Digital and Microprocessor Electronics: Theory, Applications, and Troubleshooting

Byron W. Putman

Computer Facilities and Communications
University of California, Berkeley

Formerly with the Heald Institute of Technology

Prentice-Hall, Englewood Cliffs, New Jersey 07632

Library of Congress Cataloging-in-Publication Data

Putman, Byron W.
Digital and microprocessor electronics.

Includes index.
1. Digital electronics. 2. Microprocessors.
I. Title.
TK7868.D5P87 1985 621.39 85-25632
ISBN 0-13-214354-2

Editorial/production supervision and
interior design: **Theresa A. Soler**
Cover design: **20/20 Services, Inc.**
Manufacturing buyer: **Gordon Osbourne**

Printed in the United States of America
10 9 8 7 6 5 4 3 2

ISBN 0-13-214354-2 025

Prentice-Hall International (UK) Limited, *London*
Prentice-Hall of Australia Pty. Limited, *Sydney*
Prentice-Hall Canada Inc., *Toronto*
Prentice-Hall Hispanoamericana, S.A., *Mexico*
Prentice-Hall of India Private Limited, *New Delhi*
Prentice-Hall of Japan, Inc., *Tokyo*
Prentice-Hall of Southeast Asia Pte. Ltd., *Singapore*
Editora Prentice-Hall do Brasil, Ltda., *Rio de Janeiro*
Whitehall Books Limited, *Wellington, New Zealand*

ANDERSONIAN LIBRARY
WITHDRAWN
FROM
LIBRARY
STOCK
UNIVERSITY OF STRATHCLYDE

This book is dedicated to the technicians and test
engineers of Subassembly Test in the now defunct
AMPEX short-run facility of Cupertino, California.

"Perhaps the finest group of troubleshooters
ever assembled in one lab."

My assistant Joe Yuen, Pete, Doc, Hank, Mr. Freedom,
Hung-Ta, Ernie, Big "O" and the rest.

And a special thanks to my two good friends
"Cowboy Jeff" Moschella and Dave Motoki
who taught me much.

Contents

2 INTRODUCTION TO LOGIC GATES 10

3 APPLICATIONS OF GATES 22

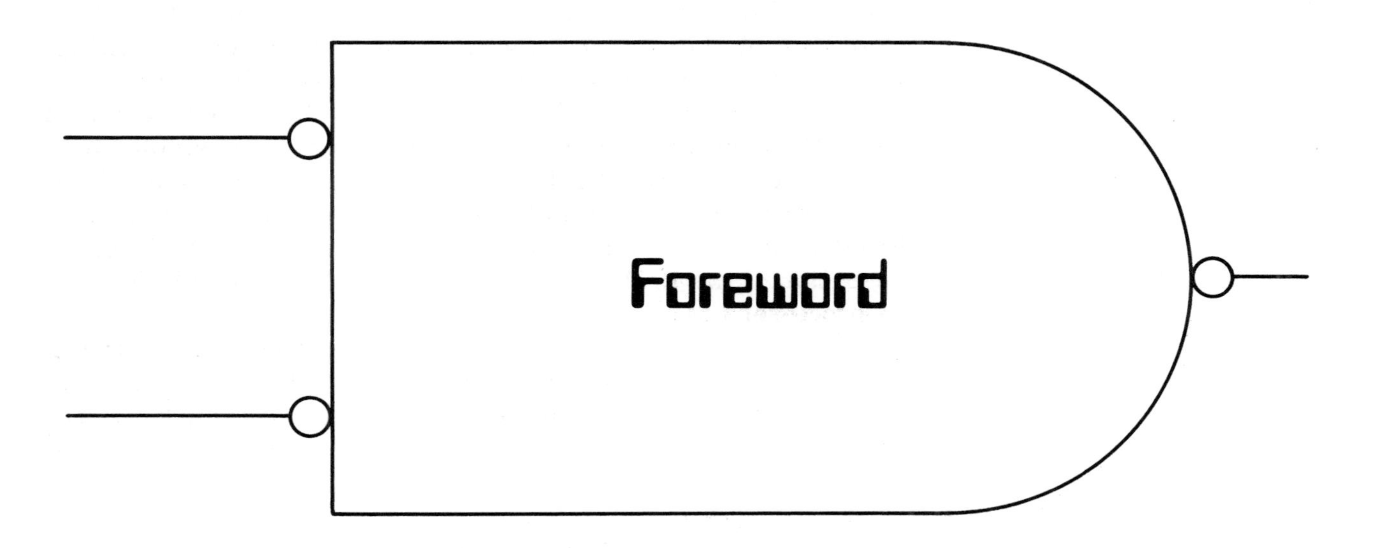

If it were necessary for one to master the chemistry and physics involved in the manufacture of molybdenum-steel piston rings before one could undertake any sort of automobile repair, there would be far fewer backyard mechanics in the world. Happily, this is not the case. After mastering some basic skills and concepts, one need only to consult the appropriate "how-to-repair" manual for the particular automobile involved to affect even complex repairs. The automobile industry is, of course, a fully mature one, so there has been time to prepare all types of helpful documentation for mechanics as well as automobile engineers. This is not the case with the emerging and dynamic field of microcomputer-based electronic equipment maintenance.

As I write this, the basic "cpu-on-a-chip" has only been commercially available for slightly longer than a decade; its low cost, flexibility, and tiny size have already enabled true computer power to be added to even such banal applications as the kitchen toaster, while at the same time a vast variety of sophisticated electronic products have become part of our daily lives. Why then, do almost all books written on the subject of microcomputer technology still treat the matter as one for research scientists and design engineers? Where are the simple descriptions of what these types of circuits look like when they're operating properly?

Until now, there has been no body of reference material available for either the newly-trained technician or the experienced technician encountering 'micros' for the first time that would simply help them fix a machine that was once working. Byron Putman in "Digital and Microprocessor Electronics: Theory, Applications, and Troubleshooting" has addressed this need. His clear and comprehensive explanations of simple logic gates lead into in-depth discussions of more complex medium-scale-integrated (MSI) and large-scale-integrated (LSI) circuits. Finally, he has included a thorough description of the internal operation of the IBM PC, which employs the powerful 8088 microprocessor.

This text has been created to provide students, practicing technicians, and other technical professionals with a valuable source that functions as both a tutorial and a comprehensive reference.

Macklin Burnham
Computer Facilities and Communications
University of California, Berkeley

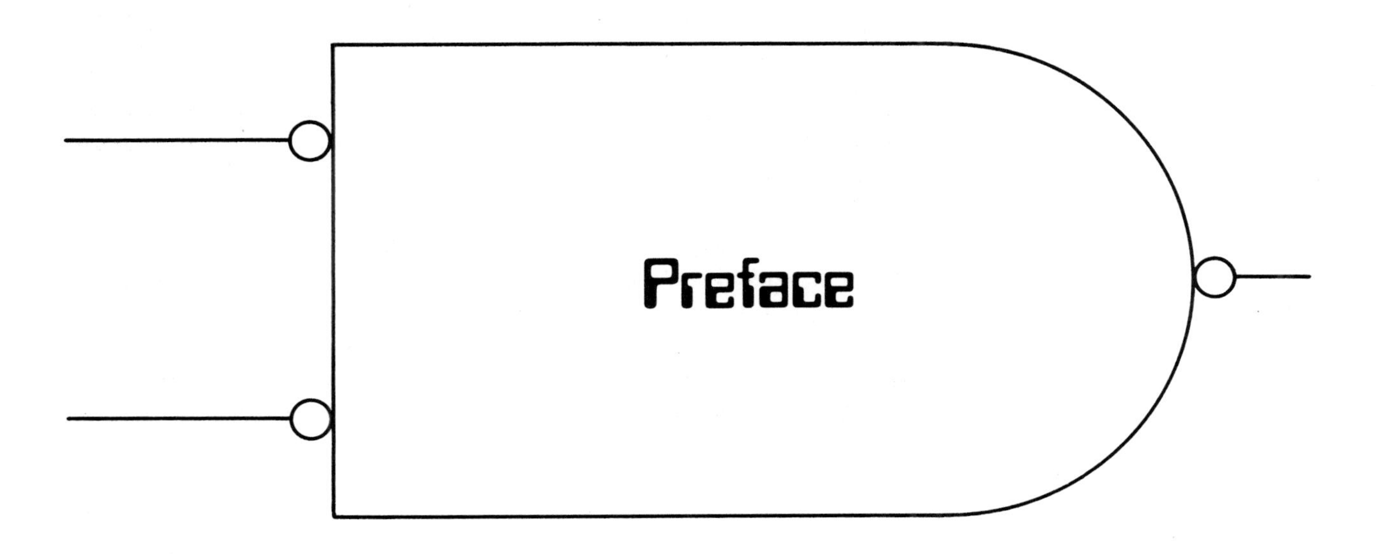

Preface

This text is the logical extension of my first book, "Digital Electronics: Theory, Applications, and Troubleshooting." It is designed for a wide range of applications; as a text for a one or two semester digital/microprocessor course in two-year colleges and technical schools, or a one semester introduction to computer hardware for BSEET and BSCS majors in four-year colleges and universities.

The emphasis of this text is on the practical investigation of SSI, MSI, and LSI devices that are the popular building blocks in modern microprocessor-based instruments and personal computers. As the title states, each device is approached in a three step manner. First the pinout and function tables are analyzed. The device is then integrated with other ICs to create practical and entertaining applications. The application is analyzed by means of a concise word description and detailed timing diagrams. The final step of the process occurs in the chapter questions and problems. Given specific symptoms or malfunctions, the student is asked to "troubleshoot" the application. This three step procedure holds the students attention, while building confidence at analyzing circuit operation; it is academically sound, yet thoroughly practical.

This book was written to be read, not just lugged to and from class or pigeonholed in a desk drawer or dusty bookcase.

It is important to note that data sheets, pinouts, schematics, and photographs are not used to "pad" the text. Each illustration is supported by an appropriate length of text. This enables students to read ahead and be properly prepared for future lectures.

The TTL logic family has been selected to illustrate SSI and MSI ICs. Although the CMOS 74HCXX family is making a strong run at the market, it is likely that LS TTL will still dominate for many years to come.

Chapter 1 introduces digital logic levels and the binary number system. Chapter 2 examines the six basic logic gates through the use of dynamic input levels and unique output levels. This technique emphasizes both the positive and negative aspects of logic gates as they relate to the gate's symbol. This enables students to effortlessly understand and accept the active-high and active-low outputs found in digital circuits.

Chapter 3 considers the concepts of Boolean algebra through the creation of simple, intuitive, and enjoyable applications of logic gates. A natural, non-mathematical approach is stressed.

Chapter 4 compares the operation of the TTL and CMOS logic gates. Input and output parameters are stressed, rather than chasing signals through the many transistors that constitute a simple gate. The elements of print circuit boards (PCBs) are examined, and troubleshooting techniques for finding opens, shorts, floating inputs, and shorted outputs are discussed.

Chapter 5 is a survey of popular MSI combinational devices. The function tables and timing diagrams of magnitude comparators, decoder/multiplexers, seven segment displays, encoders, and data selectors are examined on a line-by-line and event-by-event basis. The chapter contains many interesting applications that use simple gates and MSI ICs to provide practical functions that reinforce the need for these devices.

Chapter 6 develops sequential digital circuits by means of feedback and the concept of memory. D-type flip flops, J-K flip flops, and one-shots are examined by use of function tables, timing diagrams, and applications.

Chapter 7 introduces MSI counters that are employed as central building blocks to create complex applications that contain all of the ICs introduced in the first six chapters. This is an extremely challenging and enjoyable chapter, for instructors and students alike.

Chapter 8 describes three-bus architecture (address, data, and control/status), three-state devices and the high-z output. The microprocessor is introduced as "the master of the system bus."

Chapter 9 is a brief overview of data conversion techniques that employ the ICs introduced in the first eight chapters and op amps to construct digital to analog converters and analog to digital converts. Once again, applications and troubleshooting are stressed.

Chapter 10 uses the three-bus architecture described in Chapter 8 to develop memory systems constructed from decoders, comparators, static RAMs, ROMs, and dynamic RAMs. The mysteries of dynamic RAMs are considered in detail—row/address multiplexing, refresh, and the generation of RAS and CAS control signals. The hexadecimal number system is introduced as a shorthand method of expressing eight and sixteen bit binary quantities. Memory mapping techniques are used to analyze each memory system in the chapter.

Chapter 11 is an overview of mainframe, mini, and microcomputer systems and peripherals: the functional differences between terminals and microcomputers are considered, the modem and telecommunications are discussed, and the role of a microcomputer as an independent computing device and a channel to a mainframe computer system (using terminal emulation software) is described.

There are two standard methods of studying microprocessors. One approach examines several commercially available microprocessors. The other approach creates a "generic" model of a microprocessor that embodies the typical blocks found in commercial microprocessors. This text employs a combination of both approaches. Chapter 12 introduces generalized microprocessor architecture. This provides a student with a template in which to compare commercially available microprocessors. Chapters 14 through 17 examines three popular microprocessors—the 8085, Z80, and 6800. Each microprocessor is compared to the generalized architecture developed in Chapter 12. The concepts of memory and I/O access, interrupts, wait states and DMA are thoroughly developed.

Chapter 13 provides extremely important information that is rarely covered in microprocessor texts. The ASCII code (including control codes and flow control) is discussed in depth. The RS-232 interface, hardware handshaking, DTC, DTE, UARTS, and serial bit streams are covered in great detail. As is the Centronics parallel interface and hardware handshaking. This chapter endeavors to demystify interfacing microcomputers and terminals to printers, modems, multiplexers, port selectors, and other common communications and I/O devices.

Chapters 18 and 19 are dedicated to the presentation of the advanced LSI ICs used in the IBM personal computer. The concepts of bus systems and microcomputer architecture are developed fully. In addition to analyzing the system board of the IBM PC, the operation of video displays and character generators, and the POST (power-on self test), Chapter 19 introduces microcomputer system troubleshooting techniques, test equipment, and microcomputer preventive maintenance; including microprocessor emulators, logic analyzers, and floppy disk alignment tester/exercisers. The tradeoffs of component and board level troubleshooting are also considered.

ACKNOWLEDGMENTS

I would like to thank the following groups and persons for their invaluable help in the preparation of the manuscript:

The members of the computer facilities and communications engineering lab at UC Berkeley for their technical assistance: Jim, Mack, Sparky, Christopher, and Jack.

My good friend Roger Lanser of Lockhead Missile and Space for his expert knowledge of the IBM PC, MS-DOS and networks.

Bart Fong of the Bank of America Computer Systems Division for his insights into data communications and operating systems.

Andrew Tally of Gould Instruments Division for his feedback on test instruments, specifically logic analyzers.

Dennis Reinhardt, president of Dair Computer Systems, for his strong arguments in favor of component level repair of IBM PC system boards.

Vince Brochari of the Thomas Engineering Company for his research assistance in memory systems and DMACs.

Dr. Peter Balenger of Chevron for his invaluable feedback and his expert knowledge of microcomputer programming languages.

I would like to thank the following companies for the use of their data sheets and photographs: Texas Instruments, RCA, Data I/O, Intel, Dair Computer Systems, Lynx, Omni Logic, and Datacom Northwest.

SPECIAL ACKNOWLEDGMENT

I would like to give special thanks to Ellen Lawson, manager of the technical writing staff and customer support at the Thomas Engineering Company, for her extensive efforts as a creative consultant and editor.

An accomplished author in her own right, we who have benefitted from her superb manuals on data communications and microcomputer systems are eagerly awaiting the inevitable appearance of her first text.

Byron W. Putman

PART I DIGITAL ELECTRONICS

1 Introduction to Digital

1.1 ANALOG VERSUS DIGITAL SIGNALS

How would you describe the differences between the two waveforms in Figure 1.1? The sine wave is continually changing value while the square wave appears to change value instantaneously.

The continually changing amplitude of the sine wave gives it an infinite number of values between the positive and negative peaks. It is an *analog* signal, meaning that its amplitude changes in a continuous fashion.

The square wave is a *digital* waveform; its amplitude consists of only two values. The first half of this book is devoted to the study of digital electronics.

The term "digital" means that a numerical value can be associated with each level of the waveform. Our decimal (base 10) number system is based on the fact that human beings have 10 fingers (digits 0 through 9). In essence, a square wave has only two "fingers," so digital electronics supports a binary (base 2) number system. The digits in this system will be 0 (assigned to the low part of the square wave) and 1 (assigned to the high part of the square wave).

You are probably familiar with the term "digital" as it is commonly used—to describe a device that utilizes a numerical display: for example, watches, voltmeters, and so on. Devices that use digital displays always contain digital electronic circuitry.

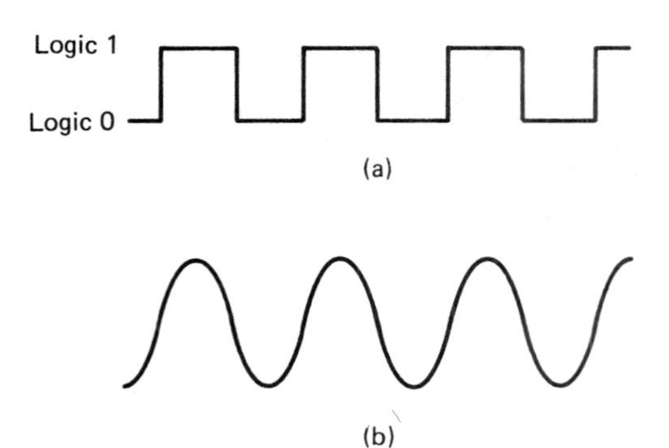

Figure 1.1 (a) Square wave; (b) sine wave.

1.2 THE BINARY NUMBER SYSTEM

As stated before, the binary (base 2) number system is used to describe digital electronics. The base of a number system describes the number of unique digits in that system. The decimal number system has 10 unique digits, 0 through 9, and the binary number system has two unique digits: 0 and 1. Most number systems operate with weighted codes, which use place values. For example, what does the number 537 in decimal really represent?

10^2	10^1	10^0
5	3	7

$$\begin{aligned} 10^0 \times 7 &= 7 \\ 10^1 \times 3 &= 30 \\ 10^2 \times 5 &= \underline{500} \\ & 537 \end{aligned}$$

The place value of the rightmost column is equal to the base raised to the 0 power ($10^0 = 1$). The next column's place value is equal to the base raised to the first power ($10^1 = 10$); the value of the last column in this example is the base raised to the second power ($10^2 = 100$). This process can continue indefinitely.

To count, we start with the first digit in the number system, which is always 0. As we increment the count, we use the next digit. When we run out of unique digits we restart the counting procedure with the number zero, and add a carry to the value in the next-highest-value column. A decimal number is called a *digit*. The term "digit" comes from counting on the fingers. A binary number is called a *bit*, which is a contraction of the terms "BInary digiT."

Notice the counting pattern in Figure 1.2. On every count the least significant bit toggles. The next column toggles every other count; the column after that, every four counts; and the most significant column, every eight counts.

The frequency with which a column toggles indicates its place value.

Counting Table

	Comments:				Decimal Equivalent
	2^3	2^2	2^1	2^0	
0000					0
+ 1					
0001				0 + 1 = 1	1
+ 1					
0010			0 + 1 = 1	1 + 1 = 0	2
+ 1					
0011				0 + 1 = 1	3
+ 1					
0100		0 + 1 = 1	1 + 1 = 0	1 + 1 = 0	4
+ 1					
0101				0 + 1 = 1	5
+ 1					
0110			0 + 1 = 1	1 + 1 = 0	6
+ 1					
0111				0 + 1 = 1	7
+ 1					
1000	0 + 1 = 1	1 + 1 = 0	1 + 1 = 0	1 + 1 = 0	8
+ 1					
1001				0 + 1 = 1	9
+ 1					
1010			0 + 1 = 1	1 + 1 = 0	10
+ 1					
1011				0 + 1 = 1	11
+ 1					
1100		0 + 1 = 1	1 + 1 = 0	1 + 1 = 0	12
+ 1					
1101				0 + 1 = 1	13
+ 1					
1110			0 + 1 = 1	1 + 1 = 0	14
+ 1					
1111				0 + 1 = 1	15

Figure 1.2 Counting table.

1.2.1 Binary Addition

From time to time you will need to add or subtract two binary numbers. Addition is as easy as counting. All you must learn are these four rules:

$$0 + 0 = 0$$
$$1 + 0 = 1$$
$$1 + 1 = 0 \text{ and a carry of } 1$$
$$1 + 1 + 1 = 1 \text{ and a carry of } 1$$

Note the following examples.

(A)	(B)	(C)	(D)
1001	1110	1010	0011
+ 0011	+ 1000	+ 0111	+ 0111
1100	1 0110	1 0001	1010

Examine addition example (D). The least significant column creates a carry. This carry is added to the next column, where two 1's are to be summed. This is a situation where the last addition rule is applied:

$$1 + 1 + 1 = 1 \text{ and a carry of } 1$$

1.2.2 Binary Subtraction

Before we attempt binary subtraction, let's review decimal subtraction. The major conceptional problem that people experience with a borrow is: "How much are we really borrowing?" Refer to the following example:

$$\begin{array}{r} 54 \\ -\ \underline{\ \ 8} \\ 46 \end{array}$$

When we attempt the subtraction it is obvious that we must borrow from the 10's column, so we are borrowing 10. Adding our borrow to the value in the 1's column yields a value of 14 and 8 subtracted from 14 gives us 6. This process may seem painfully obvious, but it is important that you consciously understand the actual mechanics involved in a borrow operation.

We have seen that one of the rules of binary addition is 1 + 1 = 0 plus a carry. The operation of borrowing will be the inverse operation of a carry. The following subtraction illustrates the manner in which you should approach a binary borrow.

$$\begin{array}{r} 10 \\ -\ \underline{\ 1} \\ 01 \end{array} = \begin{array}{r} \not{1}0^{1+1} \\ -\ \underline{\ 1} \\ 01 \end{array}$$

The following binary subtractions require elementary borrows.

$$\begin{array}{r} 1010 \\ -\ \underline{\ \ \ \ 1} \\ 1001 \end{array} \qquad \begin{array}{r} 1101 \\ -\ \underline{\ \ 11} \\ 1010 \end{array} \qquad \begin{array}{r} 1000 \\ -\ \underline{\ 100} \\ 0100 \end{array} \qquad \begin{array}{r} 1110 \\ -\ \underline{1101} \\ 0001 \end{array}$$

1.3 DIGITAL CODES

All numbers exist in some form of code. In most number systems, the place value of any given column is equal to the base raised to the number of the column. (Note that the columns are numbered from the right starting with zero.) Once again, let's examine the decimal number system to help illustrate this concept. The rightmost column of any base 10 number has the place value of 1: 10 raised to the 0 power. The next column has the place value of 10 (10 raised to the first power); the next column, 100, which is 10 raised to the second power.

1.3.1 The 8421 Code

Binary numbers can also be expressed in a form that is analogous to the decimal system. This binary code is called the *8421 code*. You should assume that all binary numbers are in 8421 code unless directed otherwise. Binary is a base 2 number system. Because 2 raised to the 0 power is equal to 1, the place value of the rightmost column is equal to 1. The place value of the next column is equal to 2 raised to the first power, or 2. The next column is 2 raised to the second power, or 4; the next is 2 raised to the third power, or 8.

1.3.2 The BCD Code

The circuitry in calculators, digital watches, or voltmeters is digital. These devices "think" in binary whereas human beings think in decimal, so a method of representing decimal numbers in a binary form was developed: *Binary-Coded Decimal* (BCD). Digital devices manipulate decimal information in an encoded binary form. Before this information is displayed, it is routed through a decoder to transform the BCD number back to its decimal equivalent. Following is a complete list of all the numbers in the BCD code.

BCD number	*Decimal equivalent*	*BCD number*	*Decimal equivalent*
0000	0	1000	8
0001	1	1001	9
0010	2	1010	Undefined
0011	3	1011	Undefined
0100	4	1100	Undefined
0101	5	1101	Undefined
0110	6	1110	Undefined
0111	7	1111	Undefined

Notice that the BCD code is the binary equivalent of the decimal digit represented. Because there are only 10 decimal digits, the last six combinations in BCD are not used and are undefined. Four bits are required for each decimal digit that we wish to represent.

Decimal number	*BCD representation*
	8 3 5
835	1000 0011 0101
	2 9
29	0010 1001
	3 3 7
337	0011 0011 0111

1.3.3 Other Codes

There are many other binary codes. If you have used a microcomputer, you have probably heard of the ASCII code, which will be introduced later in this text. Few of the remaining codes are in common use. The only other code that we will examine is the Gray code.

The *Gray code* is an unweighted code, which means that the columns in Gray code do not have a place value. Each number in the Gray code differs from the preceding number by only one bit. One application for it is interpreting the position of a mechanical shaft in binary form. Another application of the Gray code is in Karnaugh mapping. In Chapter 3 we use Karnaugh maps to analyze digital circuits.

Decimal	*Gray code*	*Decimal*	*Gray code*
0	0000	8	1100
1	0001	9	1101
2	0011	10	1111
3	0010	11	1110
4	0110	12	1010
5	0111	13	1011
6	0101	14	1001
7	0100	15	1000

1.4 CONVERTING BETWEEN DECIMAL AND BINARY

You will occasionally be required to convert between decimal and binary. Many digital devices manipulate information in groups of 4 bits. Because of this, the binary representations of the decimal numbers 0 through 15 should be memorized. You will sometimes be required to work with numbers outside this range and will need an algorithm to accomplish the conversion. (An algorithm is nothing more than a procedure for solving a mathematical problem.)

1.4.1 Binary-to-Decimal Conversion

We already know that

$$\text{place value} = \text{number's base}^{(\text{number of the column} - 1)}$$

The place values of the first eight columns of a binary number are expressed in the following table.

Column number	*Formula*	*Place value*
1	2^0	1
2	2^1	2
3	2^2	4
4	2^3	8
5	2^4	16
6	2^5	32
7	2^6	64
8	2^7	128

To convert from binary to decimal, simply sum all the place values of the columns of the binary number containing a 1. For example, convert the binary number 1001 1101 into decimal.

The binary number is broken into two groups of 4 bits to improve readability.

Column values

	128	64	32	16	8	4	2	1
Binary number to convert	1	0	0	1	1	1	0	1

$$\begin{array}{r} 128 \\ +\ 16 \\ +\ \ 8 \\ +\ \ 4 \\ +\ \ \underline{1} \\ 157 \end{array}$$

Convert 0011 0110 into decimal.

$$\begin{array}{r} 32 \\ +16 \\ +\ 4 \\ +\ \underline{2} \\ 54 \end{array}$$

1.4.2 Decimal-to-Binary Conversion

It is important to be able to perform decimal-to-binary conversion. The problem is that technicians perform these conversions so infrequently that they often forget the algorithm. There are many fast and efficient methods for accomplishing decimal-to-binary conversion, usually containing one or more "tricks" that are quickly forgotten. The conversion method that will be illustrated is slow, but it is intuitive and easily remembered.

The first step is to find the largest place value that is less than or equal to the number that we wish to convert. We will assign a one to that bit position and subtract the place value from the number, repeating the process until the number is reduced to zero. For example, convert the number 89 into binary.

We begin by listing the binary column values for reference:

Place values

128 64 32 16 8 4 2 1

Step 1. 64 is the largest column value that is less than or equal to 89. We will place a 1 in that bit position and subtract 64 from 89.

The intermediate binary result is 0100 0000.
The decimal value left to convert is 89 − 64 = 25.

Step 2. 16 is the next column value that is less than or equal to the value to convert. We place a 1 in that bit position and subtract 16 from 25.

The intermediate binary result is now 0101 0000.
The decimal value left to convert is 25 − 16 = 9.

Step 3. The next place value that is less then or equal to the value to convert is 8.

The intermediate binary result is now 0101 1000.
The decimal value left to convert is 9 − 8 = 1.

Step 4. The decimal value of 1, obviously, will fit into the first column of the binary number.

The final binary result is 0101 1001.

The other important number system used in digital electronics is called *hexadecimal* (often called just "hex"). The hexadecimal system employs a base of 16, and is used as a shorthand method of representing long binary numbers. The hexadecimal number system is discussed in Chapter 8.

1.5 CHAPTER SUMMARY

The output of a digital circuit can be only one of two possible voltage levels. The higher voltage level is assigned the numerical value of 1, and the lower voltage is assigned the numerical value of 0. A square wave is an example of a digital signal. Digital electronics is described by the binary number system.

Most students find their first class in digital electronics to be an extremely enjoyable experience. Digital circuits employ few discrete devices. A typical digital printed circuit board may contain 50 or more integrated circuits without a single discrete transistor. Because digital electronics uses so many complex integrated circuits, sophisticated systems are easily designed and constructed.

The field of electronics is constantly evolving. Any technician who desires to prosper must keep current with the present state of the technology. A characteristic that most successful technicians share is that they enjoy electronics. Do they enjoy electronics because they are good at it, or are they good at electronics because they enjoy it? In most cases, the latter is true.

QUESTIONS AND PROBLEMS

1.1 Perform the following binary additions.

(a)	(b)	(c)
1010 0111	0101 0101	1111 0001
+ 0011 0011	+ 0011 1100	+ 0100 1111

1.2 Perform binary subtractions on the problems listed in Question 1.1.

1.3 Convert the following binary numbers to decimal.

(a) 0011 1110 **(b)** 1111 0001 **(c)** 0101 1010
(d) 0001 1110 **(e)** 0011 0011 **(f)** 1111 1111

1.4 Convert the following decimal numbers to binary.

(a) 127 **(b)** 96 **(c)** 231
(d) 37 **(e)** 51 **(f)** 255

1.5 Indicate the BCD representation for the following decimal numbers.

(a) 13 **(b)** 693 **(c)** 39

1.6 What decimal numbers do the following BCD numbers represent?

(a) 1001 0010 **(b)** 0001 0101 0111 **(c)** 1100 0100

1.7 How many unique binary numbers can be created with:
(a) 3 bits?
(b) 4 bits?
(c) 8 bits?
(d) 16 bits?

1.8 Name three everyday events that are digital in nature.

1.9 Name three everyday events that are analog in nature.

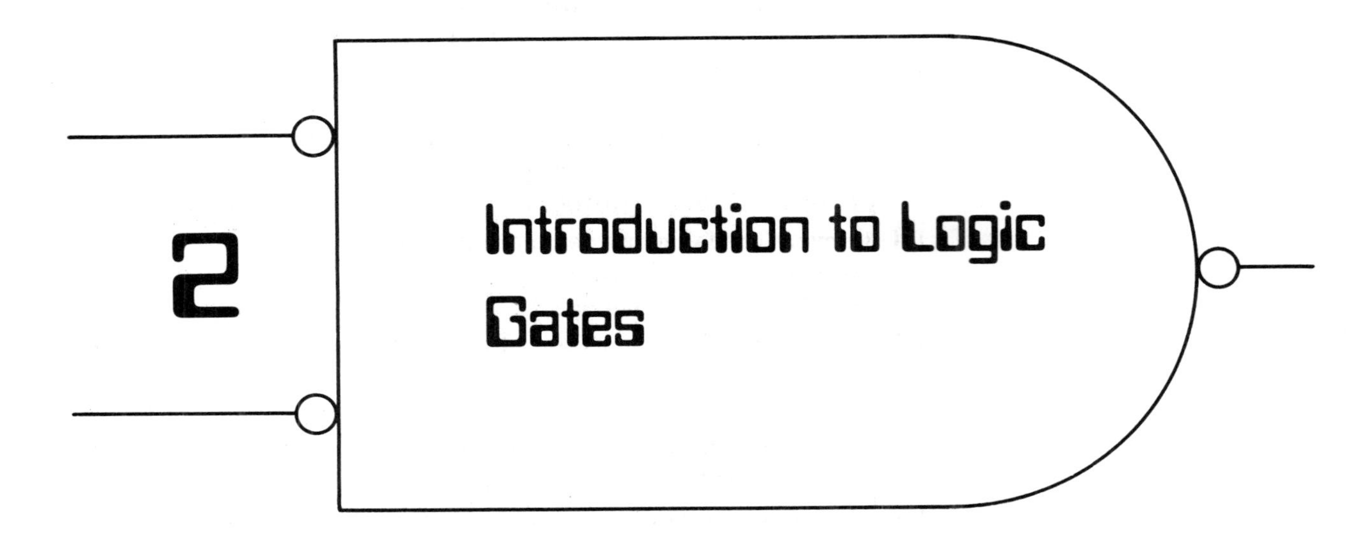

2.1 THE LOGIC GATE

The word *logic* is commonly used in conversational English and is derived from the Greek root "logs," which refers to reason, and the suffix "ic," meaning "the study of." So logic is the study of reason. In the nineteenth century, George Boole developed a branch of mathematics called *Boolean algebra* to handle logical arguments. It is an important tool for the development of digital circuits.

Gate is another common word. In your study of electronics, you have surely noticed that in technical English common words are often taken and given very specific, rigorous definitions. The common usage of "gate" is:

an opening in a wall or fence, a means of entrance or exit

The technical definition of "gate" is a natural extension of its common form:

a device that outputs a specific signal when specified input conditions are met

The logic gates examined in this book are contained in *integrated circuits* (ICs). Later, we examine the internal workings of these ICs. Until then, we will view logic gates as functional blocks. Do not concern yourself with the details of how these circuits are physically realized. Understanding the basic logic functions is more important.

Each of the gates will be introduced by illustrating the following characteristics:

1. A conversational example of how we use this logic function in everyday conversations, defining both the positive and negative approaches to each logic function.
2. A truth table that contains all the possible combinations of input values and the output value that corresponds to each input combination.
3. An explanation of the dynamic inputs and unique outputs for that gate.

   a. A dynamic input level is a certain input level that forces the output of a gate to a particular value, regardless of any other input values.
   b. A unique output level is that output level of a gate that occurs for only one combination of inputs.

Figure 2.1 illustrates some common schematic symbols that you should already know. Most logic gates have two symbols that can be used to represent them in a schematic diagram. The symbol that is chosen will be the one that best describes the function of the gate in context. Schematics should be more than just "road maps" of components; they should help the technician understand how the circuit functions.

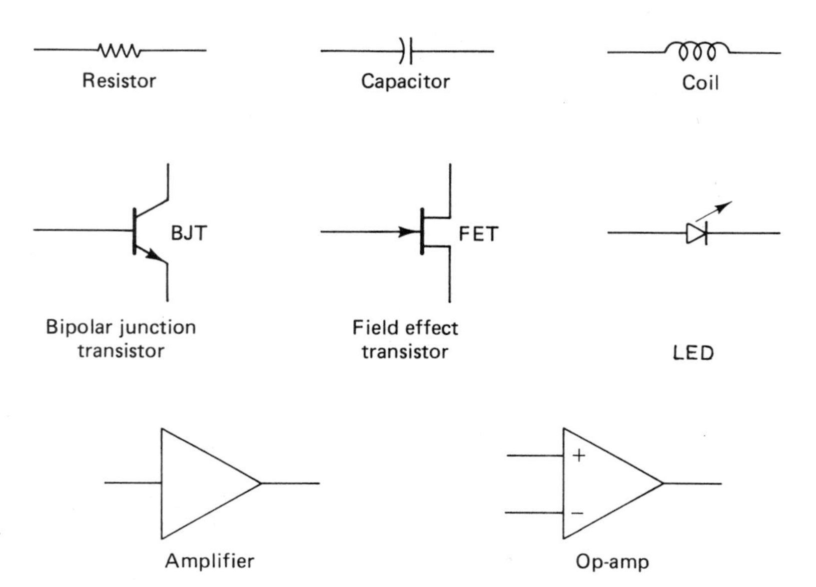

Figure 2.1 Common schematic symbols.

2.2 THE AND GATE

2.2.1 The AND Function

In conversational English we often use sentences with an AND function. For example,

If my car is running AND I get the day off, then I'll go to the beach. (1)

Both conditions must be true before the day at the beach will become a reality. We could easily add further conditions to the foregoing statement:

If my car is running AND I get the day off AND I have some money, then I'll go to the beach. (2)

Remember, in a sentence that uses the word AND to connect two or more conditions, the consequence of the sentence will be true if and only if all the conditions stated in the sentence are true.

The earlier example illustrated a positive-logic approach to the AND function. A true condition AND a true condition yielded a true consequence. The next example illustrates a negative-logic version of the AND function:

If my car is not running OR I can not get the day off, then I will not go to the beach. (3)

Statements (1) and (3) say exactly the same thing. They are two different methods of expressing an AND statement. Statement (1) has positive overtones. Statement (3) has negative overtones. Often the form of AND that we choose depends on whether we believe an event is likely to occur.

2.2.2 Dynamic Input and Unique Output Levels

Each of the four basic gates has a parameter called the *dynamic input level.* No matter what the levels of the other inputs, if the dynamic input level is applied to the gate, the output will be forced to a particular level.

The AND gate's dynamic input level is a logic 0. If a logic 0 is applied to the input of an AND gate, the output will be forced to a logic 0. Refer to the truth table of the AND gate in Figure 2.2. Whenever either of the inputs is at logic 0, the output is also at logic 0. If you are checking the inputs of a 10-input AND gate and discover a logic 0, you need not check any further. The output of the gate will be a logic 0.

Each of the basic gates also has a *unique output level.* A logic 1 on the output of an AND gate will occur only with one particular combination of inputs: all 1's. If someone told you that the output of a particular 3-input AND gate was a logic 0,

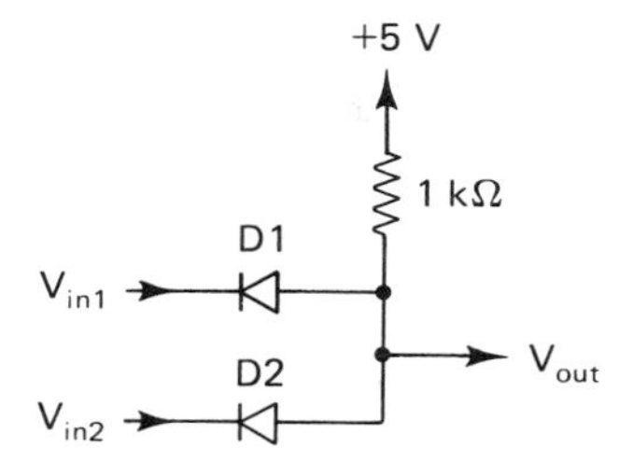

V_{in}		V_{out}
2	1	
0 V	0 V	0 V
0 V	5 V	0 V
5 V	0 V	0 V
5 V	5 V	5 V

Figure 2.2 AND gate truth table.

you could not predict the state of each input. There are seven different combinations of inputs that can result in a logic 0 on the output. Compare this to the situation where you are informed that the output of the AND gate was equal to logic 1. Could you now accurately predict the inputs? Of course! All the inputs must be logic 1. A logic 1 on the output of an AND gate is a unique event that can only result from one combination of inputs.

The dynamic input level and the unique output level of a gate are useful characteristics to know. You will see later that these two parameters actually define each of the four basic gates.

2.2.3 The Schematic Symbols for the AND Gate

We have seen that there are two approaches to the representation of an AND gate.

1. The positive approach that states: If all the inputs are true, the output will be true.
2. The negative approach that states: If any of the inputs are false, the output will be false.

Each of the different views of an AND gate has its own symbol (Figures 2.3 and 2.4). The symbol that is used will depend on the context. The symbol that looks like half of an ellipse represents the AND function. By adding more inputs we can represent AND gates of any size. The proper way to interpret this symbol is: if input *A* is high AND input *B* is high, the output will be high; otherwise, it will be low.

The negative-logic symbol of the AND function is shown in Figure 2.4. The symbol that consists of the three curved lines represents the OR function. The small circles on the inputs and output are called *bubbles*. A bubble is really a small number zero. A bubble on the input of a gate means to expect a logic 0 on this input, and a bubble on the output of a gate means to expect a logic 0 output. If input *A* is low OR input *B* is low, the

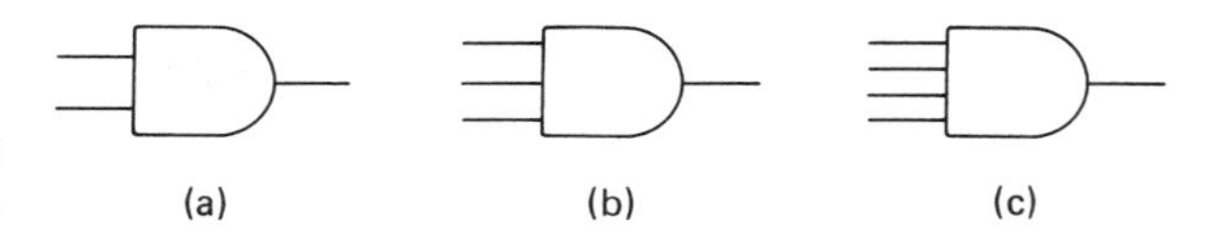

Figure 2.3 Positive logic symbols for the AND gate.

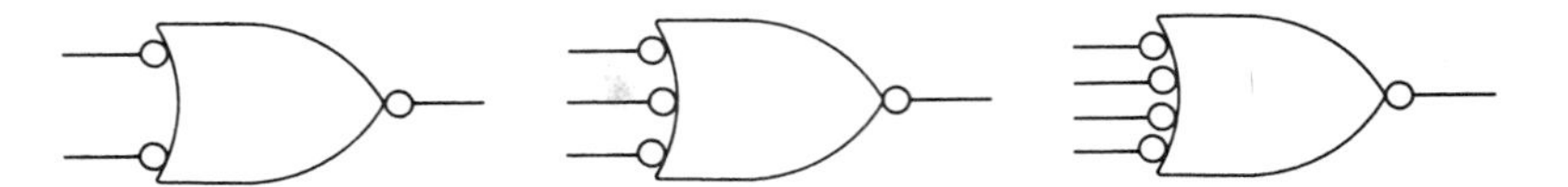

Figure 2.4 Negative logic symbols for the AND gate.

output will be low; otherwise, the output will be high. Notice that wherever we had a bubble we referred to that pin in its logic 0 state.

2.3 THE OR GATE

2.3.1 The OR Function

The other common connector that we use in compound sentences is the term OR. The following example shows the typical structure of a positive OR sentence:

If I get a raise OR the bank gives me a loan, then I can buy a new car.

If either or both conditions are true, the consequence will be true. This type of OR is called an *inclusive OR* because it includes the case when both statements are true. An *exclusive OR* does not include the case when both conditions are true. We will study the exclusive OR later in the chapter.

We can now examine the negative-logic version of the OR function.

If I don't get a pay raise AND the bank won't give me a loan, then I can't buy a new car.

Did you notice that a negative-logic AND uses an OR as a connector and a negative-logic OR uses an AND as a connector? It all depends on whether we approach a certain situation from a positive- or negative-logic point of view.

2.3.2 Dynamic Input and Unique Output Levels

Just like the AND gate, the OR also has a dynamic input level. By examining the truth table for the OR gate in Figure 2.5, it is obvious that the dynamic input level of the OR gate must be a logic 1. By applying a logic 1 to any input of an OR gate, the output will be forced to logic 1, independent of the other inputs. Again we see that the AND and OR functions have opposite

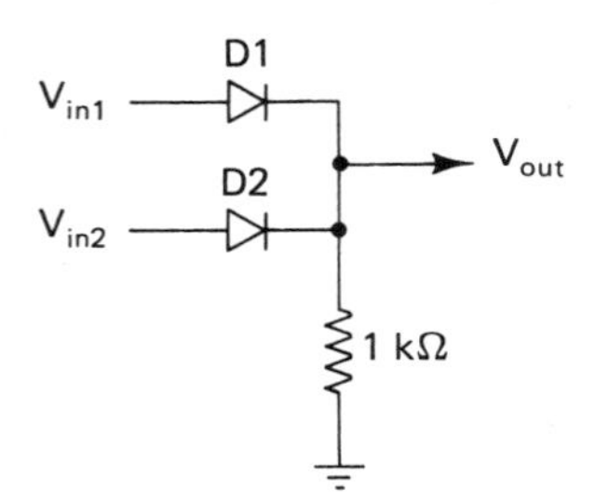

V_{in}		V_{out}
2	1	
0 V	0 V	0 V
0 V	5 V	5 V
5 V	0 V	5 V
5 V	5 V	5 V

Figure 2.5 OR gate truth table.

characteristics. As you remember, the dynamic input level of an AND gate is a logic 0, which forces a logic 0 output. The OR gate has a dynamic logic 1 input, which forces a logic 1 output.

The truth table also reveals the unique output level of the OR gate. We see that a logic 0 appears on the output for only one combination of inputs. A logic 0 is the unique output level of the OR gate. A logic 1 output can result from three separate input combinations, but a logic 0 can result from only one unique set of inputs. Again we notice that the AND and OR gates are opposite in their unique output levels.

2.3.3 The Schematic Symbols for the OR Gate

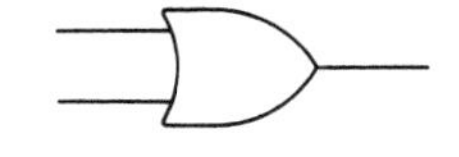

Figure 2.6 Positive logic OR gate symbol.

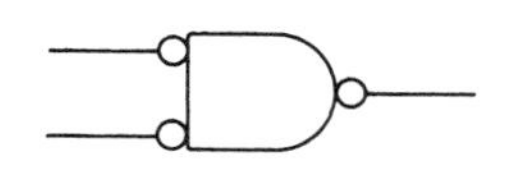

Figure 2.7 Negative logic OR gate symbol.

The OR gate also has two different symbols (Figures 2.6 and 2.7). Like the AND gate, the OR gate symbols are either positive- or negative-logic representations of the OR function. You already know the meaning of the symbol composed of the three curved lines. This is an OR symbol, representing the OR function. We have seen that the negative-logic representation of the AND function uses an OR symbol with bubbled inputs and outputs. This OR symbol has no bubbles. That means that the inputs and output are expected to be high levels for the gate to be active. You should interpret this OR symbol as saying: "If input *A* is high OR input *B* is high, then the output will be high."

You should interpret the negative-logic symbol of the OR gate as saying: "If input *A* is low AND input *B* is low, then the output will be low; otherwise, it will be high." This negative-logic OR symbol is used when we want to output a logic 0 in response to all logic 0's on the inputs.

2.4 THE INVERTER

2.4.1. The Not Function

The NOT logic function is simple. It inputs a logic level and outputs the opposite logic level. If the operand for the NOT function was logic 0, the output would be a logic 1. If the operand for a NOT function was a logic 1, then the output would be a logic 0. The truth table for the inverter is shown in Figure 2.8.

2.4.2 Dynamic Input and Unique Output Levels

Because an inverter is a single operand function, both logic 0 and logic 1 are dynamic input levels. Further, both logic 0 and logic 1 are unique output levels. If you see a logic 0 on the output of an inverter, you know that the input must be a logic 1. The same idea applies to a logic 1 on the output: you know that a logic 0 must be on the input.

The inverter is the simplest of logic functions, but is also extremely useful. Now that we have the ability to complement a logic value, we can extend our basic gates to provide us with many more useful functions.

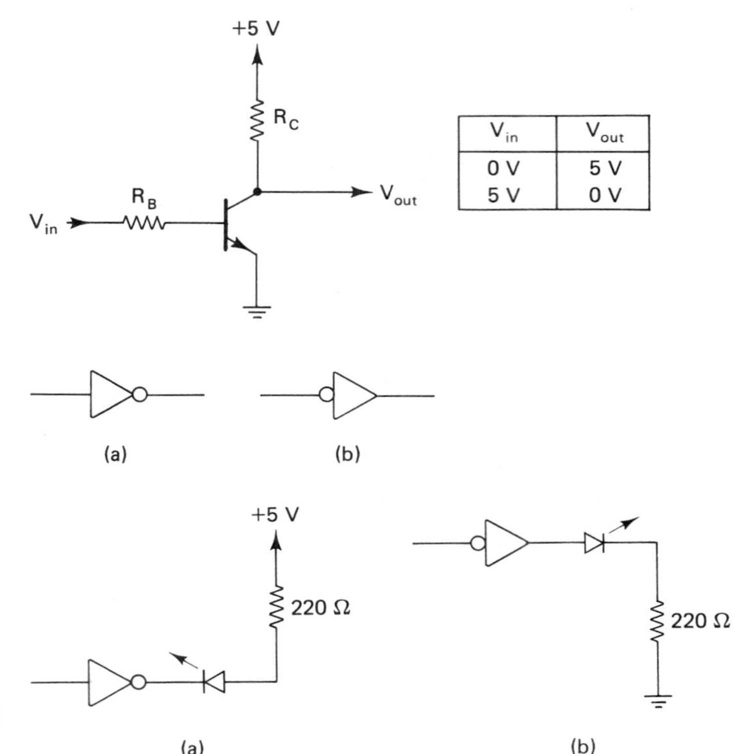

V_{in}	V_{out}
0 V	5 V
5 V	0 V

Figure 2.8 Truth table for inverter.

Figure 2.9 Logic symbols of the inverter.

Figure 2.10 Active-high and active-low inverter circuits.

2.4.3 The Logic Symbols for the Inverter

We have seen that a bubble on the input or output of a logic gate implies a low level. We can use a bubble and the universal symbol of an amplifier to create a logic symbol for the inverter. Figure 2.9a represents the positive-logic symbol for the inverter. The triangle represents an amplifier. You will discover that an inverter is often used to "buffer" an output from another logic gate. A buffer provides current gain. An amplifier symbol was chosen for an inverter to symbolize current amplification. For now, just accept the symbol at face value. A logic 1 will be input and the bubble symbolizes the inversion to a logic 0. A logic 0 can be input and the bubble will invert it into a logic 1. This is a positive-logic symbol because it "expects" to input a logic 1 and output a logic 0.

The negative-logic symbol moves the bubble to the input of the inverter. This symbol "expects" to input a logic 0 and output a logic 1. The symbol that is chosen for any particular application will depend on whether an active-high or active-low output is expected from the inverter. Figure 2.10 illustrates an active-high and an active-low output of the inverter and the proper symbol for each case.

2.5 THE NAND GATE

2.5.1 Introduction to the NAND Gate

The word NAND stands for NOT-AND. A NAND gate can be modeled as an AND with an inverter on its output. The truth

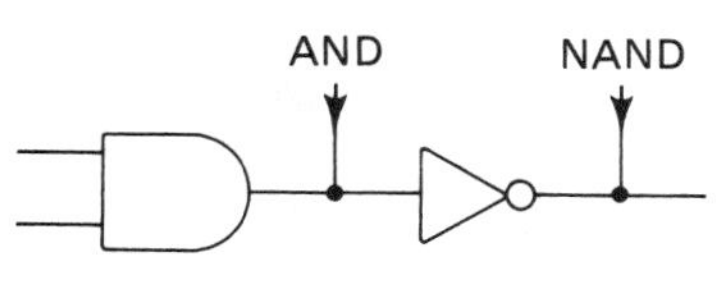

B	A	AND	NAND
0	0	0	1
0	1	0	1
1	0	0	1
1	1	1	0

Figure 2.11 NAND model and truth table.

table in Figure 2.11 compares the AND and NAND functions. There are two reasons why the NAND function is explicitly defined instead of just using AND gates and an inverter.

1. In TTL (transistor-to-transistor logic), NAND is the natural function and all gates are constructed using NAND gates. In Chapter 3 we will study the TTL logic family in depth.
2. The NAND function is a *universal building block*. All the logic functions that we have introduced so far can be built using the NAND gate.

The NAND gate may be the most widely used logic function that you will learn.

2.5.2 Dynamic Input and Unique Output Levels

Just like the AND function, the NAND function has a dynamic input level of logic 0. But unlike the AND, the NAND reacts with a logic 1 output to a logic 0 input. (Just remember that an inverter would take a logic 0 on the output of the AND and turn it into a logic 1.)

The unique output level is logic 0, which is produced only by having all inputs at a logic 1 level. This is the opposite unique output level of an AND gate. Picture an AND with all 1's on its inputs, outputting a unique logic 1. Invert this logic 1 and you will have a logic 0. The unique output level of an OR gate is also a logic 0. But a logic 0 occurs on the output of the OR gate when its inputs are all logic 0. We can see that the NAND gate shares characteristics of both the AND and the OR gates.

2.5.3 The Logic Symbols for the NAND Gate

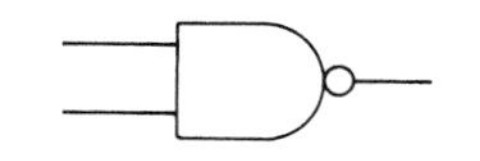

Figure 2.12 Positive-logic NAND symbol.

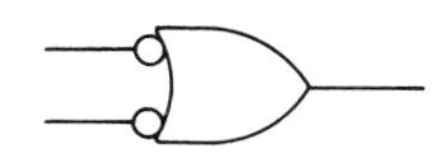

Figure 2.13 Negative-logic NAND symbol.

Like the previous three logic gates that we have encountered, the NAND gate also has both positive- and negative-logic symbols. Figure 2.12 shows the positive-logic symbol for the NAND gate. It is comprised of the AND symbol with an inversion bubble on its output. This is a shorthand version of the NAND model in Figure 2.11. The NAND gate can be thought of as an active-high-input, active-low-output AND gate.

The second NAND symbol, representing its negative-logic application, is illustrated in Figure 2.13. You can picture this symbol as having an inverter in series with each input of an OR gate. Think of it as an active-low-input, active-high-output OR gate. Another way to think of these two symbols is in the context of unique output and dynamic input levels.

2.6 THE NOR GATE

2.6.1 Introduction to the NOR Gate

The NOR gate is a combination of the OR and NOT functions: NOT OR. The NOR gate is also a universal building block. You can use the NOR gate as an inverter and an OR gate. These uses will be reserved for questions at the end of the chapter, however.

We can model the NOR gate as an OR gate followed by an inverter (Figure 2.14). Therefore, the output values of the truth tables of the OR and NOR gates will be complements of each other.

2.6.2 Dynamic Input and Unique Output Levels

The OR and NOR gates share the same relationship as the AND and NAND gates. If you remember that the dynamic input level of the NAND gate was the same as that of the AND gate, you have probably already figured out that the dynamic level of the NOR gate will be the same as that of the OR gate. But a logic 1 on the input of a NOR gate forces a logic 0 on the output, the opposite of the OR gate.

Referring once again to the truth table in Figure 2.14, we can establish that the unique output level of the NOR gate is a logic 1. This unique output level is the same for both the AND gate and the NOR gate. We have observed that the NAND gate and the OR gate both had unique output levels of logic 0, so it seems that the NOR gate is also a combination of the AND and OR functions.

2.6.3 The Logic Symbols of the NOR Gate

Figure 2.14 leads us to believe that the symbols shown in Figure 2.15 should describe the positive- and negative-logic NOR function.

B	A	OR	NOR
0	0	0	1
0	1	1	0
1	0	1	0
1	1	1	0

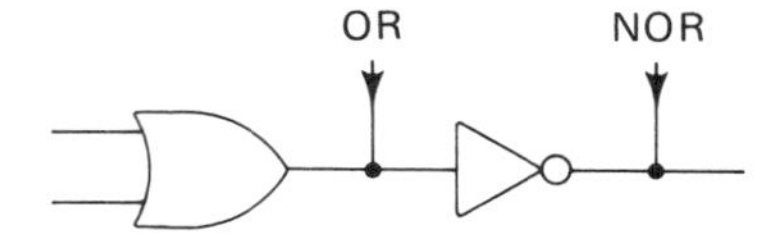

Figure 2.14 Model and truth table of the NOR gate.

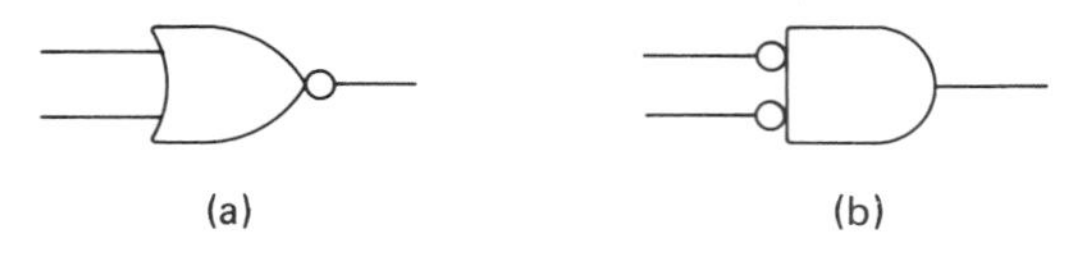

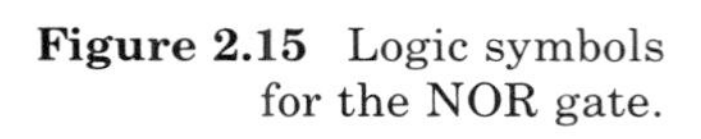

Figure 2.15 Logic symbols for the NOR gate.

2.7 THE EXCLUSIVE-OR GATE

2.7.1 Introduction to the Exclusive-OR Function

B	A	OR	XOR
0	0	0	0
0	1	1	1
1	0	1	1
1	1	1	0

Figure 2.16 Truth table of the Exclusive-OR function.

The OR function that we have studied is true if A is true OR B is true OR both A and B are true. This is formally called an *inclusive OR* because it includes the case where both A and B are true. The *exclusive-OR* function excludes the case where both A and B are true. Look at the XOR truth table (Figure 2.16). We see that the only difference in the truth table is the last line. An exclusive-OR function, abbreviated XOR, outputs a logic 0 when both inputs are the same. It outputs a logic 1 when its inputs are different.

You should notice that the XOR gate has no dynamic input level or unique output level. Other than the inverter, every gate that we have seen can be expanded to any number of inputs. The XOR gate only comes in one form, and that has two inputs.

The XOR is a special gate, used only for special applications. It is widely employed in digital circuits that perform math functions. In Chapter 3 you will learn that the XOR gate is really composed of two inverters, two AND gates, and an OR gate.

2.7.2 The Logic Symbol of the Exclusive-OR Gate

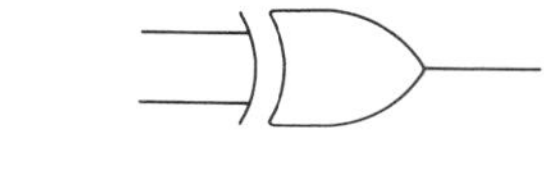

Figure 2.17 XOR gate symbol.

Because the XOR gate has no dynamic input level or unique output level, it needs only one logic symbol (Figure 2.17). It looks like the positive-logic symbol of an OR gate with a curved line across the inputs.

2.7.3 An Application of the XOR Gate

One common circuit constructed with the XOR gate is illustrated in Figure 2.18. If the control line is at a logic 0, the value on the other input of the gate passes through unchanged. If the control line is held at logic 1, the other input of the XOR gate is complemented. That is why this circuit is referred to as a *controlled complementer.*

2.8 CHAPTER SUMMARY

Digital electronics has no "positive" or "negative" nature. Give both representations of the basic functions equal time. As you proceed through this book, you will discover that the negative

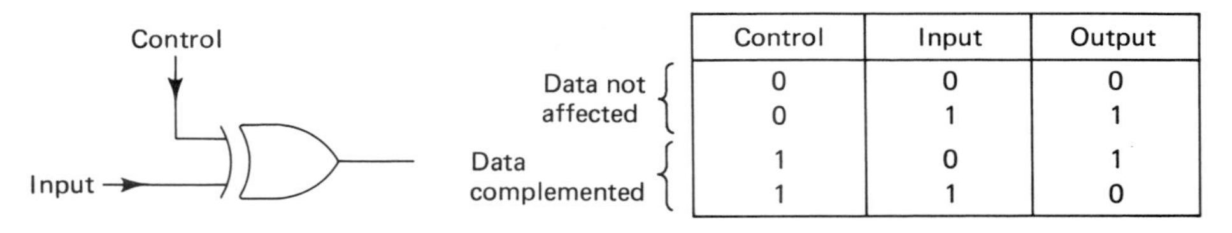

Control	Input	Output
0	0	0
0	1	1
1	0	1
1	1	0

Figure 2.18 The XOR gate as a controlled complementer.

representations are just as common as their positive equivalents, if not more so.

At times, you will need to use the basic gates in more creative ways. Some of the questions for this chapter will test your creative thinking as well as your knowledge of the basic gates. When faced with the need to build one type of gate from another, think about what the dynamic and nondynamic inputs mean. What happens if one input of a 2-input gate is tied to its nondynamic level? What happens if two inputs are tied together? Remember the relationships between the four basic gates: AND, OR, NAND, and NOR. Each one can be defined by its dynamic input levels and unique output levels.

QUESTIONS AND PROBLEMS

2.1 Define the following terms:
- **(a)** Dynamic input level.
- **(b)** Unique output level.
- **(c)** Active level of a digital input or output.

2.2 What is the dynamic input level of:
- **(a)** The AND gate?
- **(b)** The OR gate?
- **(c)** The Inverter?
- **(d)** The NAND gate?
- **(e)** The NOR gate?

2.3 What is the unique output level of:
- **(a)** The AND gate?
- **(b)** The OR gate?
- **(c)** The Inverter?
- **(d)** The NAND gate?
- **(e)** The NOR gate?

2.4 What logic body symbol is used to describe a gate:
- **(a)** In its dynamic input state?
- **(b)** In its unique output state?

2.5 Draw a circuit that illuminates an LED whenever all three inputs are low. Take care to use the proper logic symbol that best illustrates the function of this circuit. (Use a resistor in series with the LED to protect it from any power surge.)

2.6 The output of an AND gate is stuck low. What could be the possible cause of this malfunction?

2.7 The output of an OR gate is stuck high. What could be the possible cause of this malfunction?

2.8 If a particular gate has five inputs, how many possible combinations can its inputs take on?

2.9 A circuit is required that illuminates an LED when inputs A or B are high or input C is low. Illustrate this circuit.

2.10 What does a bubble on the input or output of a gate indicate?

2.11 Why are the NAND and NOR gates referred to as universal building blocks?

2.12 If the output of a NAND gate is stuck high, what could be the possible problem?

2.13 If the output of a NOR gate is stuck low, what could be the possible problem?

2.14 How does the XOR function differ from all the other logic functions?

2.15 If the output of a XOR gate is always equal to the A input, what can you conclude about the B input?

2.16 A circuit is required that illuminates an LED whenever all three of its inputs are at logic 0. Illustrate the two circuits that could accomplish this task. Remember that an LED can be illuminated on either a logic 1 or a logic 0, depending on its orientation.

2.17 A circuit is required that illuminates an LED whenever any of its three functions are at a logic 0. Illustrate the two circuits that accomplish this function.

2.18 Create two inverters from two 2-input NAND gates and two 2-input NOR gates.

2.19 Create a 2-input AND gate from two 2-input NAND gates. (*Hint:* What is the basic relationship between AND and NAND functions?)

2.20 Create a 2-input OR gate from two 2-input NOR gates.

2.21 Show how you would use a 3-input NAND gate as a 2-input NAND gate.

2.22 How would you use a 3-input NOR gate as a 2-input NOR gate?

2.23 How might you use 2-input AND gates when you needed 3-input AND gates? Can you do the same thing with OR gates?

3.1 FUNDAMENTALS OF BOOLEAN ALGEBRA

George Boole, a mathematician in the mid-nineteenth century, created the symbolic notation now used to describe digital circuitry. Boolean equations are used to represent logic circuits. Although designers work from Boolean equations to create circuits, technicians usually create the equations from the circuit and use them as tools for troubleshooting and circuit analysis.

3.1.1 The Boolean Operators

Draw the truth tables for the AND and OR functions on a piece of scratch paper. What arithmetic function describes the AND truth table? Simple multiplication. For that reason the logic AND function is represented by the multiplication symbol—a dot (·).

$$0 \cdot 0 = 0$$

$$0 \cdot 1 = 0$$

$$1 \cdot 0 = 0$$

$$1 \cdot 1 = 1$$

Note: As in algebraic equations, the dot representing the

Boolean algebra AND function is usually not used. If two Boolean algebra variables are adjacent with no separating operation symbol, the two variables are being ANDed together: for example, $A \cdot B = AB$.

Addition, with the exception of the last line of the truth table, describes the OR function. The plus sign (+) symbolizes the logical OR function.

$$0 + 0 = 0$$
$$0 + 1 = 1$$
$$1 + 0 = 1$$
$$1 + 1 = 1$$

The NOT function is indicated by a bar ($\bar{\ }$). This bar symbolizes inversion. It can extend over a single operand or group of operands.

$$\overline{0} = 1$$
$$\overline{1} = 0$$

Figure 3.1 illustrates the relationship between the truth table,

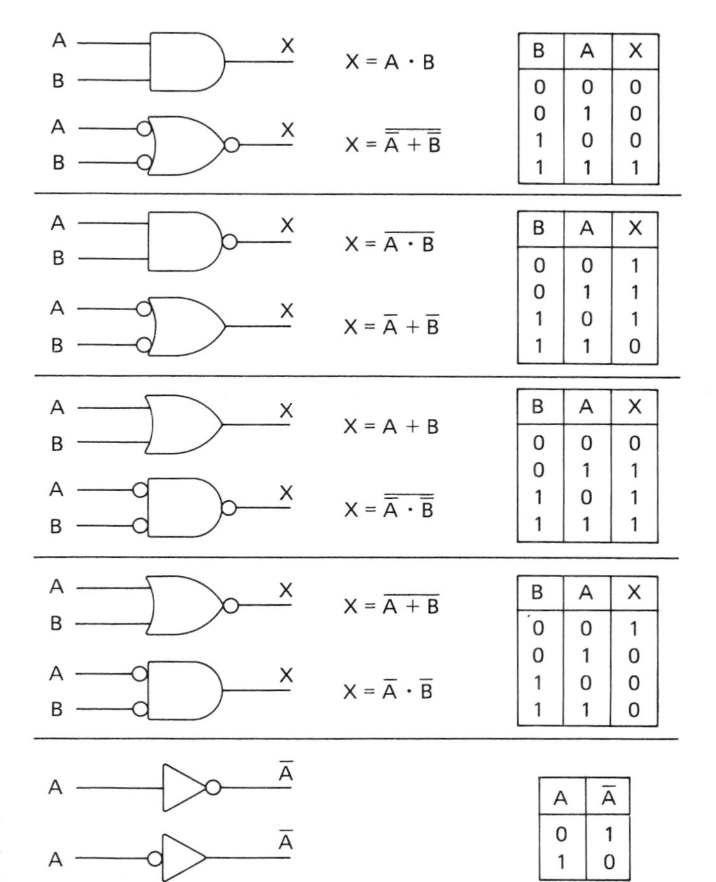

B	A	X
0	0	0
0	1	0
1	0	0
1	1	1

B	A	X
0	0	1
0	1	1
1	0	1
1	1	0

B	A	X
0	0	0
0	1	1
1	0	1
1	1	1

B	A	X
0	0	1
0	1	0
1	0	0
1	1	0

A	$\overline{A}$
0	1
1	0

Figure 3.1 Logic gates, truth tables, and their Boolean equations.

logic symbol, and Boolean equation of each function. Notice that the Boolean equations that describe the active-low output representation of the logic gates all have a long inversion bar over them.

Boolean equations follow the same rules as other algebraic equations. The inversion bar has the same grouping properties as parentheses.

The XOR function has a special symbol ($\oplus$). Later in the chapter we will discover that the XOR function is actually a composite of the AND and OR functions.

Table 3.1 contains some important Boolean relationships. These relationships, although simple, play an important role in analyzing digital circuits. It is extremely important that you thoroughly understand each identity.

Consider the first line of Table 3.1. If one input of an AND gate is tied to a logic 1 level, the output will follow the logic level applied to the other input of the AND gate because 1 is the nondynamic input level.

Because logic 1 is the dynamic input level of an OR gate, if one input of an OR gate is tied to a logic 1 level, the output will be stuck at logic 1.

Work through each line of this table so that you understand each identity. In the latter part of this chapter we use the Boolean identities illustrated in the first two lines of Table 3.1 to implement simple AND and OR gates as electronic switches.

3.1.2 De Morgan's Law

This is a simple algorithm that concerns the relationship between positive and negative logic. De Morgan's law states that if you desire to change a positive-logic symbol into a negative-logic symbol, or vice versa, execute these steps:

1. Interchange the basic logic symbol.

   An AND becomes an OR; an OR becomes an AND.

2. Add a bubble to each input and output of the new symbol.
3. If any input or output has two bubbles, they cancel each other and should be omitted.

TABLE 3.1 Boolean Identities

AND	NOT	OR
$X \cdot 1 = X$		$X + 1 = 1$
$X \cdot 0 = 0$		$X + 0 = X$
$X \cdot X = X$		$X + X = X$
$X \cdot \overline{X} = 0$		$X + \overline{X} = 1$
	$\overline{\overline{X}} = X$	

3.1.3 Creating an Equation and Truth Table from a Logic Diagram

When you are troubleshooting or analyzing digital circuits it is often helpful to create a Boolean equation describing the circuit. You can then substitute values for the inputs and find out what the output and all intermediate logic levels should be. Figure 3.2 illustrates a simple circuit composed of five gates. Because this circuit has two inputs, there are four possible input combinations. To establish what output logic level should exist for each of these four input combinations, we must follow these steps:

1. Construct a Boolean equation from the logic diagram.
2. Create a truth table and substitute the input values of each row into the equation.

Step 1. Construct the Boolean Equation. Figure 3.1 indicates the procedure for establishing the operation. Start at the inputs on the left side of the schematic. Follow each input or groups of inputs as they are applied to the various gates. At the output of each gate write the Boolean equation that describes the action of the gate. This output equation will now be used as an input to the next gate. This process will continue until you reach the final output.

Step 2. Construct a truth table. This truth table should contain the input variables, each intermediate equation, and the final equation for the output of the circuit. Fill in the truth table from left to right. Each intermediate answer will become the input value for the next equation. When the truth table is com-

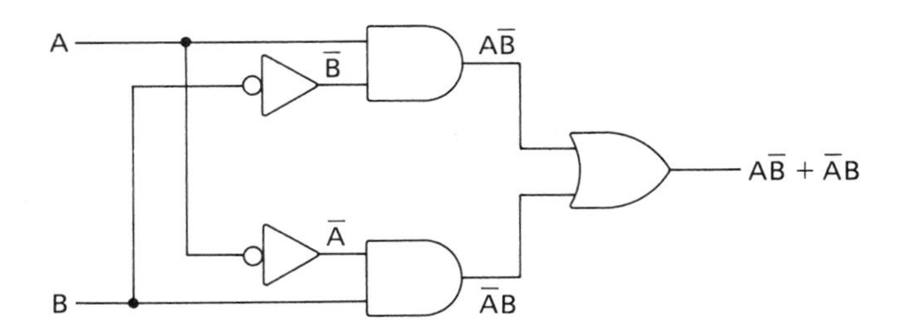

Inputs		Intermediate values				Output
B	A	$\overline{B}$	$\overline{A}$	$A\overline{B}$	$\overline{A}B$	$A\overline{B} + \overline{A}B$
0	0	1	1	0	0	0
0	1	1	0	1	0	1
1	0	0	1	0	1	1
1	1	0	0	0	0	0

Figure 3.2 Logic diagram and truth table.

plete you will know what each logic level in the circuit should be for any set of input values.

Examine Figure 3.2 closely. On a scratch sheet of paper insert each pair of input values into the equation and check to be sure that you come up with the same results. By the way, do you recognize this truth table? You should. It is the truth table for XOR. Although a designer will never use an XOR gate constructed like the one in Figure 3.2, this circuit should help you get a good feeling for how the XOR function operates.

3.2 PRACTICAL APPLICATIONS OF LOGIC GATES

This section illustrates a few applications of simple gates. It is important that you understand these simple applications before you advance to more complex devices and circuitry. To find a solution to a problem, follow the steps below.

1. Write a concise statement of the problem, defining all inputs and outputs.
2. Fill in a truth table with the proper output values for each combination of inputs.
3. Create a Boolean equation using the *sum-of-products method.* A product will be created by ANDing together the variables in any row whose output is a logic 1. After you have all the products they will be ORed together to create the final equation: thus the name "sum of products."
4. The final step is to translate the Boolean equation into the logic diagram.

3.2.1 The Three Judges: A Majority Voting Circuit

Step 1. State the problem. This circuit will analyze the votes of three judges and indicate whether a majority of the judges have voted in favor of the motion. Inputs: three, one from each judge. [The input device will be a single-throw, single-pole (STSP) switch.] Outputs: two, one light-emitting diode (LED) to indicate a majority vote, another to indicate a non-majority vote.

Step 2. Complete the truth table. Each judge will constitute one input into our circuit. With three inputs there will be eight possible voting combinations.

C	*B*	*A*	*Vote*
0	0	0	0
0	0	1	0
0	1	0	0
0	1	1	1
1	0	0	0
1	0	1	1
1	1	0	1
1	1	1	1

A logic 0 indicates a "no" vote and a logic 1 indicates a "yes" vote. Whenever two or more judges vote yes, the output will go high to indicate a majority vote of yes.

Step 3. Create a Boolean equation by the sum-of-products method. The products are:

C	B	A	*Vote*	
0	0	0	0	
0	0	1	0	
0	1	0	0	
0	1	1	1	$AB\overline{C}$
1	0	0	0	
1	0	1	1	$A\overline{B}C$
1	1	0	1	$\overline{A}BC$
1	1	1	1	ABC

The Boolean Equation is

$$\text{Majority} = AB\overline{C} + A\overline{B}C + \overline{A}BC + ABC$$

Step 4. To create a product with three input variables requires a 3-input AND gate. By inspecting the Boolean equation we can see that this circuit will require four 3-input AND gates (one for each product), one 4-input OR gate (to sum together the four products), and four inverters (to create the complements of the judges' input values and drive the nonmajority LED). The block and logic diagrams for this procedure are shown in Figure 3.3.

The SPST switches provide the digital inputs to the gates. Assume that the logic gates have an infinite input impedance, which means that current will neither enter nor exit through the inputs of these gates. (In reality, all electronic devices have a finite impedance. We cover that subject in Chapter 4.) If the switch is open, there will not be a path for current to flow; with no current flowing through the resistor, there will be no voltage dropped across it. The input to the gate will be "pulled up" to +5 V via the resistor. This resistor is called a *pull-up resistor.* When the switch is closed, +5 V is dropped across the resistor and the input to the gate will be 0 V.

3.2.2 Using Karnaugh Maps to Reduce the Number of Gates.

The circuit in Figure 3.3 accomplishes its function, but one may wonder if it can be redesigned using fewer logic gates. Circuit reduction techniques are no longer a major concern of designers because the price of gates is not as high as it used to be, but it is still an important concept to grasp.

Consider the simple Boolean equation $y = ABC + AB\overline{C}$. Does the logic value of variable C really have any effect on the

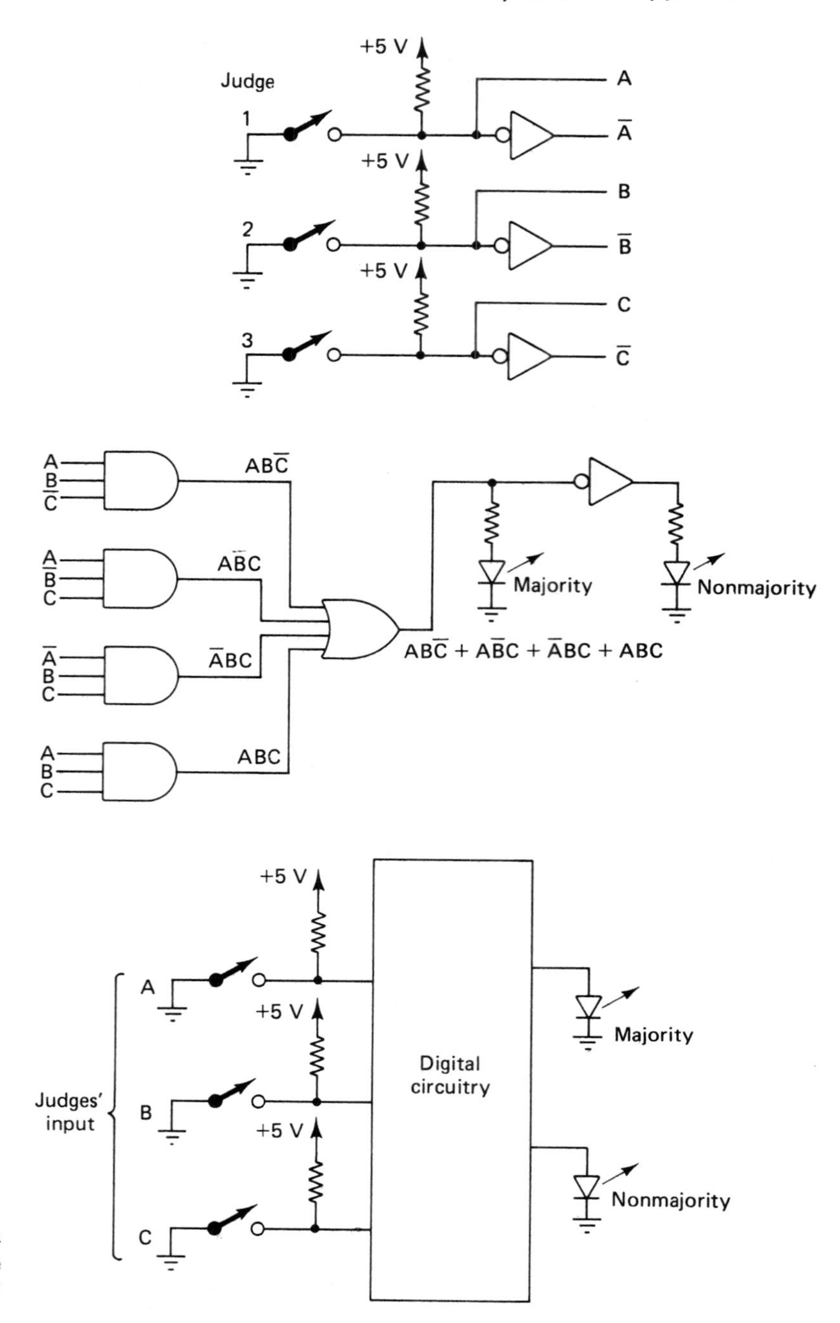

Figure 3.3 Block diagram and logic diagram for the "three judges."

value of the function? If C is high, the first product is true; if C is low, the second product is true. Therefore, we can state

$$ABC + AB\overline{C} = AB$$

We will use repeated applications of this idea to reduce the number of gates and the number of inputs required for each gate in the three-judges circuit. Our first step will be to rewrite the Boolean equation describing the three-judges circuit:

$$AB\overline{C} + A\overline{B}C + \overline{A}BC + ABC$$

$$= AB\overline{C} + ABC + A\overline{B}C + ABC + \overline{A}BC + ABC$$

Notice that the term ABC is repeated three times in our new version of the three-judges Boolean equation. This is a perfectly acceptable action because of the Boolean identity $A + A = A$. We have not really changed the meaning of the equation, only modified it to make the reduction much easier to see.

Now all we have to do is apply the concept that $ABC + AB\overline{C} = AB$ to the modified three-judges equation.

$$ABC + AB\overline{C} + ABC + A\overline{B}C + ABC + \overline{A}BC$$

$$\vee \qquad\qquad \vee \qquad\qquad \vee$$

$$AB \quad + \quad AC \quad + \quad BC$$

$$\text{vote} = AB + AC + BC$$

Whenever two judges vote yes, the vote of the third judge does not matter.

In a sum-of-products equation that has more than just a few products it can be difficult to pick out the reducible products. Karnaugh mapping provides a visual aid to assist us in finding these products.

A Karnaugh map is a modified representation of a conventional truth table and was invented by Maurice Karnaugh of Bell Labs. It uses the Gray code, so along any row or column only the state of one variable changes with each cell. Adjacent cells are arranged to contain reducible products.

Figure 3.4 illustrates the basic form for two-, three-, and four-variable Karnaugh maps. (Karnaugh maps for equations of five or more variables are extremely ungainly and complex and will not be examined in this book.) It is important to remember that these maps are just a visual aid for finding reducible products.

In Figure 3.4, the cells of the Karnaugh map are arranged to display reducible products. They do not have a sequential translation to the rows of their corresponding truth table. The cells on the left column of the Karnaugh map are adjacent to the cells in the right column and the cells in the top row are adjacent to the cells in the bottom row. If you ever question whether two cells are adjacent, examine them and determine if they are indeed reducible products.

Figure 3.5 shows the truth table and Karnaugh map for the problem of the three judges. The output logic levels from each row in the truth table have been transferred to their corresponding cells in the Karnaugh map. Notice that we can find three pairs of adjacent cells that contain logic 1 levels. The one product that is common to all three pairs is ABC. That is the product we used to accomplish the Boolean algebra reduction of this circuit.

The following list indicates each pair of reducible products and the resultant product.

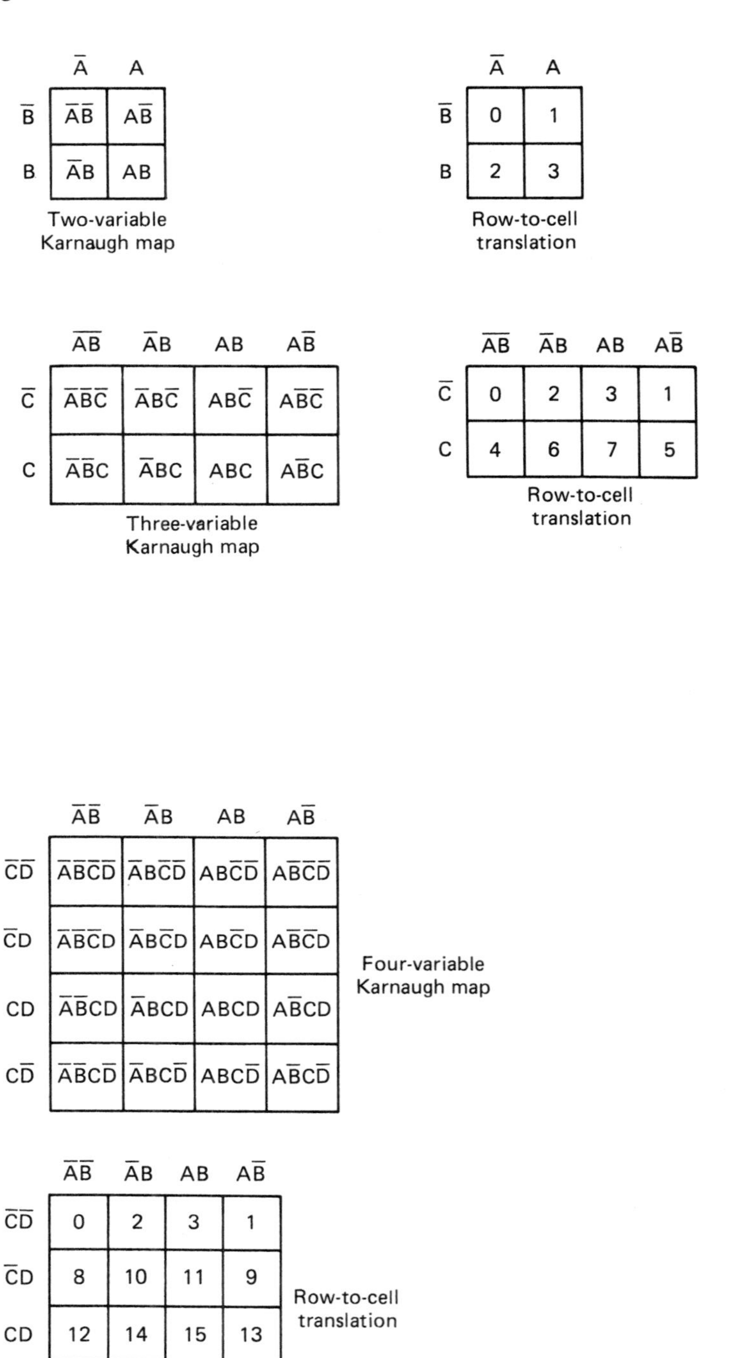

Row	B	A	Product
0	0	0	$\overline{A}\overline{B}$
1	0	1	$A\overline{B}$
2	1	0	$\overline{A}B$
3	1	1	AB

Row	C	B	A	Product
0	0	0	0	$\overline{A}\overline{B}\overline{C}$
1	0	0	1	$A\overline{B}\overline{C}$
2	0	1	0	$\overline{A}B\overline{C}$
3	0	1	1	$AB\overline{C}$
4	1	0	0	$\overline{A}\overline{B}C$
5	1	0	1	$A\overline{B}C$
6	1	1	0	$\overline{A}BC$
7	1	1	1	ABC

Row	D	C	B	A	Product
0	0	0	0	0	$\overline{A}\overline{B}\overline{C}\overline{D}$
1	0	0	0	1	$A\overline{B}\overline{C}\overline{D}$
2	0	0	1	0	$\overline{A}B\overline{C}\overline{D}$
3	0	0	1	1	$AB\overline{C}\overline{D}$
4	0	1	0	0	$\overline{A}\overline{B}C\overline{D}$
5	0	1	0	1	$A\overline{B}C\overline{D}$
6	0	1	1	0	$\overline{A}BC\overline{D}$
7	0	1	1	1	$ABC\overline{D}$
8	1	0	0	0	$\overline{A}\overline{B}\overline{C}D$
9	1	0	0	1	$A\overline{B}\overline{C}D$
10	1	0	1	0	$\overline{A}B\overline{C}D$
11	1	0	1	1	$AB\overline{C}D$
12	1	1	0	0	$\overline{A}\overline{B}CD$
13	1	1	0	1	$A\overline{B}CD$
14	1	1	1	0	$\overline{A}BCD$
15	1	1	1	1	$ABCD$

Figure 3.4 Two-, three-, and four-variable Karnaugh maps.

C	B	A	Vote
0	0	0	0
0	0	1	0
0	1	0	0
0	1	1	1
1	0	0	0
1	0	1	1
1	1	0	1
1	1	1	1

	$\overline{A}\overline{B}$	$\overline{A}B$	AB	$A\overline{B}$
$\overline{C}$	0	0	1	0
C	0	1	1	1

Figure 3.5 Truth table and Karnaugh map.

Reducible product	*Reduced term*
$ABC + AB\overline{C}$	AB
$ABC + \overline{A}BC$	BC
$ABC + A\overline{B}C$	AC

Notice that a product (in this case ABC) can be used any number of times. Now we sum these reduced products together to arrive at our new equation:

$$\text{vote} = AB + BC + AC$$

The reduced circuit is shown in Figure 3.6. Remember that within any group of reducible products in a Karnaugh map, those products that appear in both true and complemented form will be dropped from the reduced equation.

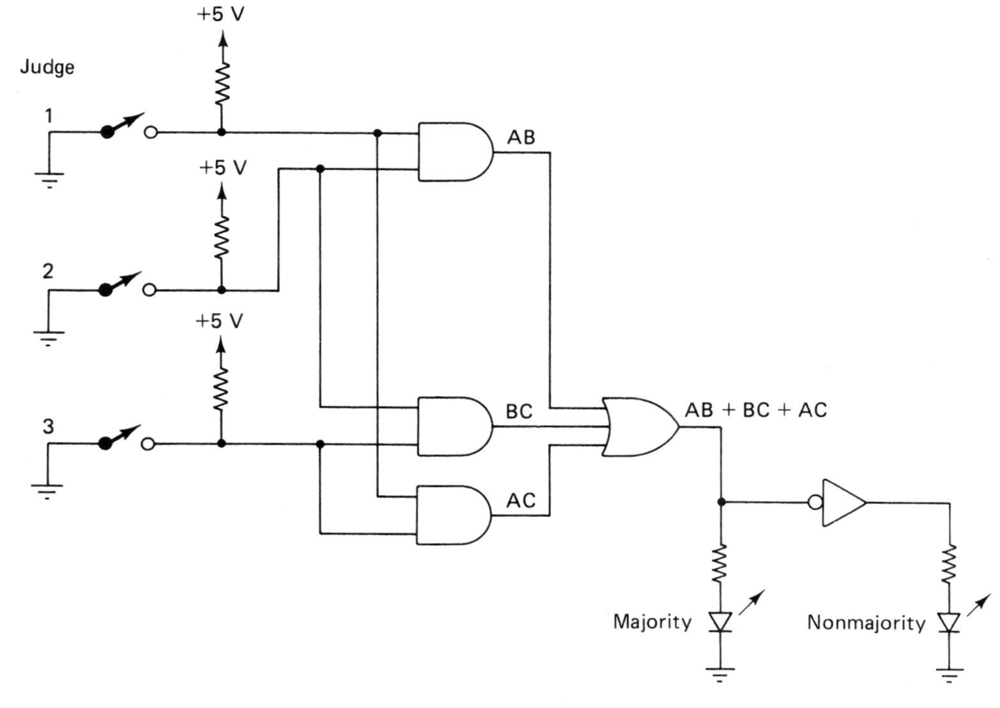

Figure 3.6 Reduced three-judges circuit.

3.2.3 The Room with Three Doors

The problem of the room with three doors is considered a classic application of logic gates.

Step 1. State the problem. A room has three doors. Next to each door is a light switch. A person should be able to enter or exit from any door and by toggling the light switch adjacent to the door change the state of the light. (By "toggle" we mean to flip a switch from on to off or from off to on.)

The inputs will be from each of the light switches. The output will turn on an LED that symbolizes the light.

Step 2. Complete the truth table. We need to make an initial assumption about the state of the light. We will assume that when all three switches are in their logic 0 state the light will be out. Whenever the variables from one row to the next change an odd number of times, the light will change states. Whenever the variables change an even number of times the light will not change states because the light is toggled once with the first change, then toggled back to its original state with the second change. The truth table below is constructed sequentially, but once understood it can be used as a look-up table. The changes listed below are provided only so that you will understand how the table was constructed.

Doors				
C	*B*	*A*	*Light*	*Number of changes from previous state*
0	0	0	Off	Assume that the light is initially off.
0	0	1	On	One change (A), so light changed.
0	1	0	On	Two changes (A and B); no change.
0	1	1	Off	One change (A), so light changed.
1	0	0	On	Three changes, so light changed.
1	0	1	Off	One change (A), so light changed.
1	1	0	Off	Two changes (A and B); no change.
1	1	1	On	One change (A), so light changed.

Step 3. Create a Boolean equation. Each row in the truth table that has a light condition of ON will be a product in the equation:

$$\text{light on} = A\overline{B}\,\overline{C} + \overline{A}B\overline{C} + \overline{A}\,\overline{B}C + ABC$$

Step 4. Construct the Logic Circuit. The circuit shown in Figure 3.7 was realized from the Boolean equation in step 3.

Step through the circuit and truth table a few times choosing different conditions and make sure that you understand exactly how the circuit was derived from the truth table and how the truth table was derived from the problem statement.

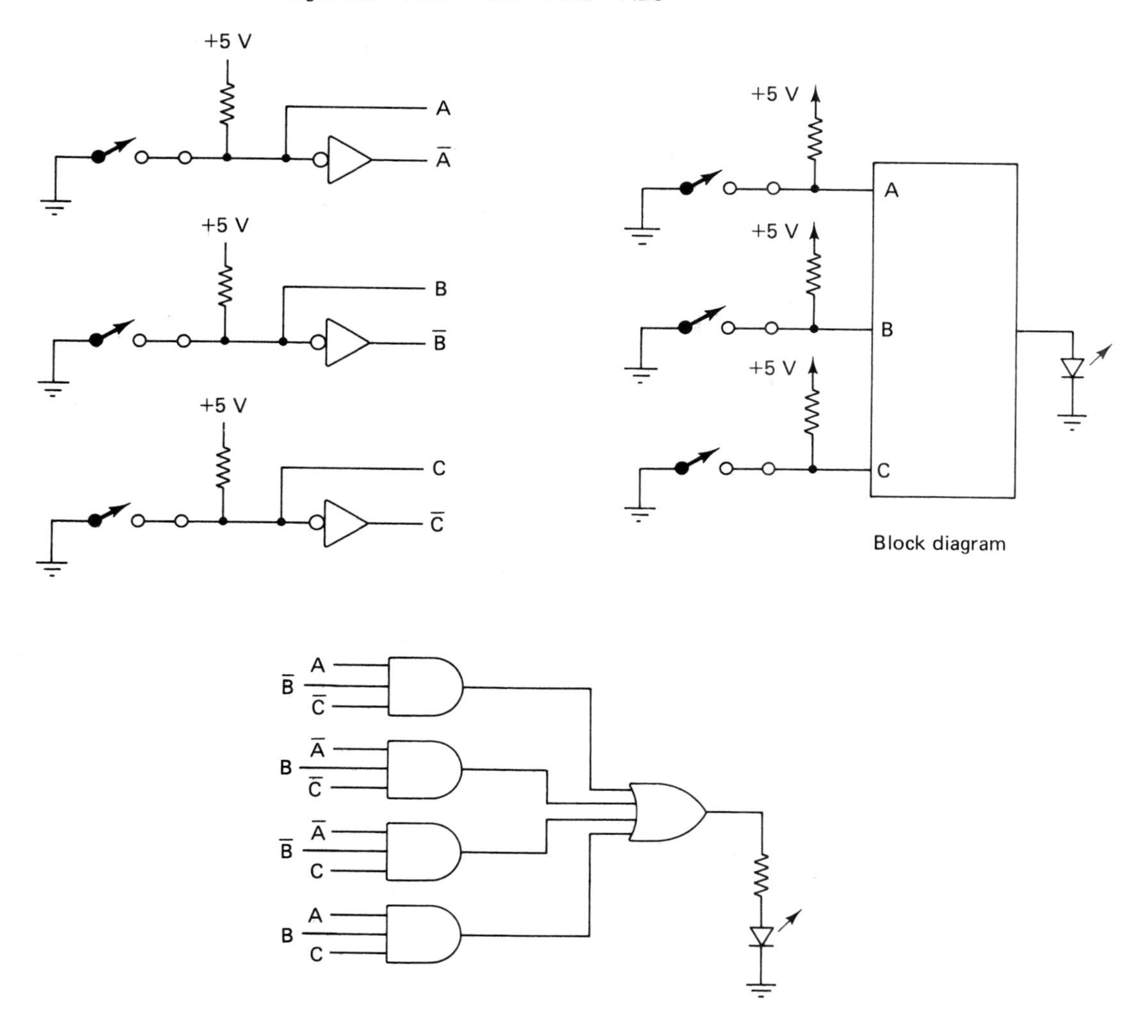

Figure 3.7 The room with three doors.

3.2.4 A 2-Line-to-4-Line Decoder

Consider the following problem.

Step 1. State the problem. There are four LEDs on a front panel. We want to be able to turn on any one of the four LEDs. The LED that will be turned on will be indicated by a 2-bit input. Two binary digits can represent four unique numbers, 0 through 3. Each LED will be assigned one of these numbers. (In digital electronics, we will always start counting with the number 0.) The first LED will illuminate whenever 00 is input; the second with 01; the third, 10; and the fourth, 11. This circuit should have an active-low output: the output indicated by the 2-bit select code will go to logic 0, and the other outputs will go to logic 1.

Step 2. Complete the truth table. This truth table will be different from the previous truth tables because it will have four outputs, one for each of the LEDs.

Inputs		*Outputs*			
B	*A*	*1*	*2*	*3*	*4*
0	0	0	1	1	1
0	1	1	0	1	1
1	0	1	1	0	1
1	1	1	1	1	0

Step 3. Create a Boolean Equation. We will create four separate Boolean equations, one for each output.

$$\text{output 1} = \overline{A}\,\overline{B} \qquad \text{output 2} = A\overline{B}$$

$$\text{output 3} = \overline{A}B \qquad \text{output 4} = AB$$

Step 4. Construct the Logic Circuit. Considering that step 3 contains four equations, one may be inclined to think that we must construct four separate circuits. Remember that only the two inputs are common to all four outputs. The circuit is shown in Figure 3.8.

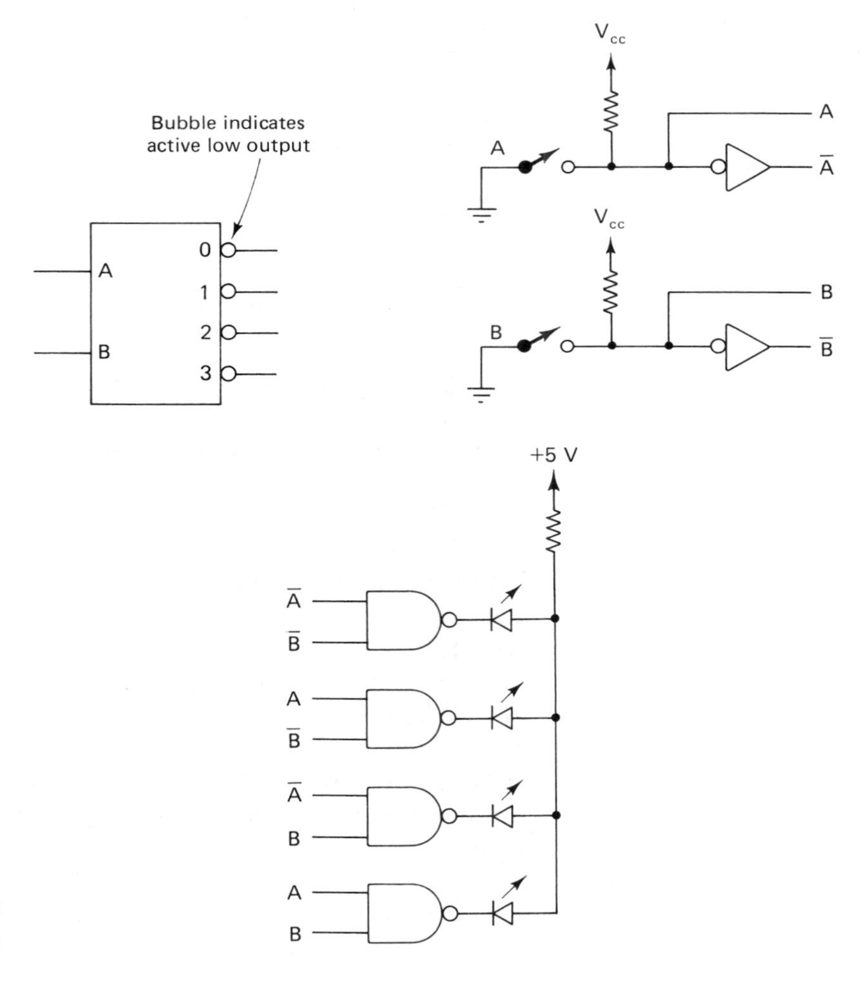

Figure 3.8 2-line-to-4-line decoder.

Decoders are useful digital circuits, but you will normally see them as a single integrated circuit rather than constructed from discrete gates as we did here.

3.2.5 A 2-Bit Magnitude Comparator

Step 1. State the problem. A circuit is required that can compare the magnitudes of two 2-bit numbers and indicate whether the first number is less than, equal to, or greater than the second number.

There will be four inputs: 2 bits for A and 2 bits for B. The three outputs will reflect the relative magnitudes of A and B: less than, equal to, and greater than.

Step 2. Complete the truth table. Remember that the inputs will be considered as weighted binary numbers and the inputs will have the decimal magnitudes of 0 through 3.

Inputs				*Outputs*		
B_2	B_1	A_2	A_1	$A < B$	$A = B$	$A > B$
0	0	0	0	0	1	0
0	0	0	1	0	0	1
0	0	1	0	0	0	1
0	0	1	1	0	0	1
0	1	0	0	1	0	0
0	1	0	1	0	1	0
0	1	1	0	0	0	1
0	1	1	1	0	0	1
1	0	0	0	1	0	0
1	0	0	1	1	0	0
1	0	1	0	0	1	0
1	0	1	1	0	0	1
1	1	0	0	1	0	0
1	1	0	1	1	0	0
1	1	1	0	1	0	0
1	1	1	1	0	1	0

Take a moment to carefully examine the outputs of the truth table. Do you agree with the value of each output?

Step 3. Create a Boolean equation. We will create one equation for each output.

$$A < B = \overline{A_1}\,\overline{A_2}B_1\overline{B_2} + \overline{A_1}\,\overline{A_2}\,\overline{B_1}B_2 + A_1\overline{A_2}\,\overline{B_1}B_2$$
$$+ \overline{A_1}\,\overline{A_2}B_1B_2 + A_1\overline{A_2}B_1B_2 + \overline{A_1}A_2B_1B_2$$

$$A = B = \overline{A_1}\,\overline{A_2}\,\overline{B_1}\,\overline{B_2} + A_1\overline{A_2}B_1\overline{B_2} + \overline{A_1}A_2\overline{B_1}B_2$$
$$+ A_1A_2B_1B_2$$

$$A > B = A_1\overline{A_2}\,\overline{B_1}B_2 + \overline{A_1}A_2\overline{B_1}\,\overline{B_2} + A_1A_2\overline{B_1}\,\overline{B_2}$$
$$+ \overline{A_1}A_2B_1\overline{B_2} + A_1A_2B_1B_2 + A_1A_2\overline{B_1}B_2$$

Step 4. Construct the logic circuit. See Figure 3.9.

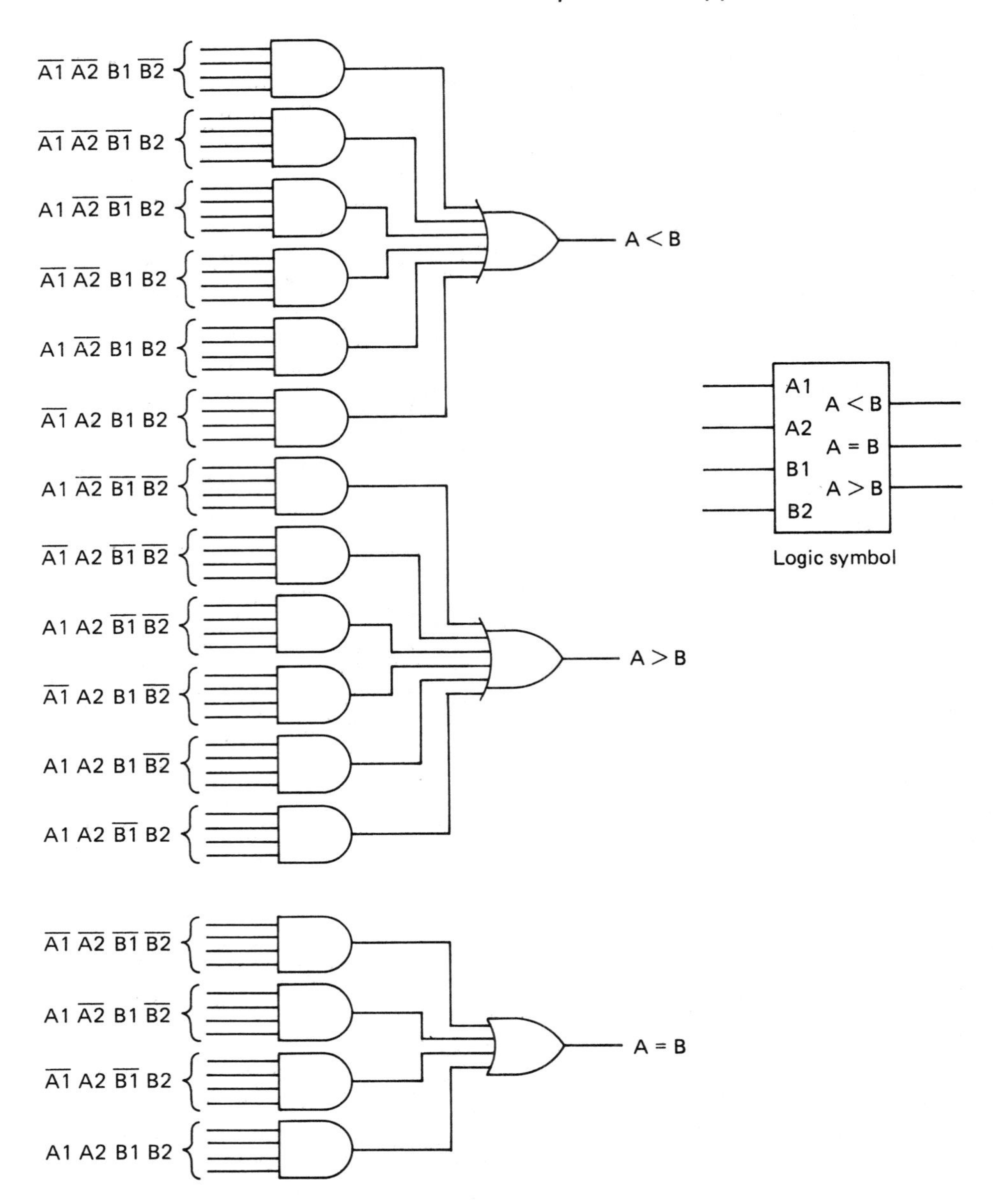

Figure 3.9 2-bit magnitude comparator.

3.3 THE GATE AS AN ELECTRONIC SWITCH

Gates are often used as switches that either pass or block input signals. Many digital circuits have enable inputs. Unless this enable bit is at the proper digital level, the circuit will not function.

3.3.1 Two-Input Gates as Simple Switches

The key to using gates as switches is dynamic input levels. Gates with two inputs are used: one input will have the data applied to it and the other will have a control level applied to it. If the control level is the dynamic input level, the output will be forced to a particular value, regardless of the data input. In Figure 3.10 data input will have a *don't-care value.* A don't-care value is denoted by an × in the truth table.

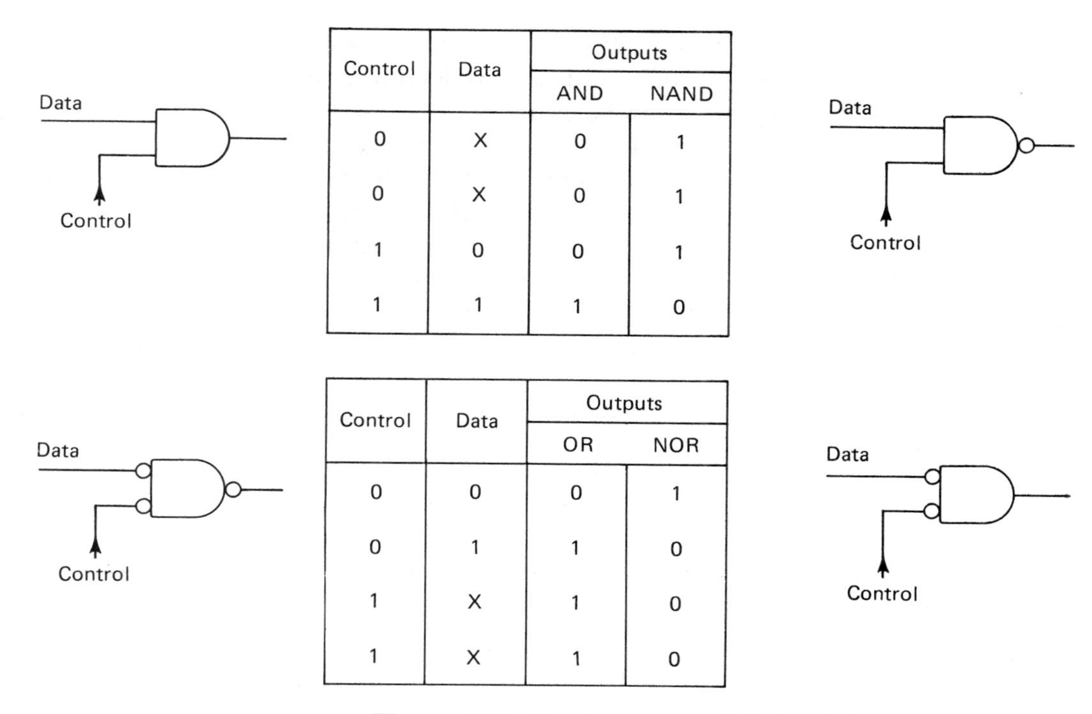

Control	Data	Outputs	
		AND	NAND
0	X	0	1
0	X	0	1
1	0	0	1
1	1	1	0

Control	Data	Outputs	
		OR	NOR
0	0	0	1
0	1	1	0
1	X	1	0
1	X	1	0

Figure 3.10 Simple gates as digital switches.

3.3.2 A 1-of-2 Data Selector

We would like to create a circuit that is the digital equivalent of a single-pole, double-throw (SPDT) mechanical switch. This circuit will have two data inputs, A and B. The select bit will steer the desired input to the output. The other input will be blocked. This application is just an extension of using simple gates as digital switches. Consider Figure 3.11.

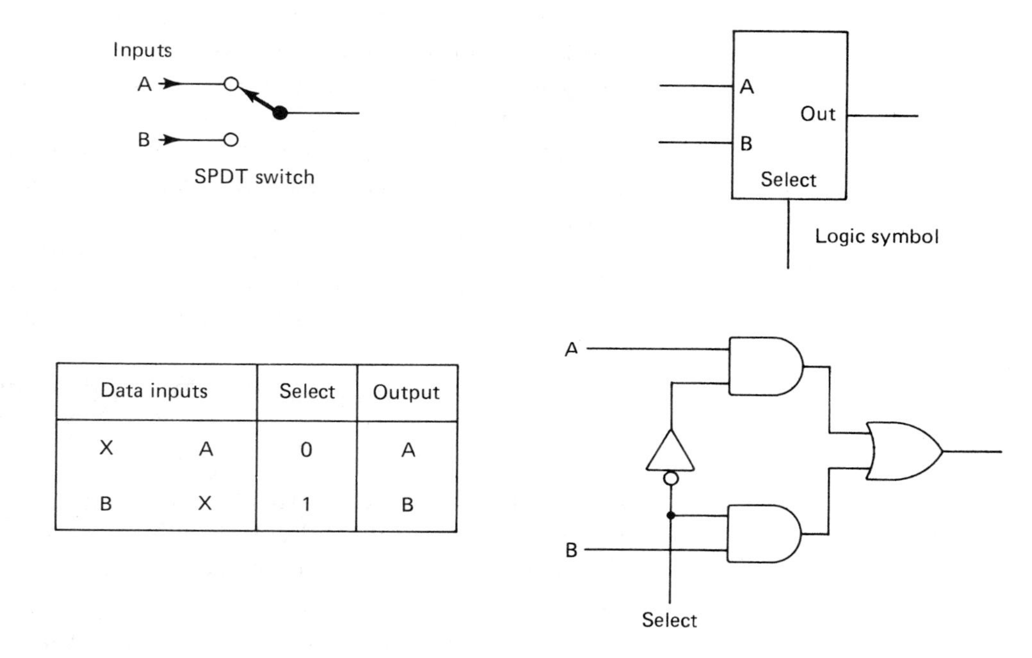

Data inputs		Select	Output
X	A	0	A
B	X	1	B

Figure 3.11 A 1-of-2 data selector.

3.4 AN IGNITION ENABLE CONTROL CIRCUIT

Step 1. State the problem. Consider a two-seat sports car. We would like to design a circuit that enables the ignition only when both doors are closed, the driver has a seat belt on, and if there is a passenger, that person must also have a seat belt on. We will approach this problem in much the same manner as the previous applications. The Boolean equation can be arrived at without the aid of a truth table.

The following table contains the five inputs and the variable names that we will use to represent them in the Boolean equation.

Input	Door 1	Door 2	Driver seat belt	Passenger seat sensor	Passenger seat belt
Variable name	D1	D2	DSB	PSS	PSB

A sensor switch that acts like a momentary SPST switch will be placed in the doors, seat belt buckles, and underneath the passenger's seat to indicate the presence of a passenger. The output will be a single active-high logic level that indicates when the ignition is enabled.

Step 2. We will skip step 2 because the Boolean equation will be easier to derive from an intuitive examination of the inputs.

Step 3. Create a Boolean equation. This is a straightforward problem except for the passenger's seat belt. If there is no passenger, the state of the passenger seat belt (PSB) switch is a don't-care condition. If the passenger seat sensor (PSS) indicates the presence of a passenger, the passenger seat belt (PSB) input must indicate that the passenger's seat belt is buckled. Using a pull-up resistor with each switch, if a switch is open, its associated logic level will be high; if the switch is closed, it will output a logic 0.

Whether or not a passenger is present, door 1 (D1), door 2 (D2), and driver seat belt (DSB) inputs must all be low. The first part of the Boolean equation should therefore be

$$\text{ignition enable} = \overline{\text{D1}} \cdot \overline{\text{D2}} \cdot \overline{\text{DSB}} \ldots$$

Now let's consider how to introduce the passenger seat sensor and seat belt buckle into the equation. If the passenger seat sensor (PSS) is high, indicating a "no passenger" condition, this Boolean equation should be true:

$$\text{ignition enable} = \overline{\text{D1}} \cdot \overline{\text{D2}} \cdot \overline{\text{DSB}} \cdot \text{PSS} \ldots$$

What if the passenger seat sensor (PSS) is low, indicating the presence of a passenger? If that is the case, the passenger seat belt must also be low, indicating that the passenger is buckled up:

$$\text{ignition enable} = \overline{D1} \cdot \overline{D2} \cdot \overline{DSB} \cdot \overline{PSS} \cdot \overline{PSB} \ldots$$

It appears that we must use an OR function to provide for the presence or absence of a passenger. The final Boolean equation is

$$\text{ignition enable} = \overline{D1} \cdot \overline{D2} \cdot \overline{DSB} \cdot (PSS + (\overline{PSS} \cdot \overline{PSB}))$$

Step 4. Create a logic function. Figure 3.12 shows the gate realization of this Boolean equation.

It is important that you understand the use of positive- and negative-logic symbols. Remember that a properly drawn schematic is an aid to understanding the manner in which the circuit operates. Step through Figure 3.12 several times until you feel that you have achieved a thorough understanding of the circuit operation.

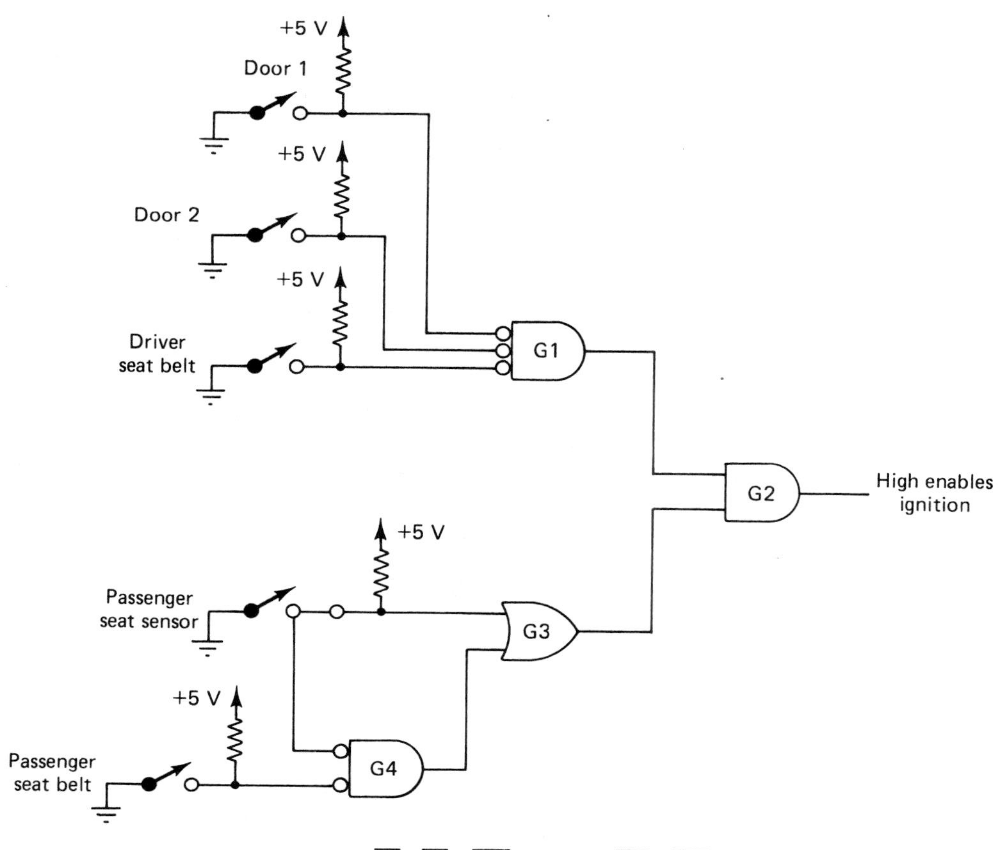

Figure 3.12 Ignition enable circuit.

QUESTIONS AND PROBLEMS

3.1. Derive the Boolean equation and complete a truth table for the circuits shown in Figure P3.1.

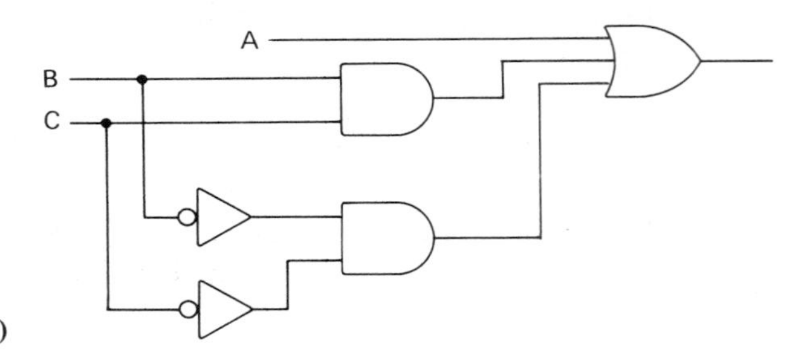

Figure P3.1(a)

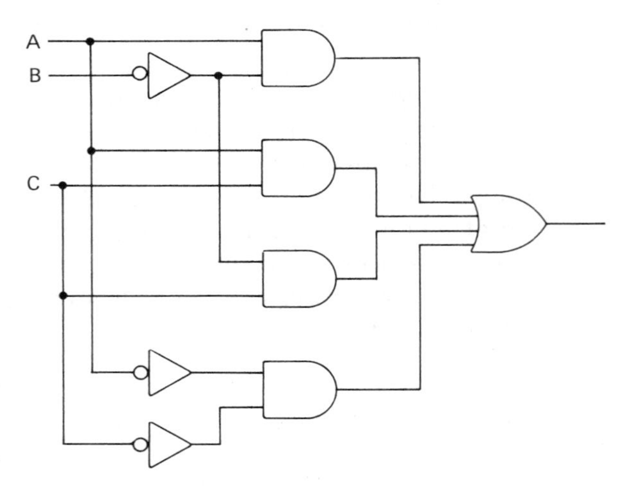

Figure P3.1(b)

3.2. Use a Karnaugh map to determine if the "room with three doors" problem can be reduced further.

3.3. Refer to the 2-line to-4-line decoder. Why can all the LEDs share the same current-limiting resistor?

3.4. Refer to the "ignition enable circuit." Why are G1 and G4 drawn in negative logic, while G2 and G3 are shown in positive logic?

3.5. Design a circuit that will emit an audible tone whenever the headlights of a car are on and the ignition switch is off. This circuit will prevent many dead batteries. The circuit will have two logic inputs: the headlight status and ignition status. A status level of logic 0 indicates that the headlight or ignition is on. The output will be an audible tone. This tone will be created will the use of a 555 timer, a driving transistor, and a speaker. Refer to Figure P3.5. The 555 timer is configured as a 1-kHz square-wave oscillator. If you are not familiar with the 555, just think of it as a black box that outputs a square wave. The transistor provides current gain to drive the speaker. If the output of the AND goes high, the transistor turns on and current flows through the speaker coil. When the AND gate goes low, the transistor turns off and the speaker coil no longer has current flowing through it.

The AND gate is to be used as an electronic switch. It is your task to design the circuitry that drives the enable input of the AND gate. Use the three-step method we introduced in this chapter.

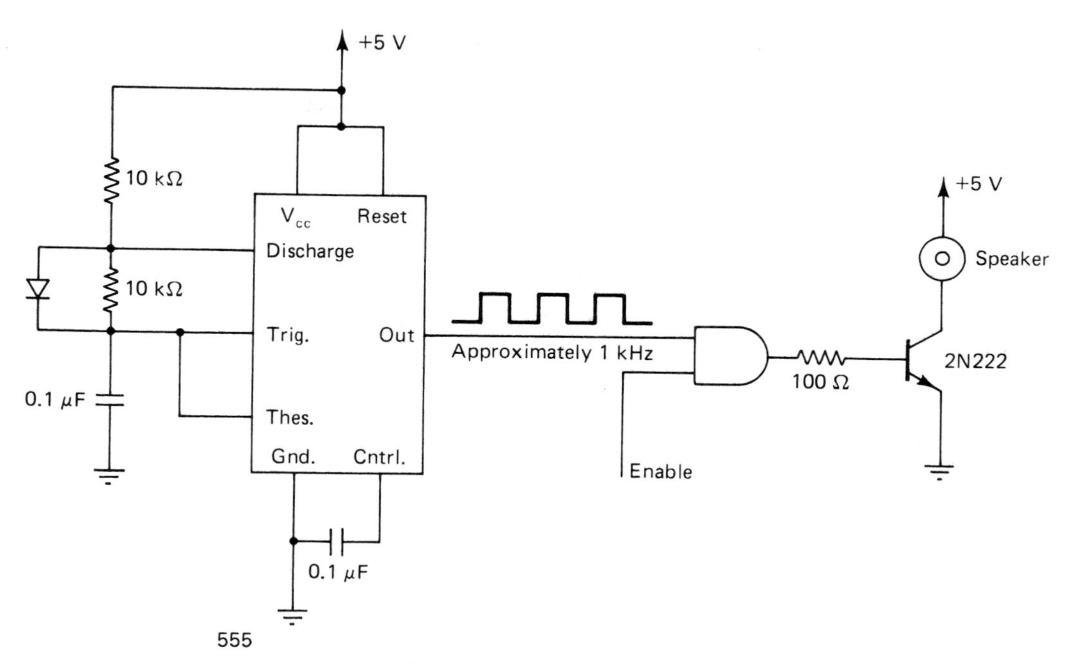

Figure P3.5

3.6. Redraw the schematic for the ignition enable circuit using only positive-logic symbols. Explain which schematic is best and why.

3.7. Design the electronics for a soda pop machine. Each soda costs 25 cents. The machine will not give change. There are four possible ways that the price of 25 cents can be paid: one quarter, two dimes and a nickel, one dime and three nickels, or five nickels. The inputs will be Q1, D1, D2, N1, N2, N3, N4, and N5.

3.8. A stairway is illuminated by one light. There are two switches that control the light: one at the top and one at the bottom of the stairway. A person should be able to control the light from either switch. *Hint:* There is a simple function that can perform this task.

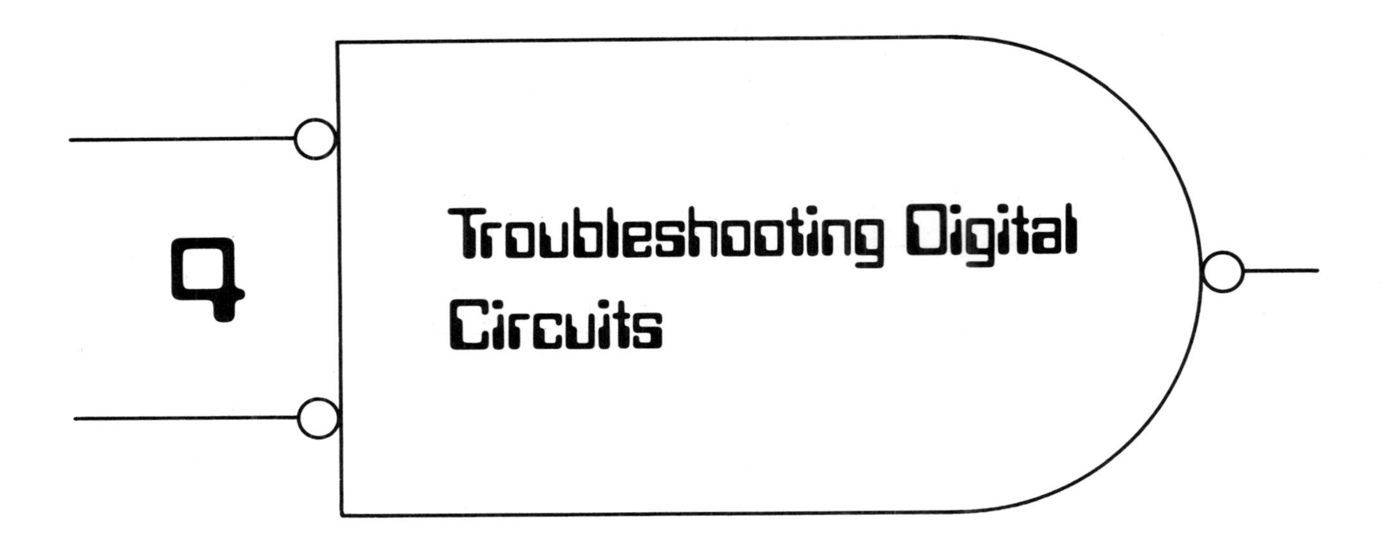

4.1 THE TECHNICIAN'S WORKING ENVIRONMENT

4.1.1 The Board Test Environment

Commonly known as *subassembly test* (SAT) or *PCB test*, this is the place where most technicians will start their careers and in many ways is the most challenging area. After printed circuit boards (PCBs) are stuffed with components, they must be tested and, if malfunctioning, repaired. Technicians work with detailed test procedures and often use special test systems to assist in the testing and troubleshooting of PCBs. Troubleshooting is down to the component level, unlike other environments, which only require the technician to find the bad PCB.

4.1.2 The Final Systems Test Environment

After PCBs are tested and repaired, they are integrated into a system. The systems test tech must check out the complete system and understand how the PCBs in the system interact. They are required to understand the system wiring and interconnect diagrams. Most systems test techs troubleshoot only down to the board level, then replace the bad board with a good unit and continue with the system test. The bad PCB is returned to board test.

4.1.3 The Field Service Environment

Field service technicians perform a wide variety of tasks. They troubleshoot many different systems to a board level, offer the customer technical assistance, and often perform a sales function. The field service tech must be capable of working with little or no supervision, and people skills are often as important as technical skills. The ability to soothe an irate customer should not be underestimated.

4.1.4 The R&D Environment

Research and development is the glamor field of electronics. Often, this glamor is more image than substance. R&D techs must work closely with high-powered, and potentially temperamental design engineers. Besides the normal electronic skills, R&D techs must also have good mathematical and mechanical ability. They build circuits and mechanical test fixtures. Long, tedious hours can be spent checking the wiring of a prototyped circuit. Weeks and months can be spent recording and analyzing data samples. The greatest attraction of the R&D environment is that R&D techs are known to quickly advance into the engineering ranks.

4.2 THE PRINTED CIRCUIT BOARD

4.2.1 The Evolution of a Design

After a design engineer has finished a preliminary schematic, the circuit must be physically realized. This is usually accomplished by a technique called *wire wrapping* (Figure 4.1). Integrated-circuit sockets are inserted into a piece of perforated fiber board (*perf board*). A wire-wrapping tool is used to wind single-strand 30-gauge wire, called Kynar wire, around the four-sided posts of the socket. The integrated circuits are then inserted into the sockets, and the design can be tested.

Wire wrapping has many advantages in prototyping and limited production applications. A wire-wrapped design can be easily modified. PCBs have a long development time and are not easily modified. R&D techs spend a good deal of time wire wrapping and debugging wire-wrapped prototyped designs.

After the prototype design has been fully debugged, updated schematics must be produced. These schematics are then given to a PCB designer to design the layout, by attaching black tape onto a Mylar sheet. The black tape represents circuit traces that will connect templates of DIP packages and other components that will reside on the PCB.

The finished tape-up is given to a photographer. A picture is taken of the taped-up sheet of Mylar. The negative or positive

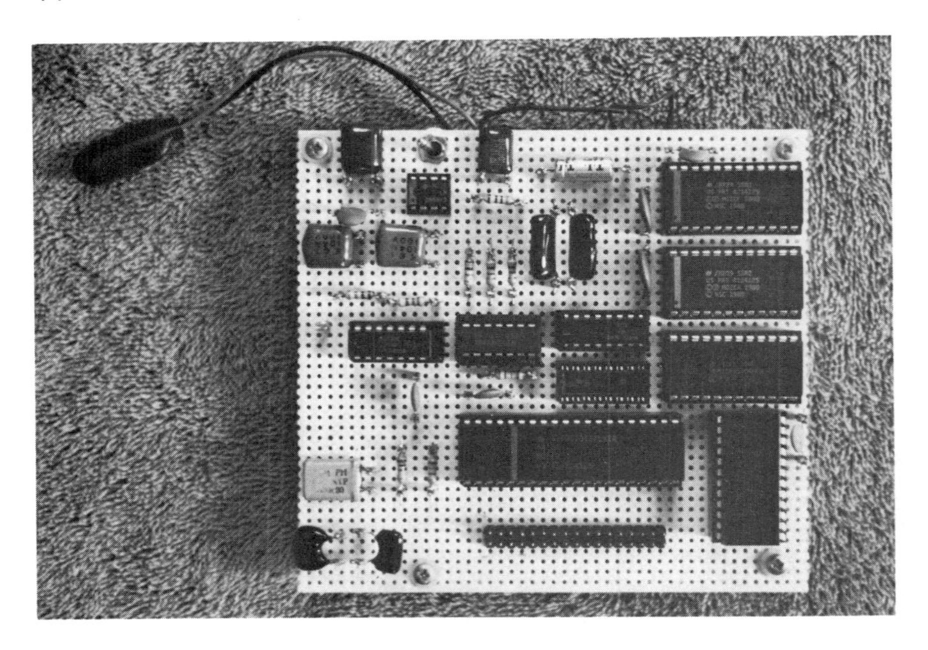

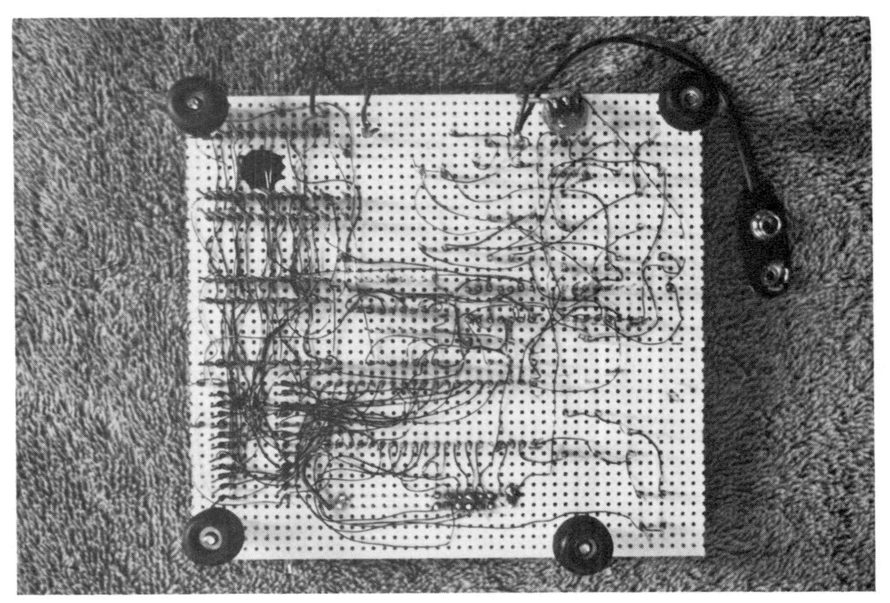

Figure 4.1 Wire-wrapped board.

of this picture is then given to a printed-circuit-board manufacturer. During a chemical process, the copper on the non-taped areas of the PCB is etched away. After the etching is complete, the PCB is thoroughly cleaned and a greenish insulating liquid called *solder mask* is applied. Finally, the PCB must have holes drilled in all the places where ICs or components will be inserted.

The ICs and all the other components are "stuffed" onto the board, either by machine or hand. The populated PCBs are run through the wave-solder machine, where all the solder connections are automatically performed. Certain plastic components such as LEDs and switches cannot be wave soldered because they would melt under the intense heat. These components must be hand soldered after the rest of the PCB is complete. The PCB is now ready to be tested in a board test environment.

The traces on the PCB are called, collectively, the *artwork*. If any mistakes are found in the artwork, the original tape-up must be modified and the whole process repeated. This is why it is important that the original wire-wrapped circuit be thoroughly debugged and a new, updated schematic be available to the PCB designer.

4.2.2 Elements of a PCB.

The bare PCB can have many manufacturing defects that will require troubleshooting in board test. It is important that the technician understand all the physical elements of a PCB. PCBs can have artwork on one side, both sides, or in layers. The majority of digital PCBs are double-sided. Multilayer PCBs are used in complicated lightweight military applications and can be difficult to troubleshoot. We will concentrate our efforts on double-sided PCBs. One side of the PCB contains the electrical components and is called the *component side*. The other side of the PCB will contain the solder connections and is called the *solder side*. Refer to Figure 4.2.

Consider the following terms describing various components that constitute a PCB:

Trace A *trace* is the conducting medium on a PCB that connects pins of ICs and other components.

Artwork As we stated previously, *artwork* refers to the PC layout. If a PCB is said to have an artwork problem, the PC layout must be modified to correct an electrical problem.

Pad ICs are plugged into holes in the PCB. *Pads* are the oblong-shaped conductors that encircle these holes on both the component and solder sides of the PCB. A square pad will designate pin 1 of an IC, the cathode of a diode or LED, the emitter of a transistor, or the positive lead of an electrolytic capacitor.

Feed-through On double-sided PCBs feed-throughs route traces from one side of the board to the other. *Feed-throughs* are plated holes through which the trace travels to the other side of the PCB.

Fingers There must be a physical interface that allows PCBs to be inserted into a system. This interface usually takes the form of an edge connector. An *edge connector* is a rectangular plastic receptor that contains an upper and lower row of spring-tension conductors. One of these conductors is called a *finger*.

Power and ground buses The traces that are wider than the rest are the power and ground buses. These buses carry considerable amounts of current, so the resistance of the trace be-

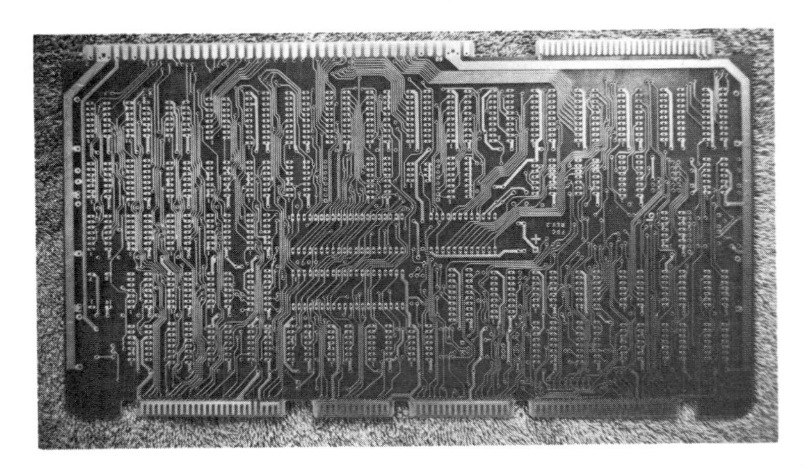

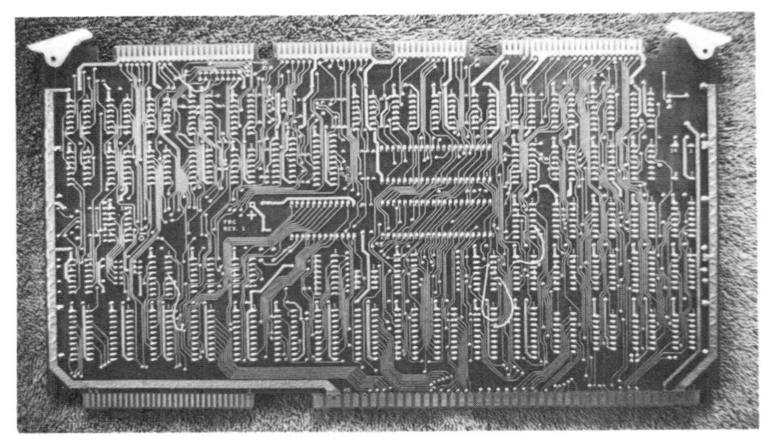

Figure 4.2 FDC: bare and stuffed.

comes an important factor. You will notice many identical capacitors on the component side of the PCB. These capacitors are called *bypass caps* and usually have a value of 0.1 μF. They are placed between the power bus and the ground bus to short any noise on the power bus to ground. Near the outside edges of the PCB you will notice large, electrolytic capacitors, located between V_{cc} and ground. These caps range in value from 22 to 100 μF. They also act as bypass caps. These larger capacitors bypass low-frequency noise. The smaller, 0.1-μF caps bypass higher-frequency noise.

Title and revision number Each PCB should have a title or a number that identifies the unit. Many PCBs even have serial numbers. No matter how well the initial design is debugged, the PCB will still go through design and artwork changes. These changes are usually denoted by a revision number or letter.

IC sockets Most ICs will be soldered directly into the PCB, but when it is not desirable to solder an IC, Dual In-line Pin (DIP) sockets are used, then the ICs are hand stuffed into these sockets. The advantage of sockets is that ICs can be quickly and easily removed and replaced without soldering. Sockets are often used for LSI devices. The process of desoldering and resoldering a 28- to 64-pin device can be frustrating and time consuming. If the IC is found not to be defective, it can be reused. CMOS devices are often socketed. Their high static sensitivity causes a high percentage of CMOS devices to malfunction during the initial checkout. Sockets can introduce problems into the system. When an IC is originally stuffed into a new socket, it may not make good electrical contact due to the tightness of the socket. It is a good habit to quickly check all socketed devices to assure that they are stuffed correctly. In electronic equipment that is subject to vibrations, ICs may work themselves loose from the sockets. Most important of all, they often cost more than the ICs that are placed into them. For these reasons the use of IC sockets is limited. In the PCB pictured, all the devices are placed in sockets because the unit is a prototype.

4.3 POSSIBLE DEFECTS IN NEWLY STUFFED PCBs

As a board test technician you will test PCBs right after they have been stuffed and wave soldered. Large companies have quality assurance (QA) people that inspect incoming PCBs. Smaller companies cannot afford this luxury. A new PCB can have many possible problems other than just malfunctioning ICs. The following is a list of common problems that you will find as a production board test tech:

Bad Components ICs, transistors, and other active devices can malfunction. They can be bad straight from the manufacturer or may be destroyed in the wave-soldering process. Capacitors can be open, shorted, or out of tolerance.

Solder Bridges A *solder bridge* is a splash of solder that shorts two or more things together. Solder bridges occur during the wave-soldering operation. The solder bridge can be between traces, between pads, or between a trace and a pad. Usually, they are visible to the unaided eye, but sometimes they are so narrow that they cannot be seen without the aid of a strong light and a magnifying glass. To remove a solder bridge, a tech can cut it with a razor knife, or heat the solder bridge and remove it with a vacuum desoldering tool or solder wick.

Unetched Traces Sometimes the chemical etching process will fail to etch away the copper between adjacent conductors. Most often, the unetched traces are between a pad and an adjacent trace. If the PCB manufacturer does not take great care in blowing the PCBs clean before the solder mask is applied, small slivers of etched-away copper may short across adjacent conductors. These shiny copper slivers will have the appearance of fine solder bridges.

Devices Stuffed Backwards It is common to find ICs, transistors, or other devices stuffed backwards. It is always a good habit to perform a quick visual check before applying power to a new PCB.

The Wrong Devices Stuffed Another common stuffing problem is inserting the wrong device. The best way to assure that all the proper components are stuffed is to compare the unit under test to a known-good unit.

Open/Cut Traces Sometimes a trace may have a small cut or break in it. This can occur at the point where the trace meets a pad, a feed-through, or a finger. An ohm meter should be used to establish if two points in question are continuous.

Open Feed-Throughs Feed-throughs are plated. During wave soldering, solder usually wicks through feed-throughs creating a good connection between the top and bottom traces. If you have an open between two points that should be continuous, be sure to check any feed-throughs for continuity.

Bent Leads under ICs When an IC is stuffed into the PCB or socket, a lead can miss the hole and be bent under the IC. It is often difficult to see these bent-under leads. If you suspect this problem, turn the PCB over to look at the soldered side. A small amount of lead should stick through each pad.

Cold Solder Joints A good solder connection should have just enough solder to cover the middle of the pad and should be bright and shiny. A connection with too much or too little solder, or a dull gray appearance, may be a *cold solder* and will need to be resoldered. By "cold" we mean that it is not forming a connective joint between the trace and the pad.

Bad Socket IC sockets can malfunction. Usually, the malfunction takes the form of an open between the IC lead and the pad. It is important to take all measurements on the lead of the IC, not the pad.

Dirty Fingers If a PCB appears to have a wide variety of problems, it is always good practice to check the fingers. Fingers should be bright and shiny. Sometimes they will corrode or be coated with a foreign insulating substance. If you suspect that a PCB may have dirty fingers, clean them with an ordinary pencil eraser until they are bright and shiny. Be sure to remove any trace of eraser bits before you reinsert the PCB into the edge connecter.

Shorted Bypass Capacitors A typical PCB may have up to 40 bypass caps. If one of these caps is shorted, the power supply will current-limit, and V_{cc} will measure as a few tenths of a volt. We discuss techniques for finding the shorted capacitor later in the chapter.

Schematic and Artwork Errors When a PCB is revised to correct a problem, another problem can be introduced. If you are troubleshooting a newly revised PCB, be alert for possible artwork problems. If your schematic and the PCB artwork do not match, this indicates a problem.

4.4 THE TTL FAMILY

TTL (transistor-to-transistor logic) is one of the two most popular integrated-circuit technologies and uses bipolar transistors. The other popular technology, CMOS, will be studied later in the chapter.

4.4.1 The 7400 Series.

The 7400 series of TTL is the most widely used group of integrated circuits in the world (see Figure 4.3). The first thing that you may notice is that Q1 has more than one emitter. Multiemitter transistors are used to provide a gate with multiple in-

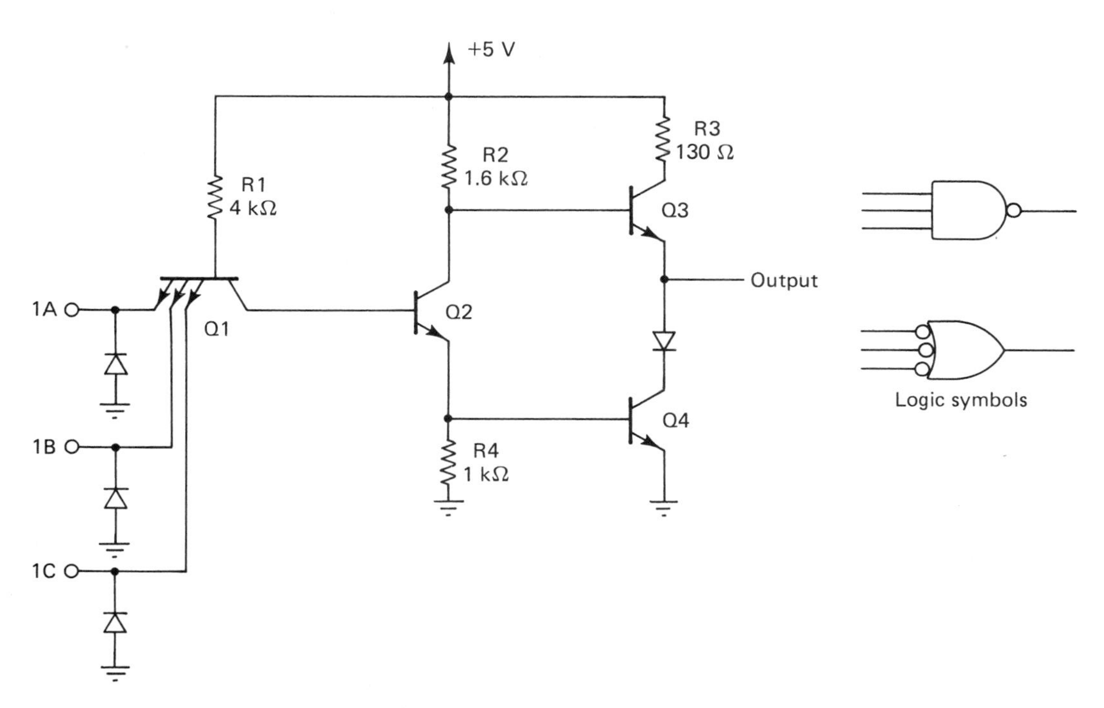

Figure 4.3 7410: three-input NAND gate.

puts. If any of the emitters have a logic 0 applied to them, Q1 will conduct and go into saturation.

The diodes on the emitters are negative-voltage-protection diodes. If a negative voltage is applied to the gate, Q1 could be damaged. These diodes clamp any negative voltage to one diode drop.

A logic 0 will forward bias the base/emitter junction of Q1 and force it into saturation, where the voltage on the emitter and collector will be approximately equal. This means that the voltage on the collector of Q1 will be 0 V, which will reverse bias the base/emitter junction of Q2. The voltage on the base of Q1 will be one diode drop (i.e., 0.6 V), leaving 4.4 V to be dropped across R1. This calculates to approximately 1.1 mA of current through R1 and the base/emitter junction of Q1, which flows into the source of the logic 0. The gate that provides the logic 0 must sink this 1.1 mA of current.

If all the emitters have logic 1's applied to them, Q1 will be cut off. Each base/emitter junction will be reversed biased. The only current that will flow in Q1 will be the base/emitter reverse diode leakage current. In a worst-case situation, this will be a maximum of 40 μA. The gate that provides the logic 1 input will not be required to source or sink any appreciable amount of current. Figure 4.4 summarizes the input circuitry of the 7400 series of TTL devices.

The standard output of the 7400 series consists of two transistors. One transistor pulls the output up toward +5 V. When the gate outputs a logic 1, the pull-up transistor will source current (i.e., the current will flow out of the gate). The other transistor pulls the output down toward 0 V. When the gate outputs a logic 0, the pull-down transistor will sink current (i.e.,

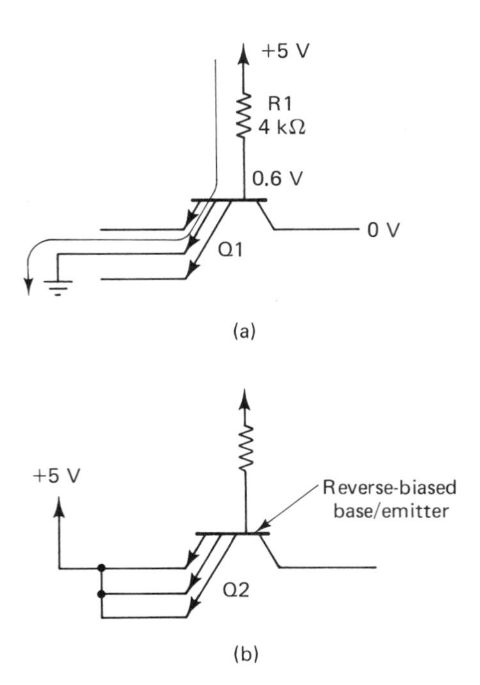

Figure 4.4 Input circuitry of 7410 three-input NAND gate.

current will enter the gate). This type of active pull-up/pull-down configuration is called a *totem-pole output.*

Consider Figure 4.5. Either Q4 will be conducting and the output will be logic 1, or Q3 will be conducting and the output will be logic 0. The diode, D1, is used to make sure that Q4 does not turn on at the wrong time. D1 can be ignored during this analysis of the totem pole.

Let's consider the current capabilities of the totem pole. If the output of the gate is shorted to ground, the 130-Ω collector resistor will limit the output current to a safe value. The trade-off in this design is that the current-limiting resistor will reduce the gate's ability to source current on a logic 1 output. TTL does not source current well.

The pull-down transistor does not have a current-limiting resistor and can sink about 40 times more current than the pull-up transistor can source. TTL sinks current well.

4.4.2 The Open-Collector Output

The totem pole is the most widely used output structure in TTL, but the *open-collector* output is used in special applications. To understand the open-collector output, we must first consider one limitation of the totem pole.

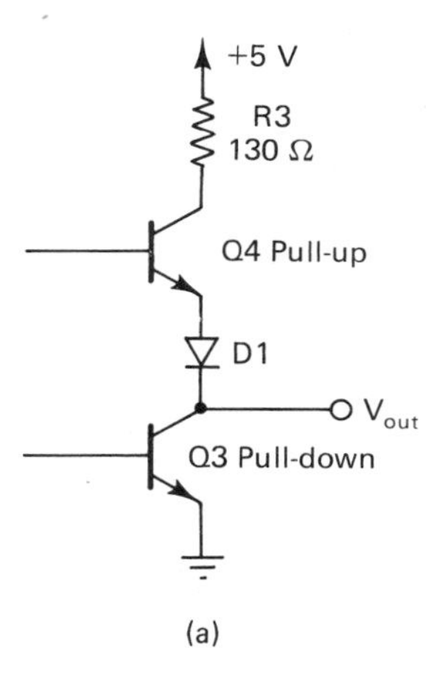

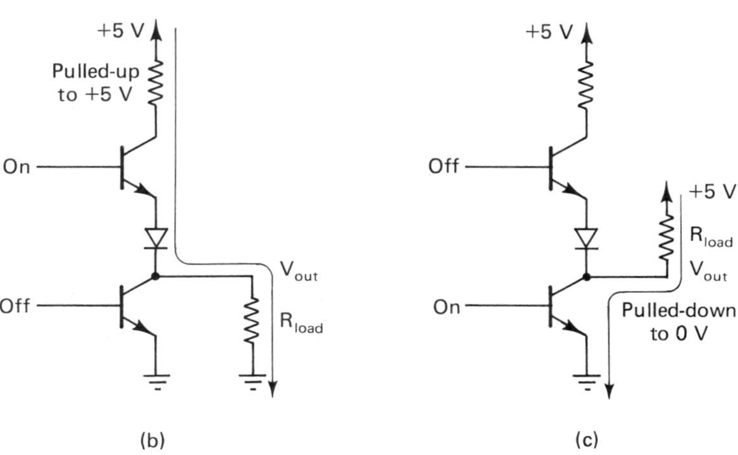

Figure 4.5 Totem-pole output.

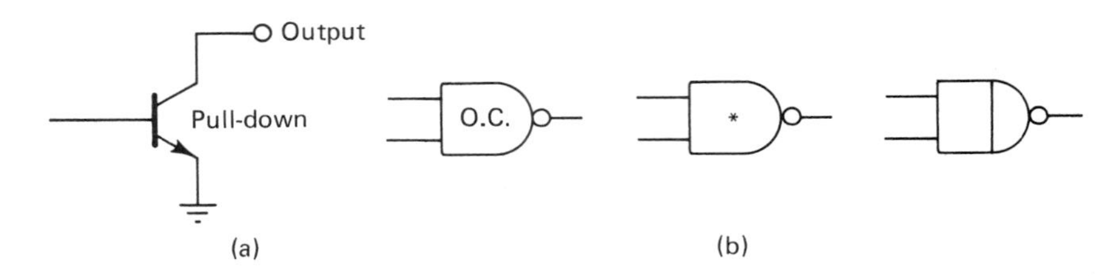

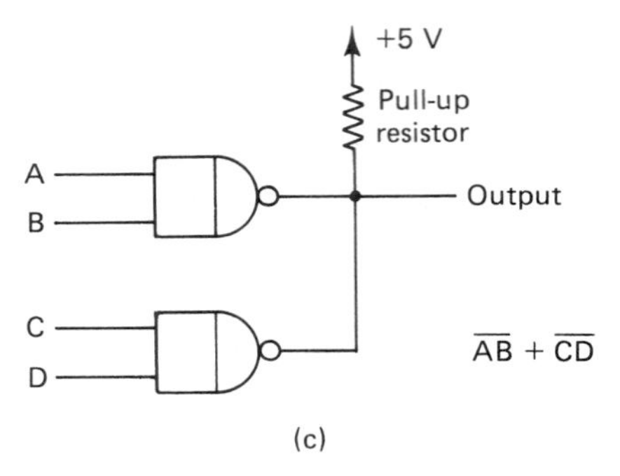

Figure 4.6 Open-collector output.

If totem-pole outputs are tied together and one gate is trying to output a logic 1 while the other gate is trying to output a logic 0, a conflict occurs which results in an indeterminate logic level. An open collector enables two or more outputs to share a common line.

Figure 4.6a illustrates an open-collector output. The output structure is simply a pull-down transistor. If the output of the gate is logic 0, the pull-down transistor is conducting. If the output of the gate is logic 1, the pull-down transistor is cut off and the output will float to an indeterminate level. The most important thing to remember about open-collector devices is that they require an external pull-up resistor.

This external pull-up resistor will have a value between 1 and 10 kΩ and provides the same function as the pull-up transistor in the totem-pole output. The advantage of an open collector is that the output of many gates can share the same pull-up resistor. Consider Figure 4.6b. (*Note:* The initials O.C. or an * or a vertical slash across the front of the gate are used in schematics to indicate an open collector.)

We can tie together an infinite number of open-collector gates sharing a common pull-up resistor; it takes only a low output on one of the gates to pull the common node low. An output that connects two or more open-collector gates together is called a wired-OR output because of the OR function provided.

4.4.3 Valid TTL Logic Levels

We have made the ideal assumption that a logic 1 is equal to +5 V and a logic 0 is equal to 0 V. Due to factors such as loading effects and nonideal transistor characteristics, there is a range of voltages for both logic 1's and logic 0's that are acceptable.

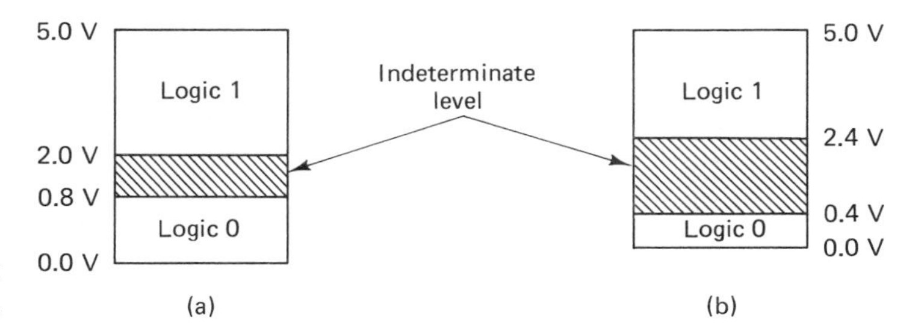

Figure 4.7 Valid TTL input and output levels.

In Figure 4.7, TTL is specified to accept an input voltage in the range 2 to 5 V as a logic 1. It is specified to accept a voltage between 0 and 0.8 V as a logic 0. Any voltage between these two ranges (i.e., 0.8 to 2 V) is neither a logic 1 nor a logic 0. These are called *indeterminate levels* and indicate a circuit malfunction.

The diagram in Figure 4.7b illustrates the specified output levels of TTL devices. These levels are guaranteed for standard TTL only if the gate is sourcing less than 0.4 mA or sinking less than 16 mA on a logic 0. Refer to Figure 4.5.

Assume that the pull-up transistor is saturated and that the totem pole is pulling the output up to logic 1. Any source current flowing out of the gate must pass through R3. As the current increases, the voltage drop over R3 will increase and the output voltage will decrease. It is specified that the gate cannot source more than 0.4 mA and still maintain an output greater than 2.4 V. If you discover a TTL logic 1 output that is less than 2.4 V, that usually indicates that the gates are being excessively loaded. Most often this is due to a short in the device that the output is driving, a solder bridge, or an unetched trace.

When the pull-down transistor is conducting, the gate is outputting a logic 0. A saturated transistor can act just like a resistor. The saturated pull-down transistor has a bulk resistance of approximately 25 Ω. As the quantity of current it is sinking increases, the voltage drop over it increases. It should be able to sink 16 mA and still maintain an output voltage that is less than 0.4 V.

4.4.4 Noise Immunity

You may have noticed that TTL output levels illustrated in Figure 4.7b are 0.4 V closer to the ideal level of +5 V and 0 V than the acceptable input levels in Figure 4.7a. This 0.4 V is the level of noise that a TTL circuit can tolerate. All circuits have a certain amount of noise. This noise is an ac quantity that amplitude modulates the signals in a circuit.

4.4.5 Driving Loads with TTL

We have seen that a TTL circuit can source only 0.4 mA of current and still meet the guaranteed logic 1 output level. Although TTL cannot source current well, it can sink up to 16 mA of current and still maintain the guaranteed logic 0 output level. Whenever TTL is required to drive any load that requires more than 0.4 mA, that load must be driven with a logic 0.

4.4.6 The Subfamilies of TTL: LS, S, L, ALS, AS

The 7400 series of TTL is a relatively old circuit technology. It has two major drawbacks: it requires appreciable amounts of bias current from the +5-V power supply and it has certain speed limitations. The subfamilies of TTL were developed to help overcome these two problems.

The 74LS00 Series This is the most popular form of TTL. The letters LS stand for *low-power Schottky*. The LS family requires much less bias current from the power supply and is also slightly faster than standard TTL. It employs Schottky transistors, which are special bipolar transistors optimized for fast switching speeds.

The 74S00 Series This series also uses Schottky transistors. Until recently it was the fastest of all TTL subfamilies. The propagation delay of Schottky TTL is about 50% less than standard TTL but requires much more power than the 74LS00 devices.

The 74L00 Series The purpose of this series is low power. The 74L00 has the lowest power requirement, but it is also the slowest of all the TTL subfamilies and has been displaced by CMOS devices.

The 74ALS00 and 74AS00 Series The advanced low-power Schottky and advanced Schottky are the most recently developed TTL subfamilies. They offer much greater speeds at lower power levels. The 74AS series is twice as fast as the 74S family.

4.5 CMOS: THE OTHER LOGIC FAMILY

CMOS stands for *complementary metal-oxide semiconductor.* CMOS employs both PMOS and NMOS transistors (FETs) in a complementary configuration. CMOS approaches the ideal characteristics of a digital device: no power dissipation, an infinitely fast switching speed, and a noise immunity of 50% of the power supply voltage. The advantages that CMOS has over TTL include low power supply requirements, excellent noise immunity, and a wide power supply voltage range.

CMOS logic gates are constructed from *N*-channel and *P*-channel enhancement-mode MOSFETs. Many transistor classes spend little or no time studying FETs (field-effect transistors). We will not try to make you an instant authority on FETs, but you should appreciate the input and output characteristics of CMOS logic devices.

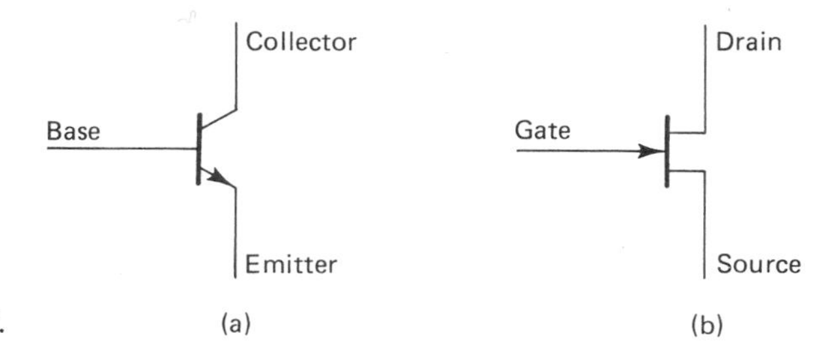

Figure 4.8 BJT and FET.

4.5.1 Comparison of Bipolar and Field-Effect Transistors

When most people mention the term *transistor* they are referring to the *bipolar junction transistor* (BJT). *Field-effect transistors* are known as FETs. BJTs and FETs are both three-terminal devices. The input that controls the operation of the BJT is called the *base.* If we source some current into the base of a *NPN* transistor, the collector current will be equal to the base current times a gain factor called *beta*. The BJT is called a *current-driven device.* It depends on base current to control the collector current. The control input of the FET is called the *gate.* The collector lead of the FET is called the *drain* and the emitter lead the *source.* The FET is a *voltage-driven device*, meaning that little current is required from the device providing the input signal. That is the reason for the extremely low power requirements of CMOS. Figure 4.8 illustrates the schematic symbols of the BJT and FET.

There are two major types of FETs: junction FETs (JFETs) and metal-oxide semiconductor FETs (MOSFETs). Because CMOS logic devices are constructed from MOSFETs we will focus our attention on them. Like TTL, CMOS requires two voltages: V_{dd} (the voltage to the drain) and V_{ss} (the substrate voltage). V_{ss} is most often referenced to ground. As we have mentioned, CMOS gates are constructed from *P*-channel and *N*-channel enhancement-mode MOSFETs, which are normally in the cutoff state and will only conduct if an appropriate gate voltage is applied.

Figure 4.9 shows the schematic symbols for these MOSFETs. The imaginary diode indicates whether the channel is *P* or *N*. If the anode of the diode is pointing at the channel, the MOSFET is a *P* type. If the diode's cathode is pointing at the channel, the MOSFET is *N* type. The FET will conduct when this diode is reversed biased. A logic 0 will cause the *P*-channel

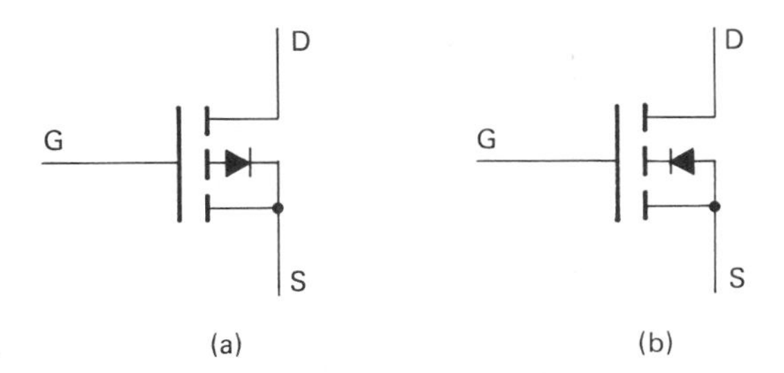

Figure 4.9 MOSFETs.

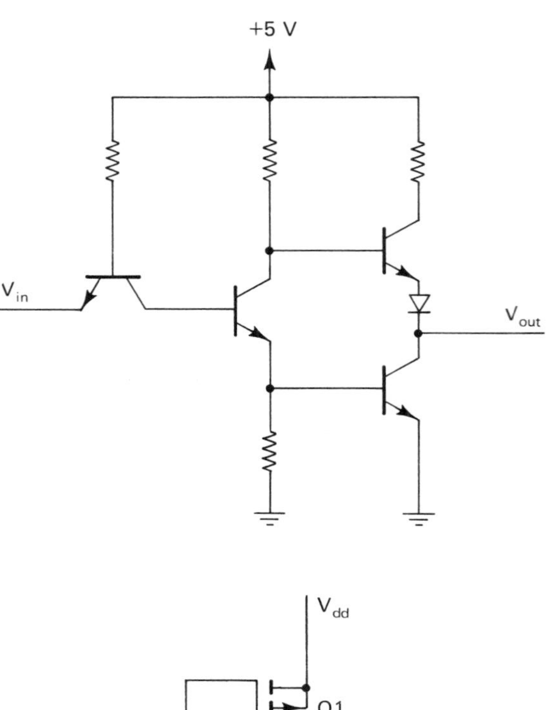

Figure 4.10 TTL and CMOS inverters.

MOSFET to saturate, and a logic 1 will cause the *N*-channel MOSFET to saturate.

Figure 4.10 illustrates a TTL inverter and a CMOS inverter. The first impression that you should get from Figure 4.10 is the simplicity of the CMOS inverter. It requires only two MOSFETs, one *P*-channel and one *N*-channel, while the TTL inverter requires four BJTs, four resistors, and one diode.

A logic 0 will drive a *P*-channel MOSFET into saturation and a logic 1 will drive an *N*-channel MOSFET into saturation. If a logic 0 is applied to the input of the CMOS inverter, Q1 will be driven into saturation and Q2 will be cut off. The output will be pulled up to V_{dd}, a logic 1. If a logic 1 is applied to the input, Q1 will be cut off and Q2 will be driven into saturation. The output will be pulled down to V_{ss}: a logic 0. Notice that Q1 and Q2 will never be in the same state. One is always saturated while the other is cut off.

4.5.2 The 4000B Series of CMOS Devices

The letter B means that this series is *buffered*. The buffer consists of a CMOS driver circuit. The 40 is often designated as 140 or 340 by various manufacturers. Thus a 4001B and a 14001B and a 34001B are all the same device. You will also discover 4000UB (*unbuffered*) CMOS devices. The major difference between the 4000B and 4000UB series is drive ability. The UB series is employed when CMOS digital devices are used as oscillators or amplifiers. UB devices are slightly faster than B devices because they do not have the extra buffer stage.

4.5.3 The 74C00 and 74HC00 Series

As the speed and drive capabilities of CMOS have increased, it has started to replace TTL in new designs. The 74C00 family is designed to be a pin-for-pin replacement for TTL. 74C00 devices are about 25% more expensive than their TTL counterparts, but their manufacturing costs are constantly dropping. Because CMOS requires so little power, other parts of the digital system, such as power supplies, are less expensive.

CMOS also has an advanced technology family, 74HC00. The letter H stands for *high speed*. These devices are 10 times faster than the 74C00 devices.

4.6 INTERFACING TTL AND CMOS

A valid logic 1 for CMOS devices with a V_{dd} of +12 V is between +12 and 8.4 V and the valid logic 0 is between 0 and 3.6 V. If we drive a CMOS gate with a TTL gate, both TTL logic 0's and logic 1's may appear to be CMOS logic 0's. On the other hand, if we drive a TTL gate with the output of a CMOS gate, the logic 1 output of the CMOS gate could destroy the TTL input transistor. We need to have some means of interfacing these two logic families.

4.6.1 The CMOS 4049/4050 Buffers–Converters

These two CMOS devices are used to convert CMOS logic levels to TTL logic levels. Notice in Figure 4.11 that the power pin is designated as V_{cc}—not the usual designation of V_{dd} for CMOS

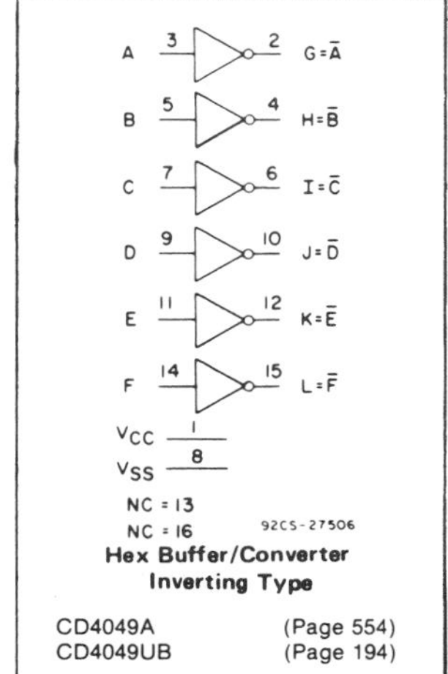

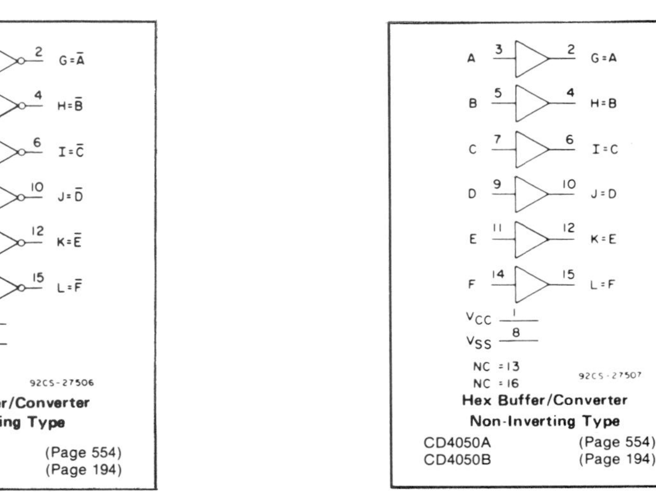

Figure 4.11 Hex buffers/converters.

HEX INVERTER BUFFERS/DRIVERS WITH OPEN-COLLECTOR HIGH-VOLTAGE OUTPUTS

06

positive logic:
$Y = \overline{A}$

See page 6-24

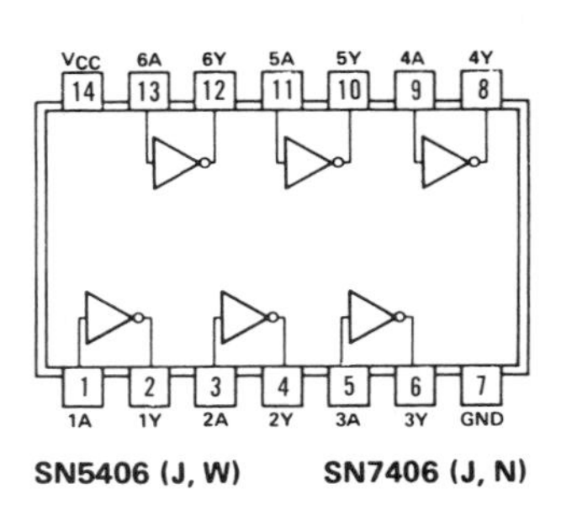

HEX BUFFERS/DRIVERS WITH OPEN-COLLECTOR HIGH-VOLTAGE OUTPUTS

07

positive logic:
$Y = A$

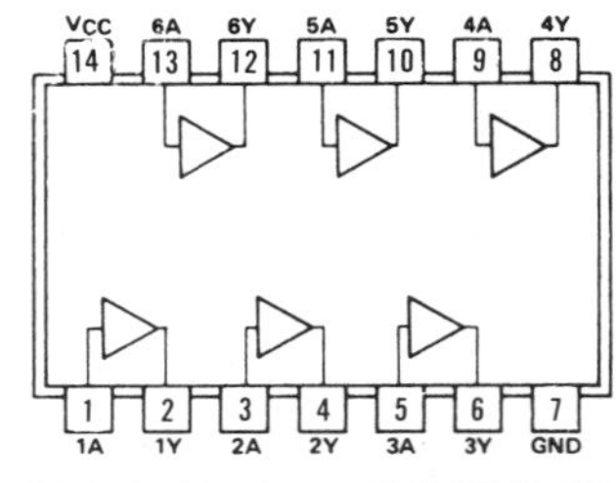

Figure 4.12 TTL-to-CMOS conversion.

devices. The 4049 converts logic levels and inverts the logic signal. The 4050 converts only logic levels. The output of these converters can drive two standard TTL inputs.

4.6.2 The 7406-7407 Open-Collector Buffer/Interface Gates

The 7406/7407 devices convert TTL logic levels to CMOS logic levels (Figure 4.12). This is accomplished by connecting the external pull-up resistor to V_{dd} of the CMOS devices. A logic 0 output will be equal to 0 V and a logic 1 output will be pulled up to V_{dd} via the external pull-up resistor. The maximum voltage the pull-up resistor can be connected to is +30 V. This is well beyond the range of CMOS V_{dd}. These higher pull-up voltages are used to drive other devices (such as incandescent displays) that require high voltages.

4.7 TROUBLESHOOTING OPENS

The questions that you should be asking yourself are:

1. What is the voltage one would measure on an open input?
2. At what logic level does a gate interpret an open input—a logic 1 or a logic 0?

4.7.1 The TTL Open Circuit

Most malfunctioning gates will be stuck at either a logic 0 or a logic 1. If you observe the output of gate switching back and forth between valid logic levels, this usually indicates that the gate is good. That does not imply that the inputs driving the gate are good.

An open between an output and an input will always result in a floating input. A floating TTL input will measure between

1.4 and 1.8 V dc. This is the single most important fact that a technician troubleshooting a TTL open can possess.

Once you have discovered a floating input by observing a dc voltage between 1.4 and 1.8 V, you must find the open. You will use your eyes and an ohm meter to find the cause of an open. Ohm meters can only be used on deenergized circuits, so power-off the circuit. The ohm meter should be placed in the most sensitive range, typically 200 Ω full scale. The ohm meter will be used as a device to test the electrical continuity between the two points. It will read infinite resistance on an open.

Does a floating input look more like a logic 0 input or a logic 1 input? A logic 0 input must be able to sink about 1 mA of current; a floating input cannot sink any current. On the other hand, a logic 1 input does nothing more than reverse bias the base–emitter junction of the input transistor; it must deliver only a minute amount of reverse diode leakage current. It is obvious that a floating input and a logic 1 input have exceedingly similar characteristics. Therefore, a floating TTL input appears to be a logic 1.

4.7.2 The CMOS Open Circuit

If a CMOS input is left floating, it will eventually destroy itself. The inputs of unused gates of TTL ICs can be left floating and no harm will come to the IC. All the unused inputs in a CMOS IC must be tied to V_{dd} via a resistor or ground. If a CMOS input is left floating, noise can cause both MOSFETs to conduct simultaneously. If this happens, neither MOSFET will have a load to current-limit it and the CMOS IC will be destroyed. Because the input impedance of CMOS inputs is in the range 10 to 50 Ω they will be affected much more adversely by noise than will the low-input-impedance TTL devices. A floating CMOS input will not exhibit a specific voltage as did the floating TTL input. An oscilloscope will usually reveal random noise transients on a floating CMOS input. Thus a floating CMOS input does not appear to be either a logic 1, as did TTL, or a logic 0—it switches with the noise.

4.8 TROUBLESHOOTING SHORTS

We are going to concern ourselves with the most common types of shorts: shorted outputs to ground, shorted outputs to V_{cc} or V_{dd}, outputs shorted to other outputs, and V_{cc} shorted to ground. Finding shorts in a board test environment may be the single most difficult task facing the technician.

4.8.1 Isolating Outputs

The output of one gate may drive the inputs of many other gates. This increases the complexity of finding the short and requires isolating the suspect output. If the gate is in a socket, bend the leg so that it does not make a connection. If it is soldered, we can cut the trace leading from it or cut the leg itself. This eliminates the possibility of the output being loaded down from an external source. If the output is still not functioning correctly, replace the chip. If it functions correctly after being isolated, the gate is good. After reconnecting the leg, all the inputs it supplies must be isolated individually, using the same techniques. If the gates all test good, the board itself must be checked. If a good visual inspection of the PCB in a well-lit area does not reveal the short, it is time to use an ohm meter. As always, whenever using an ohm meter, the first step is to deenergize the circuit. Then check for shorts between adjacent legs of the devices involved. Be patient and methodical and you will find the problem. Finding shorts of this type can be time consuming. The author has personally spent more than two hours finding one microscopic short on an extremely dense PCB.

4.8.2 Shorts in TTL Circuitry

A typical TTL low is 0.1 to 0.2 V. If you ever find a TTL logic 0 that looks like 0 V, it is probably shorted to ground. A typical TTL logic 1 is between 3.2 and 3.6 V. If you measure a TTL logic 1 that is 5 V, the gate is under a no-load condition (an open exists between the output of the gate and the input to the gate that is driving) or the output is shorted to V_{cc}.

4.8.3 The CMOS Short

CMOS logic levels approach the ideal of ground and V_{dd}. Therefore, a CMOS output short to ground or V_{dd} is more difficult to see than the same short in TTL. Just remember that if a CMOS gate is stuck high or low, it could be shorted to V_{dd} or ground, respectively.

4.8.4 Shorted TTL Totem-Pole Outputs

What happens when two totem-pole outputs are shorted together? Consider Figure 4.13. The input to G1 is a 1-kHz, 50% duty cycle, TTL-level square wave. The input to G2 is exactly the same signal, except that it has a frequency of 500 Hz. The timing diagram in Figure 4.13 is divided into four parts: T1, T2, T3, and T4. We will analyze the output of the shortened totem-poles for each time period.

T1: The inputs to both G1 and G2 are high; therefore, both active pull-down transistors in G1 and G2 saturate and the output will go to logic 0. Notice that both outputs are pulling in the same direction and actually assist each other.

T2: The input to G1 has gone low, but the input to G2 is still high. The output of G1 wants to go high; the active pull-up transistor in G1 turns on and saturates. The output of G2

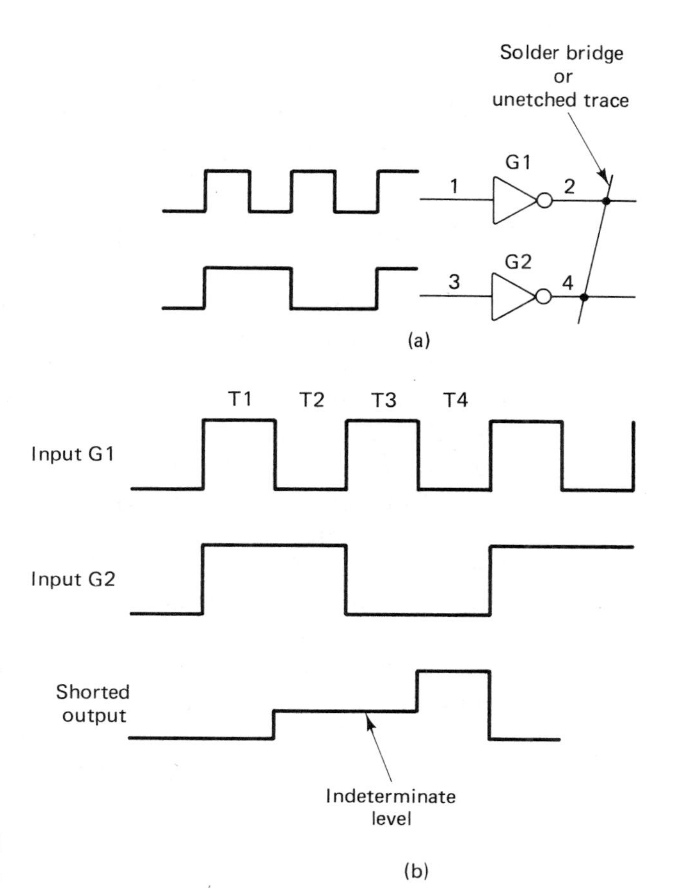

Figure 4.13 Shorted totem-pole outputs.

wants to go low; the active pull-down transistor in G2 stays saturated. It appears that because of the short between the two outputs, there is a conflict. G1 wants to go high, and G2 wants to go low. We know that TTL is much better at sinking current than it is at sourcing current. The pull-up transistor in G1 will source between 20 and 60 mA of short-circuit current. The pull-down transistor in G2 will be trying to sink all this current. The result will be a voltage level at the shorted output in the indeterminate range—somewhere between a valid logic 0 and a valid logic 1. With LS TTL this voltage level will be approximately 0.8 to 1.2 V.

T3: During this time period the input to G1 goes high, while the input to G2 goes low. Once again, there will be a conflict at the shorted outputs that results in an indeterminate voltage between 0.8 and 1.2 V.

T4: Both inputs are low, so both outputs go high. There is no conflict and the shorted node will pull up to a valid logic 1.

Review the output timing diagram in Figure 4.13. When you are observing digital outputs on an oscilloscope, the logic lows will all have the same voltage level, as will the logic highs. Whenever you discover a "third" level in the low part of the indeterminate range, this indicates a shorted totem-pole output. This is a common occurrence in a board test environment. Quickly recognizing the classic characteristic signal of shorted totem poles can save you much time and effort.

4.8.5 Shorted Bypass Capacitors

The first check you should always make on a malfunctioning PCB is to monitor the power bus with an oscilloscope. The scope should display a flat 5-V ($\pm$0.25 V) dc signal. If V_{cc} measures just a few tenths of a volt, the power supply is probably in voltage fold-back current limit (commonly called *crow-bar*), indicating a short between V_{cc} and ground.

After you have discovered what appears to be a V_{cc} short to ground, your first step should be to isolate the power supply from the PCB. Measure the +5-V output of the isolated power supply to assure that it is functioning correctly.

If the isolated power supply checks out, you should visually inspect the board. When ICs are internally shorted they will generate a significant amount of heat. Touch all the ICs, checking for one that seems to be running hotter than the others. If the visual inspection and touch tests do not reveal the problem, it is time to use the test equipment.

A shorted bypass capacitor or internally shorted IC will measure from 5 to 15 Ω, whereas a solder bridge or unetched trace short will measure a dead short of 0 Ω. Measure the resistance between the power and ground bus.

The greatest problem in finding V_{cc}-to-ground shorts is the difficulty of isolating the components. If you are using a typical digital multimeter (DMM), the most sensitive resistance scale is probably 200 Ω, which is not sensitive enough to track down the short. You must start cutting traces and lifting components. Power buses are wide traces. They are difficult to cut and resolder correctly. The most efficient way of approaching this problem is to cut the PCB in halves until you find the short. Remember that the ICs and bypass caps will be organized in rows; you should continue to isolate the short down to the bad row. At this point, you can start to desolder one leg of capacitors or isolate the V_{cc} pin on the ICs until you find the bad component.

If you are fortunate enough to have a high-resolution ohm meter, the task of finding the short will be much faster and easier. A typical high-resolution ohm meter will have a full-scale reading of 2 Ω. The high-sensitivity scale is typically labeled "Low Ohms." There are high-resolution ohm meters that measure resistances as small as 10 mΩ. Use the high-resolution ohm meter with a set of needle-pointed probes. Move the probes between the power and ground buses until you find the least resistance; this will be the point of the short.

QUESTIONS AND PROBLEMS

Use the pinouts of the TTL and CMOS devices contained in this chapter to answer the following questions.

4.1. Redraw the reduced three-judges circuit. Make a list of all the TTL ICs that would be required to build the circuit. Indicate the physical pin number of each IC on its appropriate lead in the schematic.

4.2. Repeat Problem 1 with the ignition enable circuit.

4.3. What is the typical logic 1 output of TTL? CMOS with +12-V V_{dd}?

4.4. What is the typical logic 0 output of TTL? CMOS with +12-V V_{dd}?

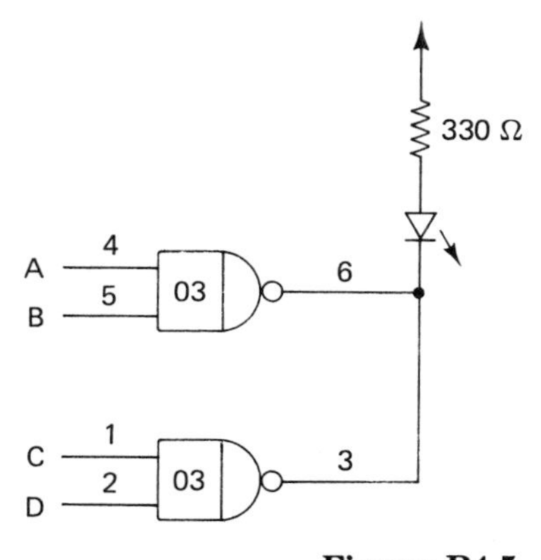

Figure P4.5

4.5. Consider the schematic shown in Figure P4.5. Under what input conditions will the LED illuminate? Redesign this circuit using gates with totem-pole outputs. What is the advantage of using open-collector gates in applications such as this?

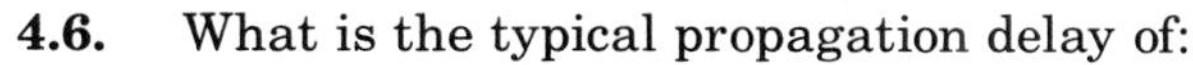

4.6. What is the typical propagation delay of:
(a) An LS TTL gate?
(b) An S TTL gate?
(c) A CMOS gate with a V_{dd} of 5 V? 10 V? 15 V?

4.7. A lumber mill wants to install digital circuits to increase the efficiency of the cutting saws. Considering the environment, should they use TTL or CMOS circuitry? Why?

4.8. The circuit shown in Figure P4.8 will output a valid logic 0, but the logic 1 output falls into the indeterminate range. What is the problem?

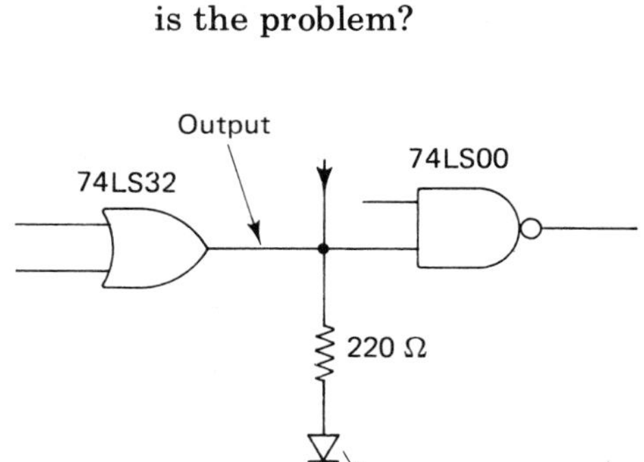

Figure P4.8

4.9. What range of voltages are indeterminate for TTL? +12-V CMOS? +5-V CMOS?

4.10. Why is LS the most popular form of TTL?

4.11. What is the greatest limiting factor of 74HC CMOS devices?

4.12. What precautions should a technician take when handling CMOS devices?

4.13. A buzzer requires a voltage of +25 V. What TTL device can be used to drive this buzzer?

4.14. In what situations do you expect to encounter wire-wrapped circuits? What are the advantages of wire-wrapped circuits? What are their disadvantages?

4.15. What are the advantages of PCBs? The disadvantages of PCBs?

4.16. What problems can occur when a PCB is being stuffed? What is the easiest way to spot these problems?

4.17. What problems can occur when a PCB is being wave soldered?

4.18. Name the situations where you expect to encounter ICs in sockets. What are the advantages of sockets? The disadvantages?

4.19. How does one distinguish a power or ground trace from other traces on a PCB?

4.20. What does a square pad designate?

4.21. What is the function of a feed-through?

4.22. What characteristics indicate a cold solder joint? Dirty fingers?

4.23. What places on a PCB are most likely to have solder bridges? Unetched traces?

4.24. How can one recognize when a lead on an IC is bent underneath the IC?

4.25. What is the function of bypass capacitors? Why don't CMOS PCBs require them?

4.26. How can one isolate a particular pin on an IC? In what situations would this be an appropriate action?

4.27. Why does a circuit have to be deenergized to use an ohm meter?

4.28. A TTL input measures 1.5 V. What is the problem?

4.29. An oscilloscope displays noise spikes on the input of a CMOS device. What is the possible problem?

4.30. Why must all CMOS inputs be terminated?

4.31. What logic level does a floating TTL input appear to be? Why?

4.32. What logic level does a floating CMOS input appear to be? Why?

4.33. What test instrument is used to find opens? Shorts?

4.34. Will a TTL gate whose output is shorted to V_{cc} be damaged? Why?

4.35. Will a TTL gate whose output is shorted to ground be damaged? Why?

4.36. What does a TTL voltage between 0.8 and 1.2 V indicate?

4.37. If a circuit fails when you first power-up, what should be your first check?

4.38. What characteristic of switching power supplies is it important to remember?

4.39. Define a high-resolution ohm meter.

5 Combinational Devices

MSI devices are used to support microprocessor, memory, input/output circuitry, and many other advanced digital functions. At this point in your digital education it is difficult to illustrate these MSI devices in meaningful circuits. Consider them as the nuts and bolts that hold together advanced LSI devices. All the ICs that constitute a PCB can be considered as building blocks. Some of the blocks, such as the basic gates, are simple; other blocks will be extremely complex. You have already mastered the most fundamental blocks. Your next task is to understand and master MSI combinational devices.

5.1 THE MAGNITUDE COMPARATOR

5.1.1 The 74LS85 4-Bit Magnitude Comparator

Specifications sheets for MSI devices include a pinout, function table, word description, equivalent gate circuit, detailed specifications, and test information. Your major focus should be on the pinout, function table, and word description.

Consider Figure 5.1. The circuit symbol in the pinout is usually the same symbol that is used to represent the device in schematics. The word descriptions provided by the manufactur-

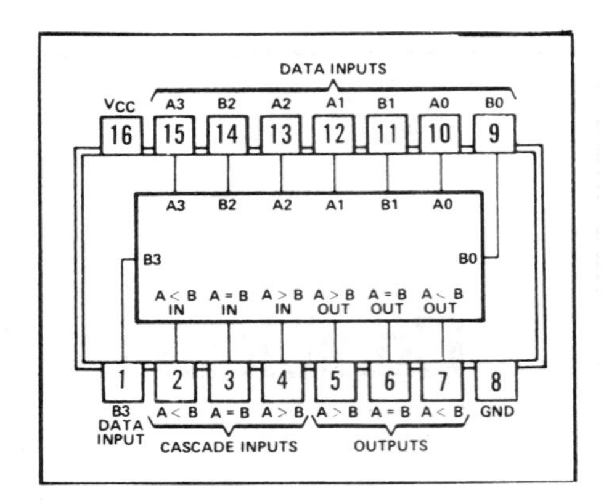

FUNCTION TABLES

COMPARING INPUTS				CASCADING INPUTS			OUTPUTS		
A3, B3	A2, B2	A1, B1	A0, B0	A > B	A < B	A = B	A > B	A < B	A = B
A3 > B3	X	X	X	X	X	X	H	L	L
A3 < B3	X	X	X	X	X	X	L	H	L
A3 = B3	A2 > B2	X	X	X	X	X	H	L	L
A3 = B3	A2 < B2	X	X	X	X	X	L	H	L
A3 = B2	A2 = B2	A1 > B1	X	X	X	X	H	L	L
A3 = B3	A2 = B2	A1 < B1	X	X	X	X	L	H	L
A3 = B3	A2 = B2	A1 = B1	A0 > B0	X	X	X	H	L	L
A3 = B3	A2 = B2	A1 = B1	A0 < B0	X	X	X	L	H	L
A3 = B3	A2 = B2	A1 = B1	A0 = B0	H	L	L	H	L	L
A3 = B3	A2 = B2	A1 = B1	A0 = B0	L	H	L	L	H	L
A3 = B3	A2 = B2	A1 = B1	A0 = B0	L	L	H	L	L	H
'85, 'LS85, 'S85									
A3 = B3	A2 = B2	A1 = B1	A0 = B0	X	X	H	L	L	H
A3 = B3	A2 = B2	A1 = B1	A0 = B0	H	H	L	L	L	L
A3 = B3	A2 = B2	A1 = B1	A0 = B0	L	L	L	H	H	L

Figure 5.1 Pinout and function table for the 74LS85.

ers are highly technical in nature. As your technical expertise increases, you will learn to read and understand these descriptions. On the job, these word descriptions will often be your sole source of information.

The Pinout The pins on the 74LS85 will fall into five groups:

1. V_{cc} and ground
2. Word A
3. Word B
4. Outputs
5. Cascade inputs

V_{cc} and Ground. All TTL devices will have pins for V_{cc} and ground. The 74LS85 is a 16-pin DIP and, as you would expect, pin 8 is ground and pin 16 is V_{cc}.

Word A and Word B. The 74LS85 compares two 4-bit binary quantities. Word *A* is the first group of four bits, while word *B* is the second group. Here the term "word" is used in its most informal sense; it designates nothing more than a group of binary inputs.

Outputs. The 74LS85 has the same three outputs as the 2-bit magnitude comparator that we designed earlier: $A = B$, $A > B$, and $A < B$. Consider the comparison as if we are subtracting the value of word B from word A: if the result is equal to 0, then $A = B$; if the result is a positive number, then $A > B$; if the result is a negative number, then $A < B$. Notice that only one output can be true for any given combination of inputs.

Cascade Inputs. You may be familiar with the definition of the term "cascade" as it applies to amplifiers. When amplifiers are cascaded, the output of one amplifier becomes the input of the next amplifier. The cascade inputs on the 74LS85 provide a means for expansion. The cascade inputs are $A = B$, $A > B$, and $A < B$. You should notice that the cascade inputs are labeled the same as the outputs. If we wish to increase the word length to compare from 4 bits to 8 bits, two 74LS85s will be cascaded. The outputs from the least significant comparator (the first comparator) will be connected to the cascade inputs of the most significant comparator. The outputs of the second comparator will be used to indicate the final result of the 8-bit compare operation.

The most significant comparator has no way of knowing that there are no other comparators upstream, so we tie the $A = B$ input high and tie the $A > B$ and $A < B$ inputs low. We are telling the comparator to assume that any upstream comparisons have produced the result that $A = B$.

The Function Table When we compare two numbers, we start our examination with the most significant digit. If one is greater than the other, our comparison is complete. There is no need to continue and examine the next digit. The only time we would continue the comparison is when the most significant digits are equal. With this in mind, examine the function table for the 74LS85 in Figure 5.1.

In the last four lines of the function table the cascade inputs become significant. If the 4 bits of word A and word B are equal, the comparator must look upstream to the next-least-significant comparator for information.

5.1.2 Applications of Magnitude Comparators

Magnitude comparators are widely employed as address decoding devices on memory boards. They are also useful devices in servomotor control circuits. By comparing a known reference quantity to a binary number that indicates the actual position of a servomotor, the outputs of the comparator can be used to drive the servomotor forward, backward, or to a halt.

5.2 THE BCD-TO-DECIMAL DECODER

A decoder converts an encoded input into an intelligible output. The BCD-to-decimal decoder inputs a BCD number and then activates one of its 10 output lines that relates to the decimal equivalent of the BCD input. The BCD decoder is also known as a 4-line-to-10-line decoder.

5.2.1 The 74LS42

Examine the pinout in Figure 5.2. This device has four inputs to accomodate a BCD number. Input *A* is the least significant bit, and input *D* is the most significant bit. There are 10 outputs, one for each possible decimal number: 0 through 9. Notice that the output lines are bubbled. When an output line is bubbled, that output is said to be *active-low*. We normally think of an active output as indicating a logic 1 level. Many TTL devices have active-low outputs. The reason for this is simple: TTL is much better at sinking current than it is at sourcing current. If the active output level is a logic 0, the TTL circuitry can provide its best current capability.

Refer to the function table. Notice that the output which is the decimal equivalent of the BCD input will go to its active-low level, while the other outputs will be at their inactive-high level. We know that the BCD codes from 1010 to 1111 are invalid. For these six invalid combinations, all the outputs will retain their inactive-high levels.

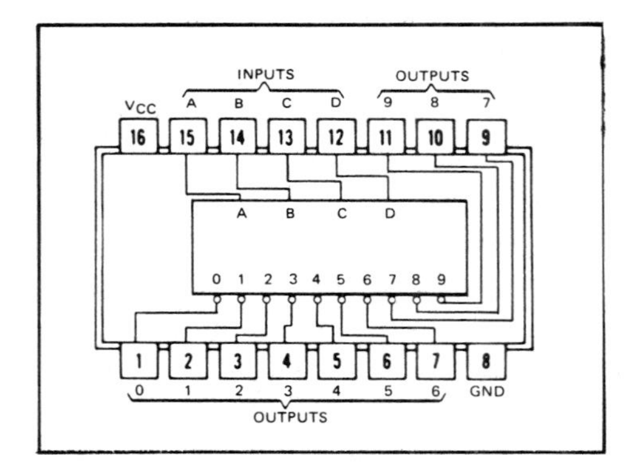

NO.	'42A, 'L42, 'LS42 BCD INPUT				ALL TYPES DECIMAL OUTPUT									
	D	C	B	A	0	1	2	3	4	5	6	7	8	9
0	L	L	L	L	L	H	H	H	H	H	H	H	H	H
1	L	L	L	H	H	L	H	H	H	H	H	H	H	H
2	L	L	H	L	H	H	L	H	H	H	H	H	H	H
3	L	L	H	H	H	H	H	L	H	H	H	H	H	H
4	L	H	L	L	H	H	H	H	L	H	H	H	H	H
5	L	H	L	H	H	H	H	H	H	L	H	H	H	H
6	L	H	H	L	H	H	H	H	H	H	L	H	H	H
7	L	H	H	H	H	H	H	H	H	H	H	L	H	H
8	H	L	L	L	H	H	H	H	H	H	H	H	L	H
9	H	L	L	H	H	H	H	H	H	H	H	H	H	L
INVALID	H	L	H	L	H	H	H	H	H	H	H	H	H	H
	H	L	H	H	H	H	H	H	H	H	H	H	H	H
	H	H	L	L	H	H	H	H	H	H	H	H	H	H
	H	H	L	H	H	H	H	H	H	H	H	H	H	H
	H	H	H	L	H	H	H	H	H	H	H	H	H	H
	H	H	H	H	H	H	H	H	H	H	H	H	H	H

Figure 5.2 74LS42 BCD-to-decimal decoder.

5.2.2 Applications of BCD-to-Decimal Decoders

Before the advent of LEDs, digital displays were implemented with neon readout tubes. Ten neon tubes would be contained in one enclosure. Each tube was in the shape of a particular decimal digit. The 10 outputs of a BCD-to-decimal decoder were used to drive the neon tubes. Only one digit per package would ever be illuminated at any instant. In this way decimal displays were created. These tubes required high voltages and moderate amounts of power. These fluorescent displays have been replaced with LED and LCD indicators, which are covered later in the chapter.

Like magnitude comparators, BCD-to-decimal decoders are also used in address decoding. Consider the situation where we wish to control 10 separate dc motors. One and only one motor will ever be energized simultaneously. All we have to do is place the BCD code of the motor that we wish to energize on the inputs of a BCD-to-decimal decoder. By changing the BCD code, the motor that is energized will change. If we desire to turn off all the motors, we must place an illegal BCD code on the inputs of the decoder.

5.3 THE 3-LINE-TO-8-LINE DECODER/DEMULTIPLEXER

With the BCD-to-decimal decoder we could drive all the outputs into their inactive state by applying an illegal BCD code to the select inputs. This would effectively "disable" the decoder. A 3-line-to-8-line decoder does not have any illegal select codes. Instead of using an illegal code to disable the outputs, 3-line-to-8-line decoders employ an enable input. When this enable input is active, the decoder will perform its normal function. When this enable input is inactive, the decoder will be disabled and all the outputs will go to their inactive level, regardless of the select code.

This enable input on decoders can also be used as a data input. In that case the decoder would be performing the function of a demultiplexer. If many data transmitters use a single, common line to send data, there must be a method of guiding the proper data to the proper receiver. That is the function of a demultiplexer. We will study demultiplexers in Section 5.7.

5.3.1 The 74LS138

The 74LS138 3-line-to-8-line decoder/demultiplexer is an extremely popular IC. It is used to decode addresses in memory systems and as a 1-line-to-8-line demultiplexer. What sets the 74LS138 apart from the other decoders that you have studied is its sophisticated enable circuitry. Refer to Figure 5.3.

You already know what a 3-line-to-8-line decoder should look like. It must have three select inputs: A, B, and C. It will also have eight outputs, 0 through 7. Look at the pinout in Fig-

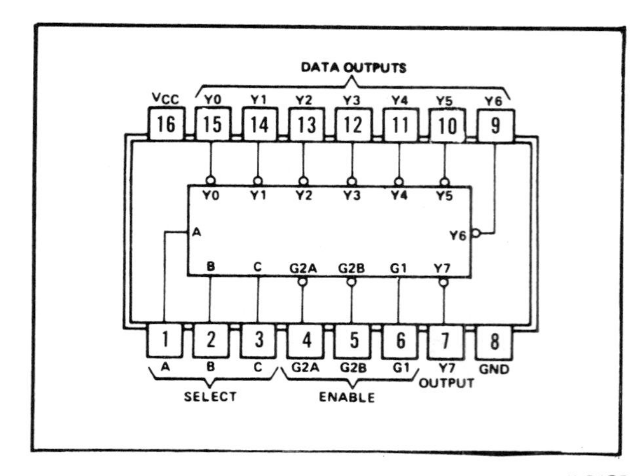

'LS138, 'S138

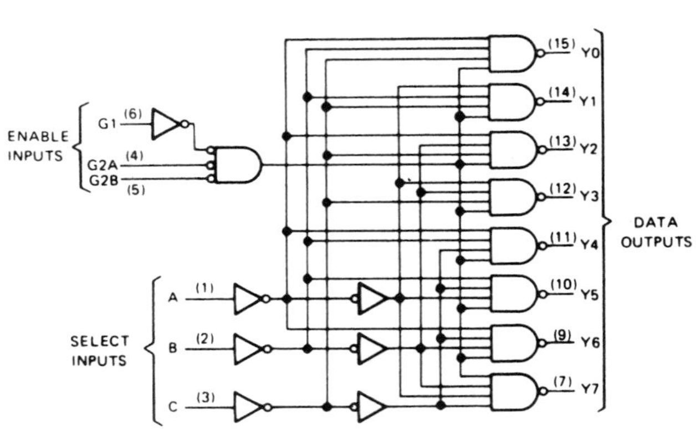

'LS138, 'S138
FUNCTION TABLE

INPUTS					OUTPUTS							
ENABLE		SELECT										
G1	G2*	C	B	A	Y0	Y1	Y2	Y3	Y4	Y5	Y6	Y7
X	H	X	X	X	H	H	H	H	H	H	H	H
L	X	X	X	X	H	H	H	H	H	H	H	H
H	L	L	L	L	L	H	H	H	H	H	H	H
H	L	L	L	H	H	L	H	H	H	H	H	H
H	L	L	H	L	H	H	L	H	H	H	H	H
H	L	L	H	H	H	H	H	L	H	H	H	H
H	L	H	L	L	H	H	H	H	L	H	H	H
H	L	H	L	H	H	H	H	H	H	L	H	H
H	L	H	H	L	H	H	H	H	H	H	L	H
H	L	H	H	H	H	H	H	H	H	H	H	L

*G2 = G2A + G2B
H = high level, L = low level, X = irrelevant

Figure 5.3 The 74LS138 decoder/demultiplexer.

ure 5.3. The pins on the 74LS138 can be divided into three functional groups:

1. Select inputs
2. Decoded outputs
3. Enable inputs

The first two groups, the select inputs and decoded outputs, should appear exactly as you expected. Notice that the decoded outputs are bubbled, indicating that they are active-low. The decoded outputs are designated by the letters Y0 through Y7 and are called "data outputs" in the pinout diagram. When this device is being employed as a decoder, the outputs will be called the decoded outputs. When this device is employed as a demultiplexer, the outputs will be called the data outputs. When we refer to a device, it will always be analyzed in terms of its present application. Because the 74LS138 is most commonly used as a decoder, it is natural to refer to it as a decoder when it is not involved in a specific application.

The only unfamiliar pins on the 74LS138 should be the en-

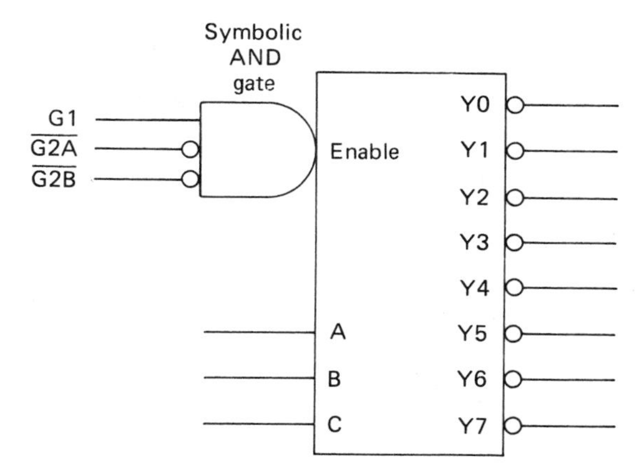

Figure 5.4 Functional schematic for the 74LS138.

ables: G1, $\overline{G2A}$, and $\overline{G2B}$. The logic diagram indicates that G1 is active high, whereas $\overline{G2A}$ and $\overline{G2B}$ are active low. These enable inputs are effectively ANDed together to form an internal enable signal. For the 74LS138 to be enabled all the enable inputs must be in their active states.

If any of the enable inputs are not at their active levels, all the inputs, Y0 through Y7, will go to their inactive-high levels, regardless of the value of the select inputs. The logic symbol in the pinout is the symbol that will be used to represent the 74LS138 in most schematics. It is easy to forget how the enable inputs operate. Because of this, the symbol shown in Figure 5.4 is gaining wide usage as the preferred schematic representation of the 74LS138. The AND gate in Figure 5.4 is symbolic; it is not a physical device. It helps remind the technician reading the schematic that the enable inputs are ANDed together to derive an internal enable.

Refer to the function table in Figure 5.3. The first two lines indicate situations where an enable input is not at its active level. When this happens the select inputs have no effect, so they are designated "don't cares" and the outputs will all be driven to the inactive-high level. Notice that the second column is labeled G2. To save space in the function table, G2 is the ORed value of $\overline{G2A}$ and $\overline{G2B}$. This says: If $\overline{G2A}$ is high OR $\overline{G2B}$ is high, the decoder will be disabled. The last eight lines in the function table are the normal decoded outputs for an enabled, active-low output 3-line-to-8-line decoder.

5.4 THE LED SEVEN-SEGMENT DISPLAY

5.4.1 The Seven-Segment Display

Seven-segment displays are the simplest method of displaying the decimal digits, 0 through 9. Each of the seven segments is constructed from a bar-shaped LED and can be lit individually. Consider Figure 5.5, where the LEDs are labeled a through g. By illuminating the proper combination of LEDs we can display any decimal digit. There are two different types of seven-segment displays: common anode and common cathode. A common-

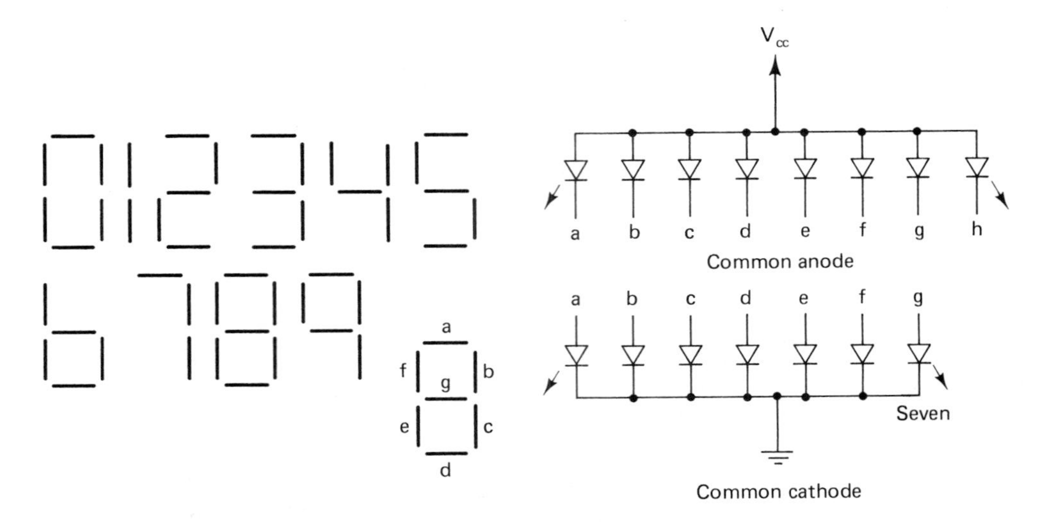

Figure 5.5 Seven-segment display.

anode display has all the anodes of the LEDs tied together and connected to V_{cc}. To illuminate any particular segment, a logic 0 must be applied to its cathode. Common-cathode seven-segment displays have all the cathodes tied together and connected to ground. To illuminate any particular segment, a logic 1 must be applied to the anode.

5.4.2 The 74LS47 BCD-to-Seven-Segment Decoder/Driver

The seven-segment display must be driven by a device that has the ability to receive a digital character in BCD form and illuminate the proper segments to create the numeral. That is the function of a BCD-to-seven-segment decoder/driver. In Figure 5.6, the provision for the BCD input is indicated on pins A through D. Notice that the outputs for the seven segments a through g are active low. This indicates that the 74LS47 is designed to drive common-anode seven-segment displays. The outputs are open collector. Each segment line will be current limited and pulled up to V_{cc} via a resistor and the associated LED segment.

There are three pins on the 74LS47 with which you are not yet familiar: lamp test ($\overline{\text{LT}}$), blanking in/ripple blanking out ($\overline{\text{BI/RBO}}$), and ripple blanking in ($\overline{\text{RBI}}$). They are all active-low signals. The best way to understand the function of these pins is by examining the function table. Refer to the first line in the function table in Figure 5.6.

The column labeled "Decimal or Function" indicates the decimal equivalent of the BCD input or a special function that the 74LS47 is capable of performing. These special functions relate to those three undefined pins. The first 16 lines in the function table will concern normal operation for the BCD codes of 0000 through 1111. You may be surprised tht the 74LS47 will

'46A, '47A, 'L46, 'L47, 'LS47
(TOP VIEW)

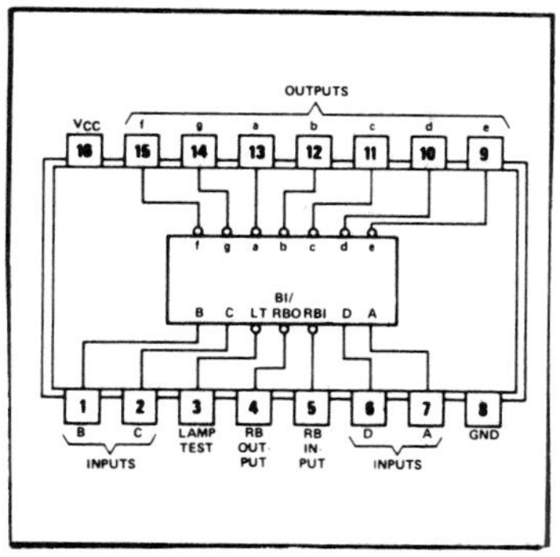

OUTPUTS

VCC f g a b c d e

16 15 14 13 12 11 10 9

1 2 3 4 5 6 7 8

B C INPUTS · LAMP TEST · RB OUT PUT · RB IN PUT · D A INPUTS · GND

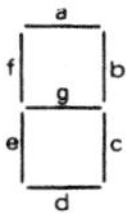

0 1 2 3 4 5 6 7 8 9 10 11 12 13 14 15

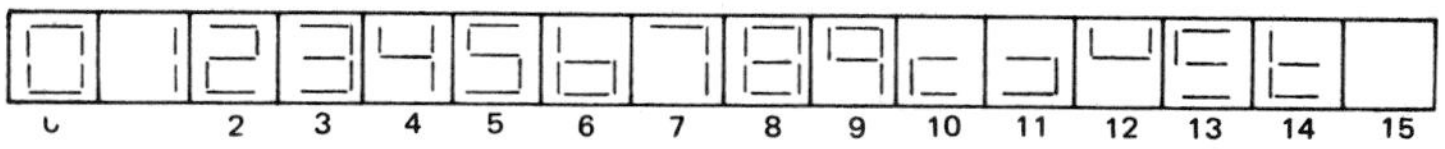

NUMERICAL DESIGNATIONS AND RESULTANT DISPLAYS

SEGMENT IDENTIFICATION

'46A, '47A, 'L46, 'L47, 'LS47 FUNCTION TABLE

DECIMAL OR FUNCTION	INPUTS						BI/RBO†	OUTPUTS							NOTE
	LT	RBI	D	C	B	A		a	b	c	d	e	f	g	
0	H	H	L	L	L	L	H	ON	ON	ON	ON	ON	ON	OFF	1
1	H	X	L	L	L	H	H	OFF	ON	ON	OFF	OFF	OFF	OFF	
2	H	X	L	L	H	L	H	ON	ON	OFF	ON	ON	OFF	ON	
3	H	X	L	L	H	H	H	ON	ON	ON	ON	OFF	OFF	ON	
4	H	X	L	H	L	L	H	OFF	ON	ON	OFF	OFF	ON	ON	
5	H	X	L	H	L	H	H	ON	OFF	ON	ON	OFF	ON	ON	
6	H	X	L	H	H	L	H	OFF	OFF	ON	ON	ON	ON	ON	
7	H	X	L	H	H	H	H	ON	ON	ON	OFF	OFF	OFF	OFF	
8	H	X	H	L	L	L	H	ON	ON	ON	ON	ON	ON	ON	
9	H	X	H	L	L	H	H	ON	ON	ON	OFF	OFF	ON	ON	
10	H	X	H	L	H	L	H	OFF	OFF	OFF	ON	ON	OFF	ON	
11	H	X	H	L	H	H	H	OFF	OFF	ON	ON	OFF	OFF	ON	
12	H	X	H	H	L	L	H	OFF	ON	OFF	OFF	OFF	ON	ON	
13	H	X	H	H	L	H	H	ON	OFF	OFF	ON	OFF	ON	ON	
14	H	X	H	H	H	L	H	OFF	OFF	OFF	ON	ON	ON	ON	
15	H	X	H	H	H	H	H	OFF	OFF	OFF	OFF	OFF	OFF	OFF	
BI	X	X	X	X	X	X	L	OFF	OFF	OFF	OFF	OFF	OFF	OFF	2
RBI	H	L	L	L	L	L	L	OFF	OFF	OFF	OFF	OFF	OFF	OFF	3
LT	L	X	X	X	X	X	H	ON	ON	ON	ON	ON	ON	ON	4

H = high level, L = low level, X = irrelevant

NOTES: 1. The blanking input (BI) must be open or held at a high logic level when output functions 0 through 15 are desired. The ripple-blanking input (RBI) must be open or high if blanking of a decimal zero is not desired.

2. When a low logic level is applied directly to the blanking input (BI), all segment outputs are off regardless of the level of any other input.

3. When ripple-blanking input (RBI) and inputs A, B, C, and D are at a low level with the lamp test input high, all segment outputs go off and the ripple-blanking output (RBO) goes to a low level (response condition).

4. When the blanking input/ripple blanking output (BI/RBO) is open or held high and a low is applied to the lamp-test input, all segment outputs are on.

†BI/RBO is wire-AND logic serving as blanking input (BI) and/or ripple-blanking output (RBO).

Figure 5.6 74LS47 BCD-to-seven-segment decoder/driver.

react to the illegal BCD codes of 1010 through 1111. Figure 5.6 depicts the resultant display for all the BCD inputs. Notice that the codes 1010 through 1110 create nonstandard characters, while the input of 1111 will cause the seven-segment display to go blank. Most applications will not use these nonstandard characters, but they are available as an option.

Under "Inputs" we see the values of $\overline{LT}$, $\overline{RBI}$, and the BCD inputs. Remember that $\overline{LT}$ and $\overline{RBI}$ are active-low inputs. BI/

$\overline{\text{RBO}}$ is an open-collector pin that serves both as an input and an output. We will discuss the function of this pin when it goes active, in the seventeenth line of the truth table.

The output columns indicate, in active-low terms, which segments will be illuminated. The four notes in the last column will be explained as they are needed. For the first 16 lines of the truth table, the $\overline{\text{LT}}$, $\overline{\text{RBI}}$, and BI/$\overline{\text{RBO}}$ pins will all be high, in their nonactive state.

Lines 1 through 16. Line 1 indicates that the BCD code for the digit 0 is input to the 74LS47. Every segment is illuminated except for g. This creates the numeral 0 on the display. Carefully examine the next 15 lines in the truth table and associate each BCD code with the resultant display.

Line 17. In this line the BI/$\overline{\text{RBO}}$ pin goes to its active-low level. Whenever this happens, all other inputs will be in don't care states and all the segments will be extinguished. To blank a display simply means to turn it off. There are three principal times when we will wish to blank a display:

1. We can control the intensity of the display by switching it on and off fast enough so that flicker is not evident, but the display will appear dimmer to the human eye. This will also decrease the power that the display consumes. A variable-frequency oscillator can drive the BI/$\overline{\text{RBO}}$ pin. By adjusting a variable resistor the intensity of the display can be controlled. Many digital alarm clocks have this feature. When this feature is being used, the BI/$\overline{\text{RBO}}$ pin is an input.
2. We may desire to blank leading zeros to make the display more readable. For example: Assume tht a calculator with seven digits is to display the number 3002. If all seven displays are enabled, the calculator will display 0003002. After the first decimal digit greater than zero is displayed, all other zeros will then be significant and must be displayed (e.g., as in 3002). The BI/$\overline{\text{RBO}}$ and the $\overline{\text{RBI}}$ pins will be used to provide this leading-zero blanking. When leading-zero blanking is used, the BI/$\overline{\text{RBO}}$ pin will be used as an output to the $\overline{\text{RBI}}$ pin on the next, less significant display.
3. A scheme called *multiplexing displays* enables one BCD-to-seven-segment decoder/driver to control many displays simultaneously. When a multiplexing scheme is being used, the BI/$\overline{\text{RBO}}$ pin is an input pin. This is discussed later.

Line 18. In this line the lamp test input is in its inactive state. The ripple blanking input is in its active-low state. This enables the display to blank the decimal digit 0. When the BCD inputs equal 0000, the seven segments will all turn off to blank the display. The command to blank a 0 is rippled from the most significant display to the least significant display.

Consider a bank of four seven-segment displays. The ripple blanking input of the most significant decoder will be tied to ground. If this decoder receives a BCD input of 0000, it will always blank the display and force its $BI/\overline{RBO}$ line to go low. This output will be tied to the $\overline{RBI}$ of the next, less significant, decoder. In this manner all zeros will be blanked until a decoder receives a nonzero input. That decoder will display the proper digit and pull its $BI/\overline{RBO}$ high to disable the blanking of any further displays.

Line 19. The lamp test (LT) input is at an active low. The $\overline{BI/RBO}$ pin will go inactive (high) and all the segments will illuminate, regardless of the condition of any other input pin. It is important to be able to test easily the integrity of all the LED segments in the system. If the $\overline{LT}$ pin is pulled low and one or more segments do not illuminate, that indicates a problem in those unlit segements, the driver outputs, or the PCB connections between the driver outputs and the displays.

5.5 THE LCD SEVEN-SEGMENT DISPLAY

5.5.1 The Liquid-Crystal Display

Seven-segment displays constructed from LEDs suffer from two major problems: they require a great deal of current and they are difficult to read in any brightly lit area. LCDs (liquid-crystal displays) were developed to overcome these limitations. LCDs consume minute amounts of power, and they are easily visible in brightly lit conditions.

LCDs do not emit light but control reflected or transmitted light. Unlike LEDs, which can be driven with a dc voltage, LCDs must be driven with a square wave. The most popular type of LCD is the field-effect LCD. Field-effect LCDs are available in a reflective version, for high-ambient-light conditions, and in a transmissive version. The transmissive type is used in low-level-light conditions and employs a backlight.

LCDs do have several disadvantages:

1. They must be used in well-lit areas, or use a backlight that defeats the advantage of low-power operation.
2. They have a limited operating temperature
3. They are more expensive than comparable LED displays.
4. They respond much more slowly than LEDs.
5. They cannot be multiplexed as easily as LEDs.

LCDs are used mainly in portable equipment, such as digital watches, calculators, and DMMs, where power consumption is a major concern.

5.6 THE ENCODER

An encoder provides the logical inverse function of a decoder. A typical encoder might be a decimal-to-BCD encoder. The name would indicate a device that has 10 inputs (decimal) and four outputs (BCD). The inputs could be active low or high, as could the outputs. Encoders are not nearly as popular as decoders. Nevertheless, there are certain situations where the encoder provides an invaluable function.

5.6.1 The 74LS148 8-Line-to-3-Line Priority Encoder

The title indicates that it will have eight inputs which can be encoded into a 3-bit binary code ($2^3 = 8$). The word "priority" is also in the name. This indicates that the inputs are arranged in some scheme, from most important (highest priority) to least important (lowest priority). If more than one input goes active, that input with the highest priority will be encoded and output on A0 through A3, while the other, lower-priority pins will be ignored.

The pinout shown in Figure 5.7 should hold few surprises. There are eight active-low inputs, 0 through 7, and the three encoded active-low outputs, A0 through A2. There is one unfamiliar input, $\overline{\text{EI}}$, and two unfamiliar outputs, $\overline{\text{EO}}$ and $\overline{\text{GS}}$. Let's tackle the explanation of these unknown pins before proceeding to the function table.

$\overline{\text{EI}}$ This is the enable input. If a high level is applied to this pin, the 74LS148 will be disabled: all the outputs will go to logic 1. A low level will enable the device to perform a normal encoding function.

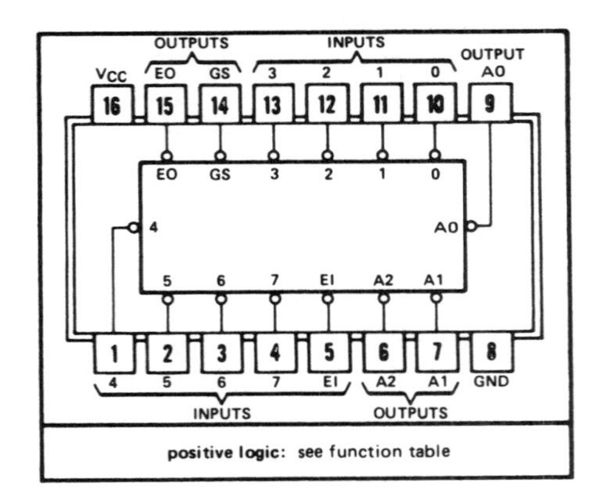

'148, 'LS148
FUNCTION TABLE

INPUTS									OUTPUTS				
EI	0	1	2	3	4	5	6	7	A2	A1	A0	GS	EO
H	X	X	X	X	X	X	X	X	H	H	H	H	H
L	H	H	H	H	H	H	H	H	H	H	H	H	L
L	X	X	X	X	X	X	X	L	L	L	L	L	H
L	X	X	X	X	X	X	L	H	L	L	H	L	H
L	X	X	X	X	X	L	H	H	L	H	L	L	H
L	X	X	X	X	L	H	H	H	L	H	H	L	H
L	X	X	X	L	H	H	H	H	H	L	L	L	H
L	X	X	L	H	H	H	H	H	H	L	H	L	H
L	X	L	H	H	H	H	H	H	H	H	L	L	H
L	L	H	H	H	H	H	H	H	H	H	H	L	H

Figure 5.7 74LS148: 8-line-to-3-line priority encoder.

$\overline{EO}$ The enable output pin will go low if the $\overline{EI}$ pin is low and none of the eight inputs are at an attractive level. The $\overline{EI}$ and $\overline{EO}$ pins are used to cascade 74LS148s. The $\overline{EO}$ pin from the highest-priority encoder will be connected to the $\overline{EI}$ pin of the next-lower-priority encoder. $\overline{EI}$ of the highest-priority encoder will be connected to ground or to the output of a gate that controls when the encoders are enabled.

$\overline{GS}$ The group select output goes low whenever the decoder is active. If the decoder is enabled ($\overline{EI}$ is low) and at least one input is at an active logic 0 level, the $\overline{GS}$ output will go low to indicate that the device has a valid encoded output on its A0 through A2 pins.

Refer to the function table in Figure 5.7. We will examine the function table on a line-by-line basis:

Line 1. The $\overline{EI}$ input is inactive, at a high level. The inputs 0 through 7 become don't-care conditions and all the outputs will go high to their inactive state. When $\overline{EI}$ is high, the 74LS148 is disabled.

Line 2. The device is enabled with a low level at $\overline{EI}$ but none of the input lines (0 through 7) are active. All outputs will be inactive except the $\overline{EO}$ line, which will go low. If another encoder is connected in cascade, this low output level will be connected to its $\overline{EI}$ input and it will be enabled.

Line 3. The encoder is enabled and input 7 is active. The rest of the inputs assume don't-care conditions. This tells us that input 7 has the highest priority; if it is active, input 7 will override all other inputs (0 through 6). Notice that A0 through A3 all go to logic 0. Remember: The outputs are active low. Here "active low" implies that the outputs will be inverted from their normal encoded values. Normally, the encoded value of decimal 7 is 111. Because the outputs are active low, the encoded value of 7 will be the inverse of 111 (i.e., 000). This will be the case for the value of each encoded output. Also notice that in line 3., when the device is enabled and at least one input line is active, the $\overline{GS}$ output will go active (low) to indicate that the 74LS148 has a valid output on lines A0 through A2.

Lines 4 through 10. Each subsequent line in the function table indicates that if the encoder is enabled and the higher-priority lines are not active, that particular active input will appear encoded and inverted on A0 through A2, and the $\overline{GS}$ will go active to indicate a valid output.

5.6.2 Applications of the Priority Encoder

Priority encoders are often used in keyboard encoding/decoding schemes. Figure 5.8 illustrates a simplified application of em-

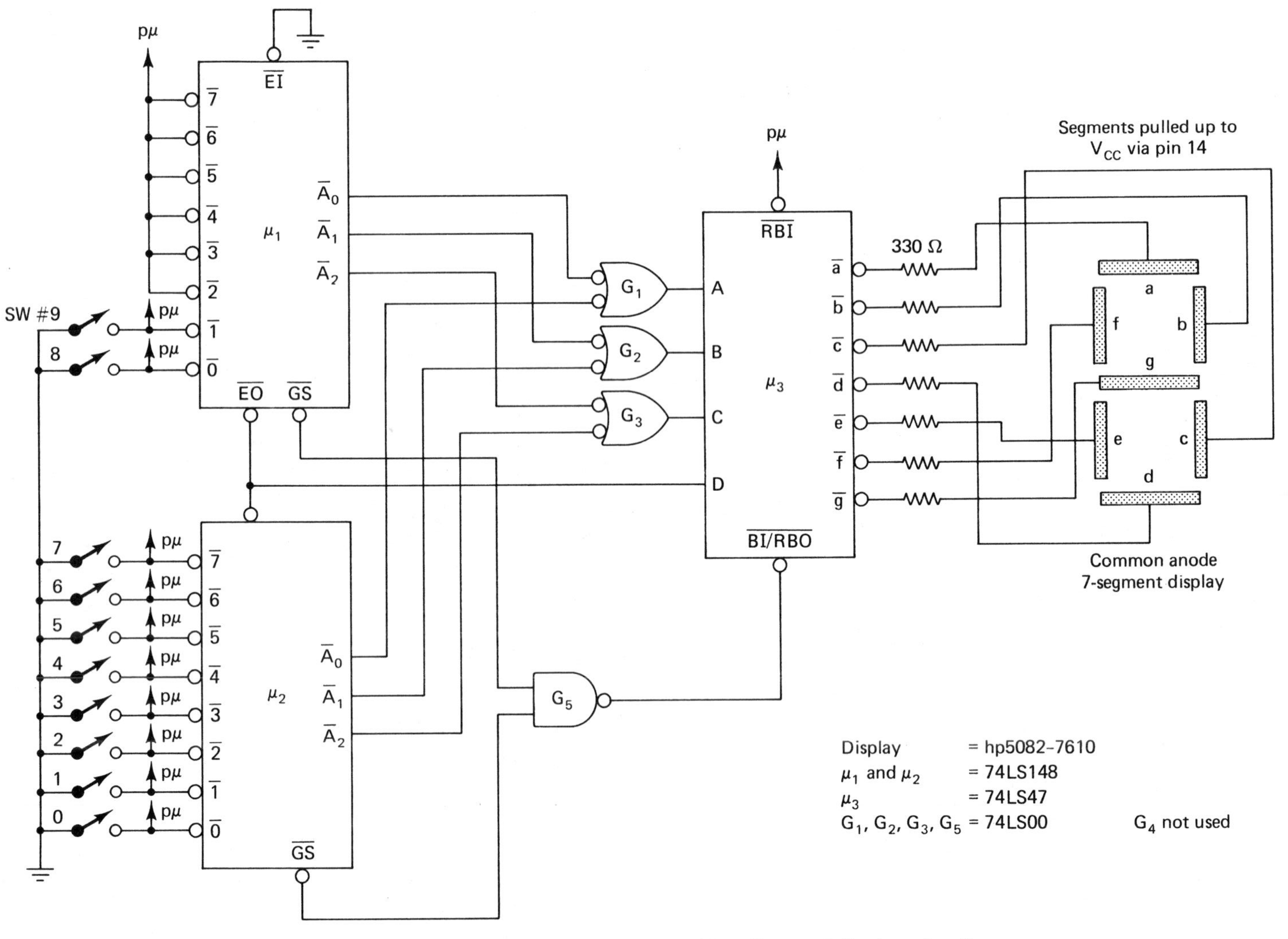

Figure 5.8 Decimal keyboard encoding/decoding and display circuit.

ploying 74LS148s to encode the depressed key in a 10-key keyboard. Switches 9 through 10 are the 10 keys that make up the decimal keyboard. U1 is the highest-priority encoder; its $\overline{EI}$ is hardwired to ground. If switches 9 and 8 are both open, U2 will be enabled via the $\overline{EO}$ pin of U1. Therefore, switch 9 is the highest-priority switch and switch 0 is the lowest-priority switch.

It is important to realize that U1 and U2 will never have simultaneous valid outputs on their A0 through A2 pins. When U1 is active, outputs A0 through A2 on U2 will all be inactive (high). When U2 is active, outputs A0 through A2 on U1 will be inactive highs. The A input on the 74LS47 will be high if A0 of U1 or A0 of U2 goes active low. The same is true for inputs B and C of U3. If the $\overline{EO}$ from U1 goes high, this indicates that either switch 9 or 8 is closed. This high will disable U2 and cause input D of the 74LS47 to go high.

If $\overline{GS}$ of U1 is high and $\overline{GS}$ of U2 is high, neither encoder is active and the seven-segment display will be blanked via the output of G5. Each segment output is current limited with a 330-Ω resistor. The segments are all pulled up to V_{cc} via pin 14 of the seven-segment display.

Redraw this circuit on a separate sheet of paper. Close each switch and derive the logic levels for each input and output of all circuit components. (Note that the letters "pu" refer to pull-up resistor.)

5.7 THE DATA SELECTOR/MULTIPLEXER

The data selector is a digitally controlled switch. In Section 4.3 we designed a 2-bit data selector. It had two data inputs and a select line that controlled which input was steered to the output. MSI data selectors are available with 4, 8, and 16 data inputs. The number of select bits required on a data selector will be a function of the number of data bits:

$$2^{(\text{number of select bits})} = \text{number of data inputs}$$

Data selectors are also used to send many different channels of data on one common transmission line.

5.7.1 The 74LS151 1-of-8 Data Selector/Multiplexer

From your previous knowledge of data selectors, you should be able to derive the basic pinout of the 74LS151. There will be eight data inputs, D0 through D7, and one data output. There must be three select bits ($2^3 = 8$): A, B, and C.

Refer to the pinout in Figure 5.9. Notice the two outputs—Y and W. Y is unbubbled and the true output. W is bubbled and is the complement of Y. There is one pin that has not been discussed: the strobe input. In digital electronics the term "strobe" has several different meanings. Most often, it refers to a fast, low-duty cycle pulse that enables a device. On the 74LS151, the

'151A, 'LS151, 'S151

FUNCTION TABLE

INPUTS				OUTPUTS	
SELECT			STROBE	Y	W
C	B	A	S		
X	X	X	H	L	H
L	L	L	L	D0	$\overline{D0}$
L	L	H	L	D1	$\overline{D1}$
L	H	L	L	D2	$\overline{D2}$
L	H	H	L	D3	$\overline{D3}$
H	L	L	L	D4	$\overline{D4}$
H	L	H	L	D5	$\overline{D5}$
H	H	L	L	D6	$\overline{D6}$
H	H	H	L	D7	$\overline{D7}$

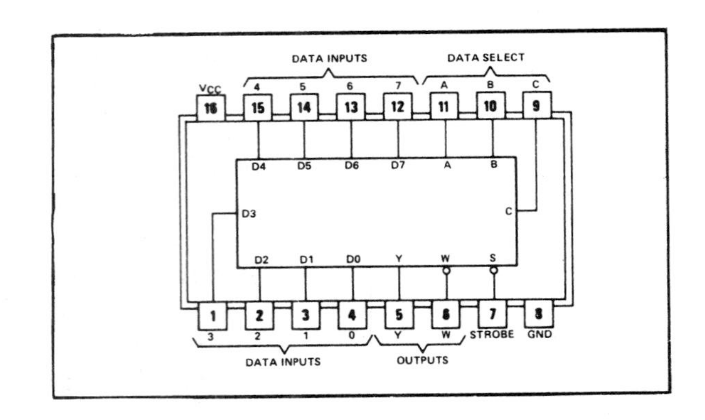

'151A, 'LS151, 'S151

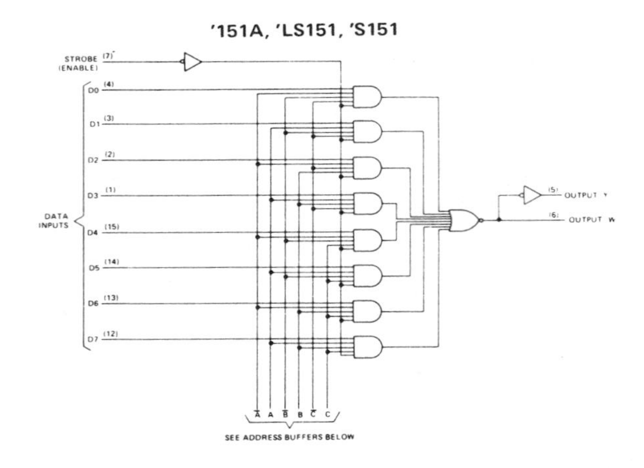

ADDRESS BUFFERS FOR 'LS151, 'S151, 'LS152

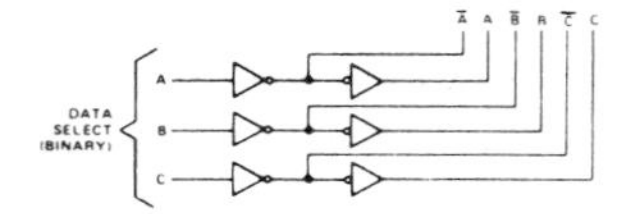

Figure 5.9 74LS151 1-of-8 data selector/multiplexer.

strobe is another name for an enable input. The strobe input is bubbled, which denotes that it is active low.

Refer to the function table in Figure 5.9. The first line indicates the situation where the strobe is at an inactive level. If the strobe input is high, the Y output will go low and the W output will go high, regardless of the values on the select or data inputs. The last eight lines of the function table illustrate each data input as it is selected and how it affects the Y and W outputs.

5.7.2 Applications of Data Selectors

Data selectors are useful in many digital applications. Like many of the other devices in this chapter, data selectors find applications in advanced circuits, such as memory controller circuits. An interesting application of data selectors is in Boolean generators.

The 74LS151 can be used to implement any Boolean equation of up to three variables. One 74LS151 can be used in place of a potentially large number of gates. Let's rebuild the "room with three doors" circuit, from Section 3.2, using a 74LS151.

C	*B*	*A*	*Out*
0	0	0	0
0	0	1	1
0	1	0	1
0	1	1	0
1	0	0	1
1	0	1	0
1	1	0	0
1	1	1	1

$$\text{Light on} = A\overline{B}\overline{C} + \overline{A}B\overline{C} + \overline{A}\overline{B}C + ABC$$

Each row in the truth table corresponds to a particular select code on the 74LS151. When the select code is 000, the *Y* output of the data selector should go low, as in row 1 of the truth table. When the select code is 001, the output of the data selector should go high, as in row 2 of the truth table. In this manner the data selector will emulate the discrete gates that made up the original circuit. To use a data selector as a Boolean function generator, follow these steps:

1. Construct the truth table.
2. Tie each input of the data selector to the logic level that corresponds to the equivalent data output in the truth table.

The following circuit shown in Figure 5.10 illustrates the 74LS151 as it implements the "room with three doors."

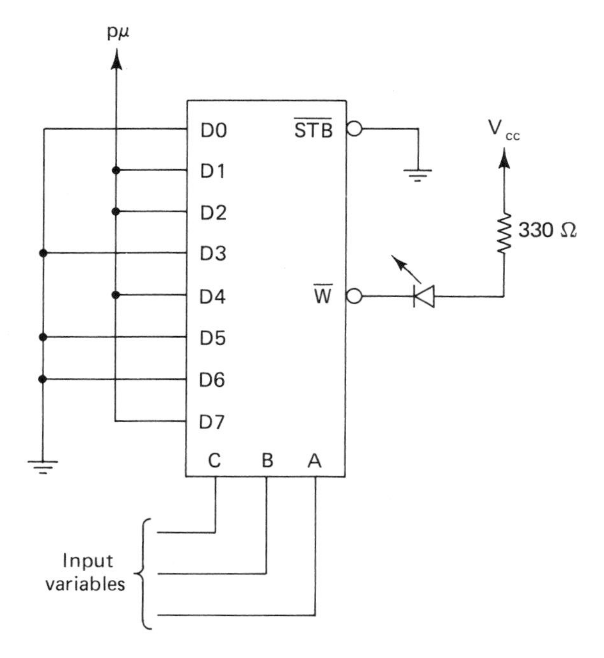

Figure 5.10 74LS151 as a Boolean function generator.

Notice that the data inputs that correspond to a row in the truth table with a logic 0 output value are tied to ground. The data inputs that correspond with rows that have logic 1 outputs are pulled up to V_{cc} via a common resistor (indicated by "pu"). Insead of requiring an inverter to drive the LED, the complemented output, W, is used.

5.7.3 Using Multiplexers and Demultiplexers

We have stated that the data selector can also function as a multiplexer. The 74LS138 was described as a decoder/demultiplexer. We are going to employ a 74LS151 as a multiplexer and a 74LS138 as a demultiplexer in a single-line digital communications circuit.

As shown in Figure 5.11, the 74LS151 functions as a 1-of-8 multiplexer. The 3-bit select code will command which of the data inputs will be steered to the *Y* output. This same 3-bit code will be applied to the select inputs of the 74LS138 demultiplexer. Instead of running eight separate data lines, this circuit employs one multiplexed data line and three lines of select code. This saves a total of four lines. The following steps describe the process of how the data is actually sent:

1. The select code of the desired data line (0 through 7) is applied to the select inputs of both U1 and U2.
2. If the selected data is low, the Y output of U1 will go low. This low is applied to the G2A enable input of U2. U2 will then be enabled and the output that corresponds to the select code will go low and match the original data from U1.
3. If the selected data is high, the Y output of U1 will go high.

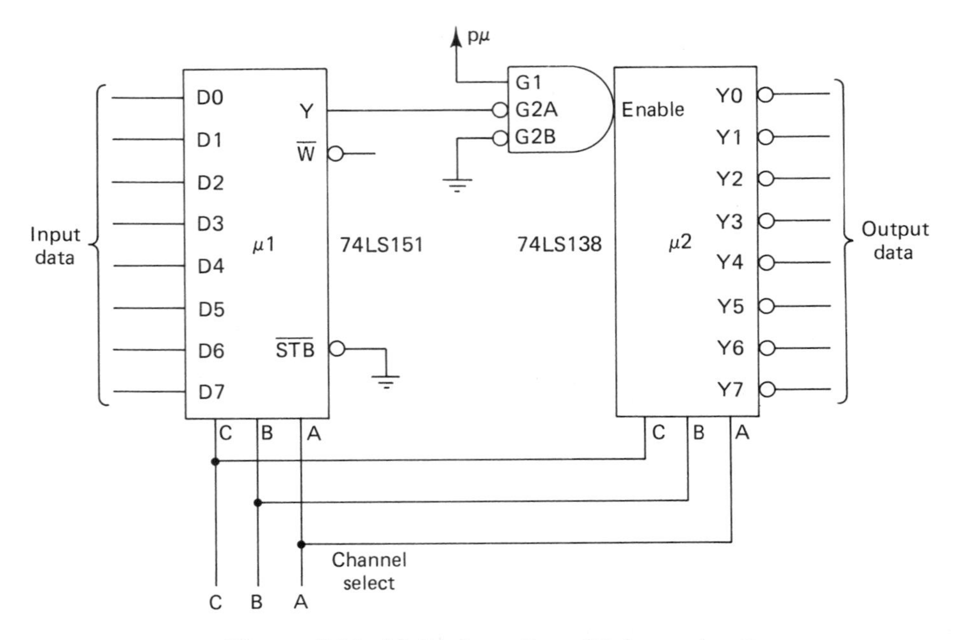

Figure 5.11 Multiplexer/demultiplexer circuit.

When this high is applied to the active-low enable input of U2, U2 will become disabled. All outputs of U2 will go high. The selected output will be high, matching the original data sent from U1.

Notice that the selected output will toggle high and low with the data sent from the multiplexer, while the nonselected outputs will stay inactive at a logic 1 level.

5.8 THE 4-BIT FULL-ADDER

In a computer-based system all the math functions are performed by special-purpose LSI devices. However, there will be occasions when two binary numbers must be added. The 4-bit full-adder can perform the addition of two 4-bit quantities, including a carry input and a carry output. Refer to Figure 5.12.

Popular forms of 4-bit adders are the 74LS83/LS283 and 4008B. The 74LS83 and 74LS283 are functionally equivalent. The 74LS83 has nonstandard power connections; the pins on the 74LS283 have been rearranged to provide the standard power terminals—ground at pin 8 and V_{cc} pin 16. The A and B inputs are arranged like the inputs on the 4-bit magnitude comparator. Instead of comparing the input values, the 4-bit adder will sum them. The symbol preceding the four outputs is the Greek letter sigma; sigma denotes the summation. There is also a carry input. A carry produced from a less significant adder will drive this input. If the carry input is not used, it must be tied to ground. The two 4-bit inputs and the carry input are added according to the normal rules of binary addition. The 4-bit adder is often used in circuits where a 4-bit offset must be added to a particular binary quantity.

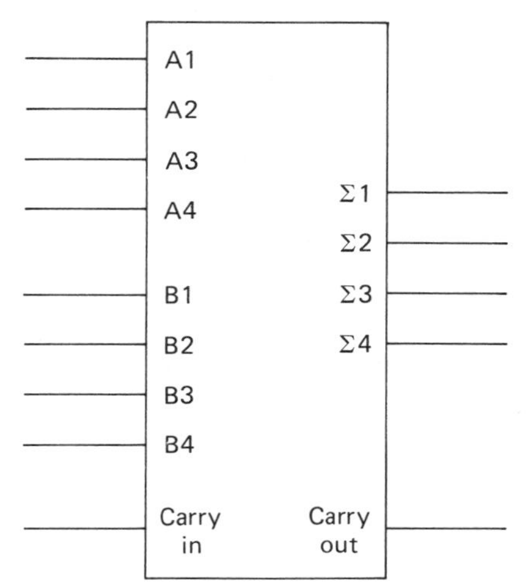

Figure 5.12 Logic symbol of a 4-bit adder.

QUESTIONS AND PROBLEMS

5.1. A BCD-to-decimal decoder functions correctly for input values from 0000 through 0111. When the BCD code of 1000 is applied to the decoder, output 0 goes active instead of output 8. Furthermore, when 1001 is applied, output 1 goes active instead of output 9. What is the cause of the problem?

5.2. How can you blank an LED seven-segment display being driven with a 74LS47 without the use of the BI/$\overline{RBO}$ pin?

5.3. How can the BI/$\overline{RBO}$ pin on the 74LS47 be used as both an input and an output? In what situations is it an input? In what situations is it an output?

Refer to Figure 5.8 for Questions 5.4 to 5.11.

5.4. When all the keyboard switches (SW 9 through SW 0) are open, the seven-segment display is illuminating the decimal digit 0. What is the problem? What are some possible causes of this problem?

5.5. The EI pin on U1 is floating instead of being tied to ground. How does this affect circuit operation? Why?

5.6. The ouput of G1 is stuck low. How does this affect circuit operation? Why?

5.7. A technician closes SW 0 and a 0 is displayed on the seven-segment display. What will happen if the technician next closes SW 1 while leaving SW 0 closed? What would happen if this was done in the opposite order (i.e., closing SW 1, then SW 0)? Why?

5.8. The 330-Ω resistors are accidentally stuffed with 33-Ω resistors. In relative terms, how would the display appear? What possible problems could occur?

5.9. The decimal digit 8 is displayed all the time, regardless of the position of SW 9 through SW 0. What is the problem?

5.10. When SW 0 is closed and all other switches are open, the display goes blank. What is the possible problem?

5.11. SW 9 through SW 5 appear to function correctly. When all the switches are open, the digit "4" is displayed. Toggling SW 4 through SW 0 does not seem to affect the display. What is the problem?

5.12. Use a 74LS151 to implement the three-judges problem from Chapter 3.

Refer to Figure 5.11 for Questions 5.13 to 5.15.

5.13. How would the circuit be affected if G2B were tied to G2A instead of ground? Which method is better? Why?

5.14. How would the output data be affected if G2A and G2B were tied to ground and G1 was connected to the Y output of the multiplexer?

5.15. How would the output data be affected if G2A and G2B were tied to ground and the W output of the multiplexer were tied to the G1 of the demultiplexer?

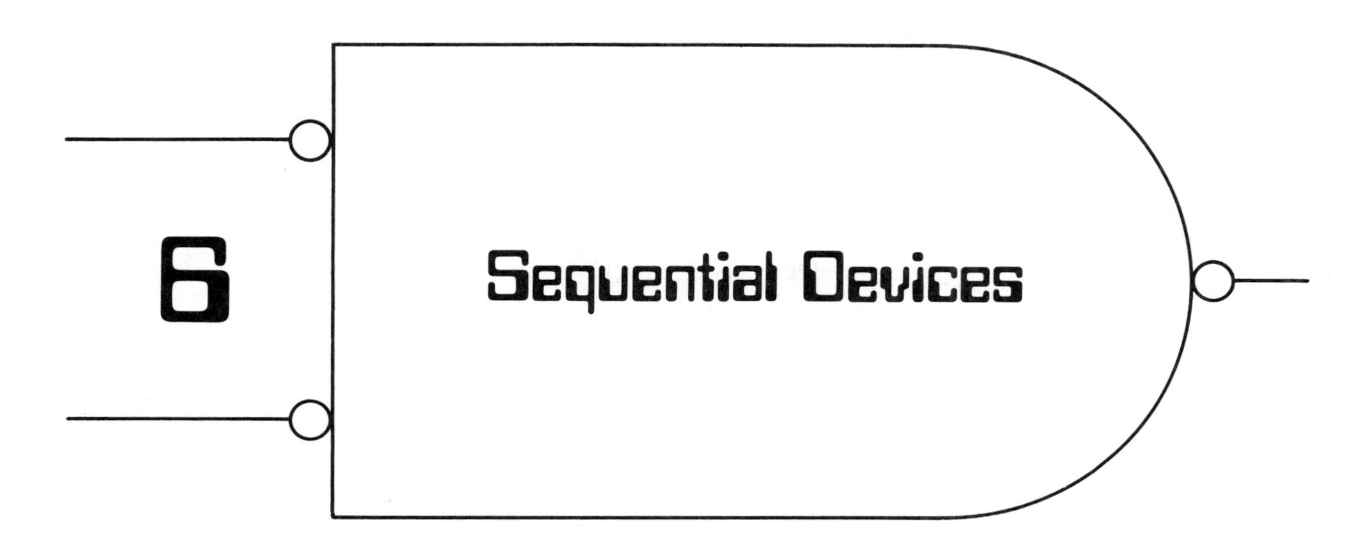

6 Sequential Devices

The outputs of logic gates and MSI combinational devices are dependent only on the present state of the inputs. This type of combinational logic operates much like the lock illustrated in Figure 6.1a.

Contrast this with the sequential lock illustrated in Figure 6.1b. This type of lock requires a sequence of events.

The most important difference between combinational and sequential devices is that sequential devices display the attribute of "memory." The sequential lock effectively remembers whether the first numbers in the sequence have been entered correctly. The combinational lock remembers nothing.

6.1 FEEDBACK AND DIGITAL CIRCUITS

You should be asking yourself: How do we design digital devices to have this capacity for memory? The answer to that question is: feedback. A simple definition of *feedback* is:

The process of sending all or part of the output signal back to the input.

6.1.1 Cross-Coupled NAND Gates

Refer to Figure 6.2, which shows our first example of a digital circuit employing feedback. The only tool we need to analyze

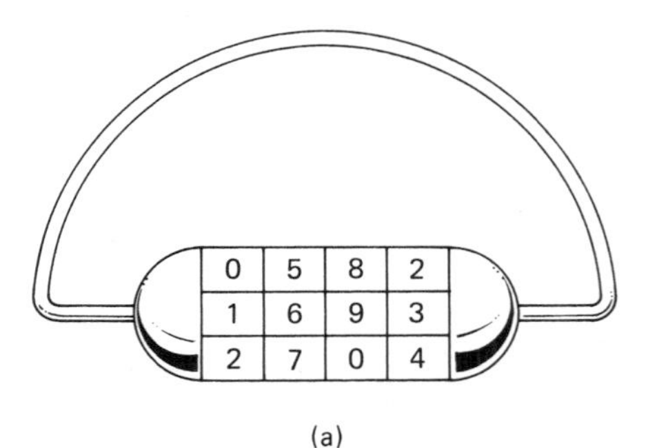

(a)

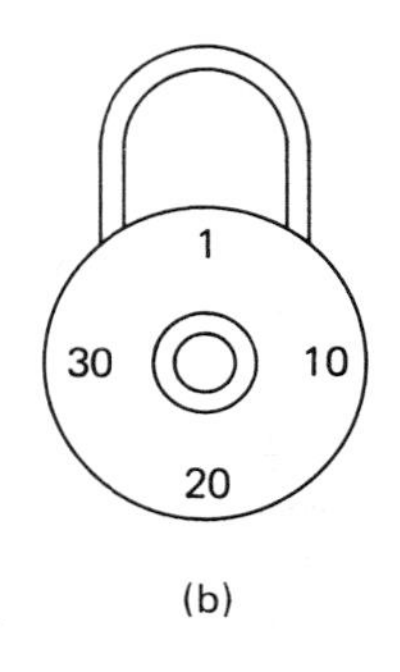

(b)

Figure 6.1 (a) Combinational and (b) sequential locks.

Input 1 — Output 1
Input 2 — Output 2

(a)

Input		Output	
2	1	1	2
1	0	?	?
0	1	?	?
1	1	?	?
0	0	?	?

(b)

Figure 6.2 Cross-coupled NAND gates.

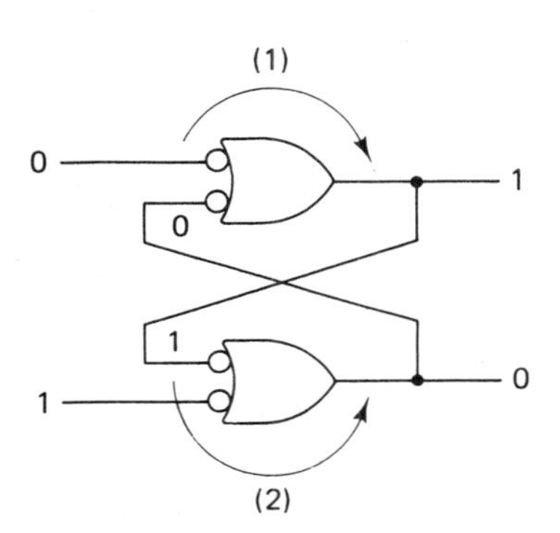

Figure 6.3 Line 1: input 1 = 0, input 2 = 1.

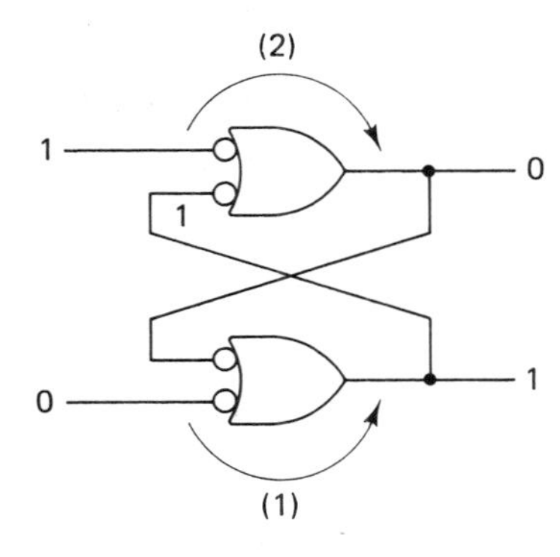

Figure 6.4 Line 2: input 1 = 1, input 2 = 0.

this circuit is the fact that the dynamic input level of a NAND gate is logic 0.

To analyze this circuit, we will complete the truth table in Figure 6.2. Normally, the case of input 1 = 0 and input 2 = 0 would appear in the first line of the truth table. To simplify the circuit analysis, we will save this input combination for the last step in our analysis.

The logic 0 on input 1 will force a logic 1 on the output of gate 1, which is fed back to an input of gate 2. Because gate 2 does not have any logic 0's on its input, its output will be a logic 0, which is fed back to an input of gate 1. Gate 1 is already outputting a logic 1, so the circuit appears stable. This is summarized in Figure 6.3.

In Figure 6.4 we find a dynamic input level on gate 2 which will force the output of gate 2 high. This logic 1 will be fed back to gate 1, whose output will go to logic 0. This is fed back to the input of gate 2. The output of gate 2 is already high, so the circuit appears to be stable.

Line 3: input 1 = 1, input 2 = 1

This is an interesting case. Neither input is at a dynamic level. We are forced to make some assumptions about the level of the outputs and analyze the circuit from that point.

Refer to Figure 6.5. Consider the first assumption that both outputs 1 and 2 are equal to logic 0. Examine Figure 6.6 carefully. Both outputs are dynamic input logic 0's. As they are fed back to the inputs of gates 1 and 2, these logic 0's will force the outputs of both gates to logic 1. Now both gates will have two

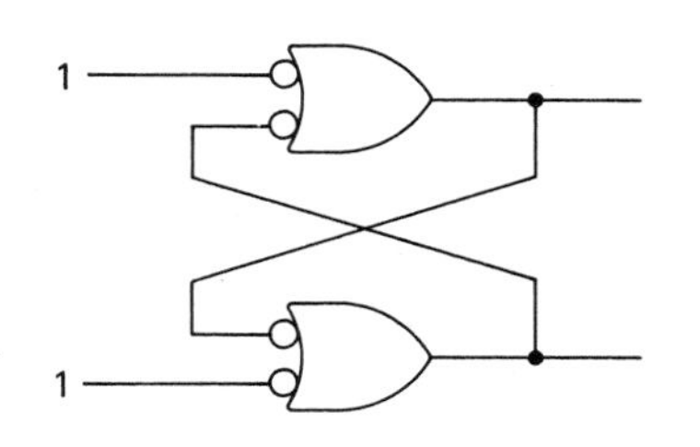

Figure 6.5 Cross-coupled NAND gates with logic 1 input values.

Outputs	
1	2
0	0
0	1
1	0
1	1

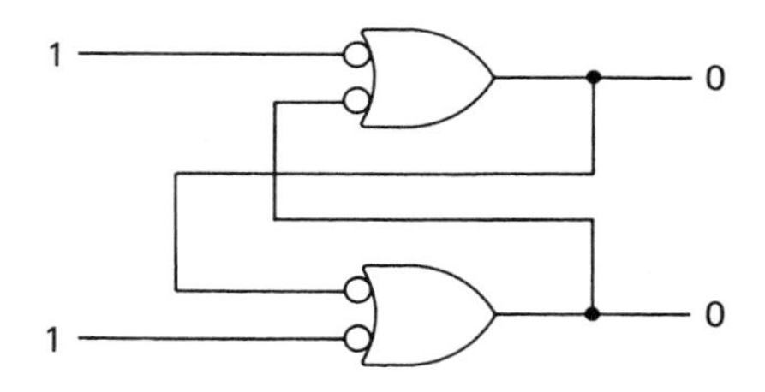

Figure 6.6 Assumption of both outputs equal to logic 0's.

logic 1's on their inputs, which will force the outputs to logic 0's. This circuit is unstable. It is impossible to predict at what level outputs 1 and 2 will finally settle.

If we had assumed that both outputs were at logic 1, the circuit action would be the same: unstable and unpredictable. This part of the analysis has yielded an important fact:

The outputs of this device must always be complements of each other.

We have now established that lines 1 and 4 of our assumption table in Figure 6.5, are unstable conditions. We are still concerned with our initial problem of analyzing the circuit when both inputs are equal to logic 1's. Let's continue examining the assumption table.

Assume that output 1 = 0 and output 2 = 1. Output 1 is a logic 0 and is fed to an input of gate 2, forcing the output of gate 2 high, which corresponds to our assumption. This high is fed back to an input of gate 1. Gate 1 has no logic 0's on its inputs, so its output must be a logic 0. Our initial assumptions concerning the state of the outputs have been proven correct.

Does this mean that when both inputs are equal to logic 1's, output 1 will be equal to logic 0 and output 2 equal to logic 1? Absolutely not! We still have not analyzed the assumption where output 1 = 1 and output 2 = 0.

Assume that output 1 = 1 and output 2 = 0. This time we will follow the logic 0 from the output of gate 2. This will force the output of gate 1 high, which is fed back to an input of gate 2. Gate 2 does not have any logic 0's on its inputs, so its output will be a logic 0. The analysis shows that our assumption of output 1 = 1 and output 2 = 0 is also true.

Refer to Figure 6.2. The last line states that both inputs 1 and 2 will be at logic 0. Can you see a problem with this? To

Inputs		Outputs	
2	1	1	2
0	0	Undefined	
0	1	0	1
1	0	1	0
1	1	Memory	

Figure 6.7 Truth table for cross-coupled NAND gates.

avoid unstable conditions in this circuit, outputs 1 and 2 must never be the same value. This combination of inputs will force both outputs to logic 1, which will eventually lead to an unstable operating condition. Figure 6.7 illustrates the completed truth table for this device.

The device that we have been examining is called an *S-R latch.* The inputs are called $\overline{S}$ and $\overline{R}$. The S and R letters stand for the operations of SET and RESET. The outputs are called Q and $\overline{Q}$. (Recall that the outputs must always be complements for proper operation.) The word "latch" is the description of this circuit's behavior. It latches or grabs inputs and remembers them on the output. The following definitions are in order.

To SET a latch means to force the Q output high and the $\overline{Q}$ output low. With the NAND type S-R latch, a set operation is accomplished by placing a logic 0 on the S input and a logic 1 on the R input.

To RESET a latch means to force the Q output low and the $\overline{Q}$ output high. On a NAND type S-R latch, the S input should be at logic 1 and the R input at logic 0.

Figure 6.8 summarizes the S-R latch.

6.1.2 Cross-Coupled NOR Gates

You know that any function accomplished by NAND gates can also be implemented using NOR gates. We are going to examine the operation of a NOR-type S-R latch (Figure 6.9).

The NAND-type S-R latch principles can easily be applied to the operation of the NOR-type S-R latch. Remember that the outputs, Q and $\overline{Q}$, must always be complements. Because a NAND gate has a logic 0 dynamic input level, the input combination where both S and R were equal to logic 0 was undefined. It should seem reasonable that the input combination

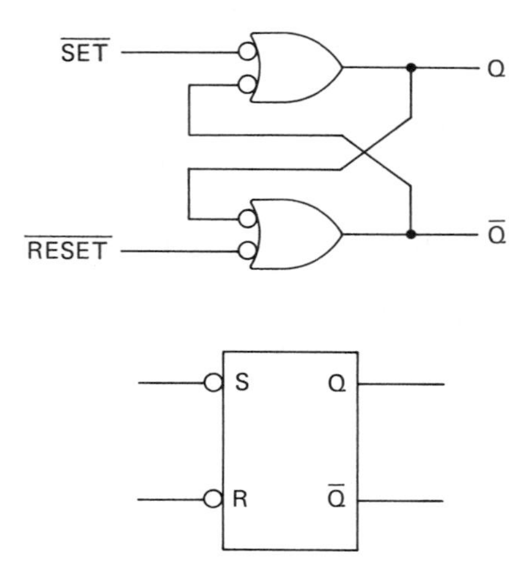

$\overline{S}$	$\overline{R}$	Q	$\overline{Q}$	
0	0	Undefined		
0	1	1	0	SET
1	0	0	1	RESET
1	1	Q_0	$\overline{Q}_0$	Memory

Figure 6.8 Summary of S-R latch.

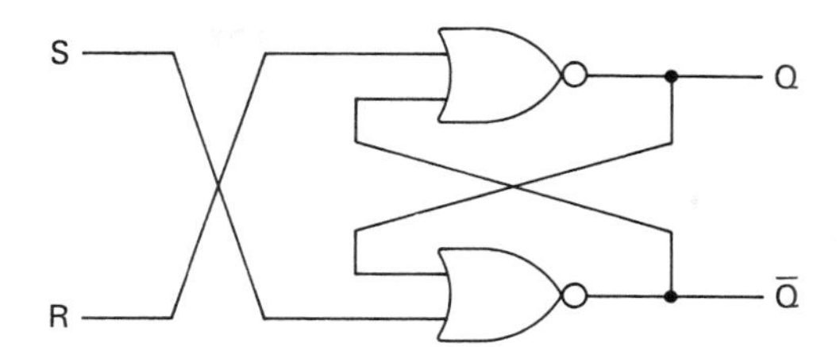

Figure 6.9 NOR-type S-R latch.

where both S and R are equal to logic 1's would be undefined for the NOR-type latch.

An undefined state in a NOR-type latch is when both S and R inputs are high.

The memory state of a NOR-type latch will be when both S and R inputs are at their nondynamic levels of logic 0.

Because the NOR has a dynamic input level of logic 1, we set the NOR latch by placing a logic 1 on the set input. As you have surely noticed, the set and reset inputs are applied to different gates in the NAND and NOR latches. The set input of the NOR latch is directed to the $\overline{Q}$ gate. A logic 1 will force $\overline{Q}$ low, which will in turn cause the Q output to go high.

Figure 6.10 summarizes the NOR-type S-R latch.

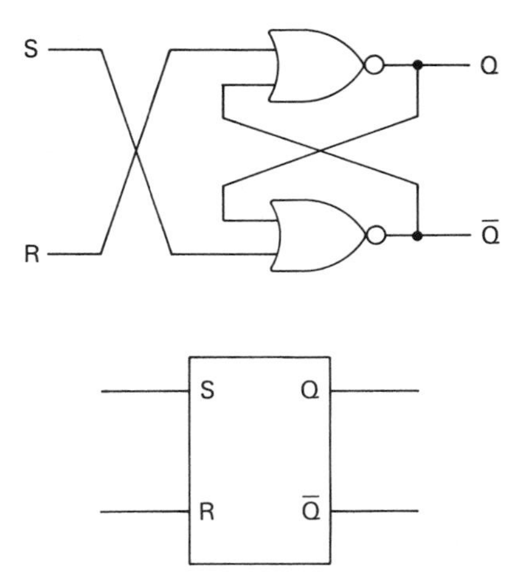

S	R	Q	$\overline{Q}$	
0	0	Q_0	$\overline{Q}_0$	Memory
0	1	0	1	RESET
1	0	1	0	SET
1	1	Undefined		

Figure 6.10 Summary of NOR latch.

6.1.3 Application of Latches as Mechanical Switch Debouncers

The most popular application for S-R latches is debouncing mechanical switches. Consider a SPDT (single-pole, double-throw) switch (see Figure 6.11). The events that will happen when the switch is toggled are:

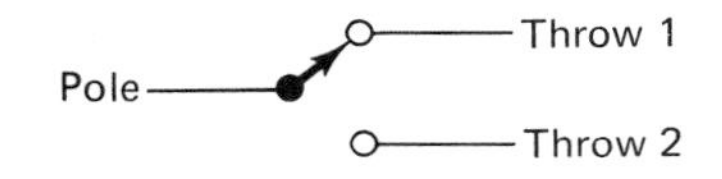

Figure 6.11 Schematic symbol for SPDT switch.

1. When the switch is first toggled, the throw will leave the pole that it was connected to and start its travel toward the other throw. There will be a short period of time when the pole is not connected to either throw.

2. The pole will connect with the second throw and because of momentum the pole will bounce off the throw, and once again it will not be in electrical contact with either throw.
3. The pole will continue to bounce on and off the throw, and finally settle down into solid electrical contact.

We must have some means of "debouncing" the output of the switch. This is a good application for the S-R latch. You should recognize the circuit in Figure 6.12 as a modified NAND-type S-R latch. Instead of having the S-R inputs we will be connected to the two throws of the switch. The pole of the switch is connected to ground. The two throws are connected to V_{cc} via 1-kΩ pull-up resistors.

When the switch is in the normally closed (NC) position, the set input will be at +5 V and the reset input will be at ground potential. The S-R latch is in a reset state when the switch is in its NC position. Let's toggle the switch into the normally open (NO) position.

1. At the first instant in time, the pole will be traveling between the two throws. Because the pole is not touching either throw, the inputs to the latch are both equal to logic 1: pulled up to V_{cc}, via R1 and R2. This is the memory state of a latch of a NAND-type S-R latch, so the latch is still in the reset condition.
2. The instant in time that the pole touches the normally open throw, the set input will go to logic 0; the reset input is already at logic 1. This corresponds with the set operation of a NAND-type S-R latch; Q will go high and $\overline{Q}$ will go low.
3. The first bounce will now occur. As the pole bounces off the normally open throw, the outputs do not change states because the latch has entered the memory state—both set and reset inputs are equal to logic 1's.
4. As the pole bounces back and forth off the NO throw, the latch will alternate between set and memory operations.
5. This process of memory state and set will continue until the bouncing finally stops.

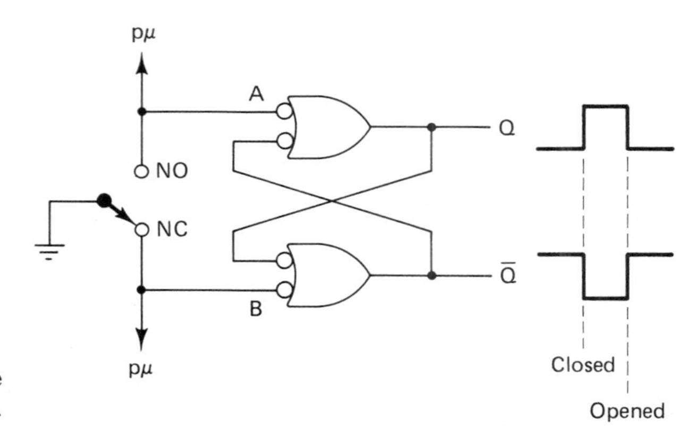

Figure 6.12 NAND-type switch debouncer.

6.2 THE CLOCK AND DIGITAL ELECTRONICS

Until this point in our study of digital electronics, the outputs of the devices have reacted immediately to the stimulus of the inputs. Circuits such as these are called *asynchronous*. This term is a combination of Greek words meaning "without regard to time." In digital systems it is important that outputs change at precise points in time. Circuits that operate in this manner are called *synchronous* circuits. Digital circuits often use time reference signals called *clocks*.

A clock signal is nothing more than a square wave that has a precise, known period. The clock will be the timing reference that synchronizes all circuit activity and tells the device when it should execute its function.

6.3 GATED TRANSPARENT LATCHES

6.3.1 A Gated Latch

We would like to develop a latch that uses a data input instead of S and R inputs. This will eliminate the problem of an undefined input combination. We will create this circuit in a short series of steps starting with the NAND-type S-R latch.

Figure 6.13 illustrates a modified NAND-type S-R latch. Instead of S and R inputs, this circuit has a D (DATA) input. The data is presented to the top NAND gate in complemented form, and to the bottom NAND gate in true form. If the data input is equal to logic 1, the latch will SET; if the data is equal to logic 0, the latch will RESET. Because it is impossible to present logic 1's to both the set and reset inputs, this latch does not have a memory state.

Figure 6.14 is a further modification of Figure 6.13. NAND gates, G1 and G2, have been added to this circuit. G1 and G2 function as electronic switches. Driven by the input called

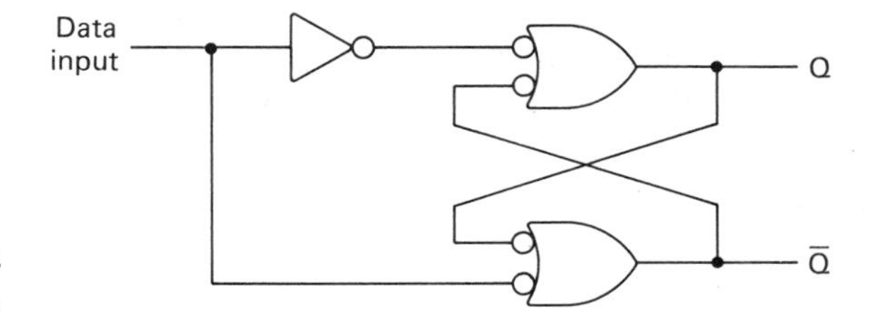

Figure 6.13 Modified S-R latch.

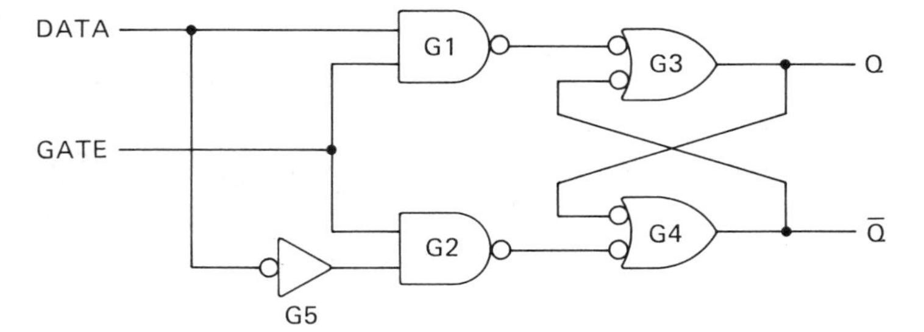

GATE	DATA	Q	$\bar{Q}$	
0	X	Q_0	$\bar{Q}_0$	Memory
1	0	0	1	RESET
1	1	1	0	SET

Figure 6.14 Logic diagram and truth table for discrete gated latch.

"GATE," G1 and G2 will either block the data input or act as inverters of the data input. Refer to the truth table in Figure 6.14.

Line 1 indicates: If the Gate input is at logic 0, the Data input is in a don't-care state and the output of the latch will be in the memory state. Notice that the Gate input goes to both G1 and G2. If this input is low, the outputs of both G1 and G2 will be forced high, making the internal S-R inputs both equal to logic 1's, which is a memory state. In this situation, the data is in a don't-care state because G1 and G2 are disabled by the logic 0 level of the Gate input.

Lines 2 and 3 show a Gate input level which is equal to logic 1. Because logic 1 is the nondynamic input level of a NAND gate, G1 and G2 will now function as inverters. That is why the location of the inverter was switched in Figures 6.13 and 6.14.

If the Gate input is at logic 1 and the Data input is equal to logic 0, the latch will RESET. Remember that G1 and G2 are acting as inverters to the Data input signal. If the Gate input is at logic 1 and the Data input is equal to logic 1, the latch will SET.

We have now accomplished the construction of a device with a Data input that can SET, RESET, and REMEMBER. The key to this device is the Gate input. The Gate is often referred to as an Enable: if it is high, the latch will be enabled to accept a data input; If it is low, the latch will be disabled and in a memory state.

6.3.2 The 74LS75: 4-Bit Bistable Latch

The 74LS75 is an example of a gated latch (Figure 6.15). Another popular name for gated latches is *transparent latches.*

Refer to the function table in Figure 6.15. If the G input (gate) is high, the logic level that appears on the D input will also appear on the Q output. The device appears to be transparent to the data input (i.e., the data flow straight through the latch). When the G input goes low, the data most recently on

4-BIT BISTABLE LATCHES

75

FUNCTION TABLE
(Each Latch)

INPUTS		OUTPUTS	
D	G	Q	$\overline{Q}$
L	H	L	H
H	H	H	L
X	L	Q_0	$\overline{Q}_0$

H = high level, L = low level, X = irrelevant
Q_0 = the level of Q before the high-to-low transistion of G

See page 7-35

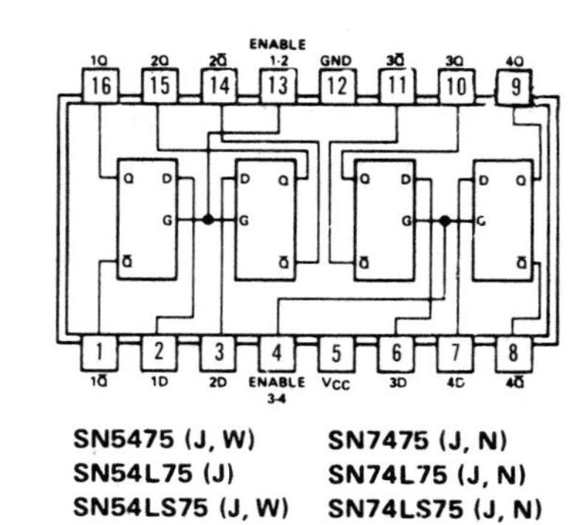

Figure 6.15 74LS75 4-bit bistable latch.

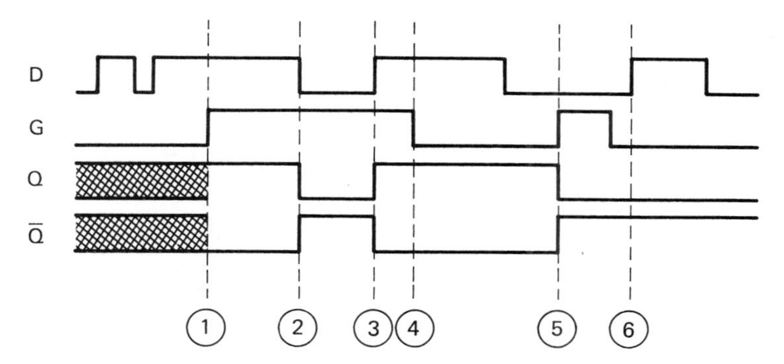

Figure 6.16 74LS75 timing diagram.

the data input before the high-to-low transition of the gate signal will be saved on the Q output. Notice on the pinout that the gate signal is called the enable. Latches 1 and 2 share one enable and latches 3 and 4 another enable.

Refer to the timing diagram of the 74LS75 shown in Figure 6.16. Previous to the gate going high, at event 1, we do not know what level and the Q and $\overline{Q}$ outputs are. (Crossed lines on timing diagrams are used to indicate an unknown state.) When the gate goes high, the latch becomes transparent. We see that the Q output follows the D input for events 1, 2, and 3. Event 4 indicates that the gate has gone to logic 0; the latch is now disabled to further data inputs and is in a memory state. The last level of data to go through the latch while the gate was still high was a logic 1. The Q output will stay at logic 1 until the gate once again goes high and some low data is presented at the D input. That is exactly what happens at event 5. Event 6 shows that even though high data is presented to the latch, the gate is disabled, so the Q output will not respond.

6.3.3 Applications of Clocked Latches

Consider the following problem: An electronic pulse meter senses and calculates pulse rates. Every 10 seconds the pulse meter outputs a narrow strobe indicating that the updated BCD code for the two decimal digits is on the output lines. We must have some method of saving the BCD codes when the strobe goes active. Refer to Figure 6.17.

While the pulse meter is busy counting the next 10-second sample of pulses, the displays are unaffected. They are driven by the two 74LS75 4-bit latches. The latches will receive the new count every time the strobe goes to a logic 1 level. Both TTL and CMOS MSI display drivers are available with on-board latches.

6.4 CLOCKS AND TRANSITIONS: RISING AND FALLING EDGES

We have talked about the clock and how it relates to digital circuits as a timing reference. Consider the transparent latches that we have just examined. Data will flow through the latch whenever the enable is active. If we wanted to catch some data

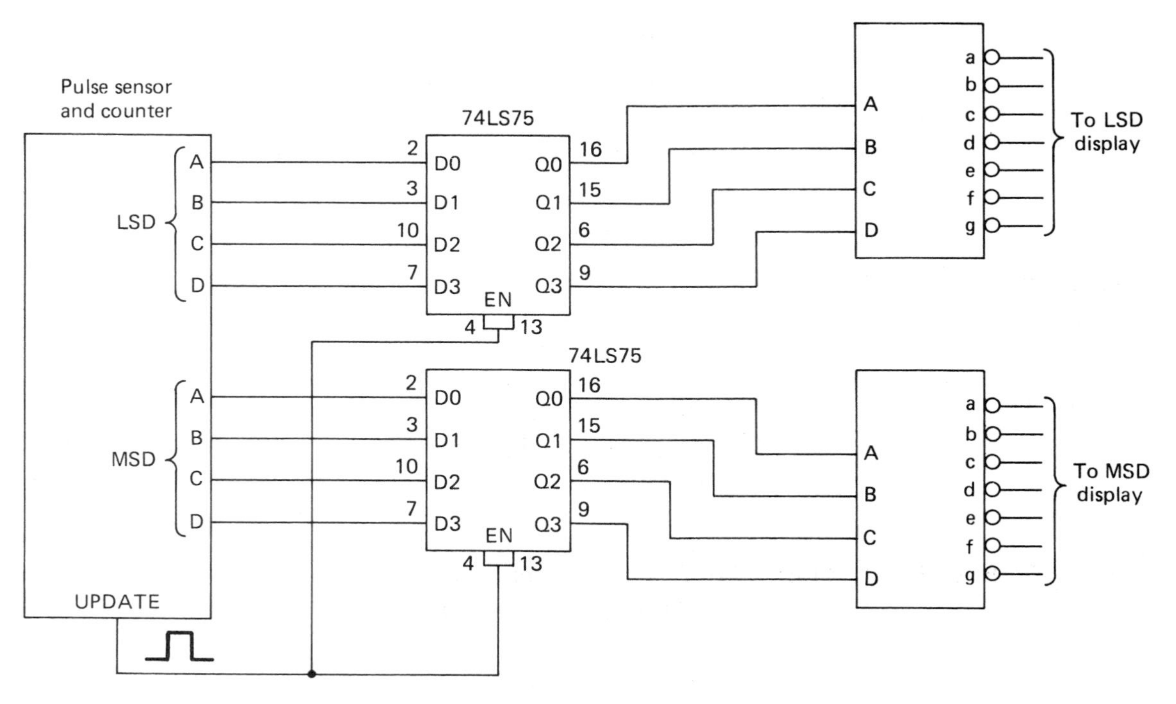

Figure 6.17 Application of 4-bit latches.

at an exact moment in time, we would have to make the enable active for a very short period.

The enable pulses shown in Figure 6.18 are narrow, but there still is a short amount of time when the latch is transparent. Event 2 shows us that the wrong data has slipped through because of the width of the gate. Instead of letting data through on the active level of the enable, why not only let the data through on the rising or falling edge of the enable? We normally assume that the rise and fall times of a square wave are infinitely fast, but of course, it takes a finite amount of time for a square wave to change levels. This time is so fast that it appears to be instantaneous compared to the speed of the other signals in the circuit. We are now going to study an important group of digital devices called flip-flops. Flip-flops have many similarities to gated latches. There will be one important difference: in

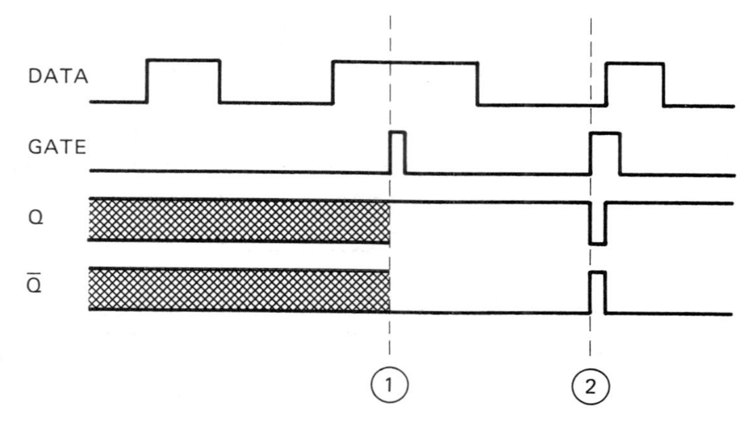

Figure 6.18 Ambiguity of level-clocked devices.

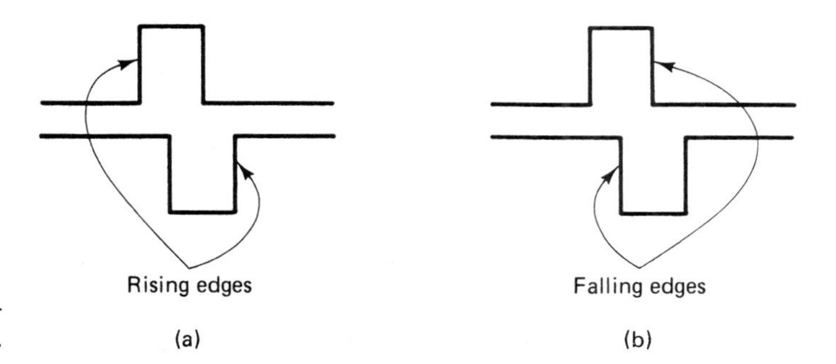

Figure 6.19 Rising edges/falling edges.

gated latches data are transferred during the active level of the enable, in flip-flops, data are transferred only during the active edge of the enable. The enables on flip-flops are called clocks. Refer to Figure 6.19 for examples of rising and falling edges.

6.5 THE D FLIP-FLOP

The "D" in "D flip-flop" stands for the word "data." The function of a D flip-flop is simple:

> *On the rising edge of the clock signal the logic level that is residing on the D input will be transferred to the Q output.*

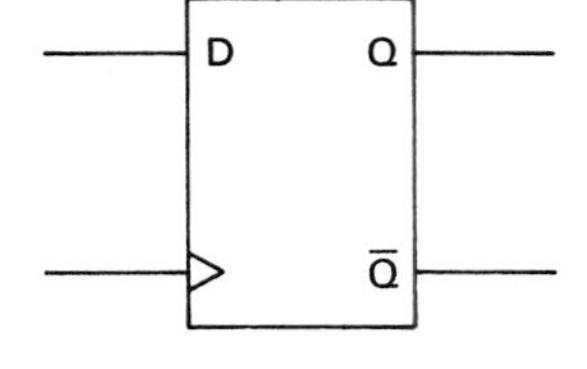

Figure 6.20 Logic symbol for data flip-flop.

Notice that this is an edge- not a level-triggered device. Figure 6.20 illustrates the standard logic symbol for a D flip-flop. The clock input is indicated by a symbol that looks like half a diamond. This symbol means that the clock is edge-active. Because this symbol is not preceded by a bubble, you should assume that it means a positive-edge-triggered clock.

Figure 6.21 illustrates a simple timing diagram for this device. The vertical dashed lines represent the times when the clock has a rising edge. It is only at these rising edges that the output will have the potential to change. Until the first rising edge, we do not know the values of Q and $\overline{Q}$. At rising edge 1, we see that the data input is equal to logic 0. This logic 0 will be transferred into the Q output, and its complement, a logic 1, will be transferred into the $\overline{Q}$ output. The outputs will stay in

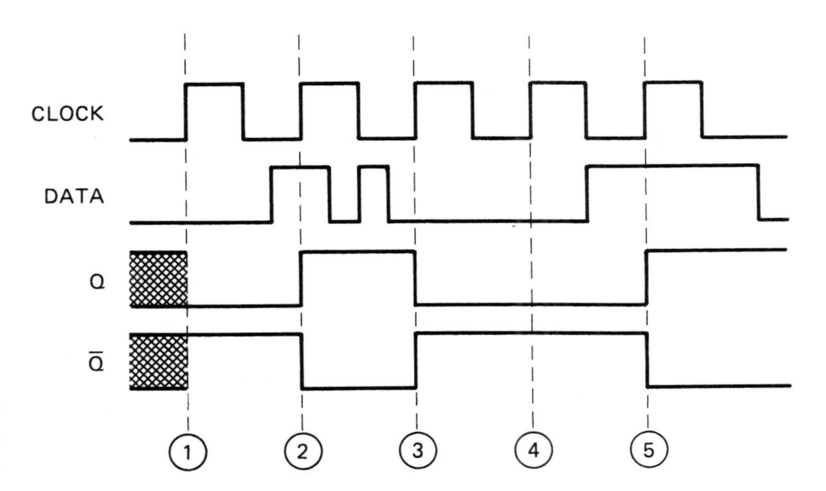

Figure 6.21 Timing diagram for positive-edge-triggered D flip-flop.

a memory state until the next rising edge of the clock. At this point the data input is logic 1; the Q output will go to logic 1. At the third rising edge the data input is logic 0 and the Q output will follow this logic 0. The fourth rising edge also occurs when the input data are logic 0, therefore, the Q output will stay at logic 0. The last rising edge in the timing diagram happens when the data are high, so the Q output will go high.

6.5.1 The 74LS74: Dual D-Type Positive-Edge-Triggered Flip-Flop

The 74LS74 is a popular D flip-flop with two added functions: a direct preset and clear. You should recognize the device shown in Figure 6.22 as a positive-edge-triggered D flip-flop. Even though you do not know what the preset and clear functions are, you can tell from the pinout and function table that they are active-low inputs.

Line 1. Preset is low, clear is high, and the clock and D inputs are don't cares. Because the preset and clear inputs are bubbled, they must be active-low. Here the preset input is at an active-low level, and the clear input is inactive high. This line represents an operation called *presetting*. This presetting operation must have a higher priority than the normal data clocking operation because the D and clock inputs are don't cares. "Preset" sounds like the term "set" that we used in describing the S-R latch. The terms "preset" and "set" are used interchangeably.

Line 2. Preset high, clear low, and the clock and D inputs are don't cares. We now see that the clear input is active. The truth table indicates that the Q output will go low and the $\overline{Q}$ output will go high, regardless of the state of the clock or D input.

DUAL D-TYPE POSITIVE-EDGE-TRIGGERED FLIP-FLOPS WITH PRESET AND CLEAR

74

FUNCTION TABLE

INPUTS				OUTPUTS	
PRESET	CLEAR	CLOCK	D	Q	$\overline{Q}$
L	H	X	X	H	L
H	L	X	X	L	H
L	L	X	X	H*	H*
H	H	↑	H	H	L
H	H	↑	L	L	H
H	H	L	X	Q_0	$\overline{Q}_0$

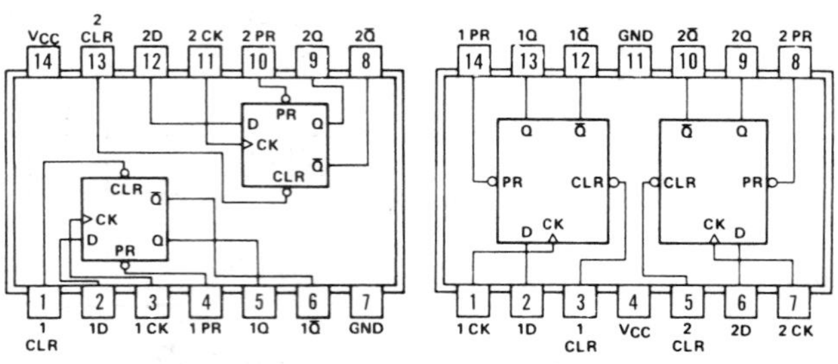

SN5474 (J) SN7474 (J, N) SN5474 (W)
SN54H74 (J) SN74H74 (J, N) SN54H74 (W)
SN54L74 (J) SN74L74 (J, N) SN54L74 (T)
SN54LS74A (J, W) SN74LS74A (J, N)
SN54S74 (J, W) SN74S74 (J, N)

See pages 6-46, 6-50, 6-54, and 6-56

Figure 6.22 74LS74 dual D-type positive-edge-triggered D flip-flop.

To clear a flip-flop means the same thing as resetting an S-R latch. The terms "clear" and "reset" are used interchangeably.

Line 3. Both preset and clear are active, the clock and D input are don't cares. This is a situation that must be avoided. Both Q and $\overline{Q}$ will both be forced high—this output state is unstable. Circuit designers will ensure that this situation will never occur during normal operation.

Preset and clear will be inactive for lines 4 through 6.

Line 4. This flip-flop will behave like the one we examined previously. With a rising edge at the clock and high data on the D input, the Q output will go high and the $\overline{Q}$ output will go low.

Line 5. The same as line 4, but with low data on the D input. Q output will go low and the $\overline{Q}$ output will go high.

Line 6. This line shows that the preset and clear are both high, and the clock does not have a rising edge: nothing is active. The symbols Q_0 and $\overline{Q}_0$ indicate that the output does not change; it is in a memory state.

6.5.2 Timing Diagrams of D Flip-Flops

When analyzing timing diagrams of D type flip-flops, you must remember that the preset and clear override the data and clock inputs. These presets and clears are often called asynchronous inputs because they occur independently of the clock.

The best approach to analyzing the timing diagram shown in Figure 6.23 is to start with the times when the preset or clear

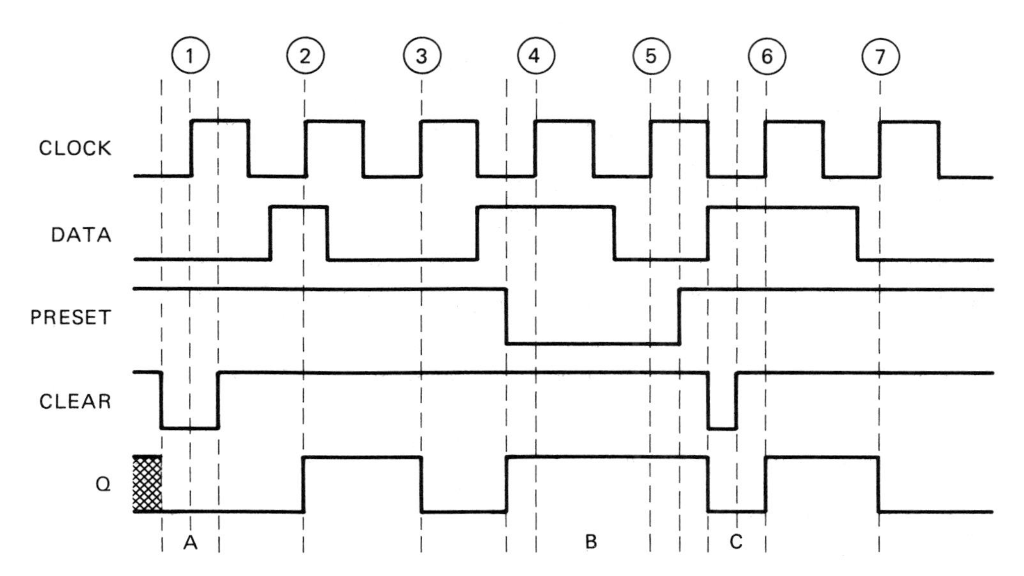

Figure 6.23 Timing diagram for D-type flip-flop with preset and clear.

inputs are active. The timing diagram is for a 74LS74 with active-low preset and clear. Whenever a preset or clear goes active, the output will automatically respond and stay at that particular level at least until the preset or clear goes inactive and another rising edge of the clock occurs. Refer to the part of the timing diagram where the clear input first goes active. The Q output responds with a logic 0. The clear stays active during the first rising edge. When the clear goes inactive, the output cannot change until the next rising edge of the clock. The Q output goes high on this second rising edge because neither preset nor clear input is active and the data input is high. This part of the Q output is labeled with an A. Examine the point where the preset input goes active. Instantly, the Q output responds with a logic 1. The fourth and fifth rising edges of the clock are overridden by the active preset. This part of the output waveform is designated by a B.

6.5.3 Setup and Hold Times

D-flip-flops have many parameters. The two most important parameters for technicians to understand are setup and hold times.

Setup time The length of time that the data on the D input must be present and stable before the active edge of the clock signal.

Hold time The length of time that data on the D input must remain stable after the active edge of the clock.

The setup time for a 74LS74 is 25 ns, and the hold time is 5 ns. The setup and hold times for CMOS flip-flops depend on the value of V_{dd}. As V_{dd} increases, the setup and hold times will decrease proportionally.

If a flip-flop circuit is displaying erratic behavior (working sometimes and other times malfunctioning), use a dual-trace oscilloscope to ensure that the setup- and hold-time specifications are being met.

6.5.4 D-Flip-Flop Applications

D-type flip-flops are used in a wide variety of applications. In practical circuits, they provide a simple means of saving a bit of data at specified time. Refer to the circuit and timing dia-

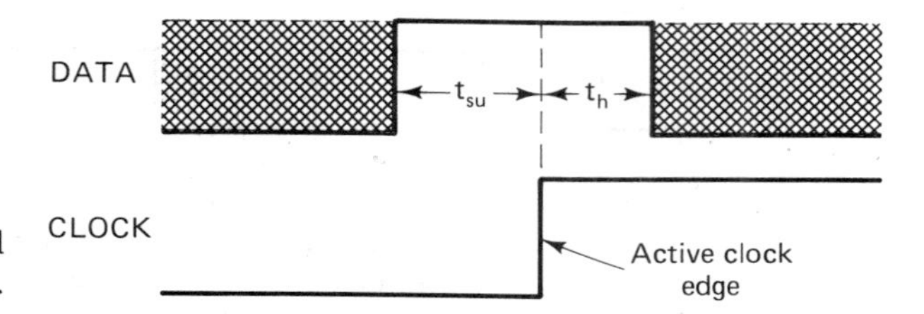

Figure 6.24 Setup and hold times.

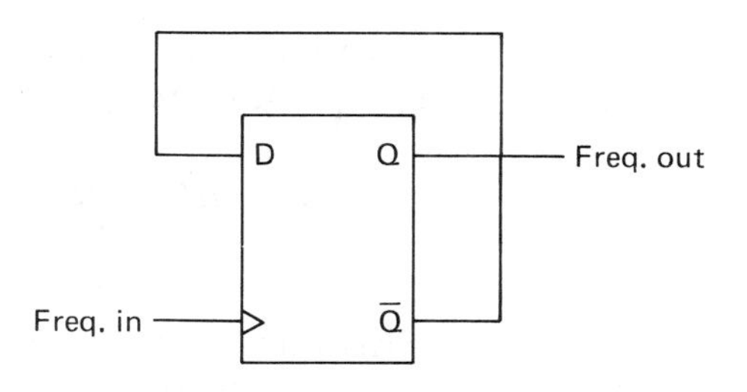

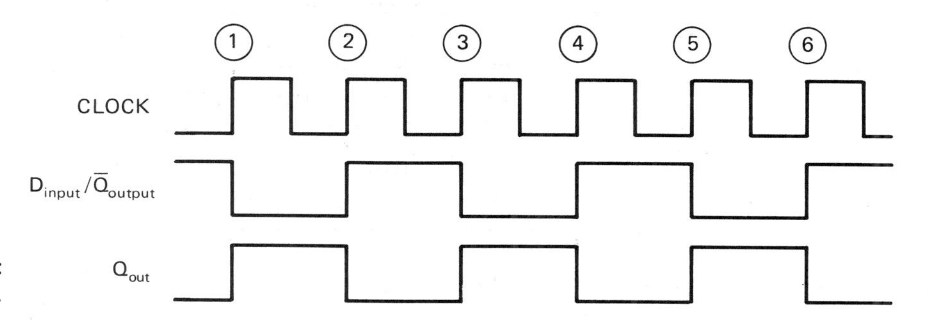

Figure 6.25 D-type flip-flop: configured to toggle.

gram in Figure 6.25. The flip-flop in this figure can be any positive-edge-triggered D type. The absence of set and clear pins indicates that these functions are not being utilized in this particular circuit. (If the flip-flop does have set and clear inputs, they must be tied to their inactive level.)

Notice that the $\overline{Q}$ output is fed back to the data input. The operation of this circuit will be analyzed by constructing a timing diagram. The timing diagram will have three lines: the clock input, Q output, and the data input/$\overline{Q}$ output. All the signals in the circuit will be referenced to the clock. [When we create a clock signal for the timing diagram its absolute frequency and duty cycle are not important. Most clock signals on timing diagrams will be a 50% duty-cycle square wave with just enough cycles to analyze the circuit operation fully. We will focus our attention on the rising edges of the clock.] Assume initially that the Q output is low and the $\overline{Q}$ output is high. The $\overline{Q}$ output is fed back to the data input of the flip-flop.

Edge 1. With a high on the data input, this rising edge will cause the Q output to go high and the $\overline{Q}$ output to go low.

Edge 2. The data input is now at a logic 0 level. This rising edge will cause the Q output to go low and the $\overline{Q}$ output to go high.

Edge 3. The data input is once again at logic 1. This rising edge will cause the Q output to go high and the $\overline{Q}$ output to go low.

By now you should realize that each time there is a rising edge of the clock, the output of the flip-flop will toggle states.

Compare the frequency of the clock signal to the frequency of the signals at Q and $\overline{Q}$.

The output frequency of this flip-flop is exactly one-half of the clock frequency.

This circuit is called a *divide-by-2 circuit.* It is the basic building block of digital counters.

Figure 6.26 illustrates a pulse synchronizing circuit. Let's analyze this circuit in functional parts.

Gates G1 and G2 are configured as an S-R switch debouncer. When the SPDT switch is toggled from the NO to the NC position, the output of G1 will be a clean, debounced rising edge.

G3, G4, G5, and G6 function as an electronic inverting differentiator. This circuit takes advantage of propagation delay.

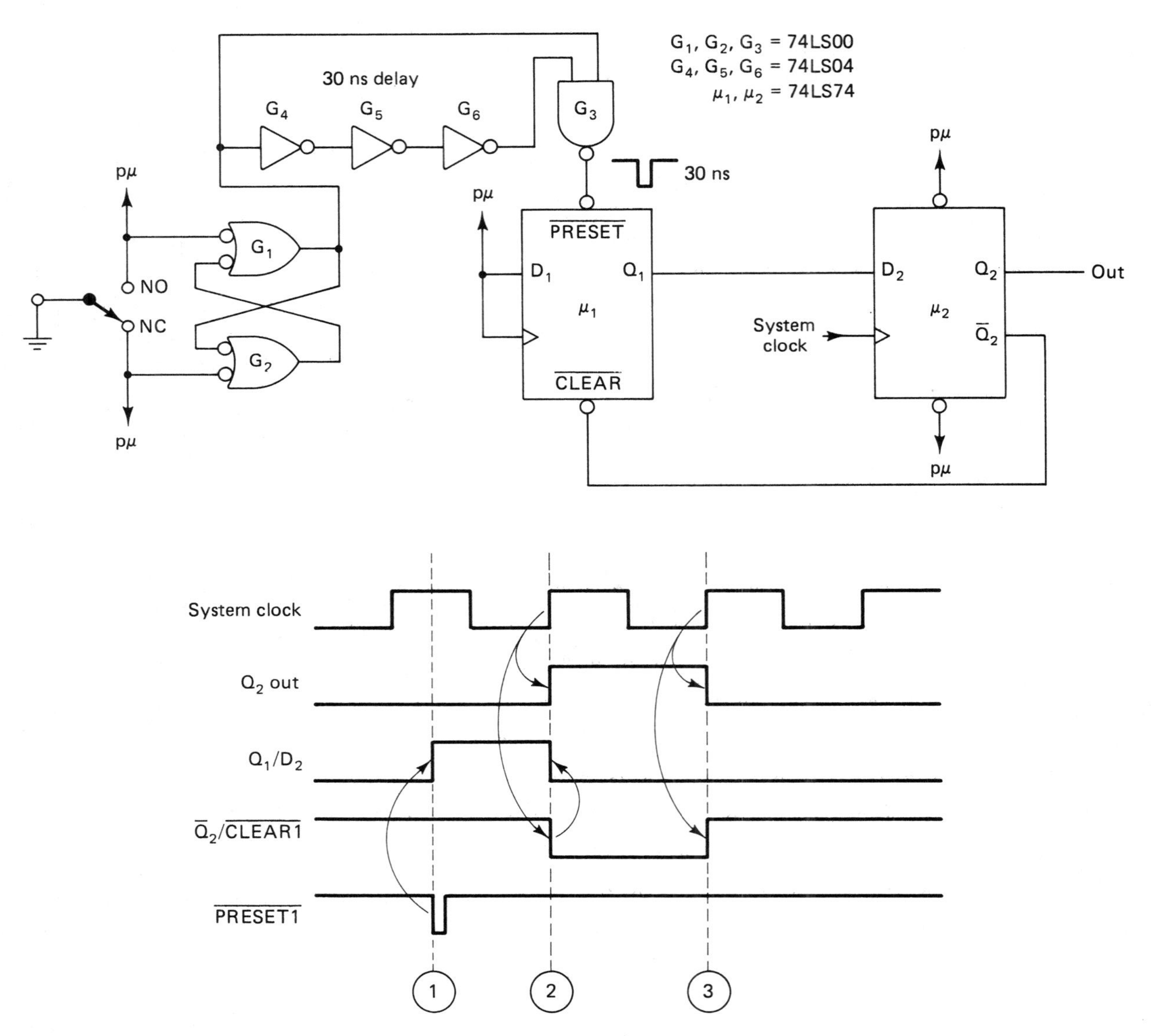

Figure 6.26 Pulse synchronizing circuit.

The typical propagation delay of a 74LS04 is 10 ns. Gates G4, G5, and G6 form a 30-ns inverting delay. With the SPDT switch in the NC position, the output of G1 is a logic 0. This logic 0 is applied to pin 2 of G3 and pin 1 of G4. A logic 0 on pin 3 of G3 will cause the preset on U1 to be high. The three inverters will cause pin 1 of G1 to be normally high. When the S-R flip flop outputs a rising edge, pin 2 of G3 will instantly go high. This will cause the preset input of U1 to go low. Approximately 30 ns after the rising edge out of G1, pin 1 of G3 will finally go low. This will cause the preset input of U1 to go high. Therefore, each time the SPDT switch is toggled from the NC to NO position, a 30-ns active-low pulse will appear on the preset input of U1.

U1 is configured as a SET/RESET flip-flop. Because the clock is tied to V_{cc}, the data input will always be don't-care condition. The only time this flip-flop will be active is when a low level is applied to the preset or clear inputs. We have already seen that the preset input is driven by the 30-ns negative pulse from the electronic differentiator. The clear input is driven by $\overline{Q}$ from U2.

Anytime Q1 is high and a rising edge from the system clock occurs, Q2 will go high and $\overline{Q}2$ will go low. This low from $\overline{Q}2$ will clear flip-flop U1.

Now that we have discussed the functional blocks in this system, it is time to examine the timing diagram. Assume that the SPDT switch is in the NC position, and both U1 and U2 are reset. There are three events of interest on the timing diagram in Figure 6.26.

Event 1. The SPDT switch is toggled to the NO position. This creates a 30-ns negative pulse on the preset input of U1. This low pulse will cause Q1 to go high.

Event 2. With D2 high, the rising edge of the system clock will cause Q2 to go high and $\overline{Q}2$ to go low. When $\overline{Q}2$ goes low, the clear input of U1 will go active, forcing Q1 low.

Event 3. The next rising edge of the system will transfer the low data onto the Q2 output. $\overline{Q}2$ will go high, releasing the active clear input on U1.

The circuit is now ready for another positive edge from G1. Notice that the Q2 output was one positive pulse synchronized with the clock.

6.5.5 A Sequential Electronic Lock.

The following circuit is an application of the D flip-flop. The electronic lock in Figure 6.27 uses both combinational and sequential logic circuits. It is an example of a sequential lock that demonstrates the capacity to remember the previous digits that

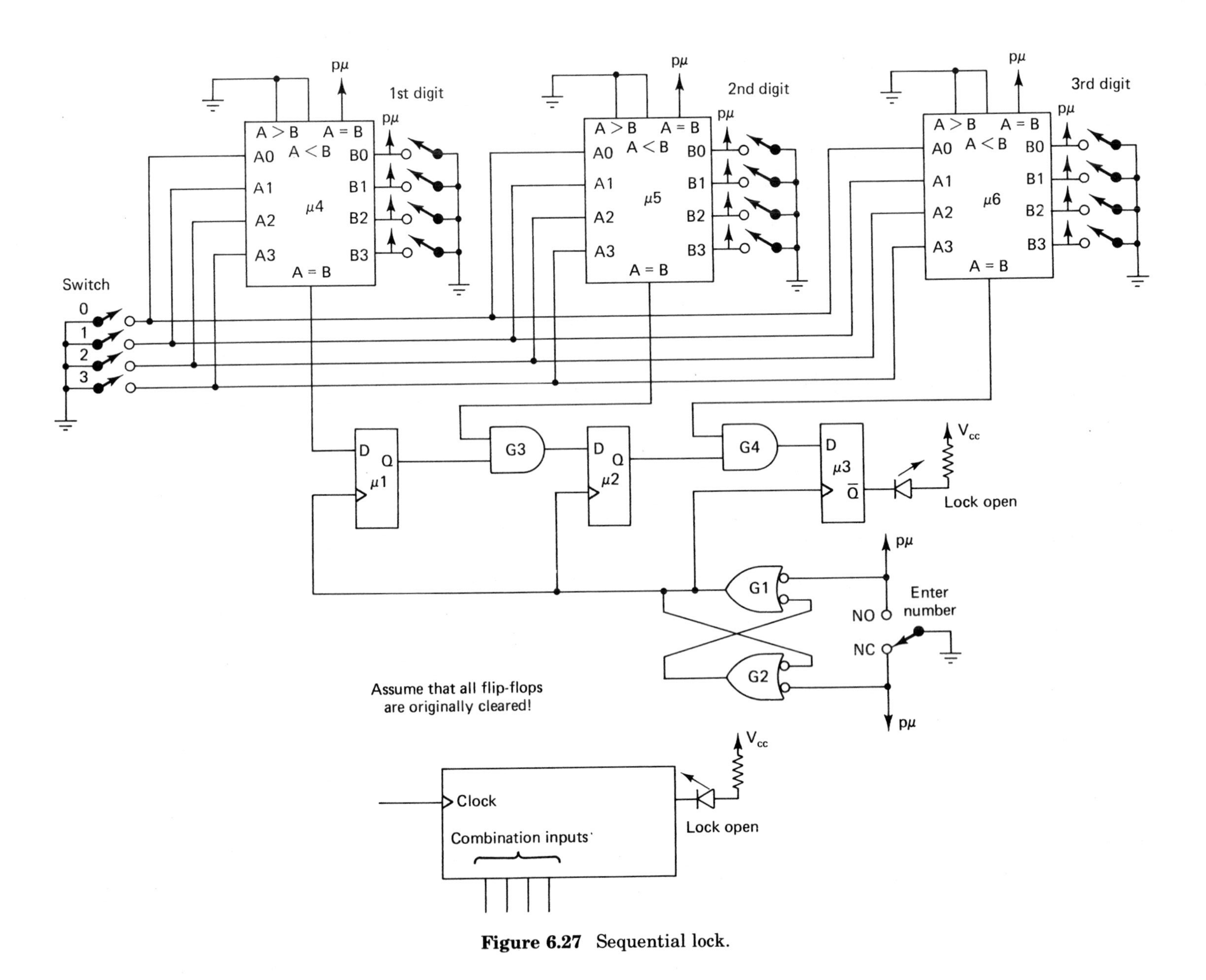

Figure 6.27 Sequential lock.

were entered. It is the electronic equivalent of the mechanical lock illustrated in Figure 6.1b.

Refer to the block diagram in Figure 6.27. The rising edge of the clock will transfer the four binary combination bits into the lock. When the LED illuminates, the lock will be open. To operate the lock, proceed as follows:

1. Set the first digit of the combination into switches SW 0 through SW 3.
2. Clock this digit into the lock by toggling the SPDT switch.
3. Repeat steps 1 and 2 for the next two digits and the lock will open.

If an incorrect number is entered, the operator must start over at step 1.

Now that you understand how to operate the lock, let's analyze how it actually functions. First, the combination is set with the three banks of switches connected to the magnitude comparators (B0 through B3). Each bank of four switches will represent a decimal number from 0 to 15. After the internal combination is set, the unit will be sealed.

The first digit of the combination will be entered into SW 0 through SW 3. If this is the correct digit, the A = B output of the magnitude comparator will go high. The A = B output is connected to the data input of the first flip-flop. Toggling the SPDT switch will create a rising edge that will transfer the status of the A = B output of U4 into the Q output of U1.

The second digit of the combination will now be entered into SW 0 through SW 3. If the Q output of U1 is high (indicating that the first digit was correct) and the A = B output of U5 is high (indicating that the second digit is correct), the data input of U2 will be high. When the SPDT switch is toggled, this value will be transferred into the Q output of U2.

The third digit is entered into SW 0 through SW 3. If the output of U2 is high (indicating that the first two digits were correct) and the A = B output of U6 is high (indicating that the third digit is correct), the data on U3 will be high. When the SPDT switch is toggled, the data will be transferred into the Q output of U3. If the Q output of U3 is high, the $\overline{Q}$ output will be low and the LED will illuminate. This indicates that the lock is now open.

Notice how each flip-flop remembers the status of the digits entered previously. If any incorrect digit is entered into the lock, the respective AND gate will force the data input of the next flip-flop low, indicating this error.

6.6 THE J-K FLIP-FLOP

The S-R latch had one major problem—one combination of S and R inputs would result in an undefined output. Gated latches and edge-triggered D flip-flops were developed to overcome this prob-

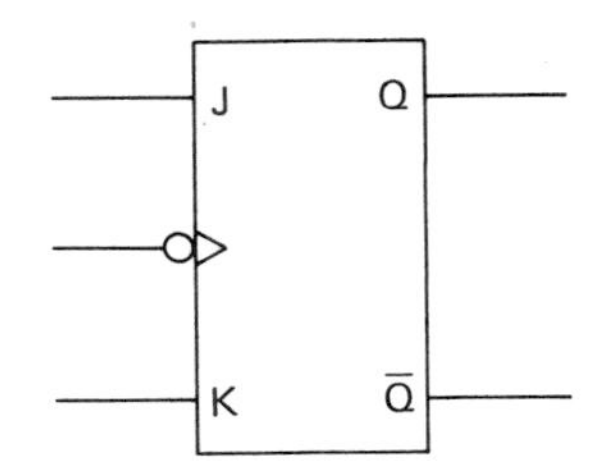

Figure 6.28 Basic logic symbol for J-K flip-flop

lem. In Figure 6.28, the logic symbol indicates a negative-edge-triggered clock. Instead of S and R inputs or a Data input. we see J and K inputs.

6.6.1 The 74LS76: Dual J-K Flip-Flop

Refer to Figure 6.29. The function table for the 7476 and 74H76 has a positive pulse symbolized in the clock column, whereas the function table for the 74LS76 has a falling edge under its clock column. The logic symbol in the pinout diagram is applicable to all three J-K flip-flops, and it indicates that the devices are negative edge-triggered. J-K flip-flops are constructed from two flip-flops, called the *master* and the *slave*. In older technology, the information on the J-K inputs was transferred into the master on the rising edge of the clock, and into the slave (output) portion on the falling edge of the clock. Because the output would react on the falling edge of the clock, these devices appeared to be negative edge-triggered. The newer technology J-K flip-flops are true edge-triggered devices. Remember that the device will appear to be negative edge-triggered.

Let's follow the function table for the 74LS76 line by line:

Lines 1–3. The preset and clear functions are the same as for the D flip-flop. Both preset and clear should never be active simultaneously.

Line 4. Both preset and clear are inactive, and the clock is showing a falling edge. With the J and K inputs both at logic 0, the output will not change. Therefore:

When J and K are both at logic 0, the flip-flop is in the memory state.

Line 5. J is high and K is low. When the falling edge of the clock occurs, Q will go high and $\overline{Q}$ will go low.

DUAL J-K FLIP-FLOPS WITH PRESET AND CLEAR

76

'76, 'H76
FUNCTION TABLE

INPUTS					OUTPUTS	
PRESET	CLEAR	CLOCK	J	K	Q	$\overline{Q}$
L	H	X	X	X	H	L
H	L	X	X	X	L	H
L	L	X	X	X	H*	H*
H	H	⎍	L	L	Q_0	$\overline{Q}_0$
H	H	⎍	H	L	H	L
H	H	⎍	L	H	L	H
H	H	⎍	H	H	TOGGLE	

'LS76A
FUNCTION TABLE

INPUTS					OUTPUTS	
PRESET	CLEAR	CLOCK	J	K	Q	$\overline{Q}$
L	H	X	X	X	H	L
H	L	X	X	X	L	H
L	L	X	X	X	H*	H*
H	H	↓	L	L	Q_0	$\overline{Q}_0$
H	H	↓	H	L	H	L
H	H	↓	L	H	L	H
H	H	↓	H	H	TOGGLE	
H	H	H	X	X	Q_0	$\overline{Q}_0$

See pages 6-46, 6-50, and 6-56

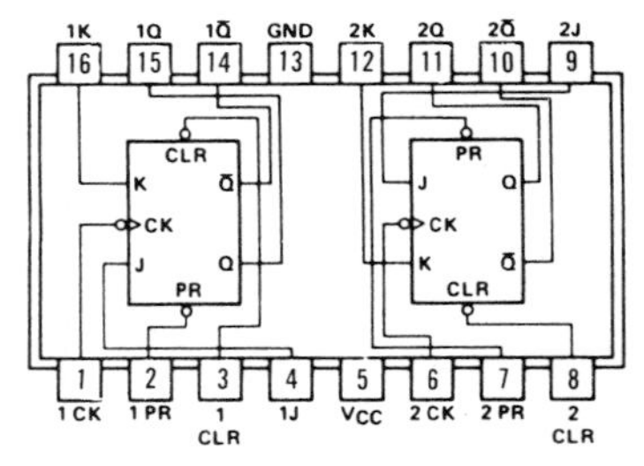

SN5476 (J, W) SN7476 (J, N)
SN54H76 (J, W) SN74H76 (J, N)
SN54LS76A (J, W) SN74LS76A (J, N)

Figure 6.29 74LS76: dual J-K flip-flop with preset and clear.

When J = 1 and K = 0, the falling edge of the clock will set the flip-flop.

Line 6. J is low and K is high. On the falling edge of the clock, the Q output goes low and the $\overline{Q}$ output will go high.

When J = 0 and K = 1, the falling edge of the clock will reset the flip-flop.

Line 7. Both J and K inputs are high. Under the outputs column we see the word "toggle." Each output, Q and $\overline{Q}$, will change to the complement of their value before the falling edge of the clock. This is how the J-K flip-flop handles the previously undefined operation.

When J = 1 and K = 1, the falling edge of the clock will cause Q and $\overline{Q}$ to toggle.

Line 8. This line symbolizes an inactive clock. The Q and $\overline{Q}$ outputs will not change.

6.6.2 Applications of the J-K Flip-Flop

J-K flip-flops are the basic building block of many advanced circuits. We could use the circuit illustrated in Figure 6.30 to sup-

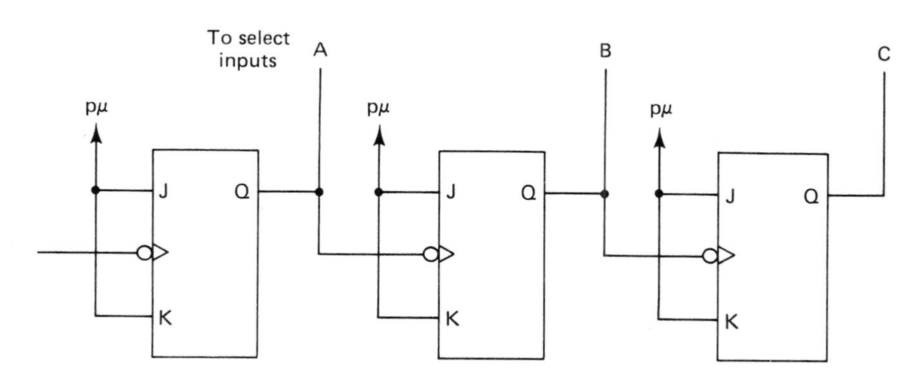

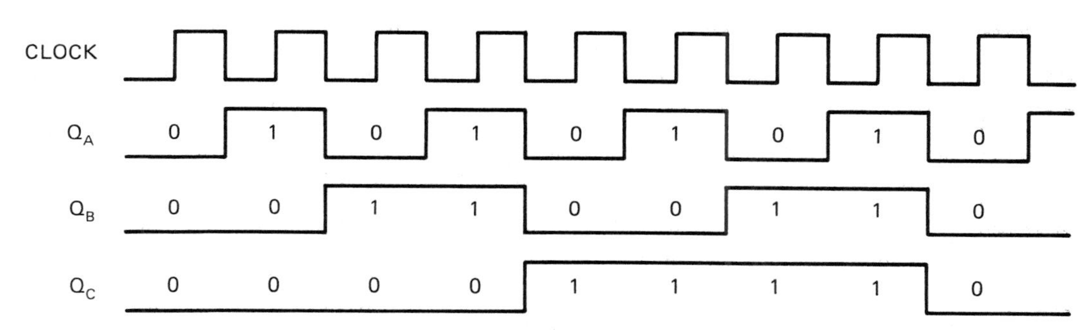

Figure 6.30 J-K flip-flop configured to supply select pulses to multiplexer/demultiplexer.

ply the select inputs to a multiplexer and demultiplexer. In this manner all eight channels will appear to be sent simultaneously.

In Figure 6.30 all three J-K flip-flops are configured to toggle. A system clock will clock the first flip-flop. The Q output from the first flip-flop will clock the second flip-flop, and the Q output from the second flip-flop will clock the third flip-flop.

If we choose Q1 to be the least significant bit and Q3 to be the most significant bit, these flip-flops will count from 0 (000 binary) to 7 (111 binary). This binary count will be applied to the select inputs of the multiplexer and demultiplexer.

Many advanced digital devices require a two-phase clock. Instead of just one clock input, these devices require two clock inputs, where the clocks are both the same frequency but are out of phase with each other.

Consider the circuit in Figure 6.31 carefully. The clock input is common to both J-K flip-flop 1 and J-K flip-flop 2. Output 1 is $\overline{Q}1$, and output 2 is Q2. Consider the feedback. When Q2 is high, the next rising edge of the clock will force flip-flop 1 to perform a reset operation. When Q2 is low, the next rising edge will cause flip-flop 1 to perform a set operation. The feedback scheme is the key to how the device operates.

To analyze this circuit we will assume that both outputs are initially at logic 0. We will freeze time just before a rising

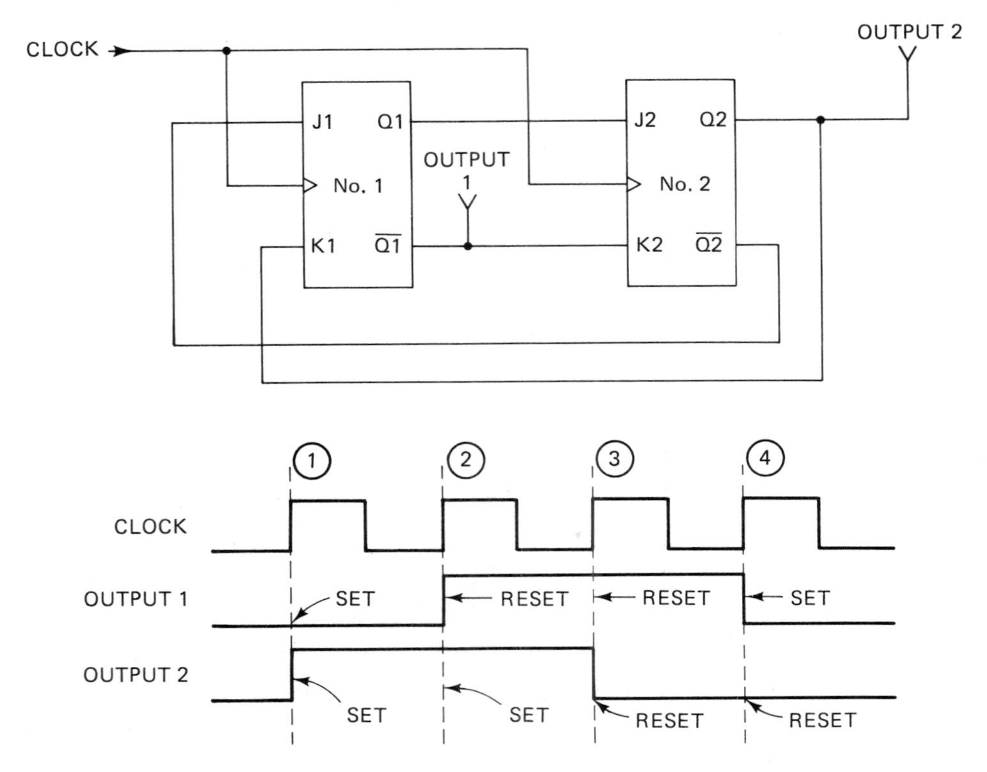

Figure 6.31 Using J-K flip-flops to create a two-phase clock.

edge of the clock, then establish what function the particular flip-flop will perform when the rising edge of the clock actually occurs.

1. Just before the first rising edge of the clock:

   J1 = 1, K1 = 0; J-K flip-flop 1 is configured for a Set operation.
   J2 = 1, K2 = 0; J-K flip-flop 2 is also configured for a Set operation.

2. On the positive edge 1:

   J-K flip-flop 1 sets and output 1 remains low,
   J-K flip-flop 2 sets and output 2 goes high.

3. Just before the second rising edge of the clock:

   J1 = 0, K1 = 1; J-K flip-flop 1 is configured for a Reset operation.
   The inputs to J2 and K2 did not change after the first clock, so J-K flip-flop 2 is still configured for a Set operation.

4. On positive edge 2:

   J-K flip-flop 1 resets and output 1 goes high.
   J-K flip-flop 2 sets once again, and output 2 stays high.

5. Just before the third rising edge of the clock:

   J-K flip-flop 1 is still configured for a Reset operation.
   J2 = 0, K2 = 1; J-K flip-flop 2 is configured to Reset.

6. On positive edge 3:

   J-K flip-flop 1 will reset for the second time in a row, and output 1 stays high.
   J-K flip-flop 2 will reset and output 2 will go low.

7. Just before the fourth rising edge of the clock:

   J1 = 1, K1 = 0; J-K flip-flop 1 is configured to Set.
   J2 and K2 have not changed and J-K flip-flop 2 is still configured to Reset.

8. On positive edge 4:

   J-K flip-flop 1 will set and output 1 goes low.
   J-K flip-flop 2 will, once again, reset and output 2 stays low.

6.7 THE ONE-SHOT

The formal name for the *one-shot* is *monostable multivibrator.* The one-shot delivers one pulse of predetermined length for each trigger pulse it receives. The one-shot can provide a delayed pulse, a shortened pulse, or a widened pulse.

6.7.1 Retriggerable and Nonretriggerable One-Shots

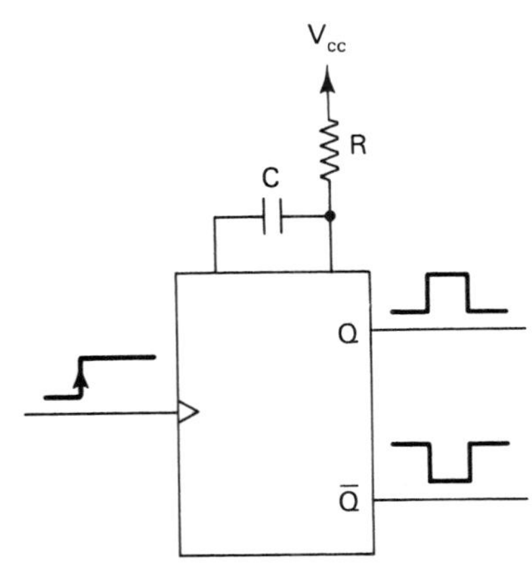

Figure 6.32 Logic symbol for one-shot.

One-shots can be divided into two different types: retriggerable and nonretriggerable. The retriggerable one-shot can continually sense input triggers.

Figure 6.32 illustrates a simplified logic diagram of a one-shot. The input of a one-shot is called a *trigger input* rather than a clock. The trigger input shown here is sensitive to rising edges. When it receives a rising-edge trigger, the Q output will respond with a positive pulse whose width is dependent on the *RC* time constant of R1 and C1. The equation that describes the output pulse width varies with each particular device. In a retriggerable one-shot, if the Q output is high and another trigger pulse occurs, the Q output will stay high for another total time period. For example, the output of a particular one-shot is set to be 10 ms; 5 ms after an initial trigger pulse is received, another trigger pulse occurs. Another 10 ms will be added to the pulse length, making it a total of 15 ms. If we trigger a retriggerable one-shot at shorter periods than its output pulse width, the Q output will stay high.

Once a nonretriggerable one-shot is fired, it will ignore any further triggers until its Q output is once again at a logic 0 level. If the one-shot in the previous example was a nonretriggerable type, its output pulse width would be 10 ms regardless of how many extra triggers occurred after the initial triggered.

6.7.2 The 74LS123: Dual Retriggerable One-Shot

Refer to Figure 6.33. The triggering schemes of one-shots are more sophisticated than a single rising edge. Examine the logic symbol for the 74LS123. Notice the symbolic three-input AND gate that is attached to the rectangle. Whenever a positive transition occurs at the output of this internal AND gate, the one-shot will receive a trigger. Examine the function table in Figure 6.33.

Line 1. This one-shot also has an active-low direct clear as in the D and J-K flip-flops. The logic diagram shows that this clear input is active low. When this direct clear is active, the A and B inputs will be don't-care conditions and the Q output will go low. This direct clear is used to effectively shorten an output pulse, or to disable the one-shot.

Line 2. Notice that the A input is active low. If A is high, it is impossible to trigger this device. A high on the A input will

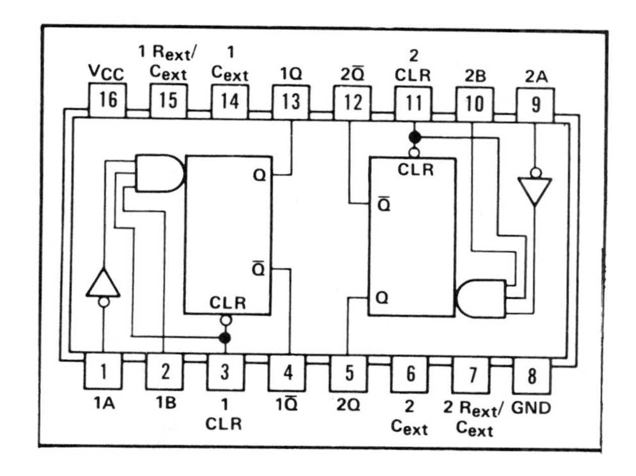

'123, 'L123, 'LS123
FUNCTION TABLE

INPUTS			OUTPUTS	
CLEAR	A	B	Q	$\bar{Q}$
L	X	X	L	H
X	H	X	L	H
X	X	L	L	H
H	L	↑	⊓	⊔
H	↓	H	⊓	⊔
↑	L	H	⊓	⊔

Figure 6.33 74LS123 dual retriggerable one-shot.

be inverted to a logic 0 (via the internal inverter) which disables the internal AND gate from having any low to high transitions.

Line 3. Input B is an active-high input. If B is low, it has the same result as we examined in line 2.

Line 4. The clear input is inactive and the A input is active low. When the B input has a rising edge, the one-shot will be fired. The positive and negative pulses in the function table symbolize the output pulses. To fire this one-shot, a positive transition must occur at the output of the internal 3-input AND gate. Figure 6.34 illustrates how the conditions in line 4 of the truth table cause this transition to occur.

Line 5. Because input A is active low, this line is the A input version of line 4. If the clear is inactive and input B is active high, a negative transition on input A will cause a positive transition on the output of the internal AND gate; this will fire the one-shot. We now see that the 74LS123 can be fired with negative or positive edge triggers.

Line 6. The third way to get a positive transition on the output of the internal AND gate is illustrated in this line. If inputs A and B are at their active levels, a low-to-high transition on the clear input will fire the one-shot. This is the least commonly used method of firing the one-shot.

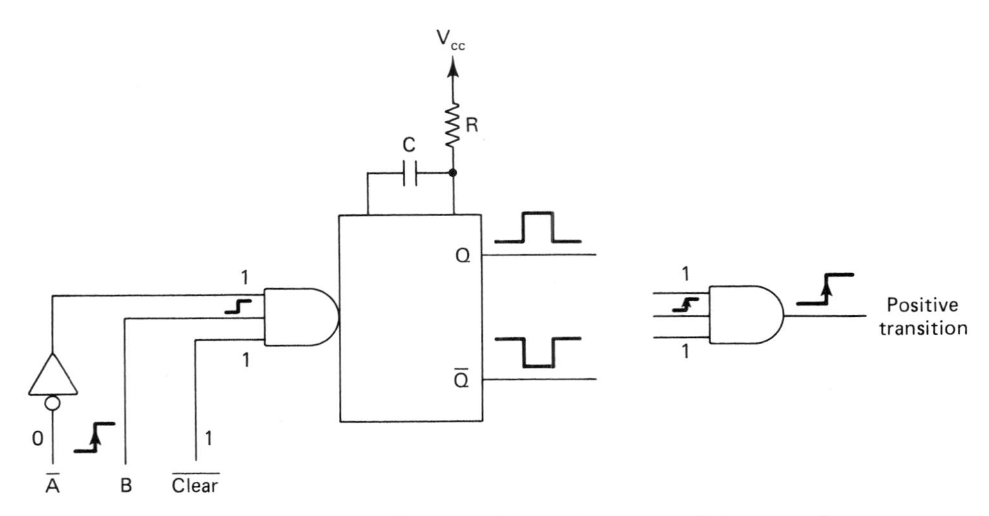

Figure 6.34 Using a rising edge on the B input to fire the one-shot.

6.7.3 The 74LS221: Dual Nonretriggerable One-Shot

The 74LS221 is an extremely popular one-shot. It can be used for many different types of applications.

Notice the cross-coupled NAND gates that form an S-R latch in Figure 6.35. The function table is almost identical to that of the 74LS123. Therefore, we can ignore the complicated logic diagram and treat this device like a nonretriggerable version of the 74LS123. There will be no indication on the logic symbol to designate whether a particular one-shot is retriggerable or nonretriggerable. It is up to the technician to refer to

FUNCTION TABLE
(EACH MONOSTABLE)

INPUTS			OUTPUTS	
CLEAR	A	B	Q	$\overline{Q}$
L	X	X	L	H
X	H	X	L	H
X	X	L	L	H
H	L	↑	⊓	⊔
H	↓	H	⊓	⊔
↑	L	H	⊓	⊔

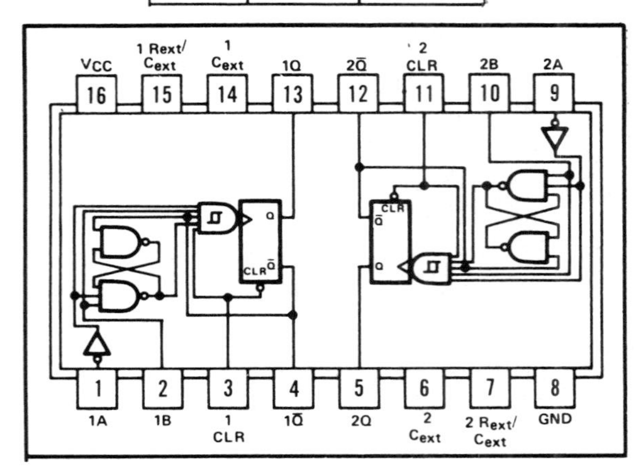

Figure 6.35 74LS221 dual nonretriggerable one-shot.

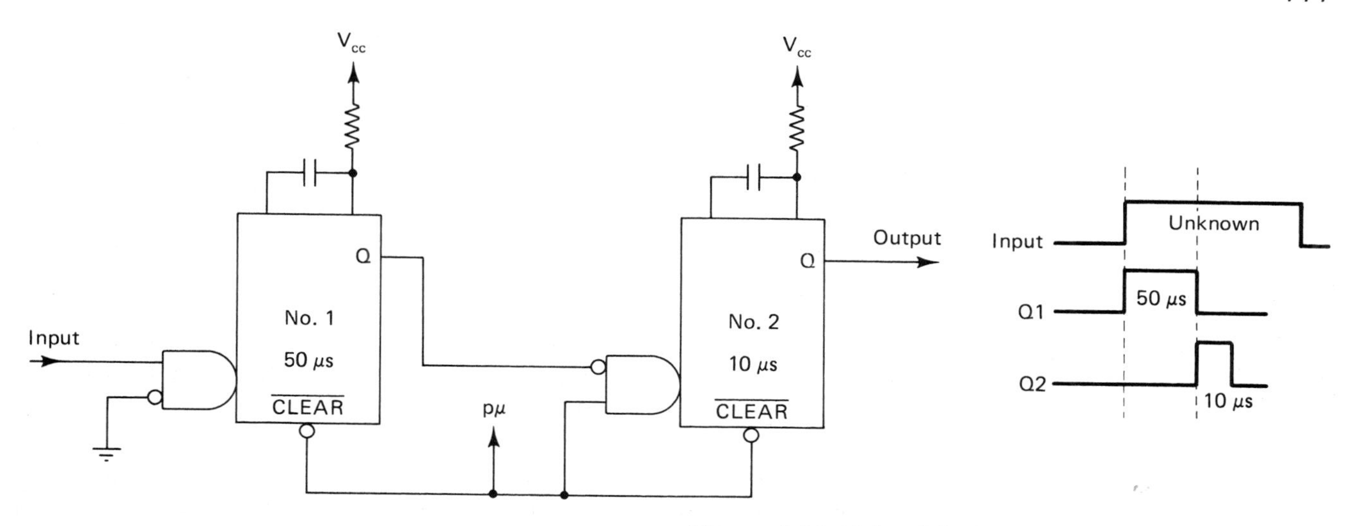

Figure 6.36 Pulse-delaying circuit.

the data sheets of the device to discover its specific characteristics.

6.7.4 Applications of One-Shots

The first application that we will examine is the one-shot as a pulse-delaying circuit (see Figure 6.36). One shot 1 is positive edge triggered and delays the input pulses by 50 μs. Notice that the pulse width of the input pulse has no effect on the outputs

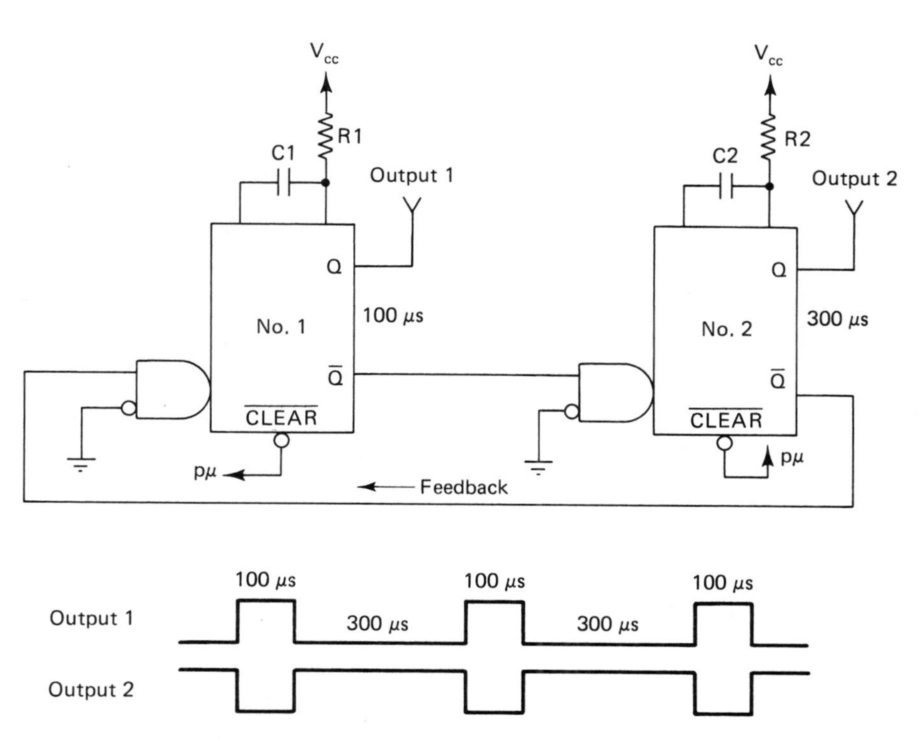

Figure 6.37 Using the 74LS221 to create a free-running oscillator.

of either one-shot; only the rising edge of the trigger input is important. Fired by the rising edge of the input signal, Q1 will go high for 50 μs. When Q1 has a falling edge, it will trigger one-shot 2. This one-shot will provide the output pulse width of 10 μs

Figure 6.37 illustrates a free-running square-wave oscillator. The output square wave can be designed for a wide range of frequencies and duty cycles, all dependent on the two *RC* time constants.

Both one-shots will be triggered on rising edges. One-shot 1 will be triggered when Q2 goes from high to low. This will cause $\overline{Q2}$ to go from low to high, which provides the rising edge. One-shot 2 is triggered in a similar manner by $\overline{Q1}$.

It looks as if we have the basis of a good oscillator; the only question is: How does this process of self-triggering begin? When V_{cc} is first applied to the one-shots, the charge on the timing capacitors is 0 V. Because of this, the action of applying power actually starts the oscillations.

QUESTIONS AND PROBLEMS

6.1. Draw a switch debouncing circuit constructed from a NOR-type S-R flip-flop.

6.2. Why can't both S-R units be equal to dynamic input levels?

6.3. Explain how the concept of feedback is used to create devices with the capacity of memory.

6.4. Complete the timing diagram for the 74LS75 shown in Figure P6.4. Complete the Q and Q-not outputs.

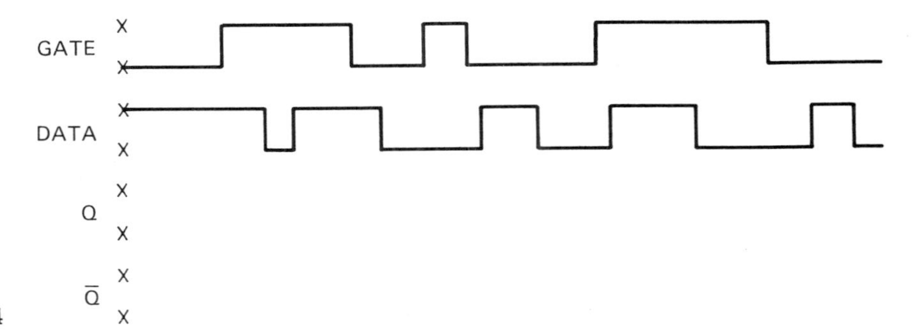

Figure P6.4

6.5. Construct a function table that describes the circuit shown in Figure P6.5.

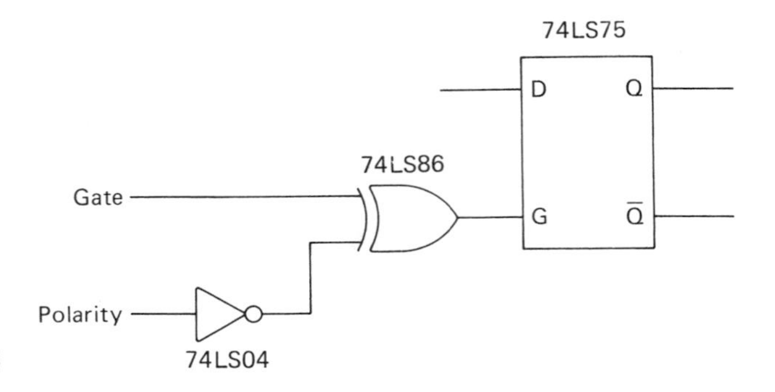

Figure P6.5

6.6. Refer to Figure 6.17. The D0 input of the top 74LS75 is floating because of a cold solder joint. How will this affect the LSD display? What would happen if the Up-Date pulse were shorted to ground?

6.7. Complete the timing diagram shown in Figure P6.7 for a positive-edge-triggered D-type flip-flop. Complete Q and Q-not outputs.

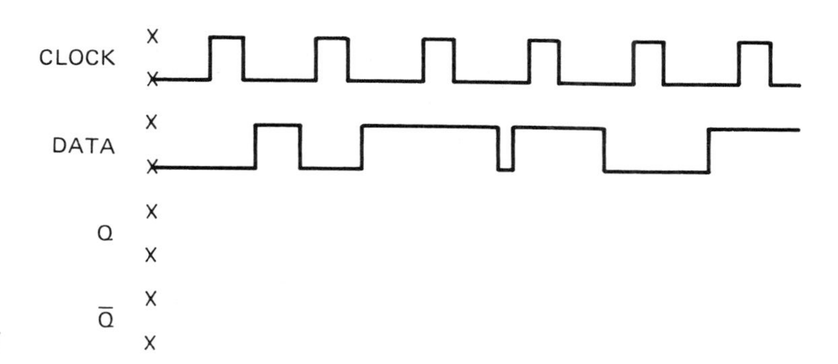

Figure P6.7

6.8. Construct a timing diagram that illustrates the operation of the circuit shown in Figure P6.8.

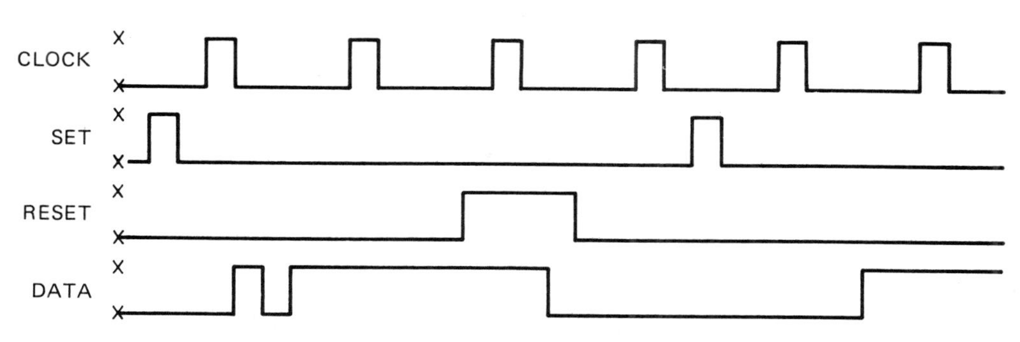

Figure P6.8

6.9. Refer to Figure 6.17. How would the circuit operation be affected if:

(a) The input to G6 were floating?
(b) D2 were shorted to ground?
(c) The frequency of the system clock were doubled?
(d) $\overline{Q}2$ were shorted to ground?
(e) D1 and the clock for U1 were left floating instead of pulled up to V_{cc}?
(f) The output of G2 were stuck internally low?
(g) There were two more inverters added to the delay circuit?

6.10. Refer to Figure 6.27. If the lock did not open after you are positive that the correct combination in the correct sequence was entered, what sequence of checks would you perform?

6.11. How would the operation of the lock be affected if Q-out of U1 was shorted to V_{cc}? Shorted to ground?

6.12. The clock input to U3 is open via an open feed-through. How does this affect circuit operation?

6.13. What checks would you perform if the lock appeared always to be open, regardless of the input combination?

6.14. Complete the timing diagram for the 74LS76 shown in Figure P6.14.

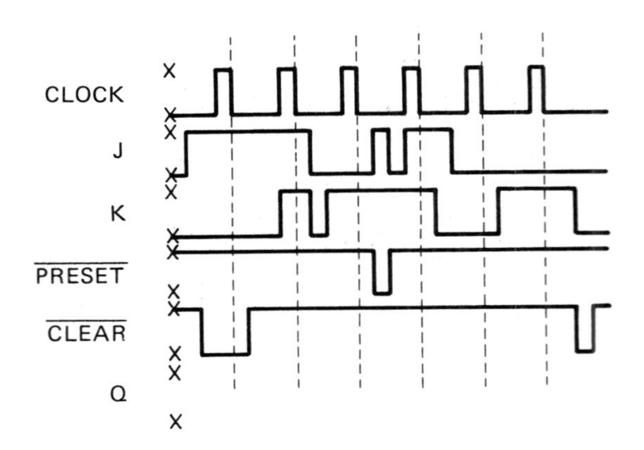

Figure P6.14

6.15. Complete the timing diagram for the circuit shown in Figure P6.15. Write a theory of operation that describes the function of this circuit.

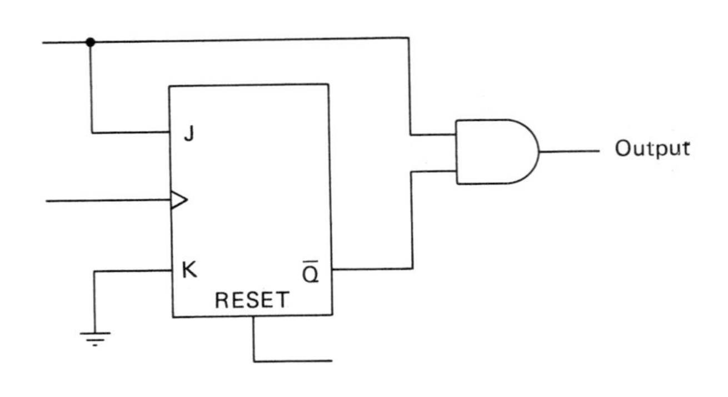

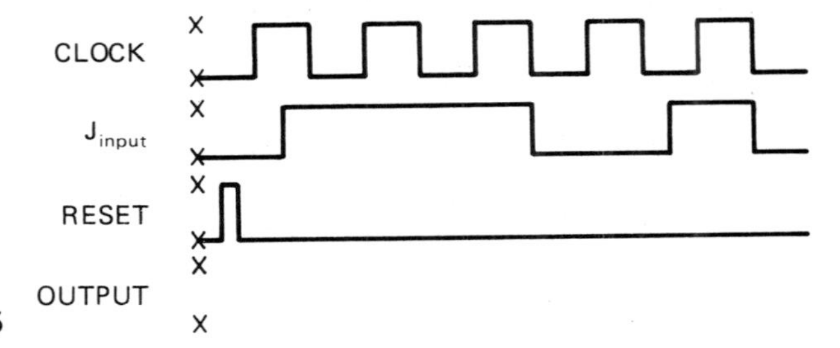

Figure P6.15

6.16. Create a timing diagram that describes the action of the circuit shown in Figure P6.16.

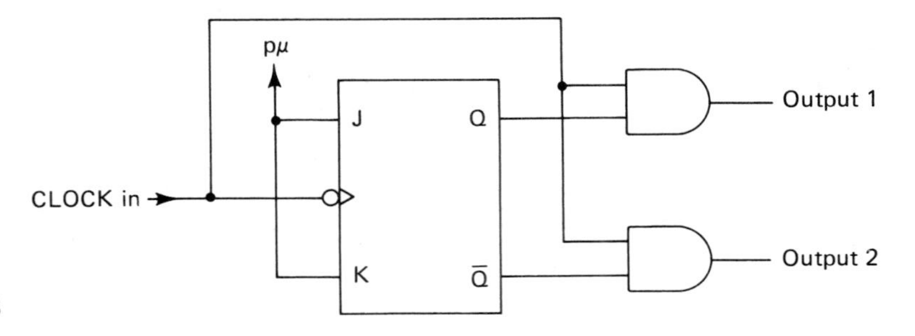

Figure P6.16

6.17. Create a timing diagram that describes the operation of the circuit shown in Figure P6.17. Be sure to include the following signals: the system clock, the pulse from the debounced switch, Q1, Q2, and $\overline{Q2}$.

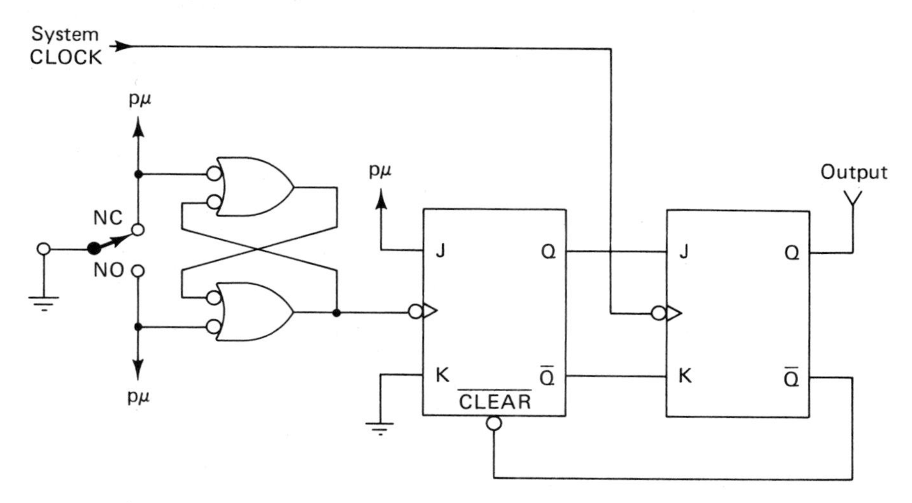

Figure P6.17

6.18. Answer the following troubleshooting questions concerning Figure 6.31.

(a) What would happen if the clock input for flip-flop 2 were floating?

(b) How would the output be affected if Q2 were shorted to ground.

(c) How would the output be affected if $\overline{Q1}$ were shorted to V_{cc}?

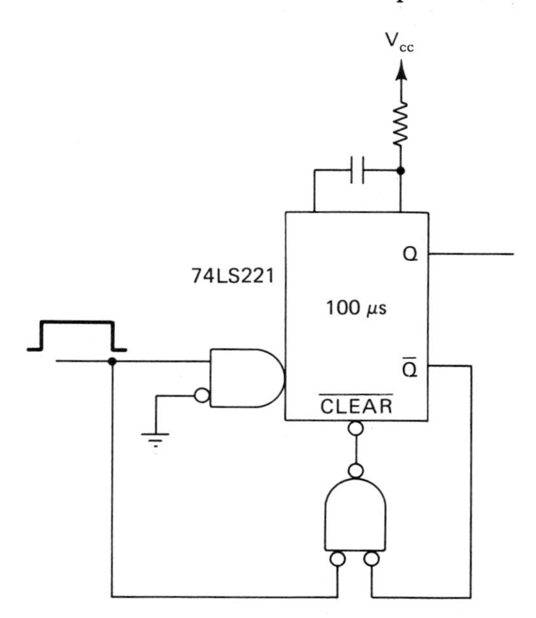

Figure P6.18

6.19. Use J-K flip-flops to construct a circuit that counts from 0 (0000 binary) to 15 (1111 binary). Be sure to label the least significant and most significant outputs. Also construct a timing diagram to describe the circuit operation.

6.20. Refer to the circuit shown in Figure P6.20, which illustrates a 74LS123 retriggerable one-shot. Create a theory of operation that describes this circuit.

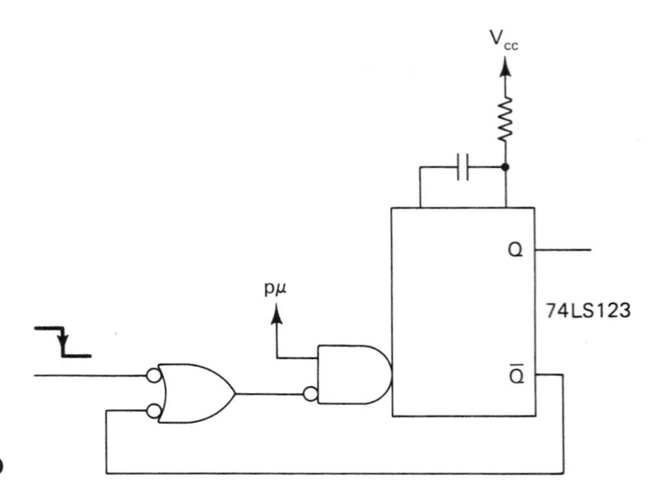

Figure P6.20

6.21. Draw a timing diagram that illustrates the operation of the circuit shown in Figure P6.21.

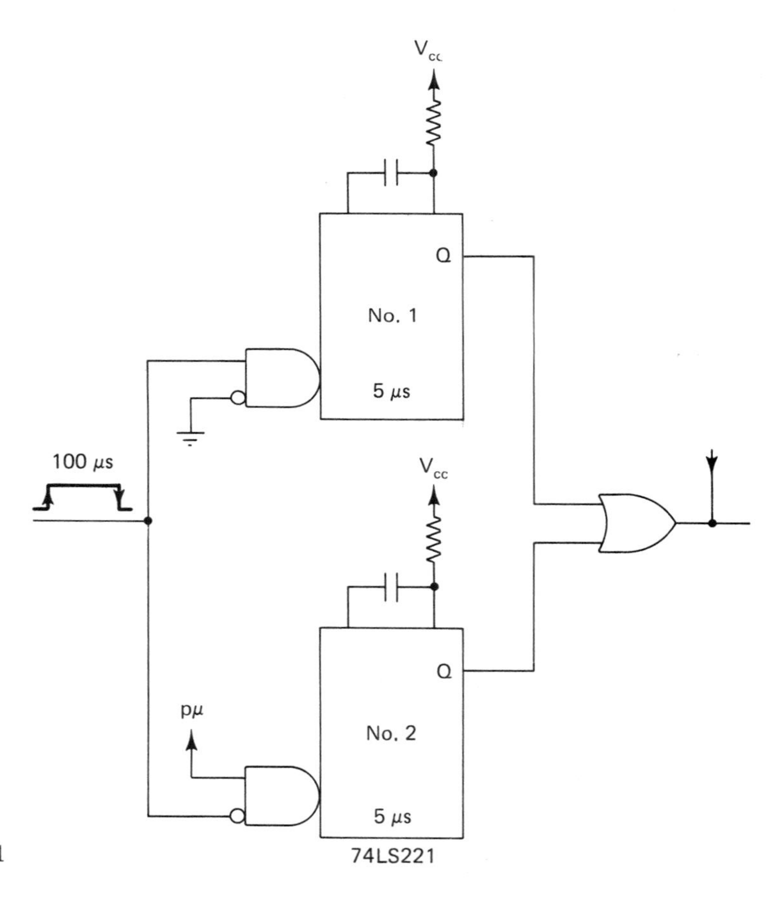

Figure P6.21

6.22. Using 74LS221 one-shots, design a circuit that delays an incoming pulse by 500 μs. The output pulse should be 1 μs long. Draw a timing diagram to summarize the circuit action.

6.23. If the Q output of a one-shot is logic 0 even though its input is being triggered, what checks would you make to find the problem?

6.24. How would the oscillator in Figure 6.37 be affected if C1 were open? If R2 were only half of its proper value?

6.25. Redesign Figure 6.37 to use the inverting trigger inputs (pull the noninverting trigger inputs to V_{cc}) instead of the noninverting trigger inputs.

7 MSI Counters and Shift Registers

Earlier we used three J-K flip-flops to construct a circuit that counted from 0 through 7. Instead of constructing counters from flip-flops, designers use MSI counters. A digital clock is really nothing more than a circuit that counts the rising or falling edges of a 60 Hz reference square wave. Another function that counters can perform is to divide a frequency by a particular integer value. A D flip-flop or J-K flip-flop configured to toggle divides the clock frequency by a factor of 2. A divide-by-N counter is a circuit that produces one output pulse for every N input clock pulses.

7.1 MODIFYING THE COUNT LENGTH

7.1.1 Modulus and Count-State Diagrams

The *modulus* of a counter refers to how many counts it will go through before repeating. A counter consisting of four flip-flops will have a modulus of 16, because it counts from 0 through 15 before starting to repeat. An alternative way to describe a counter with a modulus of 16 would be to call it a *mod-16 counter*.

A counter described as a mod-16 counter does not necessarily count from 0 through 15. A count-state diagram is used to illustrate the counting sequence of a particular counter. The

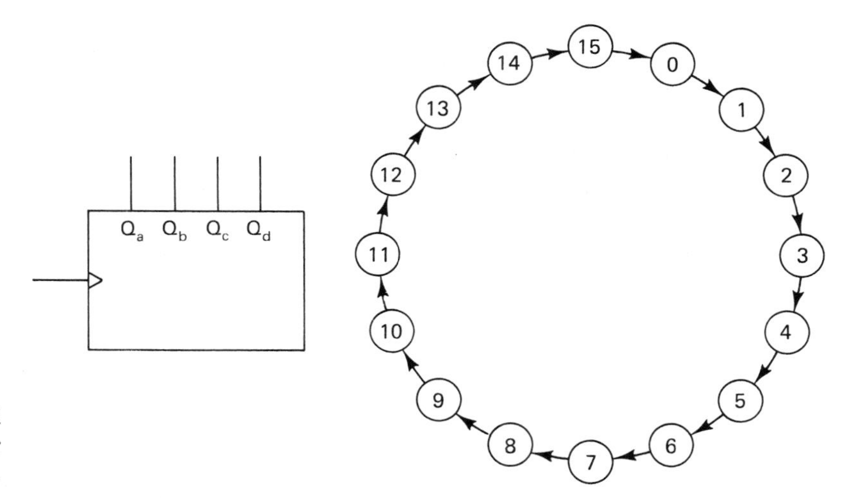

Figure 7.1 Logic symbol and count-state diagram for Mod-16 counter.

logic symbol shown in Figure 7.1 indicates a rising-edge-triggered clock and outputs Q_a through Q_d. As always, Q_a is the least significant output. The count-state diagram indicates the counting sequence for this particular mod-16 counter. Each circle represents one count state. This counter counts from 0 to 15, then begins to repeat.

7.1.2 Modifying the Count

How would we design a mod-12 counter with a count sequence that starts at 0 and proceeds through 11? If we used three flip-flops, the maximum count that could be achieved is 8; if we used four flip-flops we would jump to a modulus of 16. The answer is to use four flip-flops and modify the circuit to return to count 0 after count 11, bypassing counts 12 through 15.

The count-state diagram shown in Figure 7.2 indicates that the counter returns to count 0 after count 11. When Q_c and Q_d are high, the clear inputs to the flip-flops will go active, resetting the counter to 0. The output of the NAND gate will then go high, and the counter will resume normal operation.

When count 12 is reached, Q_c and Q_d will go high, causing the output of the NAND gate to go low; this active low is applied to the clear inputs of the flip-flops, forcing Q_a through Q_d to a logic 0 level. The counter will then continue with count 0. The events that cause the counter to reset to count 0 occur in an extremely short period of time.

All digital devices have a propagation delay. After a stimulus is applied, a certain period of time is required before the device will react. Counters of the type that we are now examining are called *ripple counters.* The term "ripple" describes how the output of one device is used to clock the next device.

7.2 MSI RIPPLE COUNTERS

7.2.1 The 749LS90 and 74LS93 Ripple Counters

The 749LS90 is a *divide-by-10 counter*. Divide-by-10 counters are often called *BCD* or *decade counters* and the 74LS93 is a

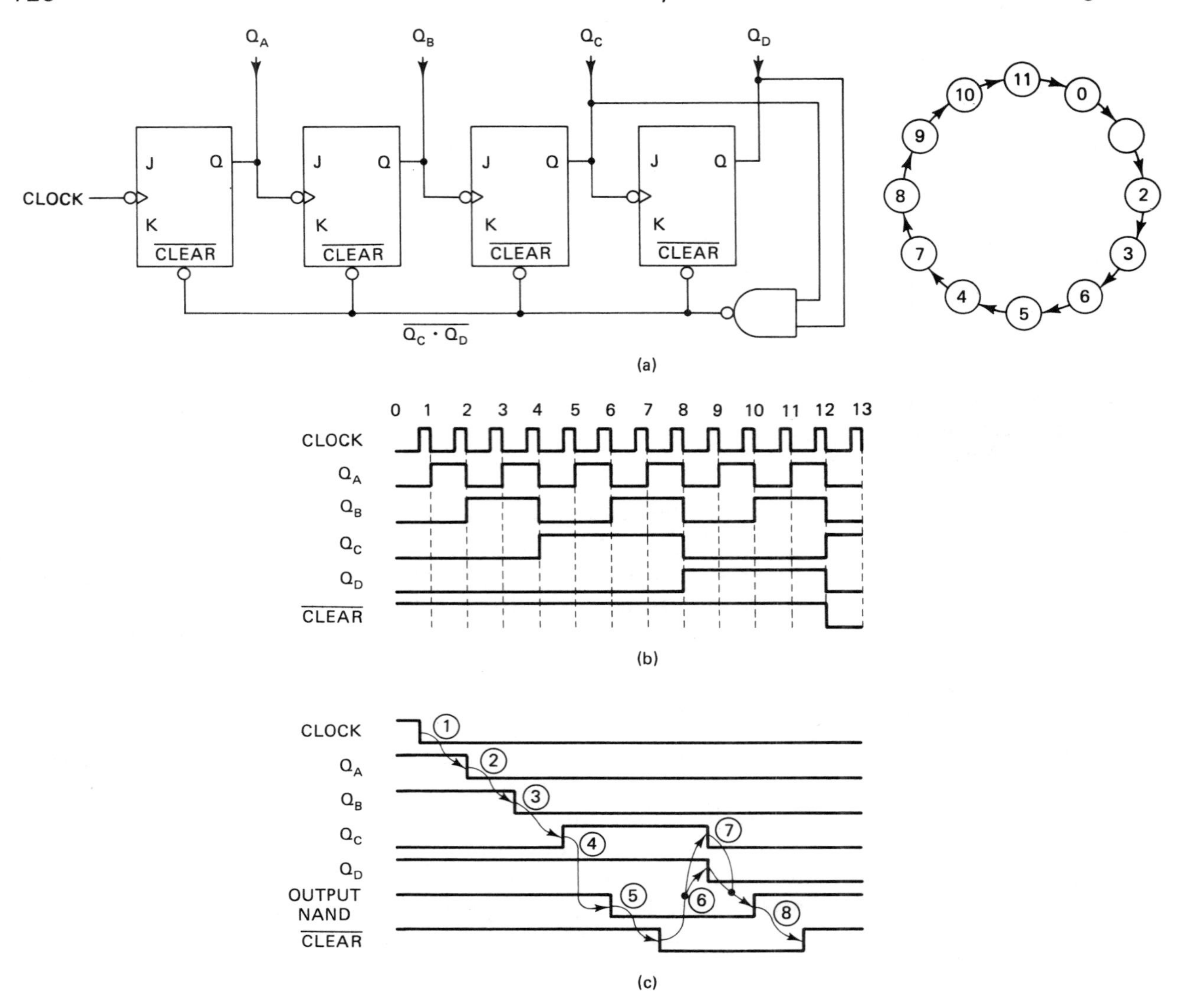

Figure 7.2 Mod-12 counter, timing diagram, and count-state diagram.

divide-by-16, 4-bit *binary counter*. Because this is an old series of ripple counters, the V_{cc} and ground pins are in nonstandard locations. V_{cc} is pin 5 and ground is pin 10. To rectify this problem, the 74LS290 and 74LS293 were introduced. The 74LS290 is the functional equivalent of the 74LS90, but its V_{cc} and ground pins are in the standard positions. The same relationship is true for the 74LS293 and the 74LS93.

Refer to the functional block diagram for the 75LS90 in Figure 7.3. When J-K flip-flops are shown with no connection to the J or K inputs, it is assumed that they are both pulled up to V_{cc} and the J-K flip-flop is configured to toggle. Notice the two NAND gates that are used to modify the count sequence of the 74LS90. When the bottom NAND gate goes active, the counter will reset to count 0; when the top NAND gate goes active, the counter will reset to count 9. There are two negative-edge-triggered clock inputs: input A and input B. The first J-K flip-flop

'90A, 'L90, 'LS90
BCD COUNT SEQUENCE
(See Note A)

COUNT	OUTPUT			
	Q_D	Q_C	Q_B	Q_A
0	L	L	L	L
1	L	L	L	H
2	L	L	H	L
3	L	L	H	H
4	L	H	L	L
5	L	H	L	H
6	L	H	H	L
7	L	H	H	H
8	H	L	L	L
9	H	L	L	H

'90A, 'L90, 'LS90
RESET/COUNT FUNCTION TABLE

RESET INPUTS				OUTPUT			
$R_{0(1)}$	$R_{0(2)}$	$R_{9(1)}$	$R_{9(2)}$	Q_D	Q_C	Q_B	Q_A
H	H	L	X	L	L	L	L
H	H	X	L	L	L	L	L
X	X	H	H	H	L	L	H
X	L	X	L	COUNT			
L	X	L	X	COUNT			
L	X	X	L	COUNT			
X	L	L	X	COUNT			

NOTES: A. Output Q_A is connected to input B for BCD count.
B. Output Q_D is connected to input A for bi-quinary count.
C. Output Q_A is connected to input B.
D. H = high level, L = low level, X = irrelevant

'93A, 'L93, 'LS93
COUNT SEQUENCE
(See Note C)

COUNT	OUTPUT			
	Q_D	Q_C	Q_B	Q_A
0	L	L	L	L
1	L	L	L	H
2	L	L	H	L
3	L	L	H	H
4	L	H	L	L
5	L	H	L	H
6	L	H	H	L
7	L	H	H	H
8	H	L	L	L
9	H	L	L	H
10	H	L	H	L
11	H	L	H	H
12	H	H	L	L
13	H	H	L	H
14	H	H	H	L
15	H	H	H	H

'92A, 'LS92, '93A, 'L93, 'LS93
RESET/COUNT FUNCTION TABLE

RESET INPUTS		OUTPUT			
$R_{0(1)}$	$R_{0(2)}$	Q_D	Q_C	Q_B	Q_A
H	H	L	L	L	L
L	X	COUNT			
X	L	COUNT			

functional block diagrams

'90A, 'L90, 'LS90

R9(1) (6)
R9(2) (7)
INPUT A (14)
INPUT B (1)
(12) QA
(9) QB
(8) QC
(11) QD
R0(1) (2)
R0(2) (3)

'93A, 'L93, 'LS93

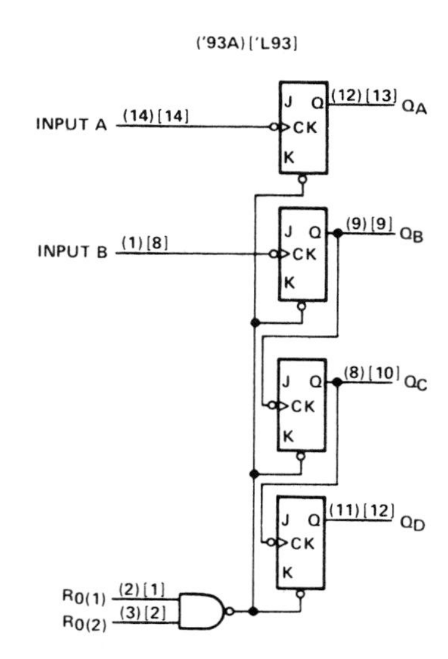

The J and K inputs shown without connection are for reference only and are functionally at a high level.

Figure 7.3 Truth tables and functional diagrams for 74LS90 and 93.

is an independent divide-by-2 block, and the last three flip-flops constitute a divide-by-5 block. To achieve divide-by-10 operation, the clock input is placed on input A and Q_a is used to clock input B.

In the reset/count function table for the 74LS90 shown in Figure 7.3, the first two lines indicate that if both reset-to-0 inputs are at an active high level and either reset-to-9 input is at an inactive low level, the output will be reset to count 0. The third line indicates that if both reset-to-9 inputs are at an active-high level, the output will go to count 9 regardless of the levels at the reset-to-0 inputs. Lines 4 through 7 indicate that if neither NAND gate is at an active-low output level, the counter will be enabled to count.

The BCD count sequence table indicates tht if output Q_a is used to clock input B, this device will exhibit a normal BCD count sequence.

Figure 7.4 depicts a circuit that uses the reset-to-0 inputs to modify the count length. Notice that Q_a is tied back to clock input B. Q_b is connected to reset-to-0 input 1 and Q_c is connected to reset-to-0 input 2. When Q_b and Q_c are both high, the counter will be reset to count 0. Q_b and Q_c will both be high, on count 6, but count 6 will never be seen. The count-state diagram in Figure 7.4 indicates the counting sequence. Count 0110 is never seen.

The 74LS93 is the divide-by-16 version of the 74LS90. You

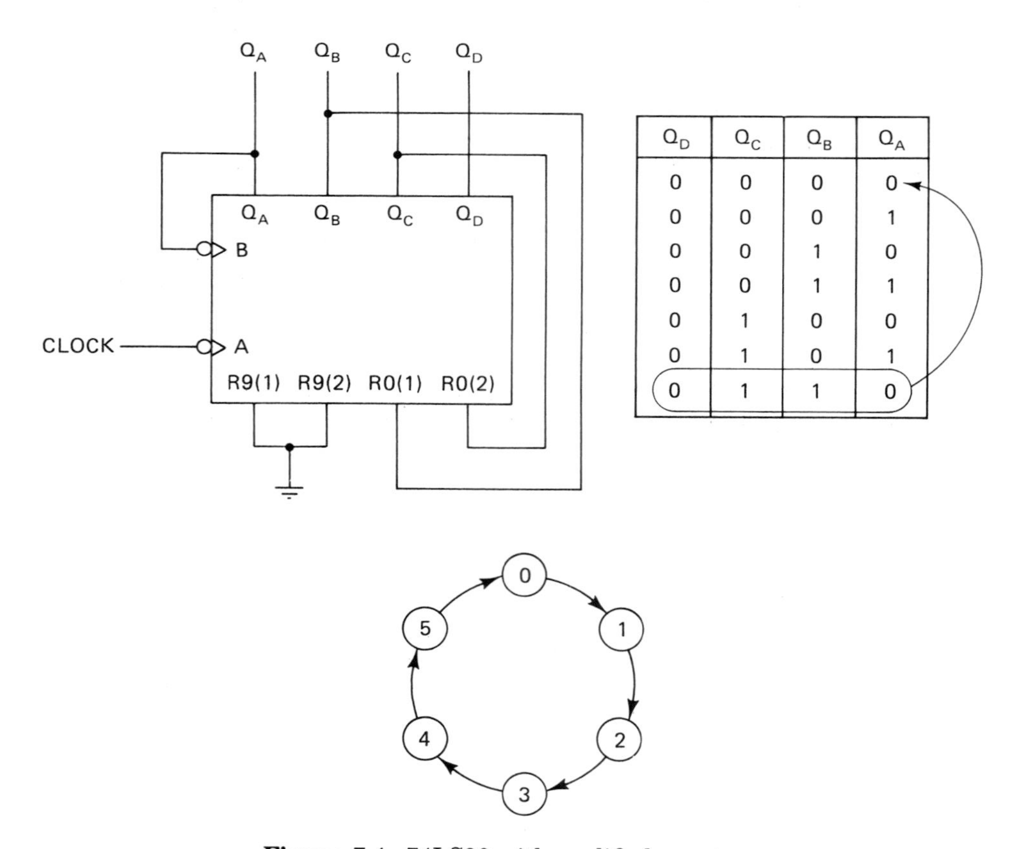

Q_D	Q_C	Q_B	Q_A
0	0	0	0
0	0	0	1
0	0	1	0
0	0	1	1
0	1	0	0
0	1	0	1
0	1	1	0

Figure 7.4 74LS90 with modified count sequence.

will find that most MSI counters come in a BCD version and a 4-bit binary version. The BCD (decade) counter is often used to drive the inputs of a BCD-to-seven-segment decoder/driver, such as the 74LS47. If the outputs of a counter are only to be used within the digital circuitry, a binary counter such as the 74LS93 is used. The 74LS93 has the same reset-to-0 inputs, but it does not have any equivalent of the reset-to-9 inputs.

7.2.2 Application of a Ripple Counter

Earlier we constructed a 10-key keyboard encoding/decoding and display circuit using two 74LS148 priority encoders. We are now going to build the equivalent circuit using a 74LS90 ripple counter, a 74LS151 data selector, and some discrete gates. This circuit, shown in Figure 7.5, is intended to illustrate the operation of a counter in a complex circuit.

We will first overview the major blocks that constitute the circuit; then we will examine how each major block functions and how it contributes to the circuit as a whole. The major blocks are:

1. The 74LS151 data selector and gates G3, G4, G5, G6, G7, and G8. These components form a decimal-to-BCD encoder.
2. The 74LS90 decade ripple counter. This counter scans the decimal-to-BCD encoder looking for a closed switch.
3. The 74LS47 BCD-to-seven-segment decoder/driver and the seven-segment common-anode display. This block takes the BCD code generated by the counter and displays it on the common-anode display.

The output of G6 is the key signal in this circuit; it is applied to two critical inputs:

1. G1 provides the function of an electronic switch; the clock is the data input and the output of G6 is the control line. If the output of G6 is high, G1 is enabled and the inverted clock will reach clock input A on the 74LS90; if the output of G6 is low, G1 will be disabled and the clock input on the 74LS90 will be held high. A logic 0 output from G6 will block the clock input to the 74LS90 and effectively cause the output count on the 74LS90 to freeze.
2. When the output of G6 is high, the output of G9 will go low and the 74LS47 will be blanked via the BI. When the output of G6 is low, the 74LS47 will be enabled and the "frozen" BCD output from the counter will be displayed.

When will the output of G6 go low? If any of its three inputs go low, the output of G6 will go low. Our remaining task is to discover how the outputs of G2, G3, and G4 will go low.

G2 is an OR gate represented in its negative-logic form. If both inputs to G2 are low, its output will go low. One input to

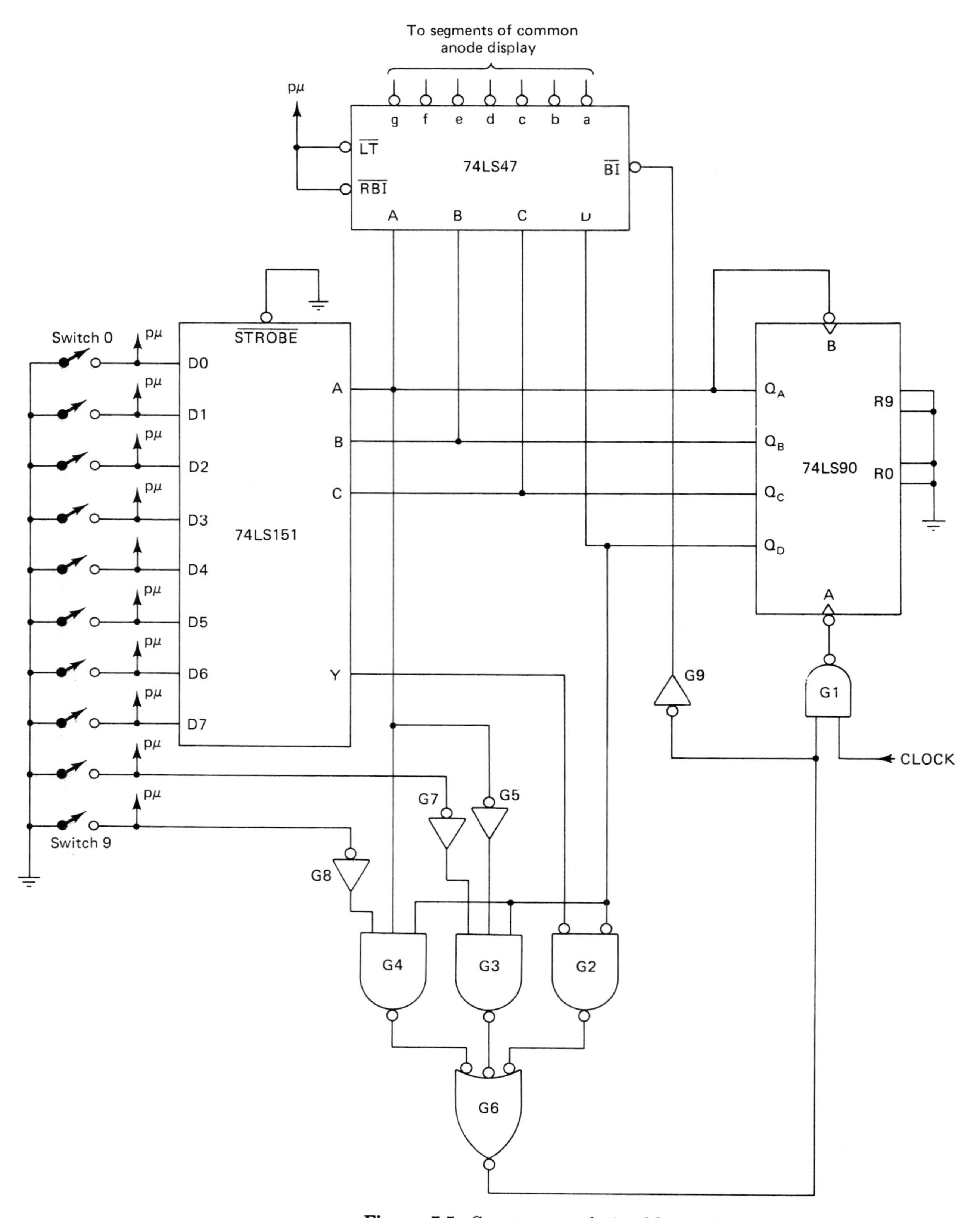

Figure 7.5 Counter-type decimal keypad encoder and display.

G2 is output Q_d from the BCD counter. If Q_d goes high, the output of G2 will be held high. This means that G2 will be enabled for counts 0 through 7, but it will be disabled for counts 8 and 9. This makes perfect sense because the 74LS151 is a 1-of-8 data selector and its data inputs are the first eight switches of the decimal keypad.

During count 0, the status of SW 0 will be placed on the second input of G2. If the switch was closed, the output of G2 will go low, forcing the output of G6 low, disabling the clock, and enabling the 74LS47 to display the digit 0. The same process will be true for counts 1 through 7.

On count number 8, Q_d will go high, disabling G2. During count 8, Q_a will be low. A high on Q_d and a low on Q_a will enable gate G3. If SW 8 is closed, the output of G3 will go low, forcing the output of G6 low, disabling the counter and enabling the 74LS47 to display the digit 8.

On count number 9, Q_d will be high and Q_a will be high. This will enable G4. If SW 9 is closed, the output of G4 will go low, forcing the output of G6 to go low, disabling the counter and enabling the 74LS47 to display the digit 9.

The counter will continue to scan the 10 switches to find a closed switch. Once a switch is closed, the counter will be disabled until it is reopened. Gates G3 through G8 are used to expand the data selector from a 1-of-8 data selector to a 1-of-10 data selector.

7.3 THE SYNCHRONOUS COUNTER

The formal name for ripple counters is *asynchronous counters*. The system clock is connected only to the first flip-flop. Subsequent flip-flops are clocked by the output of the previous stage. This asynchronous clocking method causes glitches and reduces the speed at which these counters can run effectively. Synchronous counters use a method of clocking where all flip-flops are clocked simultaneously by the system clock. All the outputs change simultaneously, eliminating glitches.

7.3.1 The 74LS160 Family of Synchronous, Presettable Counters

The 74LS160, 74LS161, 74LS162, and 74LS163 all share an identical pinout. The 74LS160 and 74LS162 are identical BCD counters except for the manner in which their clear inputs operate. The 74LS160 has an asynchronous clear. Up to this point, all the clear inputs that we have examined were called direct or asynchronous. When an asynchronous clear goes active, the outputs will instantly react. The 74LS162 has a synchronous clear. The outputs will be affected only after the clear input has gone to its active level and the next active edge of the clock occurs. The clear function will be in sync with the system clock. The 74LS161 is a 4-bit binary counter with an asynchronous

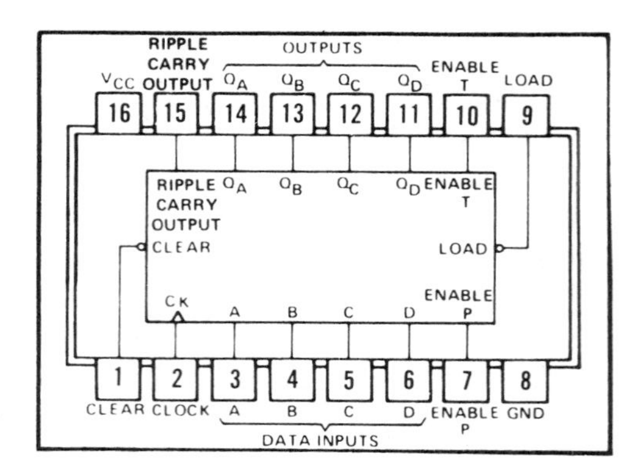

Figure 7.6 Pinout for 74LS160 series counter.

clear. The 74LS163 is identical to the 74LS161 except that it has a synchronous clear. The terms "asynchronous" and "direct" are used interchangeably.

Let's examine the pinout in Figure 7.6 in functional groups.

Clock The clock input to the counter is active on the rising edge.

Clear The clear input is active low. The pinout does not differentiate between an asynchronous and a synchronous clear.

Enables T and P T and P must both be high before the counter is enabled. These enable inputs are used to cascade counters to obtain longer count sequences.

Ripple-carry output This output will go high when the counter is on its last count. The positive pulse from the ripple-carry output will be used to enable the next significant counter in cascaded applications. The 74LS148 priority encoder has an enable output that is used to enable the next downstream encoder. You can think of the ripple-carry output on the 74LS160 series of counters as working in a similar manner.

Outputs Q_a through Q_d These are the Q outputs from the four internal counting stages.

Data inputs A through D and the load input These counters were described as *presettable*. Rather than starting at 0, it is often desirable to start the count sequence at another count. When the load input is taken to its active-low level, the next rising edge of the clock will transfer the levels on the data inputs (A through D) into the count outputs (Q_a through Q_d). This presetting operation is synchronous because it does not occur until the active edge of the clock. Another common name for presettable counters is *programmable counters.*

The timing diagram in Figure 7.7 is fairly complex. The dashed vertical lines indicate points of interest where we should focus our examination. This timing diagram represents the operation of both the 74LS160 and 74LS162 decade counters.

Event 1. The first vertical line illustrates the effect of the async clear on the 74L160. The clear input is independent of the

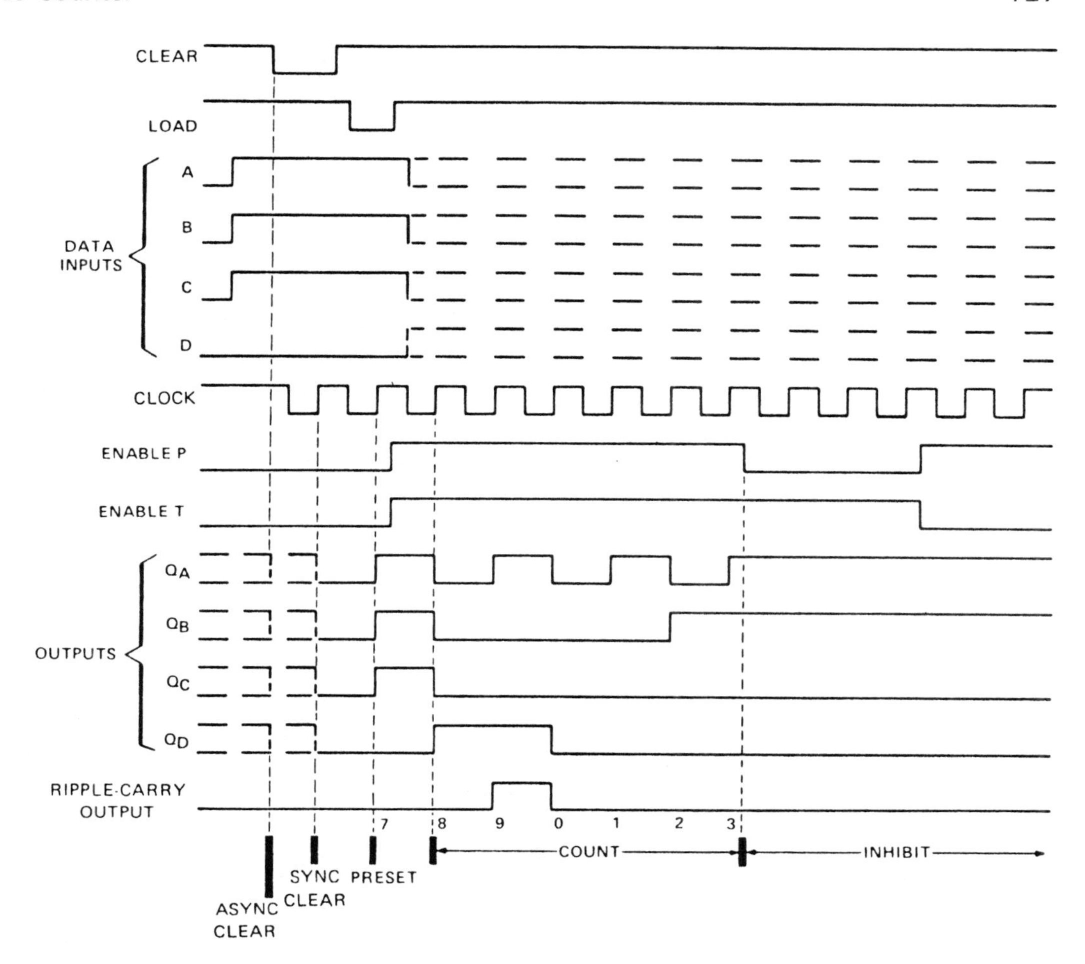

Figure 7.7 Timing diagram for 74LS160 and 74LS162 decade counters.

clock; when the clear input goes active, the outputs Q_a through Q_d will reset to logic 0. The dashed vertical lines above and below the Q_a through Q_d outputs represent their unknown states before the active async clear.

Event 2. This vertical line illustrates the effect of the sync clear on the 74LS160. The Q_a through Q_d output will not reset until the clear line is active and the clock has an active edge. The remaining events in this timing diagram are equally applicable to both the 74LS160 and 74LS162.

Event 3. This vertical line indicates the preset operation. Data inputs A through C are high and data input D is low. The load input goes active; on the next rising edge of the clock, the data inputs are transferred to the counter's outputs. The counter has been preset to the count of 7.

Event 4. The enable inputs, P and T, go to the active-high levels, the clear input is inactive, and the load input is inactive. The counter is now enabled. On the next rising edge of the clock, the counter will sequence from count 7 to count 8. The next rising edge of the clock will cause a count, then the ripple carry will go to an active-high level. The next rising edge will cause the count to return to 0 and the ripple-carry output will go to its inactive level. The counter will continue to count until some type of inhibit occurs.

Event 5. The enable P input goes inactive and the counter will be inhibited. The outputs on Q_a through Q_d will be held at the last valid count (3) until the enable P input returns high, the clear input goes active, or the load input goes active.

The timing diagram for the 74LS161 and 74LS163 is identical to Figure 7.7 except that the ripple carry will go active on count 15, instead of 9.

7.3.2 Cascading the 74LS160 Series

Figure 7.8 illustrates how the 74LS160 series of counters can be connected in cascade to produce longer count lengths. Notice that all the counters in Figure 7.8 are clocked in sync. The enable P and T on the (first) least significant counter must be taken to an active level before any counting will take place. The second counter will be enabled whenever the ripple-carry output on the first counter is high. Assume that the four counters are 74LS160 decade counters. When the first counter is at count 9, the ripple-carry output will enable the second counter. On the next rising edge of the clock, the first counter will return to 0 and the second counter will advance to 1. The third counter will be enabled at count 99. Counters 1 and 2 will return to 0 and counter 3 will advance to count 1 on the clock's next rising edge.

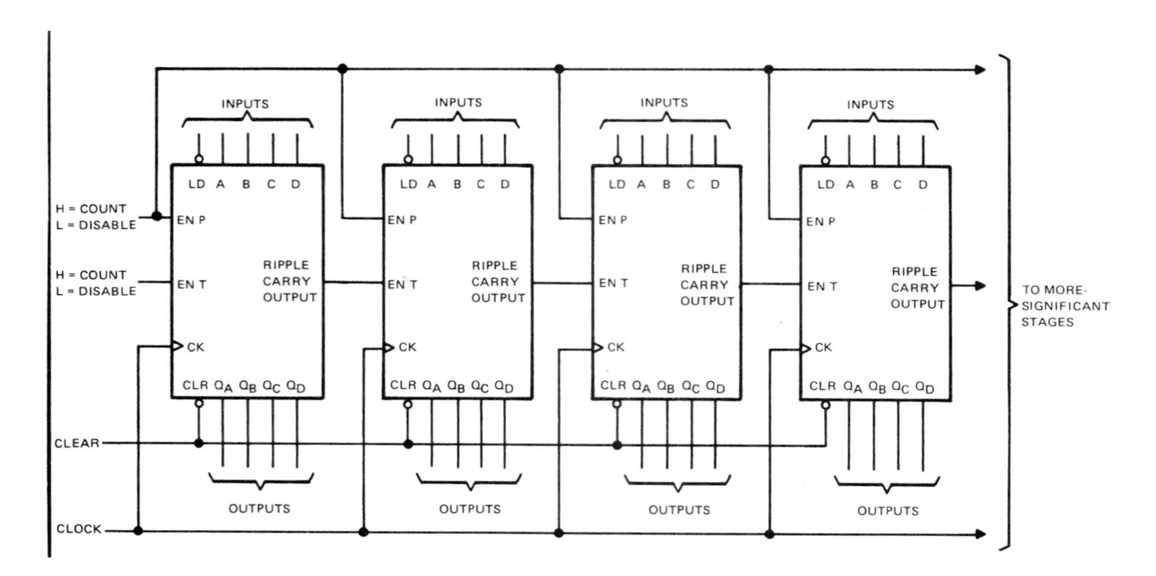

Figure 7.8 Cascading 74LS160 series counters.

At count 999 the ripple-carry outputs 1 through 3 will be high, enabling counter 4. On the next rising edge of the clock, counters 1 through 3 will return to 0 and counter 4 will advance to 1.

This process of enabling the fourth counter every 1000 pulses will continue through the count of 9999. The next rising edge will return all the counters to the count of 0.

7.3.3 Application of BCD Counters: A Digital Pulse Meter

We are going to illustrate the use of BCD counters within a pulse meter. There are three boxes in the block diagram shown in Figure 7.9 that must be accepted without detailed explanation. The box labeled "sensor" is that part of the circuit that senses a heart beat and converts it into a TTL-comparable pulse.

The box next to the heartbeat sensor will multiply the sensor's input frequency by a factor of 10. This meter will display the pulse in units of beats per minute. If we multiply the number of pulses (counted in the 6-s interval) by a factor of 10, the display can be updated every 6 s.

The third box is the 60-Hz reference. Because the power line frequency is 60-Hz, it is often used to establish a reference for timekeeping operations.

The three counters are connected in cascade to achieve a maximum count of 999. Every 6 s an update strobe will latch

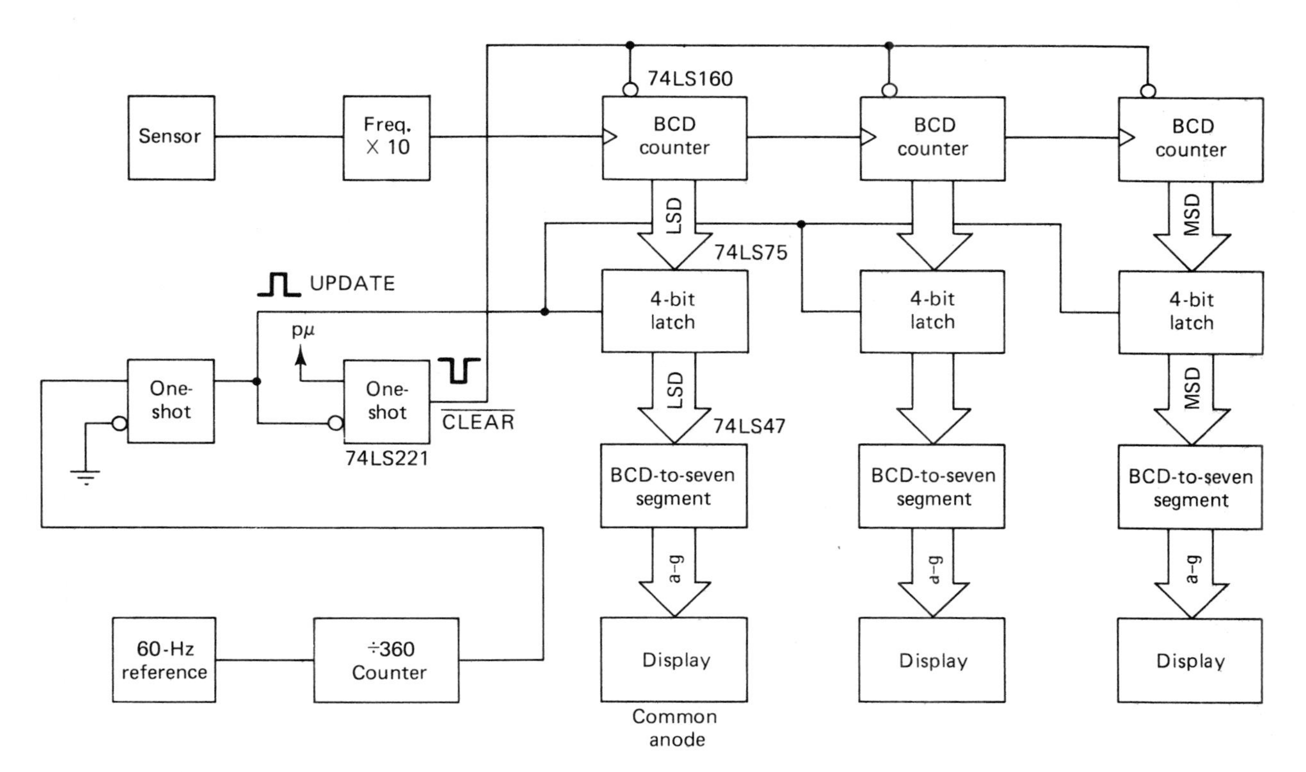

Figure 7.9 Block diagram of improved digital pulse meter.

the present count into the three 4-bit latches. These latches will drive BCD-to-seven-segment decoder/drivers.

A divide-by-360 counter will divide the 60-Hz reference frequency into 1/6 Hz. Every 6 s the positive edge from the divide-by-360 counter will fire the first one-shot, which will deliver a positive update strobe to latch the present count into the latches. The falling edge of the update strobe will fire the second one-shot. This negative pulse will clear the counter back to zero. After each 6-s sample, the process of updating the displays and clearing the counters will occur.

Figure 7.10 depicts the actual schematic for the digital pulse meter. U1 through U3 are cascaded 74LS160 decade counters with direct clear. U4 through U6 are 74LS75 4-bit latches. They will provide the temporary display memory for the 6 s between update pulses. U7 through U9 are 74LS47 BCD-to-decimal decoder/drivers. Notice that they are connected to blank leading zeros. U10 is the update one-shot. Any pulse width from 1 μs to 0.1 s will function correctly. U11 provides the clear pulse to reset the counters. Its pulse width is also not critical. Counters U12, U13, and U14 provide the divide-by-360 function; U12 is a divide-by-10 counter, U13 is a divide-by-12 counter, and U14 is a divide-by-3 counter. Notice that when counters are connected in cascade, the total factor of division is equal to the product of each counter's factor of division.

7.4 UP-DOWN COUNTERS

There are many applications that require a counter to count in both directions. The 74LS190 and 74LS191 provide this function.

7.4.1 The 74LS190 and 74LS191: Synchronous Up/Down Counters

The 74LS190 and 74LS191 are programmable, reversible up/down decade and 4-bit binary counters. The pinout for the 74LS190 and 74LS191 appears in Figure 7.11, which we will now examine according to function.

Outputs Q_a through Q_d The 74LS190 is a decade counter and the 74LS191 is a 4-bit binary counter. Other than this distinction, they are exactly alike. Q_a through Q_d, as we have seen in every MSI counter, are the count output pins.

Data inputs A through D and load These counters are presettable, just as the 74LS160 series. There is one major difference between the presetting operation of the 74LS160 series and the 74LS190/LS191. The 74LS160 counters have a synchronous preset; the data inputs appear on the Q outputs only after the load input has gone active and a rising edge appears at the clock input. The 74LS190 and 74LS191 have a direct (asynchronous) preset. When the load input goes active, the Q outputs will change instantly to agree with the data inputs.

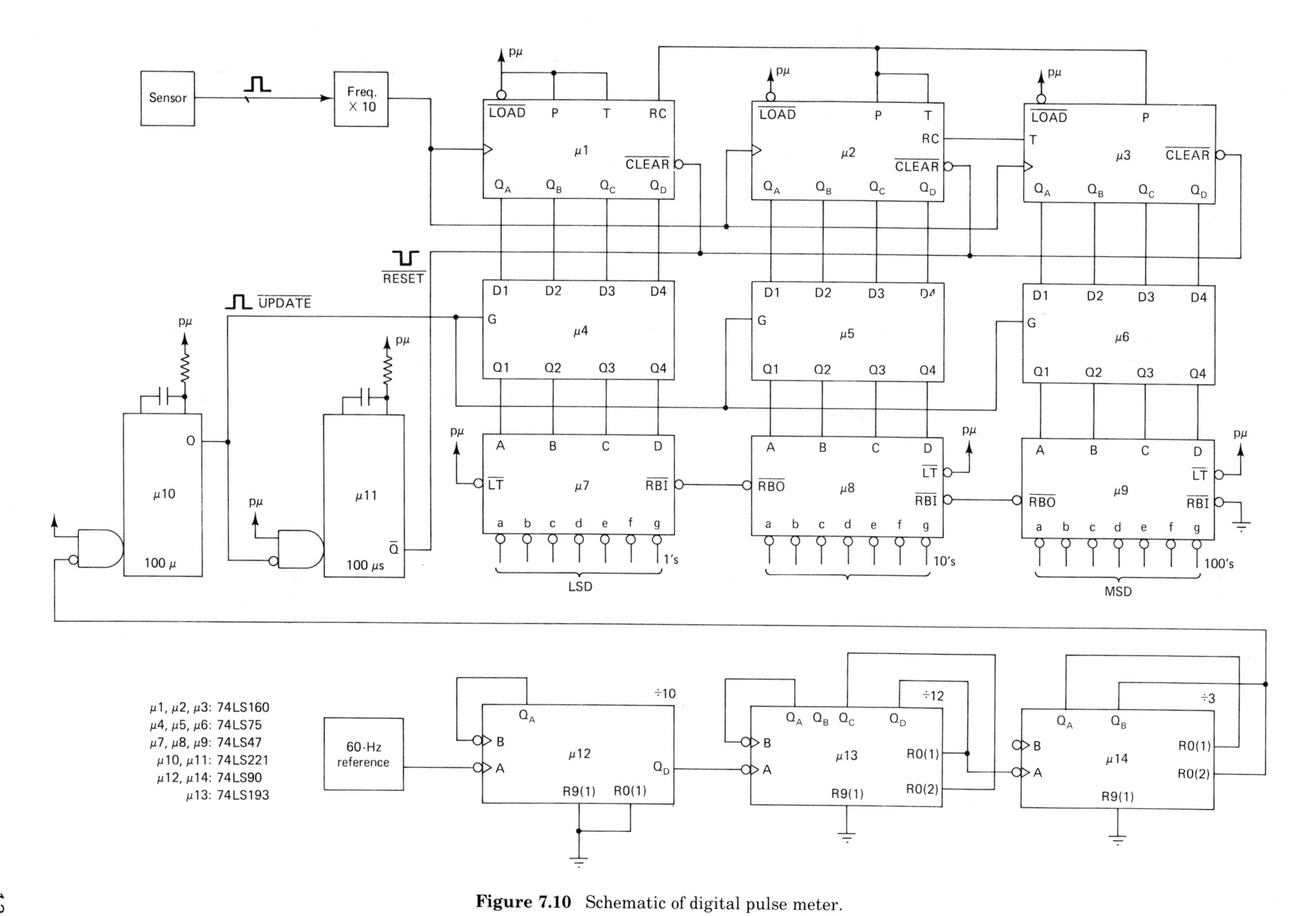

Figure 7.10 Schematic of digital pulse meter.

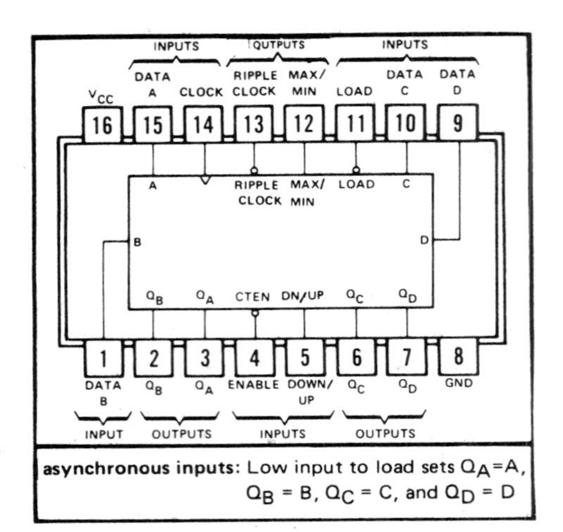

Figure 7.11 Pinout for 74LS190 and 191 up/down counters.

The 74LS190 or 74LS191 do not have a clear pin. To clear the counter, the data inputs are taken to logic 0 and the load input is pulled low.

Clock The 74LS190 and 74LS191 are positive-edge-clocked devices.

DN/UP: Down/up The logic level on this pin will determine the direction of the count. A high will cause the counters to count down; a low will cause the counters to count up.

G: Enable input The enable on the 74LS190 series is a single active-low input. Counting is inhibited until the G input is taken to a logic 0 level.

Ripple clock The ripple clock output will produce a low-level output equal to the low-level portion of the clock input whenever the counter overflows or underflows.

Max/Min This output produces a high-level pulse equal in period to the clock input whenever the counter overflows or underflows. Max/min is used to signal other devices that the counter is about to overflow or underflow.

7.4.2 The 74LS192 and 74LS193: Dual Clock Counters

The 74LS192/193 is different from other up/down counters because it has two clock inputs. One clock input is used to count up and the other to count down. Figure 7.12 illustrates the pinout for the 74LS192/193. Following are the functional groups.

Q_a through Q_d These are the four conducting outputs. The 74LS192 is a decade counter and the 74LS193 is a 4-bit binary counter.

Data A through D and load These counters are fully programmable. The present operation is asynchronous.

Count up and count down A rising edge on either clock input will cause the counter to increment or decrement. While one

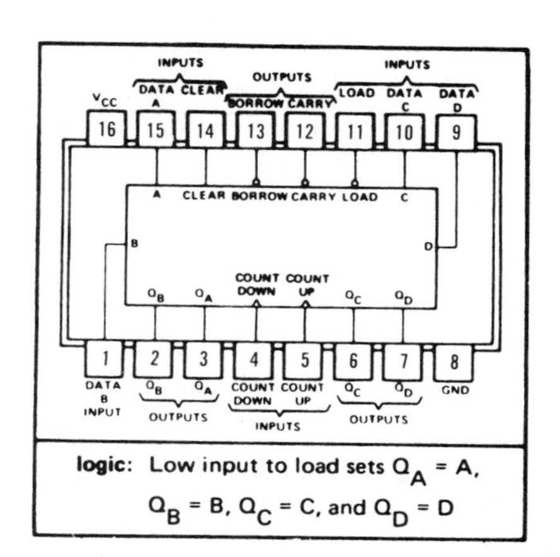

Figure 7.12 Pinout of the 74LS191/193 dual clocked up/down counter.

clock input is being pulsed, the other must be held at a logic 1 level.

Borrow and carry Because the 74LS192/193's have dual clocks, independent outputs are provided for overflow and underflow indicators. The carry output will go active to indicate an overflow condition, and the borrow output will go active to indicate an underflow condition. The counters can be cascaded by simply connecting the carry output and the borrow output to the count-up and count-down inputs of the next stage.

Clear The clear input is active high and asynchronous. When the clear input is taken to its active level, outputs Q_a through Q_d will all go to a logic 0 level, independent of either clock.

7.5 SHIFT REGISTERS

The basic building blocks of counters are flip-flops that are configured to toggle. Another application of flip-flops is shift registers. Shift registers are used for a wide variety of applications, such as serial-to-parallel conversion, parallel-to-serial conversion, sequencing circuits, and time delays.

7.5.1 The 74LS164: Serial In/Parallel Out Shift Register (SIPO)

The 74LS164 is a popular device that inputs serial data and outputs 8 bits of parallel data. In the computer world data come in two different forms: serial and parallel. Serial data are nothing more than a stream of single bits of digital information: logic 1's and logic 0's. Parallel data comprise a group of related bits. The most common size of parallel data is 8 bits. A parallel group of 8 bits is known as a *byte*. Refer to Figure 7.13. You should be able to figure out a lot about this device from the diagram even though you may not understand the details of its function.

Clock The 741S164 has a positive-edge-triggered clock.

Clear The bubble on the clear input indicates that it it an ac-

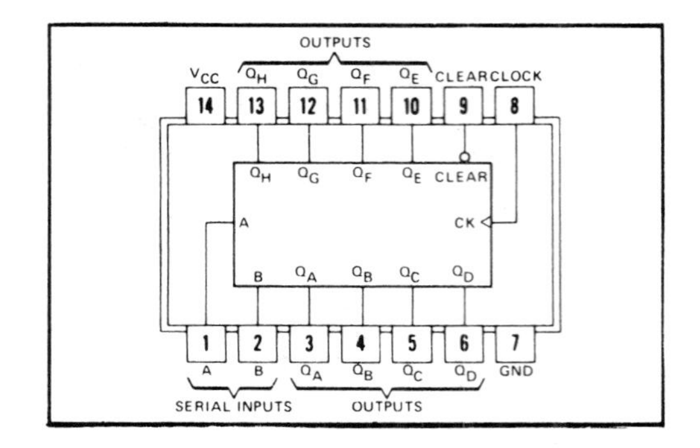

FUNCTION TABLE

INPUTS				OUTPUTS		
CLEAR	CLOCK	A	B	Q_A	Q_B ...	Q_H
L	X	X	X	L	L	L
H	L	X	X	Q_{A0}	Q_{B0}	Q_{H0}
H	↑	H	H	H	Q_{An}	Q_{Gn}
H	↑	L	X	L	Q_{An}	Q_{Gn}
H	↑	X	L	L	Q_{An}	Q_{Gn}

H = high level (steady state), L = low level (steady state)
X = irrelevant (any input, including transitions)
↑ = transition from low to high level.
Q_{A0}, Q_{B0}, Q_{H0} = the level of Q_A, Q_B, or Q_H, respectively, before the indicated steady-state input conditions were established.
Q_{An}, Q_{Gn} = the level of Q_A or Q_G before the most-recent ↑ transition of the clock; indicates a one-bit shift.

schematics of inputs and outputs

Figure 7.13 Pinout and function table for 74LS164 parallel shift register.

tive-low input. When the clear input goes active, all the Q outputs will be driven to logic 0.

Q_a through Q_h Q output pins have always represented outputs of sequential devices. These eight Q's must be the byte of parallel data that we have previously discussed.

A and B These pins are described as serial inputs. Because the 74LS164 is a serial-to-parallel shift register, the levels applied to inputs A and B must constitute serial data.

Refer to the function table in Figure 7.13. We will examine the truth table, line by line, to derive the manner in which the 74LS164 functions.

Line 1. When the clear input is at an active-low level, all other inputs become don't cares and the Q outputs will all be cleared to logic 0's.

Line 2. The clock is low. This line illustrates the situation where the clock is not active. The outputs, Q_a through Q_d, will maintain their memory states.

Line 3. The clock is active with a rising edge and the serial data inputs are both logic 1's. A logic 1 will be shifted into the Q_a output. The contents of the Q_a output, before the rising edge of the clock, will be transferred into the Q_b output. The contents of the Q_b output, before the rising edge of the clock, will be transferred into the Q_c output, and so on. The previous contents of Q_h will be lost.

Lines 4 and 5. These lines demonstrate that the A and B serial inputs are ANDed together to form the serial bit of information that is transferred into Q_a on the rising edge of the clock. If input A is low or input B is low, the result of the AND function will be a logic 0; that logic 0 will be the serial data transferred into Q_a. Only if both inputs A and B are high will the serial input data be high.

7.5.2 Application of the SIPO (Serial In/Parallel Out) Shift Register

This application of the 74LS164 will assemble 8 bits of serial information into a single parallel byte. Refer to the schematic in Figure 7.14. The components in this system are:

1. An S-R latch used to produce debounced clock pulses.
2. A 74LS164 8-bit serial in/parallel out shift register. The clear input is pulled up to the inactive level. The serial input data are controlled by the SPST switch. If the switch is open, the input data are pulled up to a logic 1; if the switch is closed, the serial input data are pulled down to a logic 0.
3. A 74LS90 counter, modified for a modulus of 8.
4. Because byte-length quantities of data are so common, many digital devices are available in groups of eight per IC. The 74LS373 is an example of such a device. The 74LS374 contains eight D flip-flops, all positive edge-triggered by a common clock input, and packaged in a 20-pin IC. The pin OE is the output enable input. When it is at an active-low level, the eight flip-flops are enabled.

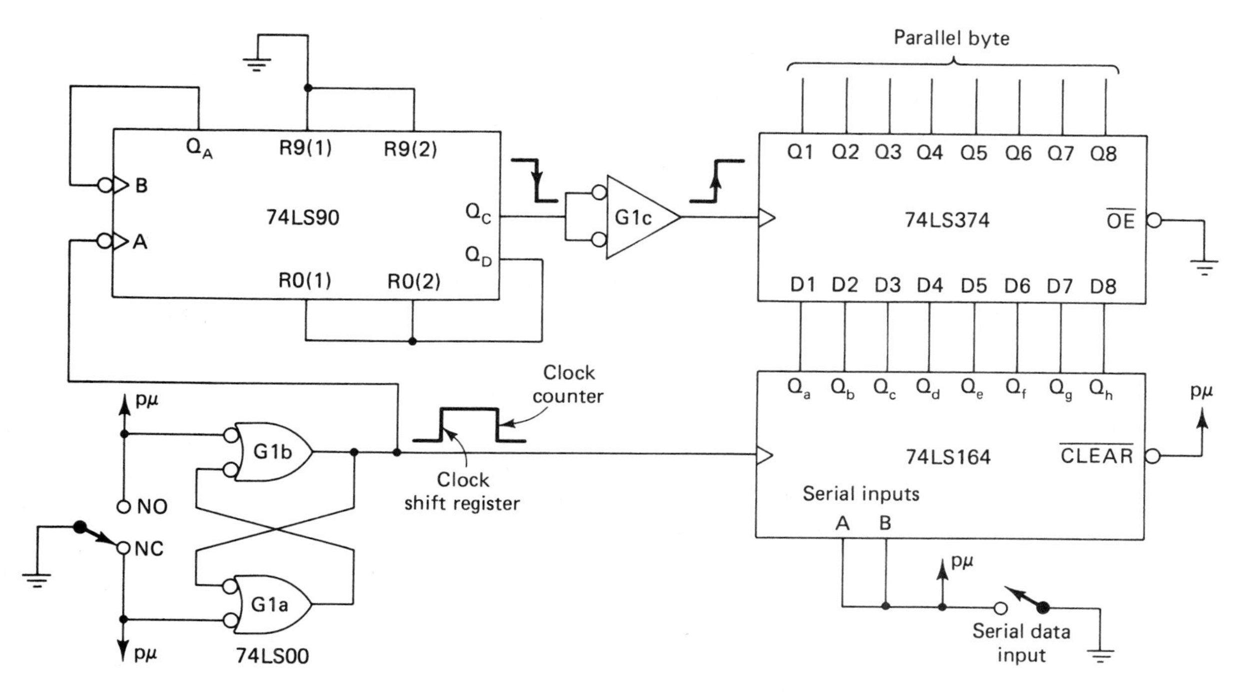

Figure 7.14 Serial-to-parallel converter.

A new notation is used to describe the three NAND gates. Schematics will sometimes refer to a particular IC as G1, G2, and so on; the gate within the IC may be designated by a letter, such as $G1_a$, $G1_b$, $G1_c$, and $G1_d$. This schematic shows that all three NAND gates physically exist in IC G1.

Let's analyze the circuit action. Assume that the counter is initially cleared to a count of 0. The most significant bit of the 8-bit word will be toggled into the serial input switch. After the data are set, the SPDT switch will be toggled. The rising edge from $G1_b$ will clock the first data bit into Q_a of the shift register. The falling edge from $G1_b$ will increment the counter from 0 to 1. This action will be repeated for the last 7 bits of the byte. On each rising edge the data will be shifted one place to the right in the shift register, and the counter will increment by one.

Picture the circuit after the first seven bits have been entered: Q_g through Q_a contain the 7 bits, where the bit in Q_g is the first entered and the most significant. The counter is on the count of 7. The last (least significant) data bit will be toggled into the serial data switch. On the rising edge from $G1_b$, the last data bit will be shifted into the shift register. The eight serial bits of data have now been reformed into a parallel byte residing on Q_h through Q_a. On the falling edge from $G1_b$, the counter will increment from count 7 to count 8. Two important events will happen:

1. When the counter increments from 7 to 8, Q_c will have a falling edge, which will be inverted by $G1_c$. The rising edge from $G1_c$ will latch the 8 bits of data on Q_h through Q_a into the octal flip-flop, where it will be saved temporarily.
2. Because the 74LS90 is a ripple counter, the action of Q_c falling from high to low will clock the Q_d stage of the counter. This will cause Q_d to go from low to high, for count number 8. Count eight will never be seen because the output of Q_d is connected to the reset to zero inputs. This high level will reset the counter back to count zero.

QUESTIONS AND PROBLEMS

7.1. Define the phrase "divide-by-N counter."

Refer to Figure 7.2 for Questions 7.2 to 7.5.

7.2. Why aren't the outputs Q_a and Q_b also required in the count modification circuitry?

7.3. In the timing diagram shown in Figure 7.2a, why do the two narrow glitches occur on Q_c and clear?

7.4. If count 12 actually occurs, why isn't it included in the count-state diagram?

7.5. Assume a propagation delay of 10 ns for all flip-flops and gates. How wide is the clear input pulse in the timing diagram shown in Figure 7.2b?

7.6. Create a count-state diagram and write a theory of operation that describes the counter circuit shown in Figure P7.6

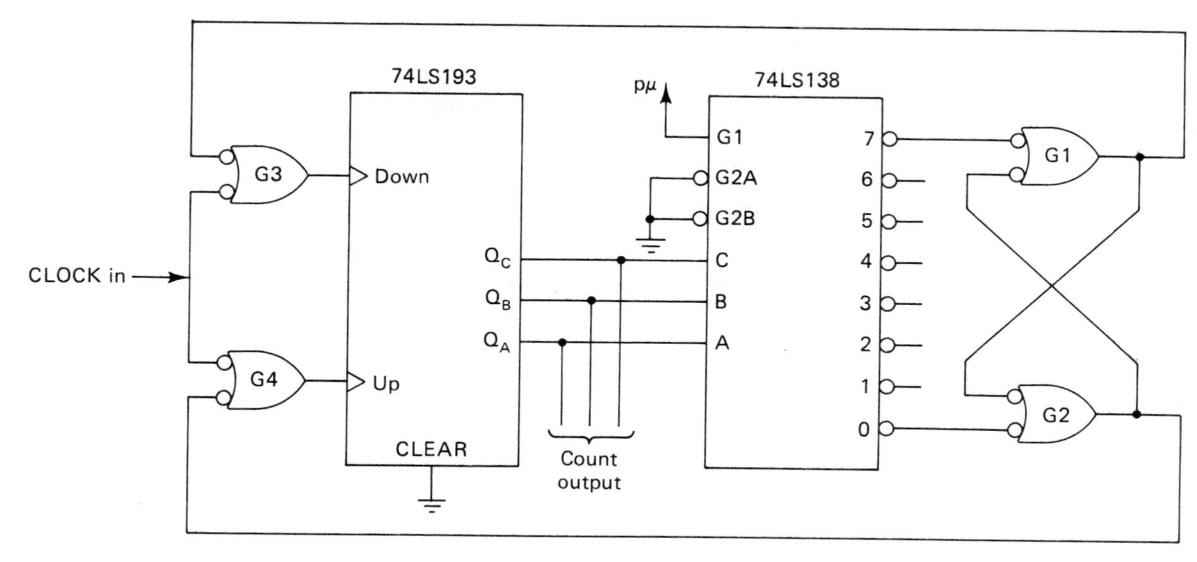

Figure P7.6

7.7. By what factor does the circuit shown in Figure P7.7 divide the input frequency?

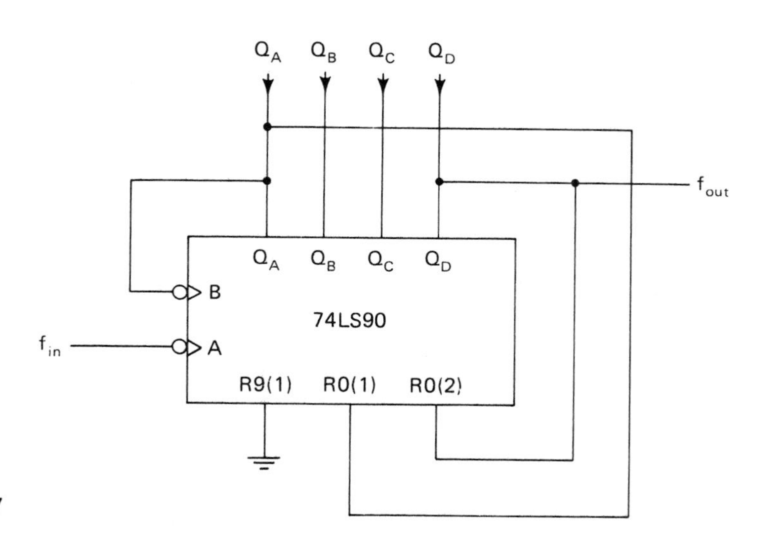

Figure P7.7

7.8. What is the major difference between ripple and synchronous counters?

7.9. Use a 74LS93 to create a mod-14 counter.

Refer to Figure 7.5 for Questions 7.10 to 7.13.

7.10. The display appears to flicker displaying the digit 8, with segments of varying intensity. What could be the possible problem?

7.11. How would the circuit operation be affected if the trace running from the output of G6 to G1 were open? Shorted to ground?

7.12. How would the circuit be affected if the output pin of G8 were bent underneath the IC during the stuffing operation?

7.13. How does the clock frequency affect the operation of the circuit?

7.14. In Figure 7.10, the output of G1 is internally stuck high. How is the circuit operation affected?

7.15. How can two 74LS75s be used to improve the operation of the circuit in Figure 7.10?

Refer to Figure 7.18 for Questions 7.16 to 7.19.

7.16. $\overline{Q}$ on U11 is stuck high. How does this affect the circuit?

7.17. The ripple-carry output of U2 is shorted to V_{cc}. How does this affect the circuit?

7.18. Latch U4 is accidentally stuffed with a 74LS76. Hoe is the circuit affected?

7.19. U12 is a divide-by-5 instead of a divide-by-10 counter. How would the circuit react?

Refer to Figure 7.14 for Questions 7.20 to 7.23.

7.20. How would the circuit be affected if the output of $G1_a$ were used to clock the shift register and counter instead of the output from $G1_b$?

7.21. The output of $G1_c$ is shorted to the output of $G1_b$. How is the circuit affected?

7.22. If the outputs of Q1 through Q8 of the octal D flip-flop are always logic 0's, what checks would you perform?

7.23. How would the circuit operation be affected if the clear input of the shift register were floating?

8 Introduction to Three-Bus Architecture

We have stated that SSI and MSI digital devices are the "glue" used to interconnect complex LSI ICs. LSI ICs are said to be *intelligent* devices. Here the term "intelligent" refers to an LSI IC's ability to execute a series of programmed instructions and alter the sequence in which the instructions are executed according to established criteria. The mircoprocessor is the most common example of an intelligent device. Detailed study of the microprocessor is reserved for the second half of this book. This chapter includes a quick peek at the microprocessor and its associated circuitry.

8.1 TALKERS AND LISTENERS

8.1.1 Three-State Logic and High-Impedance Output

We have observed two different TTL output structures: the totem pole and the open collector. If totem-pole outputs are tied together, indeterminate logic levels will result. When the output of common-collector devices are tied together, any device can take the common node to a logic 0. An open-collector connection functions like the negative-logic interpretation of an AND gate: A low on any input will pull the output low. There is a third

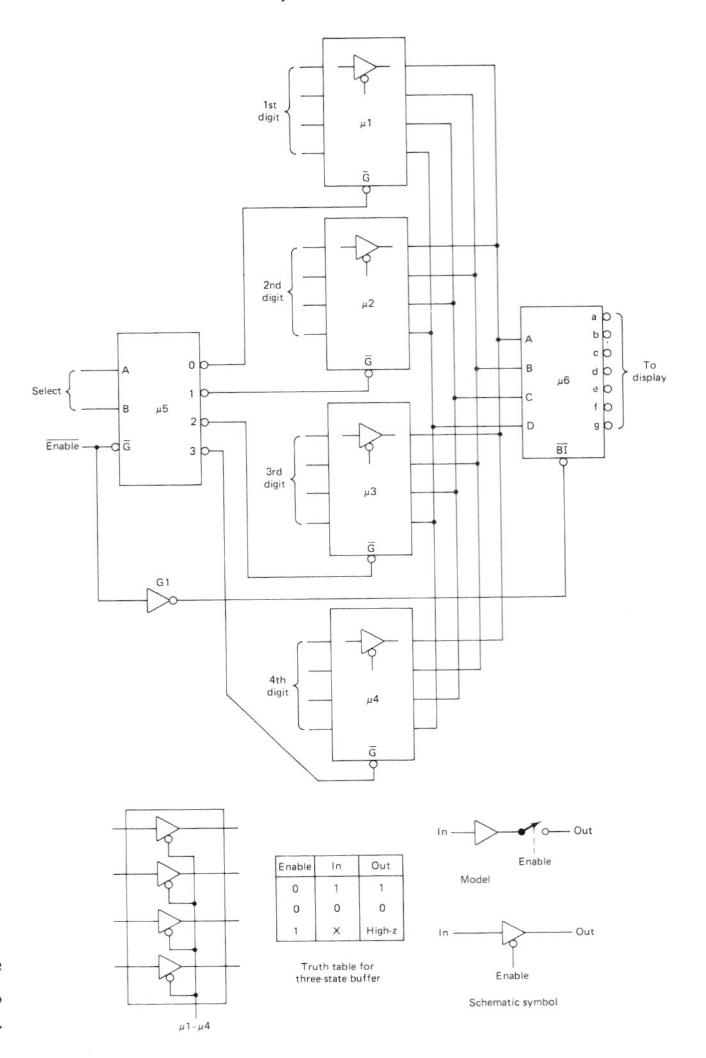

Enable	In	Out
0	1	1
0	0	0
1	X	High-z

Figure 8.1 Three-state buffer: schematic symbol, model, and truth table.

type of output structure. This structure embodies characteristics of both totem poles and open collectors.

The circuit in Figure 8.1 makes it possible for one BCD-to-seven-segment decoder and seven-segment display to be shared between four different input sources. A 2-bit select code will be applied to the 2-line-to-4-line decoder. The output of this decoder will enable one of four BCD digits to appear on the inputs of the BCD-to-seven-segment decoder. By changing the select code, any of the four BCD digits can be displayed.

You should be puzzled by the four 74LS241 devices. The four output lines of each of these devices are tied to the similar output lines on the other three devices. You are aware that totem-pole outputs cannot be tied together. Therefore, these four devices cannot have totem-pole output structures. Furthermore, these devices cannot be open collector, because there are no pull-up resistors illustrated in the schematic.

The 74LS241 is an example of the third type of output structure: the *three-state output.* A three-state device has three possible outputs: logic 0, logic 1, and *high impedance.* It is the high-impedance output that distinguishes the three-state device from totem poles and open collectors. You must assume from the schematic in Figure 8.1 that the outputs of three-state devices can be tied together with no undesirable effects.

Refer to the truth table and schematic symbol of the three-state buffer in Figure 8.1. The schematic symbol illustrates a normal buffer with an extra line. This line is the enable input. Because it is bubbled, the enabled input is active low. Lines 1 and 2 of the truth table show that when the enable is at an active-low level, the buffer passes the input level to the output. In the third line of the truth table, the enable input goes to the inactive level. The input is now designated as "don't care" and the output is labeled "high Z." *High-Z* refers to the third possible level of a three-state device: the high-impedance output level. When a three-state device is at a high-Z level, its output is effectively open or disconnected.

The model of the three-state device in Figure 8.1 shows how the output is disconnected. The enable input controls the position of a SPST switch. When the enable input is active, the switch is closed and data are passed in a normal fashion. When the enable input goes inactive, the SPST will open. The output of the buffer is now disconnected from the output of the switch.

With this new-found knowledge in hand, let's examine how the circuit in Figure 8.1 actually shares the display. U1 through U4 each contain four three-state noninverting buffers. The four buffers in each device are controlled by a common enable input. When the enable input of a particular device is active, all four buffers within that device operate normally. When the enable input of a particular device is inactive, all four buffers within that device go to a high-Z output level. The 2-line-to-4-line decoder will place an active-low level on one enable input, while placing inactive-high levels on the other three enable inputs. The device that is selected by the 2-bit select code will pass a BCD code through its four buffers; the outputs of the other three devices will be effectively disconnected.

There are two important cases that we must consider before the concept of three-state devices can be fully understood:

1. What happens if more than one three-state device, sharing the same output lines, is enabled simultaneously?
2. What happens if all the three-state devices sharing a common node are disabled?

Case 1. A three-state device can be thought of as a modified totem pole. The outputs of three-state devices sharing the same node should never be enabled simultaneously. This action would result in indeterminate logic levels. The open collector is the only output structure where common devices can be active simultaneously.

Case 2. Assume that the enable input to the 2-line-to-4-line decoder has gone to an inactive level. All the decoded outputs, 0 through 3, will be taken to the inactive-high level. None of the three-state buffers will be enabled; all the inputs on the common node are effectively disconnected. If you placed an oscilloscope probe on any BCD input of the BCD-to-seven-segment decoder, what voltage level would you observe? If all the outputs of the three-state buffers are disconnected, the entire node is floating. You would observe 1.4 to 1.8 V—a normal TTL floating input level. It would appear as if all the traces running from the three-state buffers were cut to produce opens.

When you are troubleshooting a PCB that contains three-state devices, a floating input level will not necessarily indicate a broken trace, open feed-through, or any other problem that would result in a floating input level. There will be an added possibility that the three-state devices sharing a particular node are all disabled. That in itself may not indicate a circuit malfunction. There will be many legitimate instances when all the three-state devices sharing a particular node will be disabled.

In the circuit illustrated in Figure 8.1: if the enable input of the 2-line-to-4-line decoder goes to an inactive-high level, the blanking input of the BCD-to-seven-segment decoder will go active low to blank the seven-segment display. When the BCD inputs to the BCD-to-seven-segment decoder are floating, they will appear to be logic 1's. You may remember that the 74LS47 will automatically blank the display when the BCD inputs are all equal to logic 1's. Therefore, it may appear that there is no need to blank the BCD-to-seven-segment decoder when the 2-line-to-4-line decoder is disabled. Noise glitches can momentarily pull the levels on floating inputs to logic 0. This would induce flashes on the seven-segment display. Blanking the display when the 2-line-to-4-line decoder is disabled ensures that noise glitches will not affect the seven-segment display.

Each of the quad three-state buffers can be classified as a *talker*. A talker is a device that is capable of outputting a signal. On the other hand, the BCD-to-seven-segment decoder can be classified as a *listener.* It will receive a BCD input from one talker. As in any well-mannered conversation, only one talker can be active simultaneously. If more than one talker is active simultaneously, the conversation will be garbled and the listener will not receive the correct information. Three-state output structures allow many talkers to share the same listener. If the three-state devices are enabled in the proper manner (only one active at a time), the listener can receive information from many different sources.

Three-state devices will allow us to create a circuit structure called a bus. A *bus* is nothing more than a group of signals grouped together according to a common function. In Figure 8.1, the traces that are connected to the BCD input of the BCD-to-seven-segment decoder can be called a bus. The common function of these signals is to carry a BCD digit to the BCD-to-seven-segment decoder. Using only conventional totem-pole or open-collector outputs, the circuitry required to create the bus in Figure 8.1 would be extensive and complex. The use of three-state devices and decoders makes the bus circuit almost trivial!

8.2 A SURVEY OF THREE-STATE DEVICES

Any digital device can be manufactured with a three-state output structure. Popular devices used to create buses are inverting and noninverting buffers, transparent latches, and D flip-flops.

8.2.1 The 74LS240: Octal Three-State Buffer

A three-state buffer is used to isolate the logic levels of a bus talker from the bus itself. Three-state buffers usually have the ability to sink and source much more current than the typical digital device. The 74LS240 is an example of a typical three-state inverting buffer. The first detail that you may notice in the pinout of the 74LS240 shown in Figure 8.2 is the elongated square in the center of each buffer. This is the symbol representing the property of *hysteresis*. The hysteresis symbol designates a special type of input structure that is used to reduce the effect of noise. We will discuss the concept of hysteresis in a later chapter. Until then you should treat these devices as normal three-state inverters.

The 74LS240 is an octal buffer/line driver with three-state outputs. Octal means that there are eight buffer/line drivers in this 20-pin package. You will discover that quad and octal devices are extremely popular. Earlier we discussed the idea that the quantity of 8 bits, called a byte, is the most common width of parallel data; many three-state ICs are designed to accommodate byte-wide quantities.

The term *buffer/line driver* refers to the 74LS240's ability to sink and source great amounts of current. The 74LS240 can sink 24 mA and source 15 mA. A line driver is a device that is used to drive signals through long transmission lines. This is

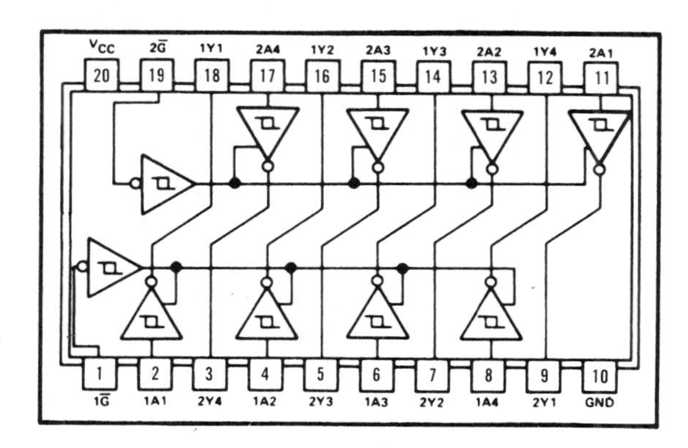

Figure 8.2 Pinout of 74LS240 octal buffer/line driver with three-state output.

where the 74LS240's good current and noise-reduction abilities are used.

Each of the eight inverting buffers has a three-state output structure. The first four buffers are controlled by the active-low enable on pin 1. The second group of four buffers is controlled by the enable input on pin 19. Used separately, each group of four buffers can control a 4-bit quantity such as a BCD digit, as was done in Figure 8.1. If pins 1 and 19 are tied together, the 74LS240 can be used to control a byte of data.

8.2.2 The 74LS363/364: Three-State Transparent Latches and D Flip-Flops

The 74LS363/364 are byte-wide (octal) transparent latches and D flip-flops (Figure 8.3). The only difference between the two devices is that the 74LS363 is level active and the 74LS346 is rising edge active. The 74LS363/364 share identical pinouts except for pin 11. The input that allows data to flow through a latch is called an *enable* and the input that causes the logic level on the D input to be transferred to the Q output in a D flip-flop is called a *clock*. The 74LS363 is a transparent latch; pin 19 is the enable. The 74LS364 is a positive-edge-triggered D flip-flop; its pin 19 is the clock.

Pin 1 of these devices is called the output control. It functions as their output enable. If pin 1 is low, the devices will perform their normal functions; if pin 1 is taken to an inactive-high level, the eight Q outputs will go to high-Z. The 74LS363/

'LS363
FUNCTION TABLE

OUTPUT CONTROL	ENABLE G	D	OUTPUT
L	H	H	H
L	H	L	L
L	L	X	Q_0
H	X	X	Z

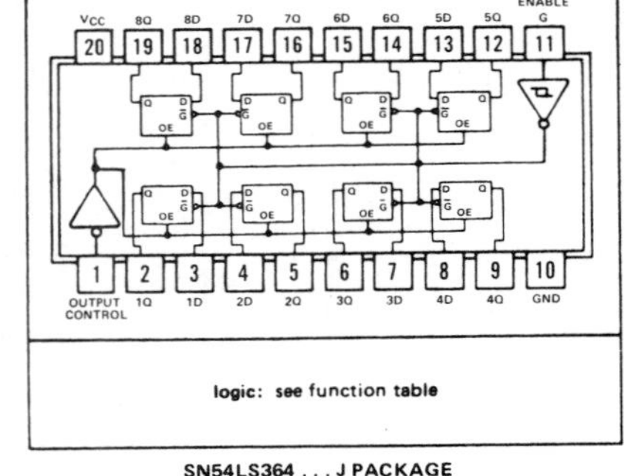

SN54LS364 . . . J PACKAGE
SN74LS364 . . . J OR N PACKAGE
(TOP VIEW)

'LS364
FUNCTION TABLE

OUTPUT CONTROL	CLOCK	D	OUTPUT
L	↑	H	H
L	↑	L	L
L	L	X	Q_0
H	X	X	Z

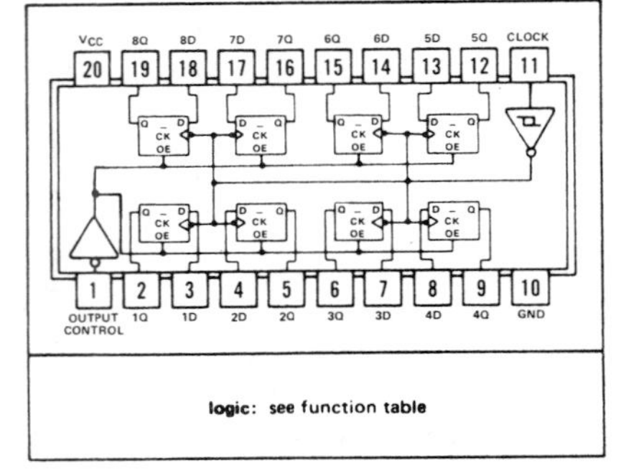

Figure 8.3 Pinouts and function tables for 74LS363/364.

364 are 20-pin packages. Because of these pin limitations, the $\overline{Q}$ outputs are not available.

The function tables shown in Figure 8.3 are identical except for the columns of enable and clock. The 74LS363 has an active-high enable and the 74LS364 is rising edge clocked. When the output control pin is at a logic 0, the octal latches and flip-flops will behave in a normal fashion. The fourth line in the function table indicates the case when the output control is at a logic 1. The eight Q outputs will go to high-Z; the enable/clock and data inputs become don't cares.

The 74LS363/364 have one other interesting characteristic. The guaranteed V_{oh} (logic 1 output voltage) is 3.45 V instead of the normal 2.4 V. This enables the 74LS363/364 to interface with MOS devices that require a higher logic 1 level than the standard 2.4 V. The 74LS373/374 that we used earlier in the serial-to-parallel converter is the same as the 74LS363/364 except that it has a standard 2.4-V V_{oh} specification.

8.3 BIDIRECTIONAL BUS DRIVERS

Every pin that we have seen on digital ICs has been designated as an input or an output. In computer systems, data must travel in both directions. The data pins on a microprocessor are both inputs and outputs. The data buses require buffers to drive them; buffers whose data pins are capable of both receiving data (as inputs) and transmitting data (as outputs). Refer to Figure 8.4.

8.3.1 A Discrete Bus Transceiver

This circuit is constructed from two three-state buffers and a conventional inverter. The input of one three-state buffer is tied to the output of the other three-state buffer. A visual inspection of the circuit does not reveal whether point A or point B is an input or an output. In fact, both points A and B are connected to inputs and outputs. Refer to the simple two-line truth table in Figure 8.4. There are two cases that we must investigate: when the control input is a logic 0 and the control input is a logic 1.

Control Input is a Logic 0 The logic 0 will enable three-state buffer 1. The inverter will place a logic 1 on the control input of three-state buffer 2. This logic 1 will effectively disconnect the output of three-state buffer 2 from the node at point A. Data can now move from point A to point B via three-state buffer 1. There is no conflict at node A because three-state buffer 2 is at high-Z.

Control Input is a Logic 1 Three-state buffer 1 will go to high-Z, effectively removing its output from node B. Three-state buffer 2 will be enabled via the inverter. The data can now move from point B to point A via three-state buffer 2. There is no conflict at node B because three-state buffer 1 is at high-Z.

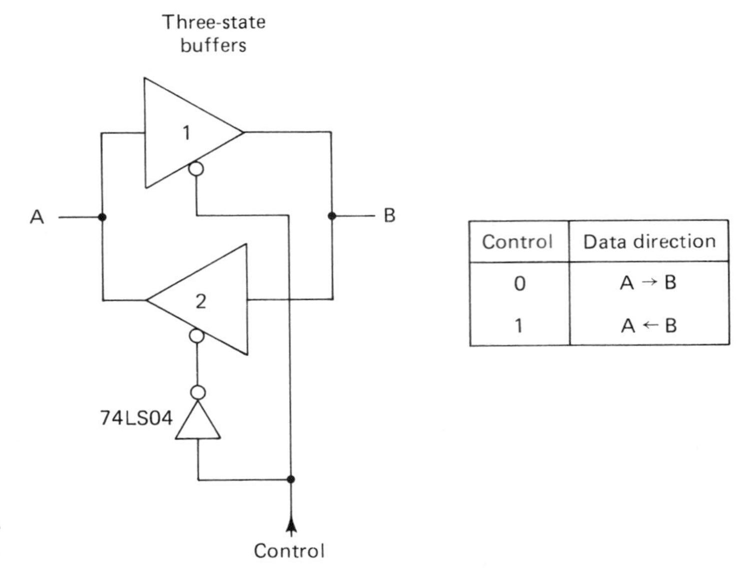

Control	Data direction
0	A → B
1	A ← B

Figure 8.4 Discrete bus transceivers.

The circuit in Figure 8.4 is called a *bus transceiver*. The term "transceiver" indicates that data can travel in either direction. (The terms "bus transceiver" and "bidirectional bus driver" are used interchangeably.) It is extremely important to realize that data do not travel in both directions simultaneously. In a correctly operating circuit, the inverter ensures that three-state buffers 1 and 2 will never be enabled simultaneously. The control input should be thought of as a directional input. If the logic level on the directional input is low, data flow from A to B; if it is high, data flow from B to A.

8.3.2 The 74LS245: Octal Bus Transceiver with Three-State Outputs

Building bus transceivers from discrete gates is impractical. The 74LS245 is an example of a byte-wide bus transceiver contained in a 20-pin package. The 74LS245 functions like the discrete bus transceiver in Figure 8.4 except it has one added mode—the ability to bring both the A and B pins to high-Z.

Refer to the function table in Figure 8.5. When the enable input is low, the 74LS245 functions like a normal bus transceiver; the direction of data movement is controlled by the direction control input. When the directional control input is a logic 0, the data flow from B to A. When the direction control input is at a logic 1, the data flow from A to B.

In the third line of the truth table, the enable input goes to an inactive-high level. Both the A side and the B side will go to a high impedance. The 74LS245 is effectively removed from the circuit.

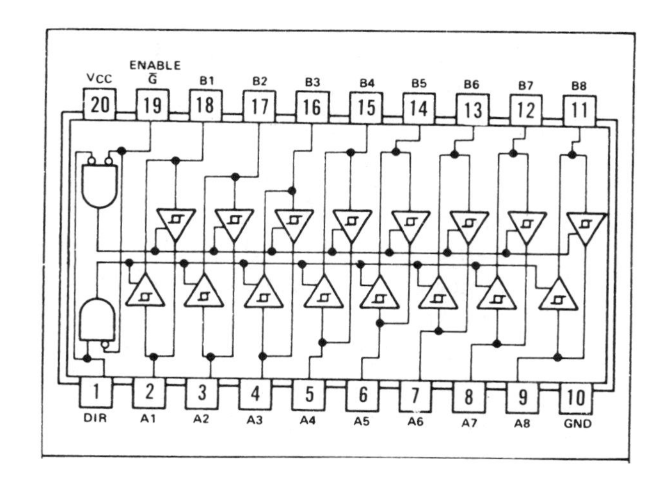

FUNCTION TABLE

ENABLE $\overline{G}$	DIRECTION CONTROL DIR	OPERATION
L	L	B data to A bus
L	H	A data to B bus
H	X	Isolation

H = high level, L = low level, X = irrelevant

Figure 8.5 74LS245 octal bus transceiver.

8.4 THE MASTER OF THE SYSTEM BUS: THE MICROPROCESSOR

8.4.1 Block Diagram of a Computer

Digital systems are composed of many complex bus structures. They require a device that can coordinate and synchronize all the circuit action in the system. This "*master* of the system bus" must be capable of retrieving and executing stored commands. It must be an "intelligent" device that can make logical decisions when confronted with choices. An LSI device called the "microprocessor" is used for such applications.

The term "computer" is extremely difficult to define. It is hard to think of an area or product where microprocessors are not used: automobile ignitions, kitchen appliances, watches, calculators, and an infinite number of other objects are controlled by microprocessors. Having a device controlled by a microprocessor does not make that device a computer. Microprocessors are used to give devices simple limited local intelligence.

The term *computer* is usually reserved to describe a general-purpose computing device that is capable of performing a wide variety of tasks. All computers spend the majority of their time reading instructions and data from memory and writing the result of calculations back into memory. Chapter 10 is about digital memory systems. Before you can thoroughly understand how digital memories function, you must have a basic idea of what microprocessors really are and how they read and write to and from memory. Do not expect to acquire a general understanding of computer concepts from this introduction to microprocessors. They will be covered in the second half of this book.

Consider the block diagram of a traditional computer shown in Figure 8.6. The diagram consists of four separate units: input circuitry, central processing unit (CPU), memory, and output circuitry. The input and output circuitry is used to interface the computer with the outside world. A typical input device is a keyboard; a typical output device is a video display (CRT). Our interest lies in the two other blocks: the CPU and the memory. Notice that arrows point in both directions between the CPU and the memory. That indicates that the CPU can read and write into the memory.

The CPU is the brains of the computer system. The CPU is divided further into two separate blocks: the arithmetic/logic unit (ALU) and the control unit (CU). The ALU performs simple arithmetic (such as addition and subtraction) and logic functions (such as NOT, AND, OR, XOR, and shifting). The CU coordinates all the activity in the computer system. A microprocessor is nothing more than a CPU implemented in a single IC package. Microprocessors can think, but they cannot hear, they cannot talk, and they cannot remember anything. That is why input/output circuitry and memory devices are required to form a complete computer system.

The microprocessor's programmed instructions are stored in memory. Along with the programmed instructions are stored data. If we program a microwave oven to cook for 30 minutes at 12:30, cook is an instruction, while 30 minutes and 12:30 are pieces of data. When you first enter the cooking program, the microprocessor takes the values from the keyboard on the oven and writes them into memory for future retrieval.

Our objective is to understand how the microprocessor reads and writes data to and from the memory.

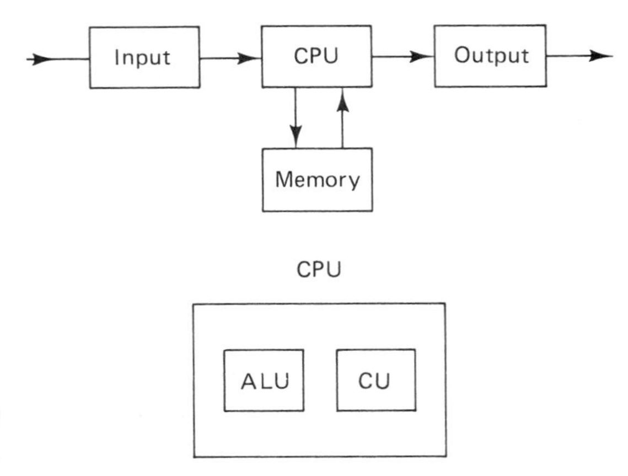

Figure 8.6 Block diagram of a computer.

8.4.2 Three-Bus Architecture

The pins of a microprocessor can be broken in three groups of buses: address, data, and control/status. Refer to Figure 8.7.

Address Bus If you want to send a letter, you must provide the mail carrier with a unique address. This unique address will guide the carrier to the correct residence. When a microprocessor needs to read from or write to memory, it must also provide a unique address. A typical microprocessor has an address bus that is 16 bits wide (A0 through A15). With 16 address lines the microprocessor can generate 2^{16} unique addresses; 2^{16} is equal to $65{,}536_{10}$. To avoid suffering under the influence of ungainly numbers, the concept of 1K of memory was developed. 2^{10} is equal to 1024. 1024 bytes of memory is called 1K of memory. Therefore, 16 address lines are said to produce 64K unique memory locations. It is understood that 64K actually means 65,536.

The address bus is output only. It does not make any sense for a microprocessor to input a memory address. Remember that the microprocessor is the master of the bus; it will provide the address of the memory location that will be accessed.

Data Bus The data bus is the path in which data enter and exit the microprocessor. Because memory locations will be read from and written to, the data bus must be bidirectional. The data bus in this example is 8 bits (one byte) wide. Each memory location will store 8 bits of data. Microprocessors are available with data buses of 8 and 16 bits.

Control/Status Bus A control line controls the interaction between the microprocessor and external devices (such as memories). Control lines are output only. Status lines are input only. The microprocessor uses status inputs to monitor the state of particular lines and sense the occurrence of important external events. A typical microprocessor has a control/status bus that is 6 to 12 bits wide.

Because of our interest in the relationship between the microprocessor and memories, we will focus our attention on

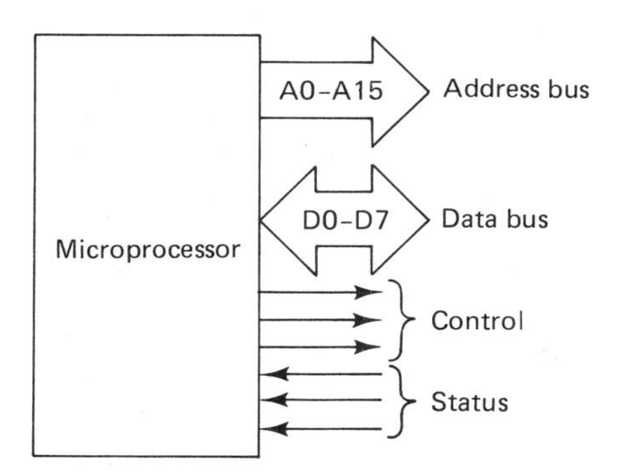

Figure 8.7 Three-bus system of a microprocessor.

control lines that a typical microprocessor might use to interface with memories: $\overline{\text{MEMRQ}}$, $\overline{\text{RD}}$, and $\overline{\text{WR}}$ (Figure 8.8). When the microprocessor needs to access memory, the $\overline{\text{MEMRQ}}$ (memory request) control line will go active low. The logic levels on the $\overline{\text{RD}}$ and $\overline{\text{WR}}$ lines will indicate whether the microprocessor is going to read from or write into memory. If the $\overline{\text{RD}}$ line is low and the $\overline{\text{MEMRQ}}$ line is low, the microprocessor is going to read from memory. If the $\overline{\text{WR}}$ line is low and the $\overline{\text{MEMRQ}}$ line is low, the microprocessor is going to write to memory. The $\overline{\text{RD}}$ and $\overline{\text{WR}}$ lines will never both be at active-low levels simultaneously. The simple two-OR-gate circuit in Figure 8.8 decodes the $\overline{\text{MEMRQ}}$, $\overline{\text{RD}}$, and $\overline{\text{WR}}$ control lines into $\overline{\text{MEMRD}}$ (memory read) and $\overline{\text{MEMWR}}$ (memory write) commands.

Refer to the timing diagram in Figure 8.8 that describes a memory-read operation. The data output of the memory device is a three-state output structure. This structure will stay at high-Z unless the microprocessor initiates a memory-read or memory-write operation.

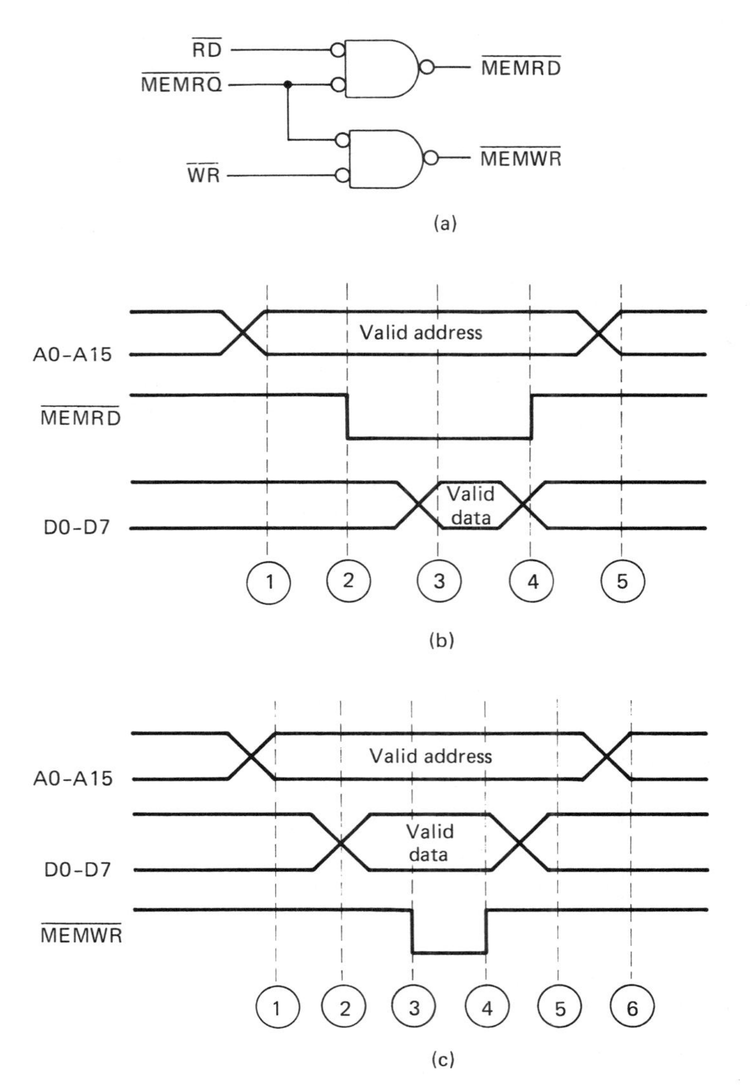

Figure 8.8 Control lines for memory read and write.

Event 1. The microprocessor will present the address of the memory location that it wishes to read.

Event 2. The $\overline{\text{MEMRD}}$ signal will go low. This indicates to the memory IC that it should enable its data bus drivers and retrieve the byte of information stored at the indicated address.

Event 3. After a specified interval, the memory will place the data byte onto the data bus. This data byte will be read into an internal storage location in the microprocessor.

Event 4. The microprocessor will take the $\overline{\text{MEMRD}}$ signal back to the inactive level. When this occurs, the memory will remove the data byte from the data bus by going to a high-Z output.

Event 5. The microprocessor will then remove the address from the address bus. The transfer from the memory device to the microprocessor is now complete.

Refer to the timing diagram in Figure 8.8 that describes the memory-write operation.

Event 1. The microprocessor will first output the address of the memory location where the data byte is to be stored.

Event 2. The microprocessor then places onto the data bus the data byte to be stored in memory.

Event 3. The $\overline{\text{MEMWR}}$ signal is brought to an active-low level. The memory device will now transfer the byte of information on the data bus into the storage location being pointed to by the contents of the address bus.

Event 4. The $\overline{\text{MEMWR}}$ signal must be held active for a minimum specified interval. The $\overline{\text{MEMWR}}$ will then be returned to an inactive level.

Event 5. The microprocessor will remove the byte of data from the data bus.

Event 6. The microprocessor will finally remove the address from the address bus.

These examples of memory-read and memory-write operations have been greatly simplified. There are many factors and specifications that must be considered in practical memory circuits. These specifications are explored thoroughly in Chapter 10.

8.5 THE HEXADECIMAL NUMBER SYSTEM

8.5.1 Introduction to Hexadecimal Notation

Consider the following typical address:

1011 0001 1111 0101

Attempting to describe an address that is 16 bits wide can be extremely confusing. In groups of 16 (or even 8) strings of 1's and 0's have a tendency to blur together. In the first chapter you learned the binary number system. The binary number system was used to describe the actual outputs of digital circuits. Because of the long data bytes and addresses in microprocessor circuitry, another number system is required that efficiently compacts these long strings of 0's and 1's into an easily readable form.

This new number system should meet the following criteria:

1. It must have the capacity to handle long groups of bits in a readily readable form.
2. Conversion between binary and this new number system must be extremely simple.

There is a number system that uniquely meets these criteria. The new number system is called *hexadecimal*. The prefix "hex" is the Greek word for the number six. The word "decimal" is equal to 10. The hexadecimal number system has base of 16 (10 + 6). Although a number system with a base of 16 may seem awkward or clumsy, you will find that hexadecimal numbers are extremely easy to understand and manipulate. If a technician is going to work with microprocessor or memory circuit boards, a fluent command of hexadecimal is required.

A number system with a base of 16, by definition, must have 16 unique counting symbols. The first 10 are borrowed from the decimal number system and the last six are taken from the alphabet.

0, 1, 2, 3, 4, 5, 6, 7, 8, 9, A, B, C, D, E, F

Binary	*Hexadecimal*
0000	0
0001	1
0010	2
0011	3
0100	4
0101	5
0110	6
0111	7
1000	8
1001	9
1010	A
1011	B
1100	C
1101	D
1110	E
1111	F

What is the simple relationship between binary and hexadecimal numbers? You know that 4 binary bits can represent 16 unique numbers. Because the hexadecimal number system has 16 symbols, any group of 4 binary bits can be represented with one hexadecimal symbol. Notice that the first 10 lines appear to be a standard binary-to-BCD conversion. Instead of having the binary combinations of 1010 to 1111 declared as illegal, they are now represented by the first six letters in the alphabet.

Reconsider the 16-bit address that we examined at the beginning of this section.

1011	0001	1111	0101
B	1	F	5

The string of 16 bits can be represented by four hexadecimal symbols. To convert any string of binary bits into a hexadecimal number, start with the least significant bit and form groups of 4 bits. If the last group has less than 4 bits, pad the missing bits with zeros. Convert each group of 4 bits into the corresponding hexadecimal symbol. The binary-to-hexadecimal conversion is now complete. To avoid confusion, the letter H is used to identify a hexadecimal number.

1011 0011 = B3H

10 1001 1111 = 29FH

1100 0101 1010 1011 = C5ABH

Converting from hexadecimal back to binary is also a simple task. Each hexadecimal symbol should be reconverted back into 4 bits.

8.5.2 Hexadecimal-to-K's Conversion

It is often required to translate an address in hexadecimal to a value in K's of memory. You already know that 1K of memory is equal to 1024 bytes. The table shown in Figure 8.9 is a convenient means of converting back and forth between hexadecimal and K's. The first column denotes the powers of 2 (from highest to lowest) for a 16-bit address. The second column indicates the decimal equivalent of each power of 2. The third column indicates the K value of each decimal number in the second column. Finally, the last column indicates the hexadecimal equivalent for each of the previous three columns.

Let's assume that you need to know the hexadecimal equivalent of a memory location at 13K.

1. Break 13K into values that are listed in Figure 8.9.

$$13K = 8K + 4K + 1K$$

Powers of 2	*Decimal*	*Value in K's*	*Hexadecimal*
2^{15}	32,768	32	8000
2^{14}	16,384	16	4000
2^{13}	8,192	8	2000
2^{12}	4,096	4	1000
2^{11}	2,048	2	0800
2^{10}	1,024	1	0400
2^{9}	512	1/2	0200
2^{8}	256	1/4	0100
2^{7}	128	1/8	0080
2^{6}	64	1/16	0040
2^{5}	32	1/32	0020
2^{4}	16	1/64	0010
2^{3}	8	1/128	0008
2^{2}	4	1/256	0004
2^{1}	2	1/512	0002
2^{0}	1	1/1024	0001

Figure 8.9 Hexadecimal-to-K's conversion table.

2. Sum the hexadecimal equivalents of the individual values derived in step 1.

$$(13K) = 2000H + 1000 + 0400 = 3400H$$

The hexadecimal equivalent of 13K is 3400H. In Chapter 10 we will examine the topic of memory mapping. An important part of the memory-mapping process is converting between hexadecimal and K's.

8.5.3 Hexadecimal to Decimal Conversion

Consider the place values of a four-symbol hexadecimal number:

Place values

X	X	X	X
$16^3 = 4096$	$16^2 = 256$	$16^1 = 16$	$16^0 = 1$

To convert a hexadecimal number to its decimal equivalent you must first convert the hexadecimal symbol in each column to its decimal equivalent. Multiply the decimal equivalent of each symbol by the place value of the column. Finally, sum up all the products from the multiplication and the result is the decimal equivalent of the hexadecimal number.

Convert A3F9H to decimal.

1. The decimal equivalent of each column is:

10 3 15 9

2. Multiply each of these column values by the proper place value.

$$10 \times 4096 = 40960 \qquad 3 \times 256 = 768$$

$$15 \times 16 = 240 \qquad 9 \times 1 = 9$$

3. Sum the products from 2.

$$\begin{array}{r} 40960 \\ +\ 768 \\ +\ 240 \\ +\ \ \ 9 \\ \hline 41977_{10} \end{array}$$

This procedure is identical with the binary to decimal conversion procedure. The only difference is the place values.

8.5.4 Decimal to Hexadecimal Conversion

This procedure will also be identical to the decimal to binary procedure that we studied in Chapter 1.

Convert 45,346 to hexadecimal

	Intermediate value	*Hexadecimal sum*
Step (1)	45,346	
	−32,768	8000H
	12,578	
	− 8,192	+2000H
	4,386	A000H
	− 4,096	+1000H
	290	B000H
	− 256	+0100H
	34	B100H
	− 32	0020H
	2	B120H
	− 2	+0002H
	0	B122H

Check: Reconvert B122H to decimal.

$$\begin{array}{r} 11 \times 4096 = 45056 \\ 1 \times 256 = 256 \\ 2 \times 16 = 32 \\ 2 \times 1 = 2 \\ \hline 45{,}346_{10} \end{array}$$

Figure 8.10 Decimal-to-hexadecimal conversion and check.

We will use Figure 8.10 as an aid in the conversion process. The first step in the conversion process is to find the largest decimal number in column 2 of Figure 8.10 that is smaller than the number to be converted. This value in column 2 will be subtracted from the decimal number that we are converting. The hexadecimal equivalent in column 4 will become the first value in a hexadecimal sum. We will continue to subtract decimal numbers from the intermediate value to be converted and sum their hexadecimal equivalents until the intermediate value is equal to 0. Refer to the following example of a decimal to hexadecimal conversion. It requires six subtractions to reduce the decimal number to 0. Notice how the hexadecimal column is summed to produce the converted hexadecimal number. The second part of Figure 8.10 is a reconversion from the hexadecimal answer to decimal. The reconversion checks that our hexadecimal answer is equivalent to the decimal number that we wanted to convert.

8.5.5 Adding Hexadecimal Numbers

The hexadecimal addition during the decimal to hexadecimal conversion will not generate any carries. A hexadecimal carry is generated when an addition causes a hexadecimal result which is greater than F. All carries, in any number system, are handled in the same manner. The base of the number system is subtracted from the result that produced the carry. The result of this subtraction is the new answer for the column that produced the carry and a one is added to the next significant column.

It is difficult to add directly in hexadecimal. When performing an addition, the hexadecimal symbols A through F are converted into their decimal equivalents, then each column is added. If any column results in an answer greater than 15, then 16 (the base value of the hexadecimal number system) is subtracted from that answer. The result of the subtraction is the answer for that particular column and a carry is added to the next significant column. If any columns have a result of between 10 and 15, the number is converted into the proper hexadecimal symbol.

8.5.6 Subtracting Hexadecimal Numbers

The most important part of hexadecimal subtraction is the borrow. When a borrow occurs in decimal subtraction, the value of 1 is subtracted from the next significant column and the value of 10 is added to the column that required the borrow. In a subtraction of any base, the value that is added to the column that requires the borrow is always equal to the base of the number system. Therefore a hexadecimal borrow will result in a value of 1 being subtracted from the next significant column and the value of 10H (16 decimal) being added to the column that required the borrow.

To make the numbers easier to handle, when we added hexadecimal numbers, hexadecimal symbols were converted to their decimal equivalents. We will perform hexadecimal subtraction in the same manner. You must practice until you feel confident with the hexadecimal number system. In the microprocessor world, hexadecimal is the most widely used number system. Data bytes and memory locations will be displayed in hexadecimal. When a group of memory locations are *dumped* onto a video display they will be shown in hexadecimal form. Experienced microprocessor technicians can convert back and forth between binary and hexadecimal in their heads. Hexadecimal to K's is also an extremely common conversion, as is hexadecimal addition and subtraction. The least used conversion skill is converting between hexadecimal and decimal. Remember that binary and hexadecimal are closely related, whereas in the digital world, decimal is an unnatural number base.

QUESTIONS AND PROBLEMS

Refer to Figure 8.1 for Questions 8.1 to 8.5.

8.1. A scope probe is placed on input A of U6. The oscilloscope displays the following waveform.

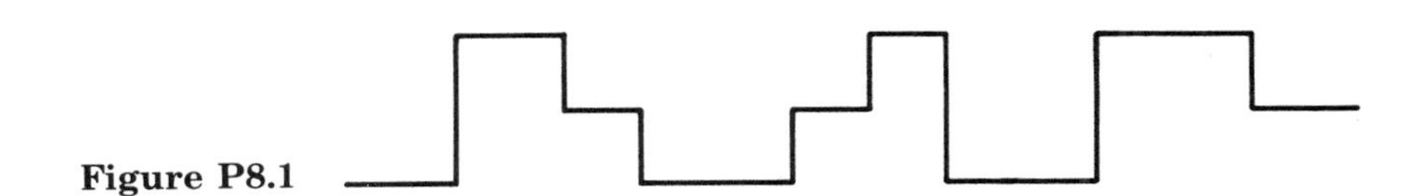

Figure P8.1

Discuss each voltage level as it pertains to the circuitry.

8.2. A scope probe is placed on input D of U6. The oscilloscope displays the following waveform.

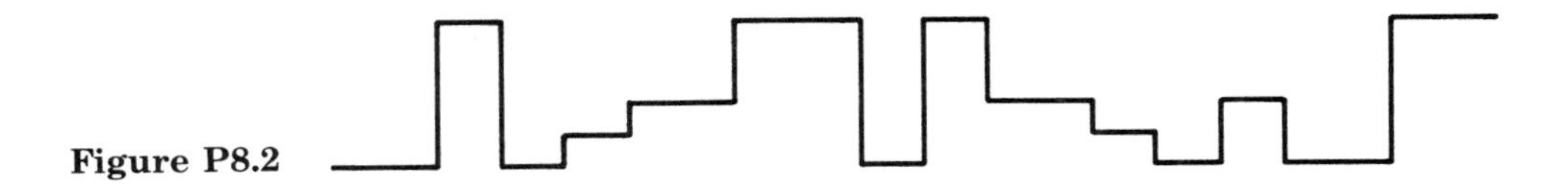

Figure P8.2

Discuss each of the four voltage levels. Explain the circumstance in which each voltage will occur.

8.3. The circuit in Figure 8.1 works correctly for select codes 00 and 01. When select code 10 is applied to the decoder the first BCD digit is displayed instead of the third BCD digit. When select code 11 is applied the second digit is displayed instead of the fourth digit. What is the most likely cause of this circuit malfunction?

8.4. None of the segments of the display will illuminate. What is the most likely cause of such a malfunction?

8.5. Segment *a* of the seven segment display burns out each time it is replaced. What is the most likely problem?

8.6. Define the term "buffer with 3-state output."

8.7. Create a truth table for each of the following circuits. Explain why the common node of an open collector circuit is called a "wired-OR" or "wired-AND" connection.

8.8. In what applications do you expect to encounter each of the following output structures?

a. totem poles
b. open collectors
c. three-state outputs.

8.9. What could be a possible reason that the Q outputs of a 4502B are always logic 0's or floating?

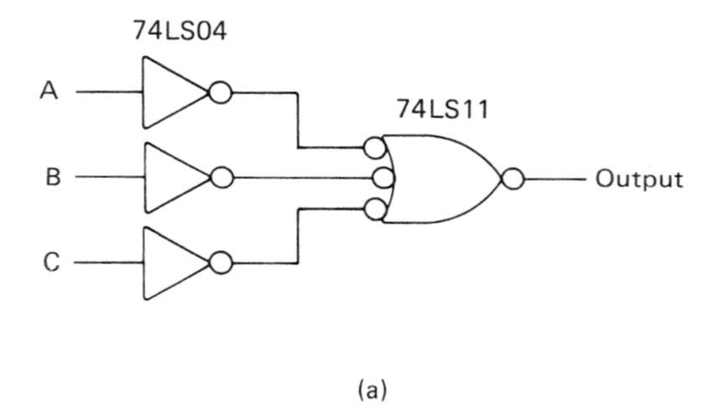

(a)

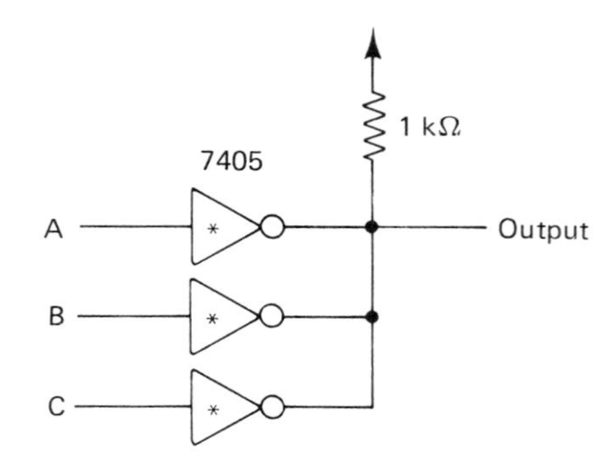

Figure P8.7 (b)

8.10. Is there any danger in using three-state devices in CMOS circuitry?

8.11. Create a truth table that describes the following circuit. The two control inputs should be renamed. Their new names should indicate the functions that they provide.

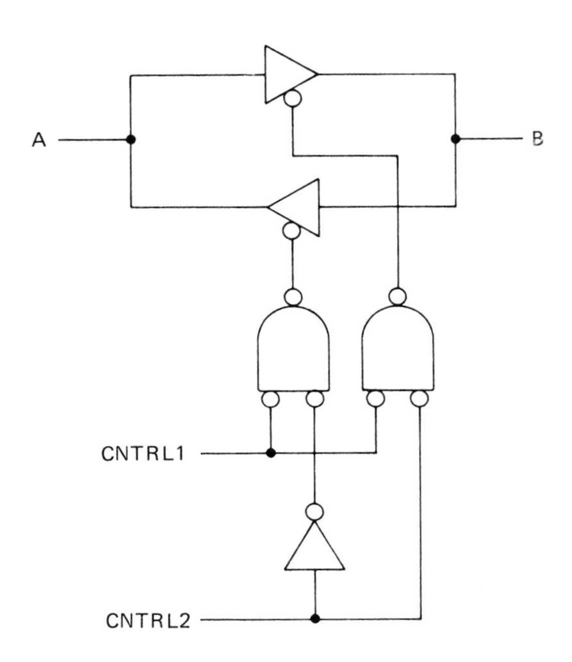

Figure P8.11

8.12. What is a simple definition of "microprocessor"?

8.13. Draw the block diagram of a computer.

8.14. What is the definition of "Bus"?

8.15. Name and describe each bus in a microprocessor based system.

8.16. Describe the manner in which a microprocessor:
a. fetches a byte of data from memory.
b. writes a byte of data into memory.

8.17. What advantages do hexadecimal numbers have over decimal and binary numbers?

8.18. Convert the following values:
(a) 40K = _______H **(b)** 56K = _______H
(c) 64K = _______H **(d)** 32K = _______H
(e) FFFFH = _______K **(f)** C000H = _______K
(g) 0C00H = _______K **(h)** A800H = _______K

8.19. Convert the following hexadecimal values to decimal:
(a) 00FFH = _______ **(b)** F0A4H = _______
(c) 01FFH = _______ **(d)** 07FFH = _______

8.20. Convert the following decimal values to hexadecimal:
(a) 225 = _______H **(b)** 510 = _______H
(c) 1044 = _______H **(d)** 2560 = _______H

8.21. Perform the following additions:

(a)	(b)	(c)	(d)
10BH	9F8H	998H	F0FDH
+ 88H	+ 8H	+ 42H	+ 704H

8.22. Perform the following subtractions.

(a)	(b)	(c)	(d)
FA0H	770H	4000H	F000H
− 3FH	− F7H	− 7FFH	− 800H

9 Introduction to Data Conversion

Digital circuits are often used to measure, process, and control analog quantities. The process of converting analog quantities to their digital equivalents is called *analog-to-digital conversion*. The process of converting a digital number into its analog equivalent is called *digital-to-analog conversion*.

A whole book could easily be devoted to the topic of data-conversion techniques. This chapter provides an overview of data-conversion techniques. Other than in the section on Schmitt triggers, no real-world ICs will be examined. All circuits will be modeled with familiar components. This will enable you to acquire a solid understanding of data-conversion techniques, without getting bogged down with specific devices.

9.1 THE SCHMITT TRIGGER

The *Schmitt trigger* is a digital device that converts slowly rising or falling signals, or noisy signals, into clean, precise digital waveforms. We often need a 60-Hz reference for an input to a circuit. The power-line frequency in the United States is a fairly accurate 60-Hz. Many digital clocks use the 60-Hz line frequency as a reference.

The half-wave rectifier in Figure 9.1 produces a 4-V pulse with a frequency of 60 Hz. (Resistor R_S is a series current-lim-

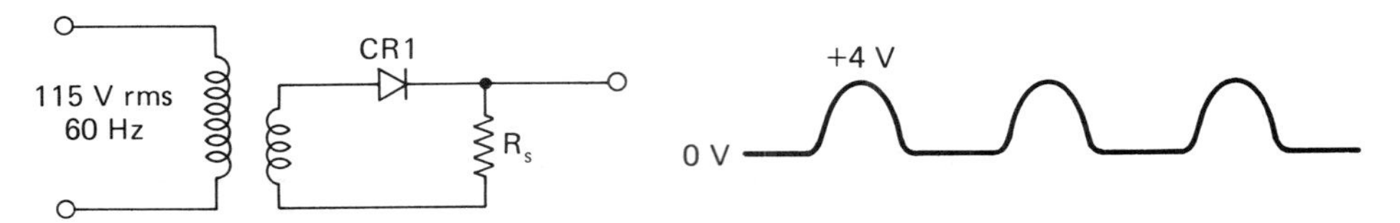

Figure 9.1 Half-wave rectifier with step-down transformer.

iting resistor that protects the diode.) Is the output of the rectifier a digital or an analog signal?

It has characteristics of both digital and analog signals: The two predominate levels could pertain to a logic 0 and a logic 1, but it changes between these two levels very slowly. Actual digital signals closely approach infinitely fast rise and fall times.

The voltage of a sine wave is continually changing. A half-wave rectified sine wave will have a stable logic 0 level, but the logic 1 level will have long rise and fall times. If this signal was used to drive the input of a digital device, the slowly rising and falling edges would cause the digital device to be biased in its linear region of operation. This would cause the output of the digital circuit to oscillate spuriously.

We need a device that will transform this quasi-digital signal into a true digital signal with fast rise and fall times. That is the function of the Schmitt trigger. Schmitt triggers use a phenomenon called *hysteresis*. In the study of physics, hysteresis describes the lagging of magnetization after the magnetic field has been removed. This concept can be implemented in electronic circuitry to form the Schmitt trigger (Figure 9.2). Consider each of the specifications illustrated in Figure 9.2.

Positive-Going Threshold Voltage, V_{T+} (Figure 9.2a). As the input volt into a Schmitt trigger increases, it will reach a

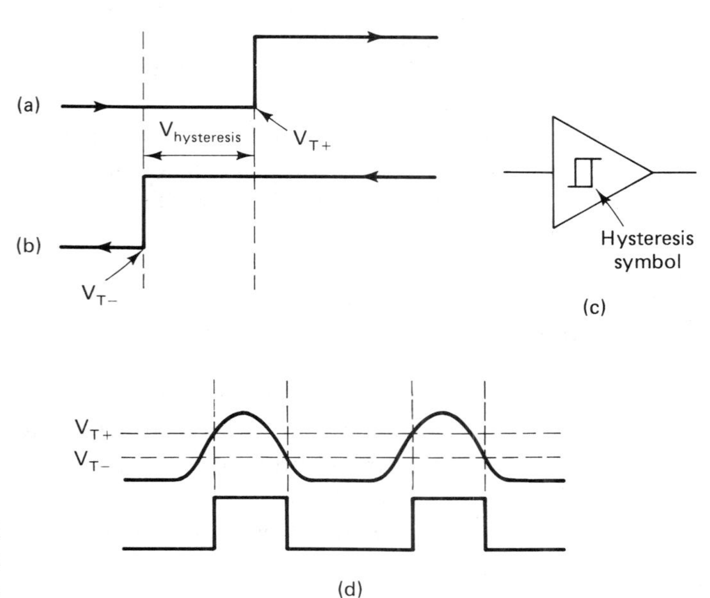

Figure 9.2 Upper, lower, and hysterisis voltages of a Schmitt trigger.

point where the output toggles from a logic 0 to a logic 1. This voltage is called the positive-going threshold voltage. Typically, for TTL Schmitt triggers it is equal to 1.7 V.

Negative-Going Threshold Voltage, V_{T-} (Figure 9.2b). The input voltage is now decreasing from above the positive-going threshold voltage. As the voltage decreases under 1.7 V, the output does not change to a logic 0. The input voltage must go below the negative-going voltage threshold before the output toggles to logic 0. For TTL Schmitt triggers this voltage is typically 0.9 V.

The difference between the negative- and positive-going voltage thresholds is called the *hysteresis voltage.* Consider the output of a Schmitt trigger circuit just as the input voltage reaches 1.7 V—the output snaps to a logic 1 level. If noise on the input causes the voltage to dip as low as 0.91 V, the output will still stay at logic 1. The hysteresis voltage indicates the noise immunity of the Schmitt trigger.

Schmitt Trigger Noninverting Buffer (Figure 9.2c). Notice the symbol placed in the center of the buffer. This symbol denotes a Schmitt trigger device. The hysteresis symbol is a composite of the upper and lower trigger point graphs. This is how a normal buffer and a Schmitt trigger buffer are distinguished. The 74LS14 is a hex Schmitt trigger that is pin-for-pin compatible with a 74LS04. The 74LS132 is a quad Schmitt trigger NAND gate that is pin-for-pin compatible with the 74LS00. Any logic function can be constructed with Schmitt trigger circuitry.

Half-Wave-Rectified Sine-Wave Input and the Digital Output of the Schmitt Trigger Buffer (Figure 9.2d). Notice how the output snaps to logic 1 when the input voltage becomes greater than the V_{T+} and then snaps to logic 0 when the input goes below the V_{T-}.

Schmitt triggers are used to receive digital signals that may have been exposed to noise or that may have degraded rise and fall times because of long transmission lines.

9.2 DATA ACQUISITION AND CONTROL

A Schmitt trigger is used to "clean up" waveforms that are already digital in nature. Physical parameters such as temperature, acceleration, pressure, position, and flow are analog quantities. Figure 9.3 illustrates a typical data acquisition and control system. Devices 1 through 5 are the data acquisition components, and devices 6 through 8 are the components of the control system. Let's investigate each device in Figure 9.3.

1. *Transducer.* A transducer is formally defined as a device that converts energy from one form to another. An audio speaker is a common example of a transducer; it converts electrical current into sound waves. The transducers used in a data acquisition system convert a physical parameter into an elec-

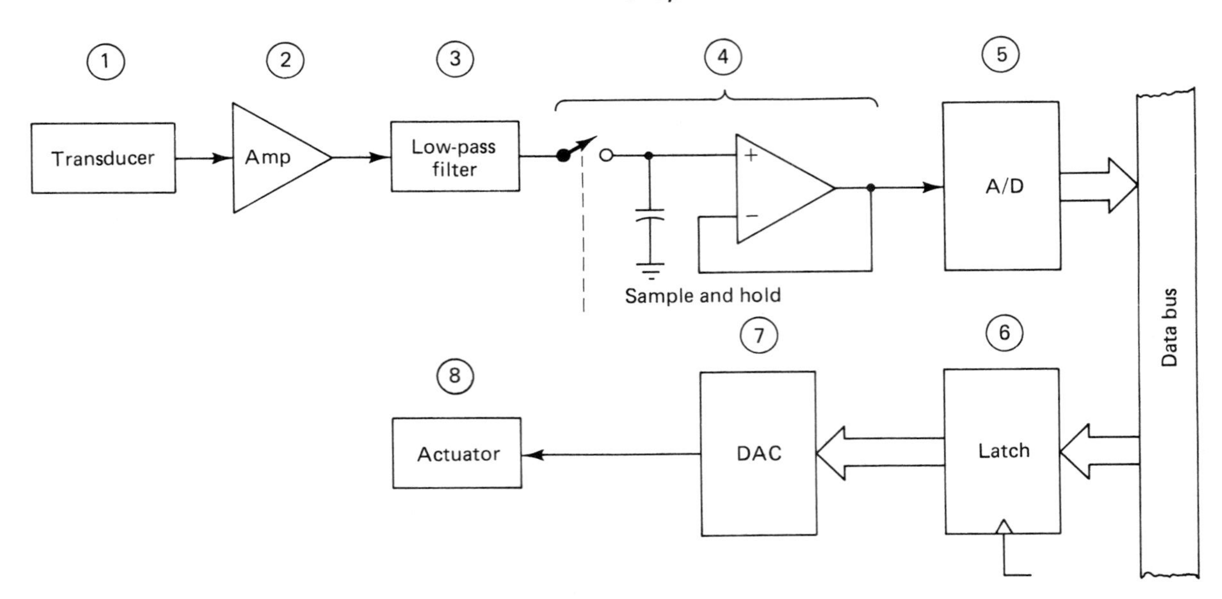

Figure 9.3 Data acquisition and control system.

trical voltage or current. Typical transducers convert temperature, pressure, flow, acceleration, or position into a voltage or current. Good-quality transducers are often the most expensive element in a data acquisition system.

2. *Amplifiers.* A high-gain operational amplifier (op amp) is commonly used as a voltage or current amplifier. This amplifier must have an extremely high input impedance, which will ensure that the output of the transducer will not be excessively loaded. Excessively loaded transducers will produce inaccurate outputs.

3. *Low-pass filter.* This device is used to filter high-frequency noise from the output of the amplifier. This filter is typically realized with an op amp and passive components. Active filters have sharp response curves and do not attenuate the signal like filters constructed exclusively from passive components.

4. *Sample-and-hold circuit.* This block is constructed from a CMOS analog switch, capacitor, and op amp. When the microprocessor is ready to sample the output voltage from the transducer, it will issue a narrow pulse that closes the CMOS switch for a short period. The capacitor will quickly charge up to the output voltage of the low-pass filter. The op amp is used to buffer the voltage on the capacitor. Like the transducer, the capacitor can be easily loaded. The capacitor functions as an analog memory device. It holds the voltage sample while the microprocessor is performing a conversion. This will ensure that the voltage does not change during the analog-to-digital conversion.

5. *A/D (Analog-to-Digital Converter).* This is the actual device that converts the sampled voltage into a digital quantity. The digital quantity is then output onto the microprocessor's

data bus. A/Ds have a command input that is used by the microprocessor to start a conversion, and a status output that signals to the microprocessor when the conversion is complete.

6. *Latch.* When the computer has processed the input data, it may be ready to issue a control command to the analog system. Consider a home heating and cooling system that is controlled by a microprocessor. It will continually sample the temperature until an out-of-range error occurs. The microprocessor then must either turn on the air conditioner or the furnace to bring the temperature back into the acceptable range. A system such as this is called a *closed-loop system.* "Closed loop" indicates that the microprocessor is continually given feedback information (from the data acquisition devices) that indicates the result of the control commands. Compare this with an *open-loop system*, which operates without the benefit of feedback.

The latch is used to hold the microprocessor's output control word. This will free the microprocessor to perform other tasks, such as sampling the temperature from the transducer. Notice that the microprocessor must output the command onto the data bus and then send a rising edge to the latch to store the command.

7. *DAC (Digital-to-Analog Converter).* The DAC is used to convert the digital control command into an analog voltage or current. We have now closed the loop from the analog world to the digital world and back again! The DAC provides the inverse function of the A/D.

8. *Actuator.* The actuator converts a current or voltage into a mechanical output. In the heating/cooling system analogy, the actuator would be the control on the air conditioner or furnace.

Many of the devices illustrated in Figure 9.3 are optional. A weather monitoring system that measures temperature and rain fall will use only the data acquisition circuitry; output control circuitry would not be required. A robot arm that is moved by the output of the control circuitry may not require the data acquisition circuitry. If the microprocessor knows the starting position of the robot's arm, it can calculate the resultant position after each command. That would be an example of an open-loop system.

If the output of the transducer is a clean, slowly varying voltage, the low-pass filter and sample-and-hold circuits may not be required. One A/D converter may be shared by many transducers. An analog multiplexer is used to steer the selected transducer output onto the analog input of the A/D converter.

DACs and A/Ds require an external precision reference voltage. Special-purpose ICs are designed to supply temperature and load stable reference voltages. The DAC's reference voltage will define the largest analog input voltage that can be converted to an accurate digital value. The reference voltage on an A/D defines the maximum output of the device when the digital inputs are all logic 1's.

Resolution refers to the number of digital inputs on a DAC or the number of digital outputs of an A/D. It is also expressed in parts of the full-scale output: An 8-bit DAC is said to have a resolution of 1 part in 256.

DACs and A/Ds are described by many complicated specifications. Those specifications will not be discussed in this text. The aim of this chapter is to familiarize the student with the fundamental concepts of data conversion.

9.3 THE DAC

Digital-to-analog conversion is not a complex process. A binary word will be converted into an equivalent analog voltage. The concept of *equivalence* between the digital and analog worlds is interesting. The equivalent analog output is equal to the reference voltage multiplied by the ratio of the input binary word to the full-scale binary word. For this reason, DACs that use external reference voltages are called *multiplying DACs.*

9.3.1 Basic Operational Concepts of DACs

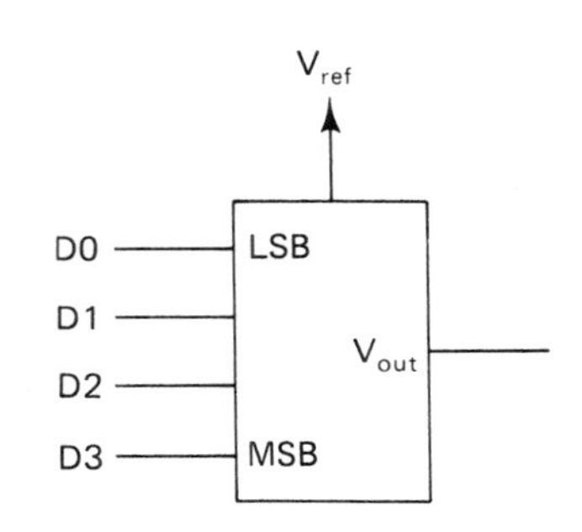

Figure 9.4 Pinout of a 4-bit DAC.

The inputs to a DAC will be a weighted digital word. Depending on the DAC, this word will be 4 to 16 bits wide. Let's examine the pinout of a 4-bit DAC (Figure 9.4). D0 through D3 are the four data bits that will be converted to an analog voltage. The output will be the analog voltage that is equivalent to the data inputs. This range of the analog voltage will be set by the pin labeled V_{ref}. Assume that the reference voltage is tied to 7.5 V. When the digital input is equal to 0000, the output voltage will be equal to 0 V; when the digital input is equal to 1111, the output voltage will be equal to the reference voltage of 7.5 V. A digital input between the minimum (0000) and the maximum (1111) will be equal to a proportional part of the reference voltage.

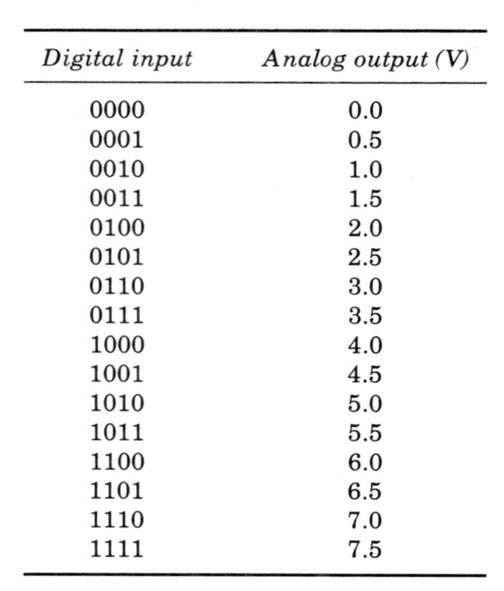

Digital input	*Analog output (V)*
0000	0.0
0001	0.5
0010	1.0
0011	1.5
0100	2.0
0101	2.5
0110	3.0
0111	3.5
1000	4.0
1001	4.5
1010	5.0
1011	5.5
1100	6.0
1101	6.5
1110	7.0
1111	7.5

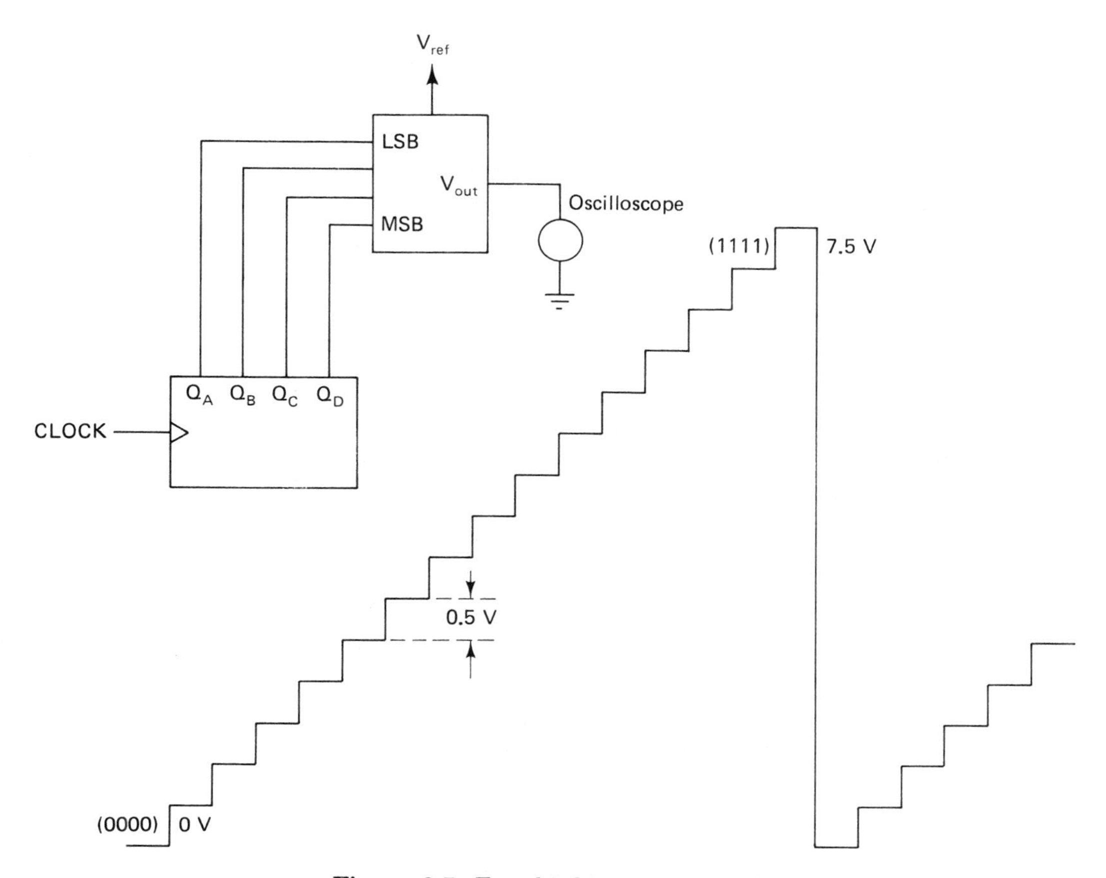

Figure 9.5 Four-bit binary counter driving the inputs of the DAC.

The resolution of a DAC is defined as the smallest incremental change of the analog output. A 4-bit DAC has 16 possible input states. The first input state of 0000 will output 0 V. That leaves 15 input states to be divided between the reference voltage of 7.5 V. The resolution of this 4-bit DAC with a reference voltage of 7.5 V is 0.5 V. Notice that the analog output voltage is proportional to the binary-weighted input. In Figure 9.5, the output of a 4-bit binary counter is driving the digital inputs of the DAC. An oscilloscope connected to the analog output will display a stair-step waveform. When the count begins at 0000 the analog output voltage will be 0 V. Each time the counter is incremented, the analog voltage will increase by 0.5 V. This process will continue until the count of 1111, when the reference voltage of 7.5 V is reached. The next rising edge of the clock will reset the counter to 0000 and the analog output voltage will follow by returning to 0.0 V.

9.3.2 A Summing Op-Amp DAC

A simple DAC can be constructed using an op amp in a summing configuration. Let's quickly review the concept of a summing op amp (refer to Figure 9.6). The most important fact to

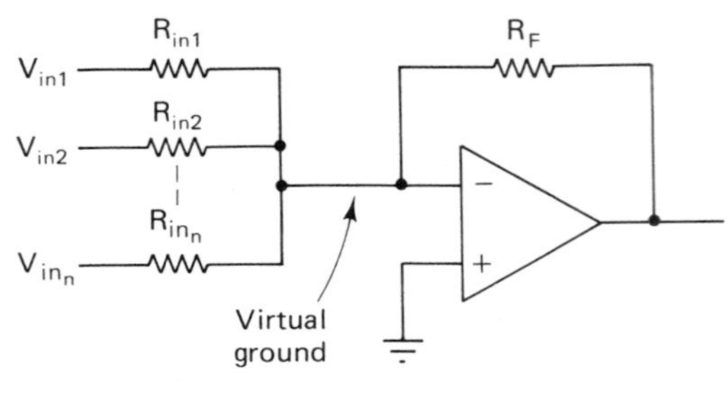

$$I_{sum} = \frac{V_{in1}}{R_{in1}} + \frac{V_{in2}}{R_{in2}} + \cdots + \frac{V_{in_n}}{R_{in_n}}$$

$$V_{out} = -(I_{sum} \times R_F)$$

Figure 9.6 Review of summing-Op-Amp configuration.

remember about the operation of op amps is:

The output will feedback sufficient voltage to the inverting input to assure that the differential voltage between the inverting and noninverting inputs is equal to 0 V. An op amp operating as a linear device must employ negative feedback.

Because the noninverting input is tied to ground, the output of the op amp will source just enough current through the feedback resistor to hold the inverting input at ground potential. The inverting input is said to be at *virtual ground*. That means that the common node of the input resistors appears to be tied to ground.

The input currents will be summed at the virtual ground at the inverting input of the op amp. Because one end of each resistor appears to be tied to ground, the individual currents will equal the input voltage divided by the input resistance. The summed current will then be passed through the feedback resistor, providing a negative feedback path and functioning as a current-to-voltage converter. The output voltage is equal to the voltage drop across the feedback resistor: $I_{sum} \times R_f$. Because this op amp is configured as an inverting amplifier, the polarity of the output voltage will be negative. The summing op amp is useful in applications where many signals must be mixed.

We can use the summing op amp with binary-weighted resistor values to construct a simple DAC. Consider a 4-bit binary number. The place value of the least significant bit of a binary number is equal to 1. The place value doubles for each successive bit. In a summing op amp DAC, the least significant bit should provide the least amount of current, and each successive bit should provide twice the current of the previous bit. Remember that the current is being summed, not the voltage. Because the input current of each bit will be inversely proportional to resistance, the input resistance of the least significant bit should be the greatest; the input resistance of each successive bit should be half of the previous bit. This will double the current for each successive bit. In Figure 9.7, the reference voltage of 7.5 V and an input resistor will create a constant-current source.

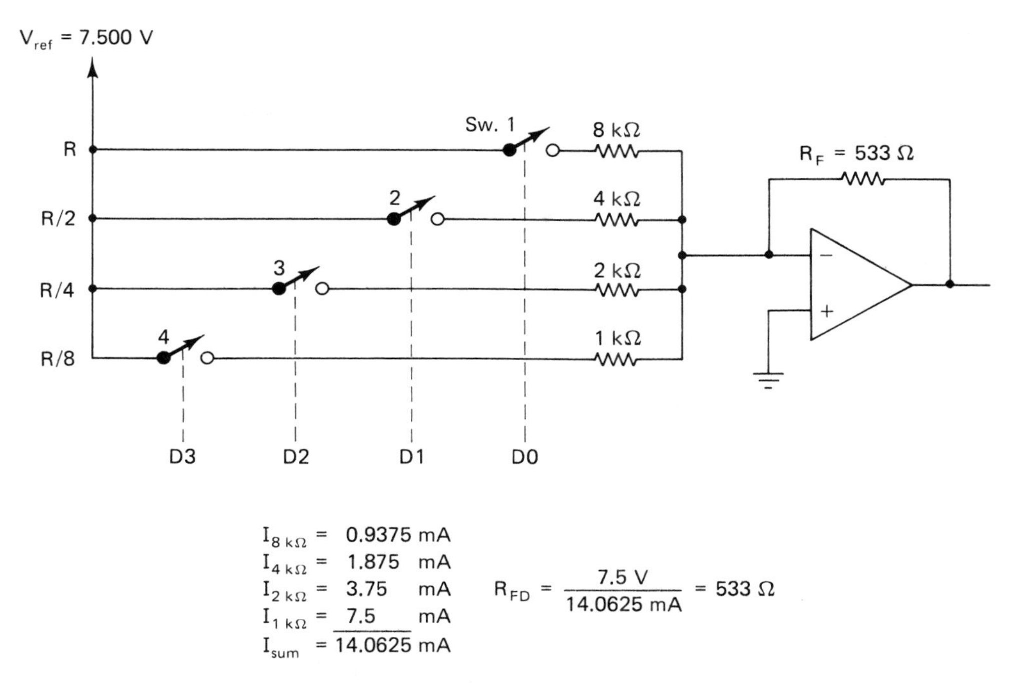

Figure 9.7 Four-bit summing-Op-Amp DAC.

Switches 1 through 4 are CMOS transmission gates. Each input bit will control one constant-current source. A logic 1 will close the CMOS switch and the constant-current source will feed the summing node; a logic 0 will open the CMOS switch. Data bit 0 controls the least amount of current; each subsequent data bit controls twice as much current as the previous data bit.

When the input value of 0000 is placed on the controls of the CMOS switches, no current will flow and the output of the op amp will be 0 V. When the input value of 1111 is placed on the controls of the CMOS switches, the maximum current will flow (14.0625 mA) and the op amp will output −7.5 V. Another op amp can be used to invert the negative output to a positive output.

9.3.3 An R-2R Ladder-Type DAC

The accuracy of the DAC in Figure 9.7 is dependent on the precision of the input resistors and the reference voltage. A 16-bit DAC will require 16 different-value input resistors. This creates a problem: It is not practical to build DACs from discrete parts and precision resistors cannot be manufactured within ICs. We must discover a method of creating constant-current sources that do not require a large number of different-value resistors.

We can construct a DAC of any input width using only two different resistor values (Figure 9.8). The absolute value of each resistor is not important; the critical factor is the 2:1 ratio between the resistors. The mathematical analysis of the R-2R ladder is long and tedious. What you must understand is that the

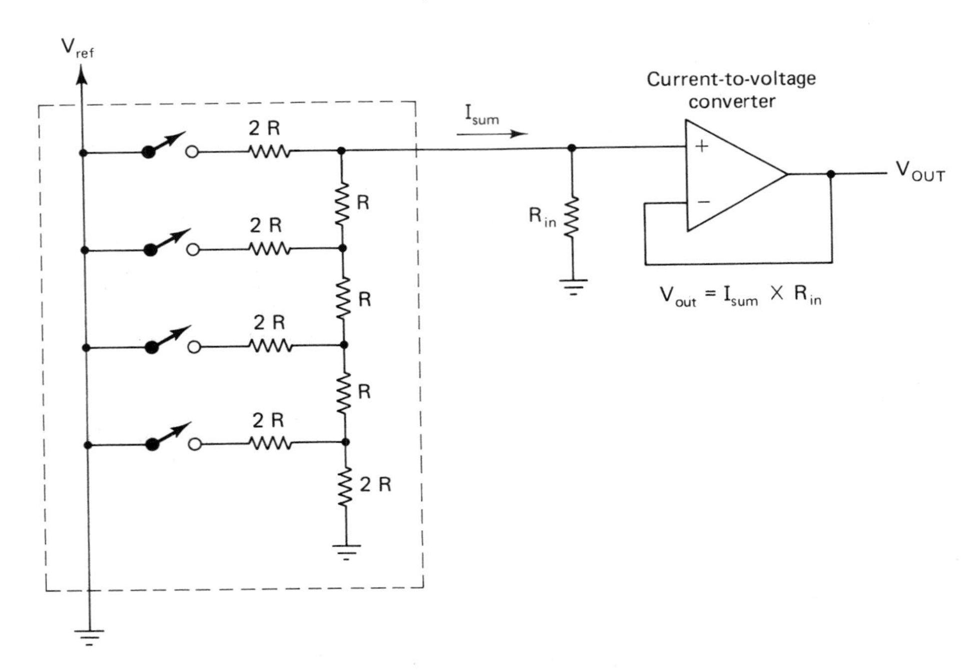

Figure 9.8 R-2R ladder-type DAC.

R-2R ladder functions exactly as the binary-weighted resistors did in the previous DAC. The major advantage of the R-2R ladder is that it requires only two different resistor values. All monolithic DACs (DACs contained in an IC) use an R-2R ladder.

The dashed lines in Figure 9.8 indicate the part of the DAC that is actually contained in an IC—the CMOS transmission gates and the R-2R ladder. The output of most DACs is not voltage, but current. An external op amp with an appropriate feedback resistor must be added to the circuit. Most DACs do not contain an internal voltage reference, so this must also be added to the circuit. DACs that do not contain voltage references are called *multiplying DACs.* This is because the output is always equal to the reference voltage multiplied by a fraction denoting the input digital word.

9.4 THE A/D (ANALOG-TO-DIGITAL CONVERTER)

Analog-to-digital conversion is much more complex than digital-to-analog conversion. A *digital word* is a known quantized value. Here the word "quantized" means a quantity that is composed of discrete values, unlike an analog voltage, which is continuous. Converting a digital word to an analog equivalent is a simple process. Analog-to-digital conversion begins with a continuous voltage. The A/D converter must quantize this continuous voltage into an exact digital value.

There are four popular methods of converting analog voltages to digital values. Their trade-offs are speed, accuracy, and cost. Each method of A/D has an appropriate area of application.

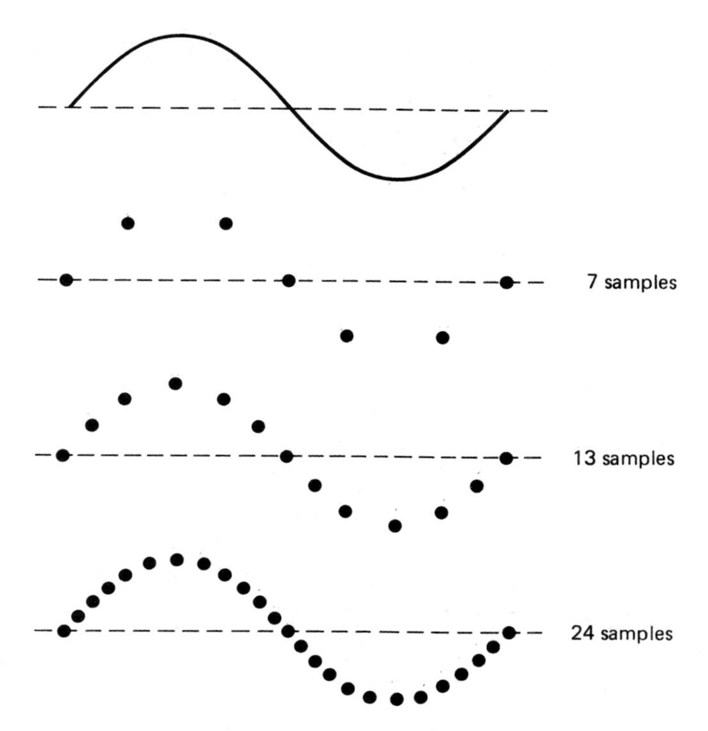

Figure 9.9 Digitized sine waves.

9.4.1 A Digitized Waveform

A sine wave is a continuous function. One cycle of a sine wave is constructed from an infinite number of points. When we digitize a waveform (convert it from analog to digital), we are assigning a digital value to selected points of the waveform. If we sample enough points, the digitized waveform will be a recognizable copy of the original. If we sample too few points, the digitized waveform will appear as nothing more than a series of unrelated points. Refer to Figure 9.9.

The intelligibility of a graph is proportional to the number of points sampled. Because the sine wave is a smooth and continuous function, even the graph with only seven samples is fairly recognizable. A graph depicting the acceleration of a revolving body would be highly complex and impossible to reconstruct with only a few samples. In general, the sampling frequency should be at least 100 times greater than the frequency of the waveform that is being digitized.

Every analog sample must be assigned a numeric value. That is the function of an A/D. An A/D will have one analog input and a group of digital outputs (the number of which depends on the particular A/D). The precision of an A/D is directly proportional to the number of digital outputs.

After a sample has been digitized, it can be stored into memory or manipulated in many other ways by the microprocessor. Conventionally, music is recorded in an analog format and degrades with use. Music stored as a sequence of digital words cannot get noisy or degrade; the digital words are a quantized value, not an analog representation of the original music.

9.4.2 A Counter/Ramp-Type A/D

A DAC is often one of the components in an A/D system. The counter/ramp type of A/D uses a counter, voltage comparator, and a DAC to convert an analog sample into a digital word (Figure 9.10). You have accumulated enough electronics knowledge to completely analyze the operation of the circuit shown in Figure 9.10. Before continuing, take a moment to try your hand at creating a theory of operation that describes the counter/ramp type of A/D converter.

Let's analyze the function of the individual devices before attempting to describe the system interaction.

Voltage Comparator When the analog input voltage is greater than the voltage output of the DAC, the output of the comparator will be a logic 1. This logic 1 will enable the 74LS161 counter. When the voltage out of the DAC becomes greater than the analog input voltage, the output of the comparator will go to a logic 0 and the counter will be disabled. The falling edge of the voltage comparator will also fire the one-shot. The output of

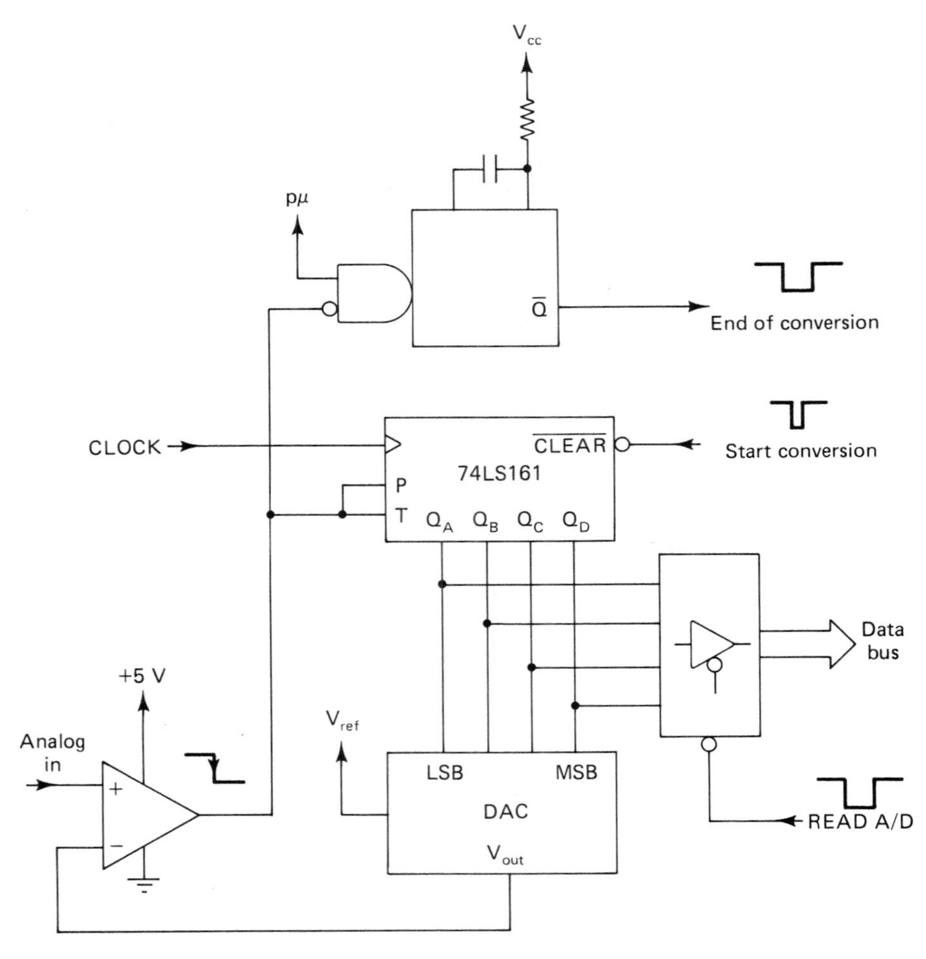

Figure 9.10 Counter/ramp-type A/D converter.

the one-shot is a negative pulse which indicates that the A/D conversion is complete.

Counter The counter will be enabled when the output of the voltage comparator is a logic 1. The 4-bit output of the counter is applied to the inputs of the DAC and the three-state buffer.

DAC The 4-bit DAC will reconvert the digital output of the counter into an analog voltage. The output of the DAC and the analog input voltage will be compared. This is an example of a closed-loop system. The DAC feeds the output of the A/D converter back to the input. The counter will continue to count until the output of the DAC becomes slightly greater in amplitude than the analog input voltage.

Three-State Buffer The output of the three-state buffer is connected to the microprocessor's data bus. Like any bus talker, the output of this three-state buffer will stay at high-Z until it is enabled by the microprocessor.

One-Shot As indicated in the description of the voltage comparator, the one-shot will output a negative pulse when the comparator outputs a falling edge. The negative pulse will indicate to the microprocessor that the A/D conversion is complete.

Let's consider the system interaction as an analog-to-digital conversion is processed.

Event 1. Assuming that the analog voltage to be converted is on the noninverting input of the voltage comparator, the microprocessor will issue a negative-going "start conversion" pulse. This pulse will clear the Q outputs of the counter.

Event 2. When the outputs of the counter are cleared, the analog output of the DAC will go to 0 V. This will cause the output of the comparator to go high.

Event 3. The counter is now enabled. At each rising edge, the counter will be incremented. This updated count will be converted to analog by the DAC, and then compared with the analog input. The counter will continue to increment until the output of the DAC becomes greater than the analog input voltage.

Event 4. When the output of the DAC becomes greater than the analog input, the output of the comparator will go low. This action will freeze the counter and fire the one-shot. The output of the counter is the digital equivalent of the analog input voltage.

Event 5. The one-shot will fire a negative pulse to alert the microprocessor that the A/D conversion is complete. Remember that the output of the counter is frozen. As long as the analog input voltage does not increase, the counter will stay frozen with the digitized value.

Event 6. The microprocessor will place a negative pulse on the enable input of the three-state buffer. This will momentarily pass the outputs of the counter onto the data bus. The microprocessor will store this value in a memory location.

The A/D conversion cycle is now complete, and the circuit is ready to perform another conversion.

Now that we understand how the system functions, we must consider some of its operational concepts. The conversion time is directly proportional to the amplitude of the analog input. If the amplitude of the analog input is greater than the full-scale output voltage of the DAC, the counter will never be disabled and the A/D-conversion one-shot will never fire.

The counter/ramp A/D derives it name from the output of the DAC. The DAC will output a linear voltage ramp whose peak amplitude is slightly greater than the analog input. If more resolution is required, another 4-bit counter and an 8-bit DAC can be used to increase the output of this A/D to a full byte.

9.4.3. Dual-Slope A/D Conversion

One of the most accurate (but also slowest) A/D conversion methods is the dual-slope A/D. It is widely used in products that do not require fast conversion speeds, such as digital voltmeters.

The key to dual-slope A/D is the integrator. An integrator circuit continuously sums a current and stores it as a charge across a capacitor. The op amp is the major component in an integrator. Figure 9.11 is a review of the op-amp integrator. The integrator illustrated is an inverting integrator. A positive V_{ref} will drive the output voltage negative, and a negative V_{ref} will drive the output voltage positive. A reference voltage and resistor are used to create a constant-current source. The output of this constant-current source is used to charge the summing capacitor in the feedback loop.

The voltage on the capacitor is proportional to the magnitude of the constant current and the length of time that it has been charged, and is inversely proportional to the value of the capacitor. That means that the greater the magnitude of current from the constant current source and the longer the capacitor is charged, the greater the voltage on the capacitor. Similarly, the larger the capacitor, the more time it requires to charge up to a particular voltage.

Refer to the graph in Figure 9.11. The slope of the graph is equal to the constant current divided by the capacitance. The slope can be easily modified by varying the current—if the current increases, the slope gets steeper; if the current decreases,

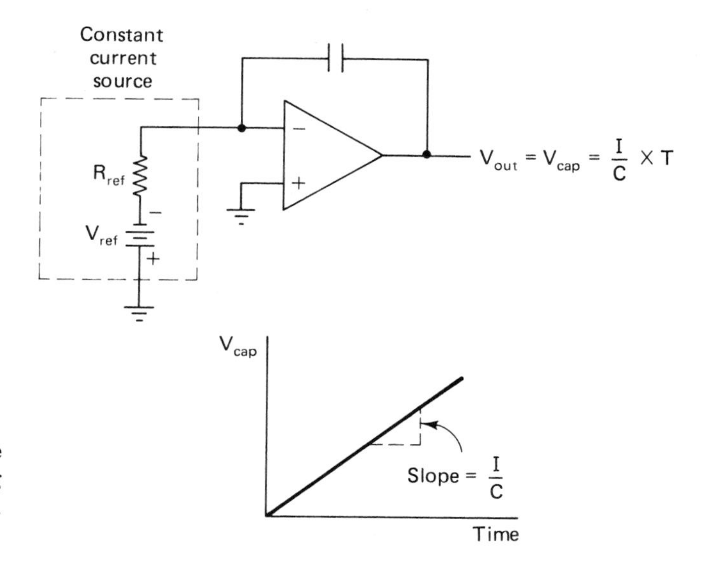

Figure 9.11 Review of the Op-Amp inverting integrator.

the slope becomes more gradual. This is the basis of the dual-slope A/D. The first half of the graph in Figure 9.12 illustrates three different constant currents integrated for a fixed period of time. The second half of the graph illustrates the output voltage as it is discharged by a known reference current until V_{cap} is equal to 0 V.

Remember that the slope of the charging graph is proportional to the magnitude of the constant-current source. The largest constant current is labeled 1 and the smallest current is labeled 3. At the end of the fixed charge period, the voltage across the capacitor is proportional to the magnitude of the constant charging current.

What would happen if we discharged the capacitor with a known reference current of the opposite polarity? The length of time required to discharge the capacitor to 0 V would be proportional to the magnitude of the original charging current. The graph indicates that the discharge slope of all three lines is equal. The largest-magnitude constant current required the

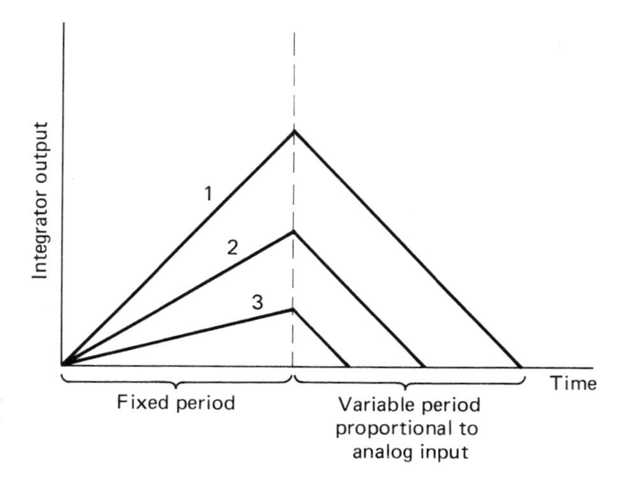

Figure 9.12 Integrator with fixed charge and variable discharge intervals.

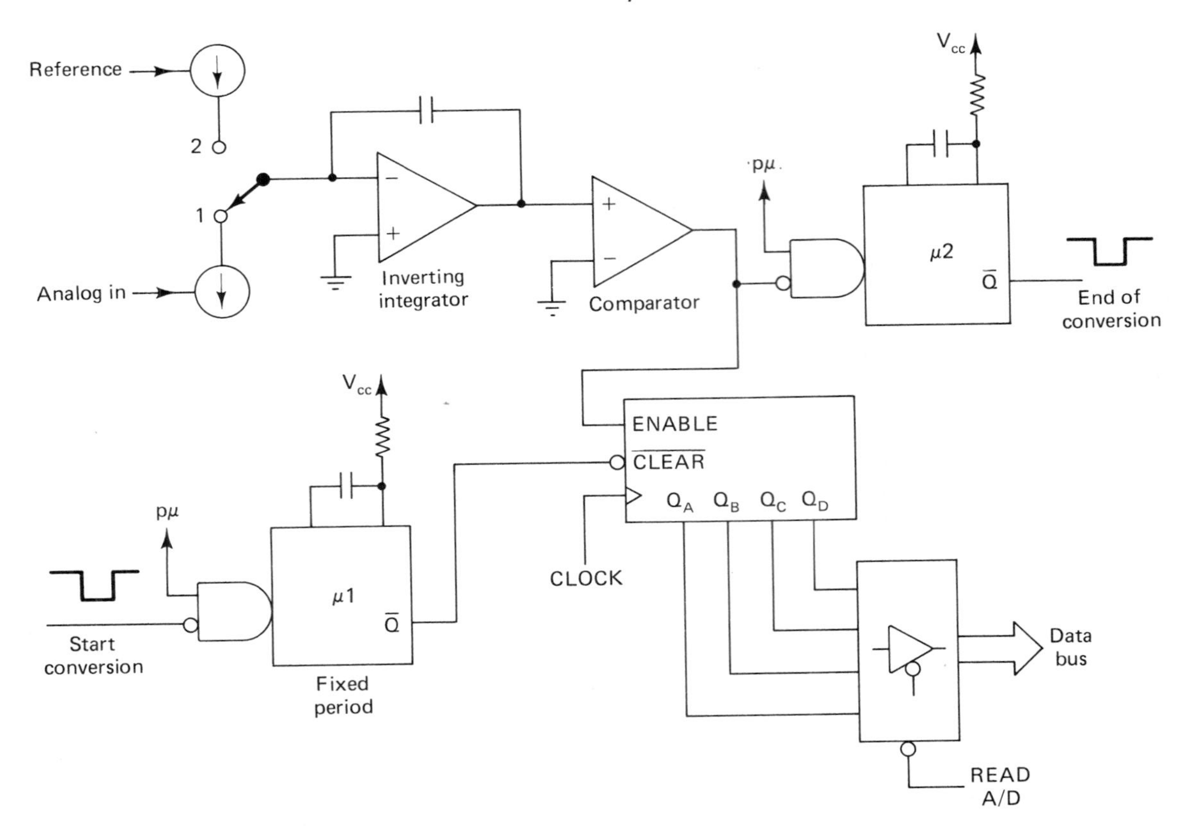

Figure 9.13 Model of dual-slope A/D.

longest discharge time, and the smallest-magnitude constant current required the shortest discharge time.

Finally, consider what would happen if we enabled a counter from the beginning of the discharge time until the capacitor was discharged to 0 V. The digital output of the counter would be proportional to the original constant current. We have just completed an analog-to-digital conversion. Figure 9.13 illustrates a highly simplified model of a dual-slope A/D. The schematic symbol for a current source is an arrow enclosed in a circle. Figure 9.13 depicts two current sources: the output of the top current source is proportional to a reference voltage; the output of the bottom current source is proportional to the analog input voltage.

Let's consider the operation of this dual-slope A/D.

Event 1. The microprocessor will initiate the conversion by outputting a start-conversion pulse. The falling edge of the pulse will trigger the one-shot. The $\overline{Q}$ of the one-shot will go low for a specified interval. The length of the negative pulse will be the fixed interval during which the integrator will sum the current from the variable current source. When the $\overline{Q}$ output of the one-shot is low, the counter will be cleared and inhibited from counting.

Event 2. $\overline{Q}$ of the one-shot will go high at the end of the fixed period. The CMOS switch will toggle to position 2, and because the integrator was charged to a positive voltage, the output of the comparator will be a logic 1. The counter is now enabled.

The reference current source will discharge the integrator to 0 V. The counter will convert the period required to discharge the integrator 0 V into a digital value.

Event 3. When the noninverting input of the comparator goes slightly negative, the output will toggle to a logic 0. This falling edge will trigger the end-of-conversion one-shot, and the logic 0 level will freeze the output of the counter.

Event 4. The microprocessor will issue a read A/D command, and the output of the counter will be passed onto the data bus, thus concluding the dual-slope A/D operation.

9.4.4 Successive-Approximation A/D Conversion

Consider a simple balance scale. It consists of two metal plates supported in the center by a vertical post, and a set of calibrated weights. For this example the balance scale will have calibrated weights of 1, 2, 4, and 8 ounces. Notice that the calibrated weights increase in powers of 2 (just like the binary number system).

The object to be weighed will be placed on one of the metal plates. The largest calibrated weight will then be placed on the opposite plate. If the scale tips in the direction of the calibrated weight, it is too large and must be removed; if the scale stays tipped in the direction of the unknown weight, the calibrated weight will be left on the scale. This process is repeated for each of the calibrated weights, starting with the heaviest and continuing to the lightest.

To find the weight of the unknown object, the calibrated weights that were left on the plate must be summed. With the calibrated weight set of 1, 2, 4, and 8 ounces, an object up to 15 ounces can be weighed with a resolution of 1 ounce.

The act of weighing an object can be thought of as an analog-to-digital conversion! That means we are assigning a numerical (digital) value that describes the weight (analog parameter) of the object. The block diagram in Figure 9.14 illustrates a successive-approximation A/D converter based on the concept of the balance scale. It is the electronic version of the balance scale. Like the counter/ramp type of A/D, the successive-approximation method uses a DAC and a voltage comparator. Operation of the successive-approximation A/D centers around the SAR (successive approximation register). The SAR contains a shift register and control circuitry.

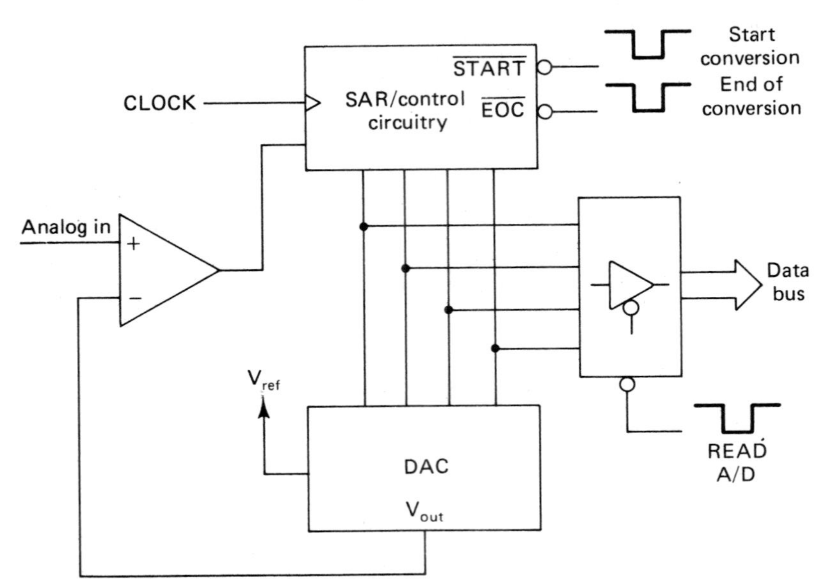

Figure 9.14 Successive-approximation A/D converter.

After the microprocessor initiates a conversion by pulsing the start-conversion input, the SAR will shift a logic 1 into the most significant bit of the output. This digital value (1000) will be converted by the DAC and compared to the analog input. If the output of the voltage comparator stays at a logic 1 level, the SAR will leave the most significant bit set; if the output of the voltage comparator toggles to a logic 0, this indicates that the digital output is too large and the SAR will reset the most significant bit.

This process will be repeated for the remaining 3 bits. The SAR will then pulse the EOC (end-of-conversion) output low, which indicates to the microprocessor that the conversion is complete. The microprocessor will respond by taking the read A/D line low. The three-state buffer will pass the outputs of the SAR onto the data bus. The conversion cycle is then complete.

Unlike the counter/ramp-type or dual-slope-type A/Ds, the successive-approximation A/D will always complete a conversion in the same period of time, regardless of the amplitude of the analog input. Successive-approximation A/D is fast and finds applications where many samples must be taken in short periods of time.

9.4.5 The Flash (Simultaneous) A/D Converter

The fastest A/D converter is the flash type. The other methods of A/D require a number of passes or clock cycles to complete the conversion. The flash A/D operates in parallel, so all bits are converted simultaneously. Figure 9.15 illustrates a 3-bit flash A/D. For an output of N bits, 2^N resistors and 2^{N-1} voltage comparators are required. The 3-bit flash converter in Figure

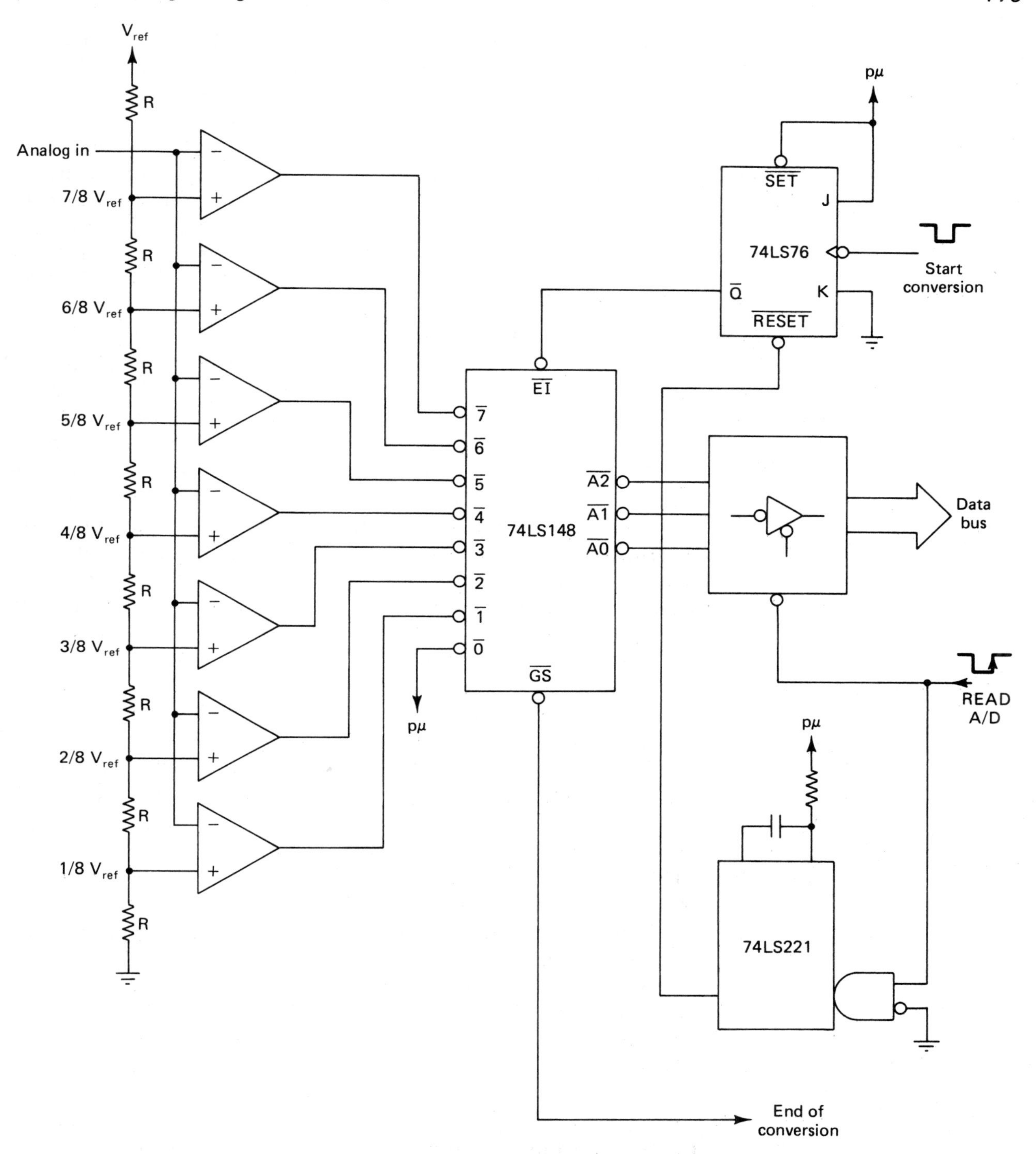

Figure 9.15 Flash-type A/D converter.

9.15 has eight resistors and seven voltage comparators. The output of the seven voltage comparators drive the inputs of an 8-line-to-3-line priority encoder. The active-low outputs of the priority encoder are inverted by the three-state buffer.

The eight resistors are all equal. The absolute value of each resistor is not important, but it is extremely important that all the resistors be closely matched within a few tenths of a percent. The eight resistors constitute a precision voltage divider. The noninverting input of each voltage comparator will be a multiple of one-eighth of the reference voltage.

The analog input voltage is connected to the inverting input of each voltage comparator. If the analog input voltage is greater than the reference voltage of a particular voltage comparator, that comparator (and all other comparators with smaller reference voltages) will go to an active-low output level. The highest-priority active low will then be encoded into a 3-bit binary output by the 74LS148.

Let's assume that the reference voltage is 8.000 V. The divided-down reference voltages will be 1.000 (at comparator 1) through 7.000 (at comparator 7). We will apply a voltage ramp to the analog input of the flash A/D. The ramp will start at 0 V and rise to 7.999 V. *Note:* Theoretically, the inputs to the voltage comparator can never be equal. For this reason, the previous table does not indicate any ramp voltage that is equal to a reference voltage. Notice that the output of the inverting three-state buffer is the digital equivalent of the analog input. A priority encoder is needed because the output of a less significant voltage comparator does not toggle back to an inactive state when the next significant comparator goes active low.

	Output of comparators							*Output onto data bus*		
Ramp voltage	*7*	*6*	*5*	*4*	*3*	*2*	*1*	*B2*	*B1*	*B0*
0.000–0.999	0	0	0	0	0	0	0	0	0	0
1.001–1.999	0	0	0	0	0	0	1	0	0	1
2.001–2.999	0	0	0	0	0	1	1	0	1	0
3.001–3.999	0	0	0	0	1	1	1	0	1	1
4.001–4.999	0	0	0	1	1	1	1	1	0	0
5.001–5.999	0	0	1	1	1	1	1	1	0	1
6.001–6.999	0	1	1	1	1	1	1	1	1	0
7.001–7.999	1	1	1	1	1	1	1	1	1	1

A conversion will be initiated by the microprocessor pulsing the clock input of the J-K flip-flop. The J-K flip-flop is configured to set on the falling edge of a clock pulse. The $\overline{Q}$ output will go low, driving the enable input of the 74LS148 to an active-low level. The outputs of the comparators will be encoded into a 3-bit active-low code by the 74LS148. After the inputs are encoded, the group select output will go low to indicate that the conversion is complete. The microprocessor will then pulse the read A/D line. The reinverted binary code will appear on the data bus and the rising edge of the read A/D signal will fire the one-shot that resets the J-K flip-flop.

9.4.6 Summary of A/Ds

Practical A/Ds are not constructed from discrete devices. They are available in two different packages: monolithic (integrated circuits) and hybrids. Hybrid devices are sealed packages that contain a combination of integrated circuits and discrete components. Let's summarize each type of A/D converter.

Counter/Ramp Practical counter/ramp A/Ds use an up/down counter. This enables the device to continually track the analog input voltage. Tracking counter/ramp A/Ds are used in medium speed/medium performance circuits and are usually manufactured in hybrid packages.

Dual Slope Although dual slope is the slowest form of A/D conversion, it is also one of the most popular. Dual-slope A/Ds are simple, accurate, and inexpensive. They also have exceptional noise immunity. Most digital voltmeters use a dual-slope A/D.

Successive Approximation Successive-approximation and dual-slope A/Ds constitute over 90% of A/Ds in use today. The conversion speed of successive-approximation A/D conversion is limited by the settling time required for the feedback DAC. High-performance, two-step successive approximation A/Ds change 2-bits at a time. This effectively doubles the conversion speed. Successive-approximation A/Ds are used in moderate-to-high-speed applications.

Flash (Known Also as Simultaneous or Parallel A/Ds) An 8-bit flash converter requires 255 voltage comparators, 256 resistors in the precision divider, and the equivalent of a 255-line-to-8-line priority encoder. This complexity results in such a high cost that flash converters are only used in super-high-speed video applications in excess of 20 MHz.

QUESTIONS AND PROBLEMS

9.1. Complete the timing diagram for the noninverting Schmitt trigger shown in Figure P9.1.

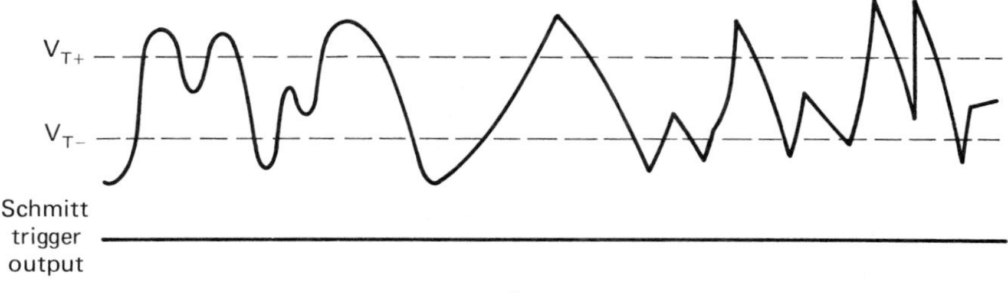

Figure P9.1

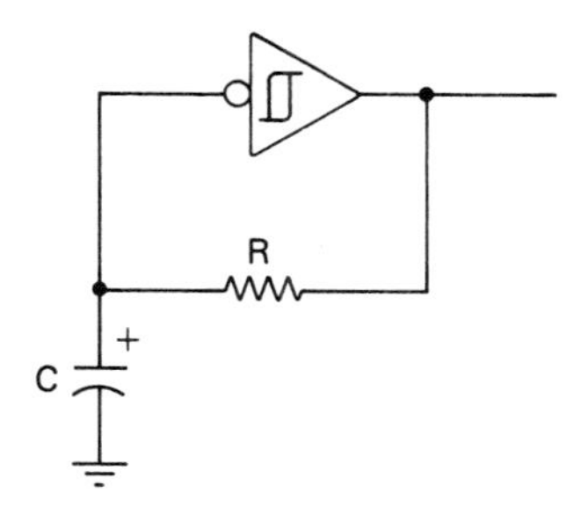

Figure P9.2

9.2. Describe the output of the circuit shown in Figure P9.2. (*Hint:* Remember that the capacitor will appear as a short to ground when V_{cc} is first applied.)

9.3. Describe the output of the circuit shown in Figure P9.3 when SW 1 is closed.

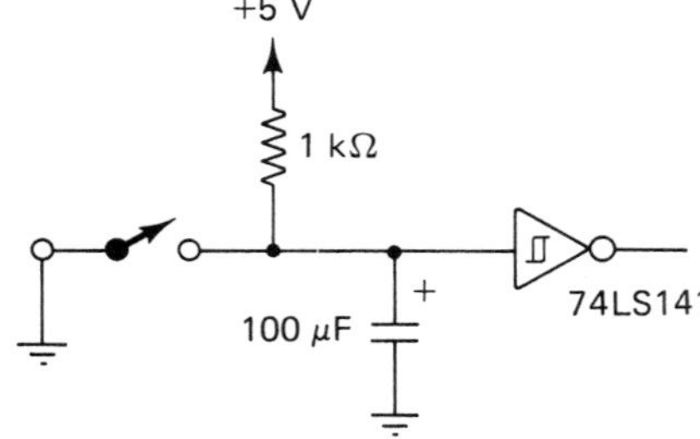

Figure P9.3

9.4. The diode in Figure 9.1 is installed backwards. What is the output of the circuit? The diode is shorted. What is the output now? What would the output be if the diode was open?

9.5. Define the specifications V_{T+} and V_{T-} as they pertain to the Schmitt trigger.

9.6. In a data acquisition circuit, how would a thermistor be classified?

9.7. Name two devices in which you have seen a capacitor used as a storage element.

9.8. Refer to Figure 9.5. Q_d is shorted to ground. Draw the output of the DAC.

9.9. Explain the concept of a virtual ground as it pertains to op-amp circuits.

9.10. How is the op amp in Figure 9.8 performing a current-to-voltage conversion?

9.11. A square wave with an amplitude of V_{ref} is applied to the input of an A/D converter. What is the output waveform? A triangular waveform with a peak amplitude of V_{ref} is applied to the input of an A/D converter. What is the output waveform?

Refer to Figure 9.10 for Problems 9.12 to 9.14.

9.12. Draw a timing diagram that describes the events that constitute an A/D conversion cycle.

9.13. The microprocessor issues a start conversion command, and the end-of-conversion pulse never occurs. What is the most likely problem?

9.14. The enable input of the three-state buffer is shorted to ground. How does this affect system operation?

9.15. Where does the dual-slope A/D derive its name?

9.16. Refer to Figure 9.9. The one-shot that provides the fixed period has too much drift for this high-precision circuit. Given a 1-MHz clock, design a circuit that provides a 256-μs fixed time period. (Hint: Consider a counter.)

9.17. Describe how a successive-approximation A/D functions.

9.18. What limits the speed of a successive-approximation conversion cycle?

Refer to Figure 9.15 for Problems 9.19 to 9.22.

9.19. The output of comparator 6 is stuck low. How does this affect circuit operation?

9.20. The EI of the 74LS148 is shorted to the A2 output. How does this affect circuit operation?

9.21. EI is shorted to ground. How does this affect circuit operation?

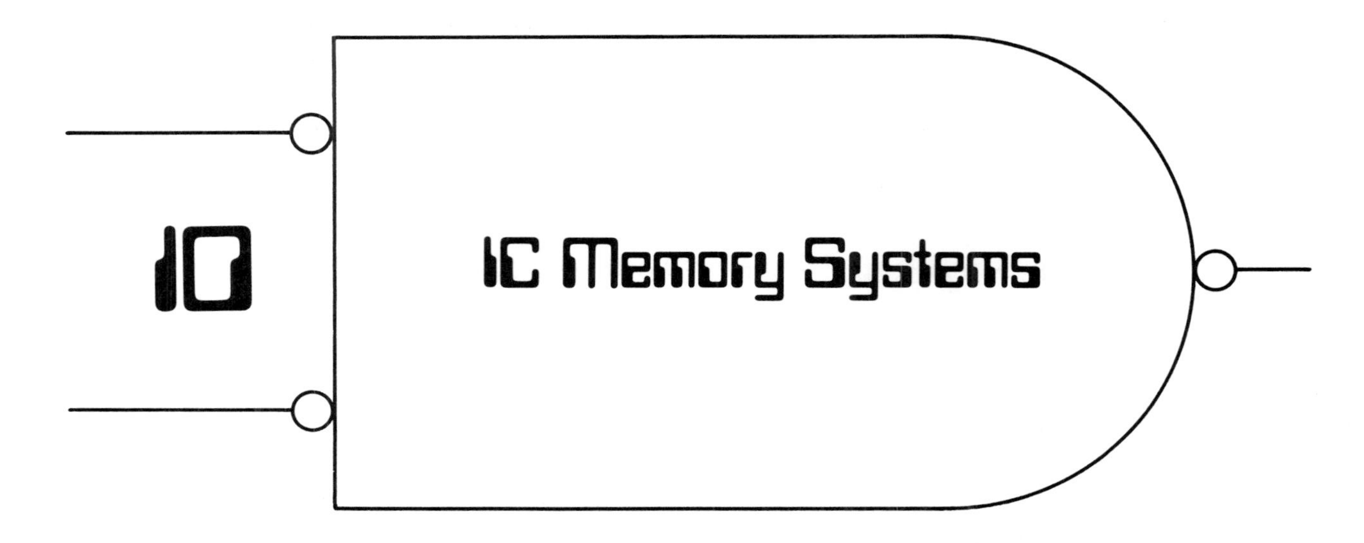

In the block diagram of the compter we looked at earlier, one block was labeled "memory." We stated that memory was used to store programs and data. Earlier, we discovered that sequential devices have the capacity to "remember" a digital bit of information. We have also used latches and D flip-flops as memory devices in many different applications.

Computers employ many different types of memory devices. Some memory devices are designed to hold large quantities of data. These devices are mechanically driven and store the digital information on some type of magnetic media. IC memories are constructed from D flip-flops or MOS capacitors.

10.1 A 16-BIT READ/WRITE MEMORY

Let's consider each of seven input and output lines that interface with the circuit in Figure 10.1. The data line is bidirectional. During the write operation, it carries a bit of information from the microprocessor to be stored in the memory; during the read operation it carries a bit of information retrieved from the memory to the microprocessor. The D inputs of each flip-flop and the output of three-state buffer G2 are connected to the data line.

The $\overline{RD}$ and $\overline{WR}$ lines come from the control bus of the microprocessor. When $\overline{RD}$ goes low, a memory-read operation is des-

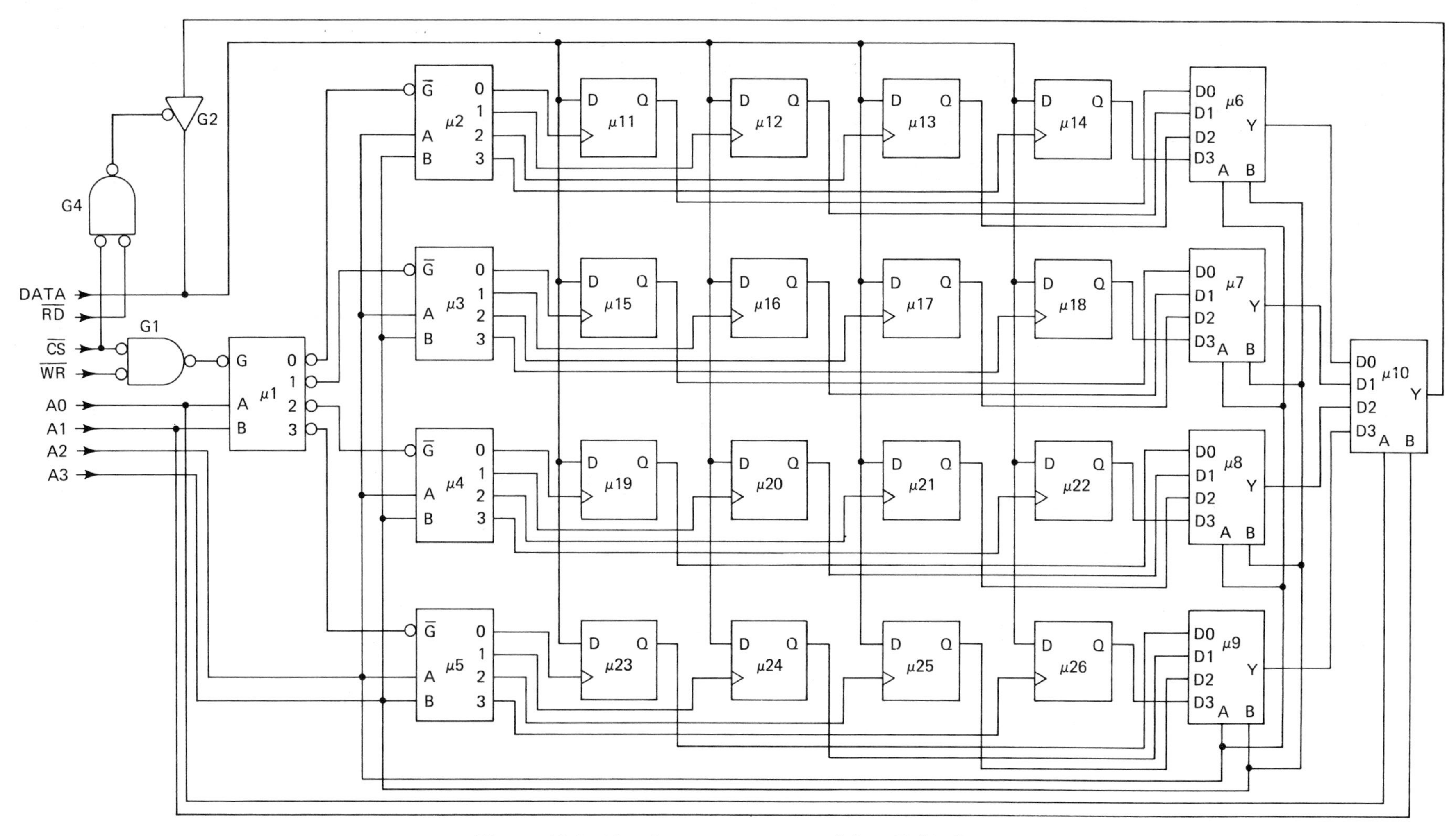

Figure 10.1 16 × 1 memory constructed from D flip-flops.

ignated; when $\overline{WR}$ goes, low a memory-write operation is designated. The $\overline{RD}$ line is connected to an input of G4. If the $\overline{RD}$ line is at a logic 1 level, the output of G2 will be high-Z. $\overline{WR}$ is connected to an input of G1. If $\overline{WR}$ is at a logic 1 level, U1 will be disabled.

The line designated $\overline{CS}$ stands for "chip slect." This active-low signal functions like an enable input. If chip select is at an inactive-high level, the memory is disabled; it cannot be accessed by the microprocessor for either a read or a write operation. If the chip select goes active low, the memory is selected and can be accessed by the microprocessor. The $\overline{CS}$ is applied to gates G1 and G4. If the $\overline{CS}$ is high, G2 will be at high-Z and U1 will be disabled.

Earlier you learned that a typical microprocessor has 16 address lines. In this example, only the first four are used. Four address lines can uniquely address 16 memory locations. A0 and A1 are supplied to a 2-line-to-4-line decoder U1 and 1-of-4 data selector U10. A2 and A3 are connected to the 2-line-to-4-line decoders U2 through U5 (with active-high outputs) and 1-to-4 data selectors U6 through U9.

This circuit has 16 memory locations; each D flip-flop (U11 through U26) constitutes one memory cell. Notice that the 16 flip-flops are arranged in a square matrix of four rows by four columns. The rows are labeled 0 through 3. Flip-flops U11 through U14 constitute row 0; U15 through U18 row 1; and so on. The columns are also labeled 0 through 3. Flip-flops U11, U15, U19, and U23 constitute column 0; U12, U16, U20, and U24 column 1; and so on.

Each of the 16 flip-flops can be uniquely identified by a row-and-column designation.

Column	*Row*	*Flip-flop*	*A3 A2 column*	*A1 A0 row*
0	0	U11	00	00
0	1	U15	00	01
0	2	U19	00	10
0	3	U23	00	11
1	0	U12	01	00
1	1	U16	01	01
1	2	U20	01	10
1	3	U24	01	11
2	0	U13	10	00
2	1	U17	10	01
2	2	U21	10	10
2	3	U25	10	11
3	0	U14	11	00
3	1	U18	11	01
3	2	U22	11	10
3	3	U26	11	11

The last two columns, labeled "column" and "row," designate the actual binary address of each memory cell. Notice that the lower 2 bits are the row address and the upper 2 bits are the column address.

The function table of a D flip-flop indicates that if the set and preset inputs are held at inactive levels, the Q output can only change on the rising edge of the clock input. During a write operation, the data that will be stored in memory is placed on the input of every D flip-flop in the memory matrix. The row and column decoders will steer a rising edge to the clock input of the addressed memory cell. In this manner, only the addressed memory cell will be affected.

During a write operation, the output of three-state buffer G2 must be held at high-Z. This is because during the memory-write operation, the memory device is a bus "listener" and must not output any logic levels onto the bus. The only time a memory device will become a bus "talker" is during a read operation. The $\overline{\text{RD}}$ input must be held at an inactive-low level during the write operation.

Let's examine a typical write operation. We will analyze each event in the write-cycle timing diagrams (Figure 10.2) as it is carried out in the circuitry of Figure 10.1. Assume that the $\overline{\text{RD}}$ input is held high during the write operation.

Event 1. The 4-bit address designating a unique memory location is placed on the address bus (A0 through A3). A0 and A1 designate the row address; A2 and A3 designate the column address.

Event 2. The $\overline{\text{CS}}$ input goes active low. This indicates that the memory is now selected.

Event 3. The microprocessor places the data bit to be stored into memory on the data input.

Event 4. The $\overline{\text{WR}}$ input goes active low. The active-low $\overline{\text{CS}}$ and $\overline{\text{WR}}$ inputs cause the output of G1 to go active low; U1 is now enabled. U1 is called the *row decoder.* Each of the four decoded outputs of U1 are associated with a particular row of

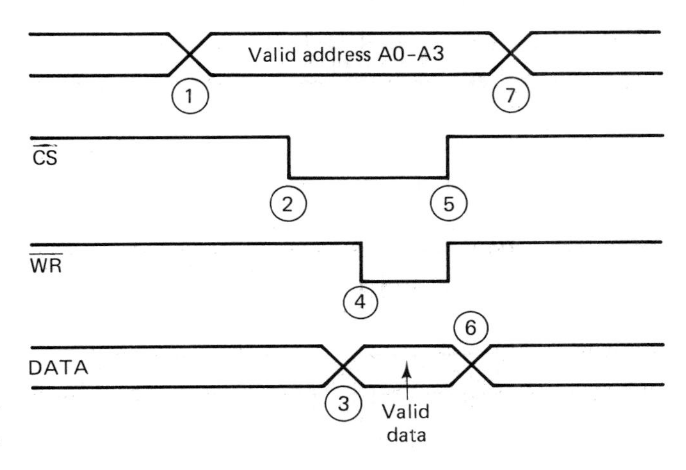

Figure 10.2 Typical write-cycle timing diagram.

memory cells. The row address of A0 and A1 will force a decoded output of U1 to an active-low level.

U2 through U5 are the *column decoders.* Now that a row has been selected by the output of U1, a column within that row must be selected. The column address is applied to column decoders. Notice that these 2-line-to-4-line decoders have active-high outputs. When the selected column output goes to an active-high level, the rising edge on the clock input of the addressed memory cell will transfer the bit on the D input onto the Q output. The data bit has now been stored in memory.

Events 5 through 7. These show the address, data, and control lines returning to inactive states.

Notice that the data must be given a short period to stabilize before the $\overline{WR}$ input goes active low. If this period is too short, the wrong logic level may be stored. The data input also has to be held for a short time after the falling edge of $\overline{WR}$. This is due to the setup and hold times of flip-flops, which we examined in Chapter 8.

After the falling edge of $\overline{WR}$, the propagation times of the OR gate (G1), row decoder, column decoder, and a flip-flop must occur before the data are actually written. Remember these delays when we encounter actual IC memory devices.

Let's consider a typical memory-read operation. Assume that the $\overline{WR}$ input is held at an inactive level during the read cycle. Refer to Figure 10.3

Event 1. The address of the memory cell to be read is placed on A0 through A3 by the microprocessor.

Event 2. The chip select goes to an active-low level. U1 will be disabled throughout the read operation. Flip-flops need not be clocked to be read.

Event 3. The $\overline{RD}$ input goes active low. The active-low $\overline{CS}$ and $\overline{RD}$ inputs will cause the output of G4 to go active low. G2 is now enabled to pass data.

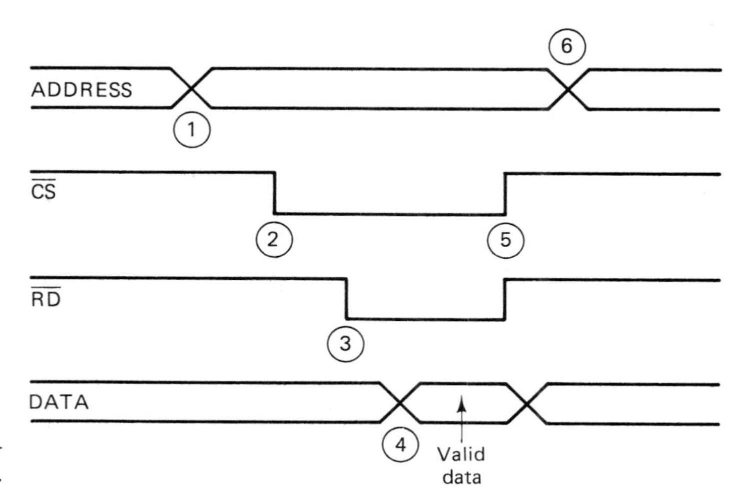

Figure 10.3 Typical read-cycle timing diagram.

G2 is driven by the output of data selector U10. U10 is the memory-read counterpart of U1. U1 is the row writer decoder, and U10 is the row read data selector. The select inputs of both U1 and U10 are driven by the row address A0 and A1.

U6 through U9 are the counterparts of U2 through U5. U6 through U9 are the column read data selectors. The selects of U2 through U5 and U6 through U9 are driven by the column address bits A2 and A3.

The Q output of the selected memory cell will first be transferred through a column read data selector. Then it will be transferred through the row read data selector and placed on the input of G2. This read process will be subject to the propagation delay of a column read data selector, the row read data selector, and G2.

Event 4. After the sum of all the propagation delays, the valid data appear on the output of G2.

Event 5. The $\overline{CS}$ and $\overline{RD}$ inputs are taken inactive high. After a slight delay the output of G2 will return to high-Z.

Event 6. The microprocessor changes the address on A0 through A3.

Constructing memories from discrete flip-flops is impractical. This 16-bit memory has been used to illustrate some important points concerning IC memories. Memories are constructed in square matrixes, where the lower half of the address bits represent a row address and the upper half of the address bits represent a column address. You can now appreciate the complex timing parameters that describe the operation of IC memory devices.

10.2 OVERVIEW OF MEMORY DEVICES

The memory that we have just examined is constructed from D flip-flops. There are many other types of memory devices used in computer systems. Some memory devices are fast but have limited storage capacity. Others have vast storage capacities but are exceedingly slow. Yet others are a compromise between these two extremes.

10.2.1 Mass Memory

The 16-bit memory that we constructed with D flip-flops is said to be a volatile. When V_{cc} is turned off, all the information stored in the D flip-flops is lost. *Mass memory devices* are nonvolatile. They retain information without the need of V_{cc}. Mass memory devices store digital information on magnetic media. As the name implies, mass memory devices are used to store large quantities of data. Because mass memory devices are mechanical in nature, they operate too slowly to be accessed directly by microprocessors. The digital information stored in mass memory devices must first be loaded into semiconductor memory before it can be used by the microprocessor.

Many inexpensive home computers use audio cassette recorders as mass memory devices. They store logic 0's and logic 1's as different tones. Cassette storage is inexpensive, but it is also extremely slow. Many manufacturers build high-speed digital cassette drives designed specifically for computer applications, but they are much more expensive than standard audio cassette drives.

Cassettes are an example of serial memories. Data are stored in a long line of sequential bits. One cassette may have 10 programs stored on it. If you desire to place the eighth program into the computer's memory, you must first read sequentially through the first seven programs. This can be a slow and frustrating process.

The 16-bit memory that we constructed from D flip-flops is called a *random-access memory.* If we want to read the memory cell in the second row, third column, we can access it directly without having to read through the other memory cells. Every location in a random-access memory takes the same length of time to access. The period required to access information in a serial memory device depends on the location of the information.

Large computers use magnetic tape mass memories. You have surely seen pictures of large reel-to-reel magnetic tapes in computer rooms. Magnetic tape is an old, but still useful form of mass memory. Its use is limited to large computers in the data processing industry.

The *hard disk* is a high-speed mass memory. A metal disk is coated with a ferrite material. Several of these metal disks are mounted on a central spindle. Each disk is called a *platter.* The platters spin at extremely high speed. Digital information is stored on the platters by magnetic write heads. The information is read with magnetic read heads. The read/write heads never touch the surface of the platters. They ride on a few microns of air. Hard disks are available for both large computers and microcomputers.

The most popular form of mass storage for microcomputers is the *floppy disk.* Instead of being constructed from a metal plate, a floppy disk is a circular piece of plastic coated with a ferrite material. After a hole is punched out of the center of the floppy disk, it is inserted into a protective cover. This cover has a slot where the read/write head can access the surface of the floppy disk. Unlike hard disks, the read/write head actually contacts the surface of the floppy disk. Standard sizes of floppy disks are 8, 5.25, and 3.5 in.

Hard and floppy disks are pseudo-random access memories. The disk is formatted in a series of ciruclar tracks. Each *track* is broken down into further blocks called *sectors.* A particular track can be accessed without reading all the previous data; that is the random-access portion. To reach a particular sector within the track requires that the whole track be read. This is the serial-access portion. The speed at which disks operate is limited by their mechanical drives. IC memories speeds are limited only by the propagation delay of their circuitry.

Like cassettes, floppy disks are a removable medium. A typical microcomputer user will have many floppy disks. Each can contain several programs and files of data. To be accessed by a computer, the floppy disk must be inserted into a floppy disk drive.

10.2.2 Main Memory

Microprocessors operate at high speeds. The memories that the microprocessor accesses must be capable of operating at the speed of the microprocessor. They contain the current programs and data that are being processed. These memories are called *main memory.*

Each type of mass memory was driven by a mechanical means that limited their response time. Main memory devices are constructed from high-speed semiconductors. In the early days of computers, main memories were constructed from small iron doughnuts called *magnetic cores.* The modern use of magnetic core memory is limited to extremely specialized applications where IC memories cannot perform reliably, such as radiation, hot, or mechanically destructive environments. To this day the main memory in computers is still often called *core memory* even though it has long since been replaced by IC memories.

We have already examined the most common form of main memory: the random-access read/write memory. Read/write main memory devices are called RAMs. As we stated previously, RAMs are volatile memory devices. When V_{cc} is turned off, the contents of the RAM are lost. RAMs are used to temporarily store programs and data loaded into the computer from mass memories.

There are two types of RAMs: static RAMs and dynamic RAMs. RAMs constructed from flip-flops are called *static RAMs.* Once a bit of information is written into a static RAM, no further action (other than maintaining V_{cc}) is required to assure that it is not lost. *Dynamic RAMs* are constructed from MOS capacitors. When a logic 1 is stored in a dynamic RAM cell, an MOS capacitor is charged to a logic 1 level. When a logic 0 is stored in a dynamic RAM cell, an MOS capacitor is discharged to 0 V. All capacitors suffer from leakage current. In a matter of a few milliseconds the logic 1 stored on the MOS capacitor may discharge to a logic 0 level. Because of these leakage currents, dynamic memories must be *refreshed* at least once every few milliseconds. The refresh process will read the logic level stored on a MOS capacitor memory cell, and rewrite it. This will prevent logic 1 level charges from leaking down to logic 0 levels. Although refreshing dynamic RAMs may sound inconvenient, you will discover that they have many advantages over static RAMs.

Another type of main memory device is the read-only memory. Read-only memories are called ROMs. A ROM cannot be written into by the microprocessor. ROMs are nonvolatile; they are used to store programs, data, and tables of information that must not be lost when V_{cc} is turned off.

There are many different types of ROMs. Some must be ordered from the factory, whereas others can be programmed by the user. (Here the term "program" describes the process of burning information into the memory cells of the ROM.) Some ROMs can be programmed only once; others can be erased and reprogrammed many times.

10.3 THE STATIC RAM

Because static RAMs do not require refreshing, they are often used in simple microprocessor applications. Static RAMs have two major disadvantages:

1. They are limited to small storage capacities. D flip-flops require much space on the surface of an IC. This greatly limits the number of memory cells that be can integrated onto a single chip. Two or three circuit boards of static RAMs must be used to achieve the same storage capacity as that of eight dynamic RAM ICs.
2. Because more ICs are required to gain the same storage capacity, static RAMs tend to consume great amounts of power. As power supplies grow in capacity, they also increase in price, size, and weight.

Even with these limitations, static RAMs are ideal for small memory systems.

10.3.1 The 6116: 2K × 8 High-Speed CMOS RAM

The 6116 uses 16K D flip-flops for its storage cells. These flip-flops are organized in 2K groups of 8. That is exactly what the description 2K × 8 states.

In general, the description of the storage capacity of memory ICs will take the following form:

unique locations × memory cells at each location

The product of the number of unique locations and the number of memory cells at each location will yield the total number of storage devices in the memory IC. By observing the number of address inputs, you should be able to calculate the total number of unique storage locations within a memory device. The pinout in Figure 10.4 indicates that the 6116 has 11

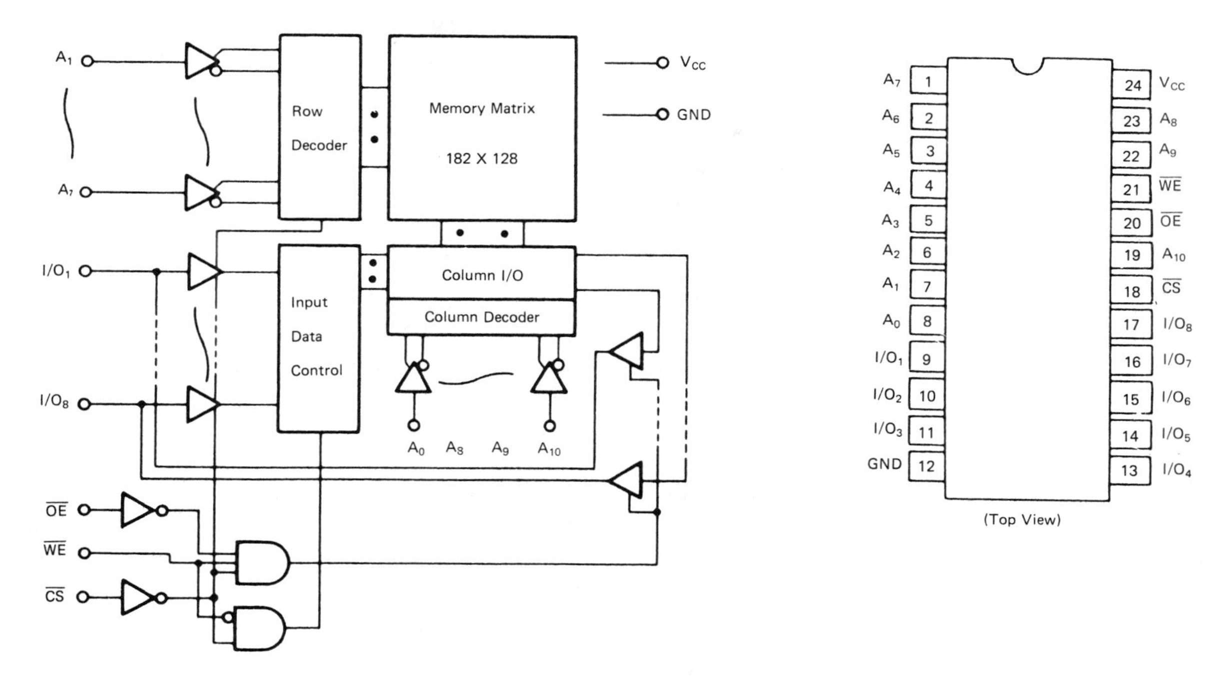

Figure 10.4 Functional block diagram and pinout of a 6116 static ram.

(A0 through A10) address pins. 2^{11} = 2K unique memory locations. This agrees with the description of the 6116.

Counting the number of data inputs or outputs will tell us the number of memory cells at each unique location. You will not find any pins on the 6116 labeled as data in or data out. Instead, there are eight pins labeled "I/O." This is the method used to indicate data pins on memory devices that are bidirectional—that have both inputs and outputs. Because the 6116 has eight I/O pins (I/O 1 through I/O 8), there must be eight storage cells at each unique location. Once again this agrees with the 2K × 8 description of the 6116.

Refer to the functional block diagram in Figure 10.4. Notice the odd inverter-like devices that the 11 address lines are driving. This is a way of illustrating that the address lines are buffered and require only minimal drive currents. The inverter-like drivers imply that both the true and inverted values of the address lines are applied to the row and column decode circuitry. (Remember the discrete 2-line-to-4-line decoder that we built? Both the true and complemented forms of all the inputs were required to drive the internal circuitry.) The memory matrix (D flip-flop storage cells) is described as 128 × 128. This calculates to a total number of 16K flip-flops. This agrees with the product of 2K × 8.

The data I/O lines are buffered on the input and driven by three-state drivers on the output. This tells us that the I/O lines are capable of going to high-Z. You know that any talker on a microprocessor bus must be capable of going to high-Z.

The only power requirements of the 6116 is +5 V (V_{cc}) and ground. This makes it compatible with TTL and microprocessor

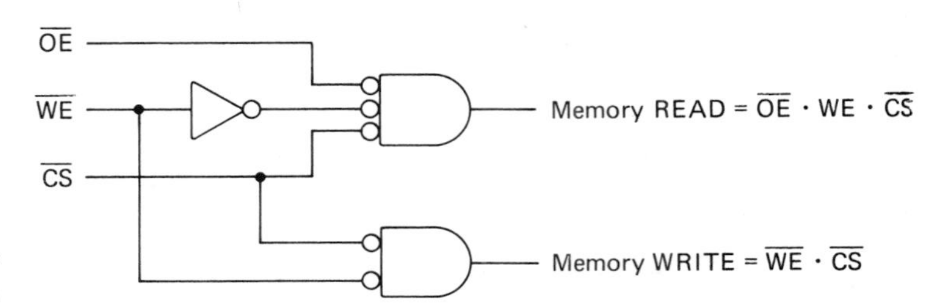

Figure 10.5 Internal decode circuitry of the 6116.

circuitry. Although the 6116 is a CMOS device, this fact will usually be transparent to the design engineer and the technician.

The last part of the functional diagram that must be analyzed is that of the control bus inputs: $\overline{\text{OE}}$ (output enable), $\overline{\text{WE}}$ (write enable), and $\overline{\text{CS}}$ (chip select). Notice that these three signals are barred; they are active low. The output of the top AND gate appears to control the enable lines on the three-state data output buffers. If the output of the top AND gate goes active high, the three-state data output buffers are enabled. This AND gate must control the memory-read operation.

The active-high output of the bottom AND gate is connected to the "input data control" block. It must control the memory write operation. The control circuitry is redrawn in Figure 10.5. Notice that the only signal common to both NOR gates is the chip select input. The chip select must be at an active-low level before the RAM will respond to either read or write commands.

The read operation will be true when chip select and output enable are at active-low levels and the write enable input is at an inactive-high level. The write operation will occur when both the chip select and write enable inputs are active low. The only signal that is not applied to both AND gates (in true or complemented form) is the output enable. In many applications the output enable pin is hardwired to ground. When this is true, the 6116 will either be reading or writing whenever chip select goes active (depending on the status of the write enable input). In our application of the 6116, the output enable input will be driven by a memory-read signal.

10.3.2 Timing Parameters of the 6116

The timing parameters describing memory ICs are complex. A typical read or write cycle has at least 10 specifications. These specifications describe minimum pulse width and setup and hold times for the address, data, and control signals. Design engineers must consider each specification carefully when they are designing memory systems. Technicians only need to know the most basic of these specifications. The words and phrases used to describe these timing parameters are not standardized. They differ greatly between manufacturers. After you study and understand the timing parameters of the 6116, you should be able to read and understand to some degree the timing parameters

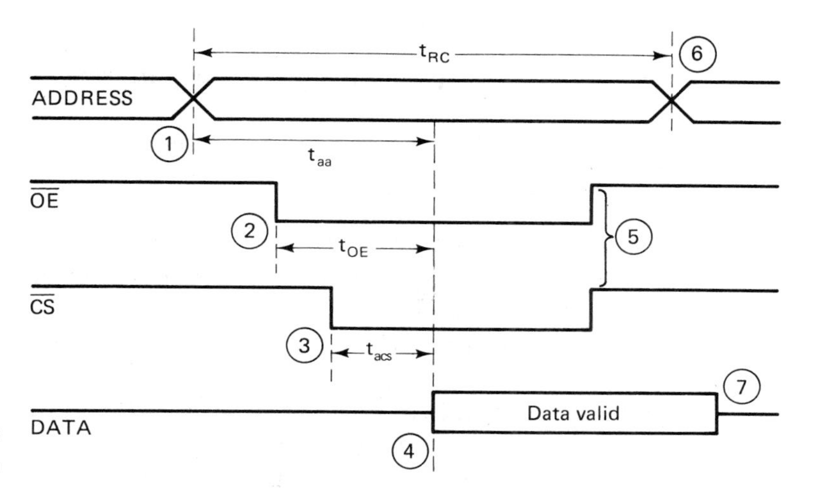

Figure 10.6 Read cycle of the 6116.

for any static RAM. The microprocessor performing the read and write operations will have an 8-bit bidirectional data bus, a 16-bit address bus, and a control bus with these active-low signals: $\overline{\text{MEMRQ}}$ (memory request), $\overline{\text{WR}}$ (write), and $\overline{\text{RD}}$ (read).

Figure 10.6 describes a typical read cycle of the 6116. Assume that the $\overline{\text{WE}}$ signal is held at the inactive-high level throughout the read cycle.

Event 1. The microprocessor will provide a 16-bit address denoting the memory location to be read. Address bits A0 through A10 are connected directly to the 6116. Address bits A11 through A15 and $\overline{\text{MEMRQ}}$ will be used to create the chip select pulse. A complete schematic and explanation of the chip select derivation is presented later in this chapter.

Event 2. The $\overline{\text{OE}}$ signal goes active low. The output enable of the 6116 is driven by the $\overline{\text{RD}}$ output of the microprocessor.

Event 3. The chip select goes active low. Our discussion of the internal 6116 control circuitry indicated that this combination of signals ($\overline{\text{OE}}$ and $\overline{\text{CS}}$ active, $\overline{\text{WE}}$ inactive) enabled the three-state output buffers for a memory-read operation.

Event 4. After a short delay, the data output is available on pins I/O 1 through I/O 8. Notice that the line before and after the data valid indication is neither high nor low but appears at a floating level. This indicates that before and after the data valid time, the I/O pins are held at high-Z. Unless a bus talker is selected, its outputs will remain at high-Z. The I/O lines are the only outputs of the 6116; the other signals—address and control—are inputs; only output lines need to be taken to high-Z when a device is de-selected.

Event 5. The $\overline{\text{OE}}$ and $\overline{\text{CS}}$ lines are taken to inactive levels.

Event 6. The address bus is taken to another address. The read operation is now complete.

Event 7. There will be a delay after the address bus changes and the I/O outputs return to the high-Z levels.

Let's consider the major timing parameters indicated in the timing diagram.

Read-cycle time (t_{RC}) This is the most important specification of a memory IC. It specifies the minimum time in which a read cycle can be completed. When an engineer or technician references a particular RAM, two numbers must be indicated: the part number and the read access time. For example: 6116s are available with 120-, 150-, and 200-ns read access times. When you say that you need a 6116, the read access time must also be specified: for example, 150-ns 6116. Faster-memory ICs are proportionally more expensive than slower-memory ICs.

Address access time (t_{AA}) This is the maximum time required between a valid address and a valid data output.

Output enable to output valid (t_{OE}) This is the output enable equivalent to the address access time; t_{OE} is the maximum time required between an active output enable and valid data.

Chip select access time (t_{ACS}) Like t_{AA} and t_{OE}, chip select access time is the maximum time required between an active chip select and valid data.

All three of the preceding specifications must be met before the data output is valid.

Figure 10.7 illustates a typical 6116 memory-write cycle.There are two possible ways to perform a 6116 write. We are going to assume that the $\overline{OE}$ is held at an inactive logic 1 level. This will assure that the three-state output buffers in the 6116 are never enabled when the microprocessor is outputting data onto the data bus. Another method of performing a write operation is to hardwire the $\overline{OE}$ to ground. When the chip select goes active, the three-state buffers will be enabled momentarily until the write enable goes active. An overlap of ac-

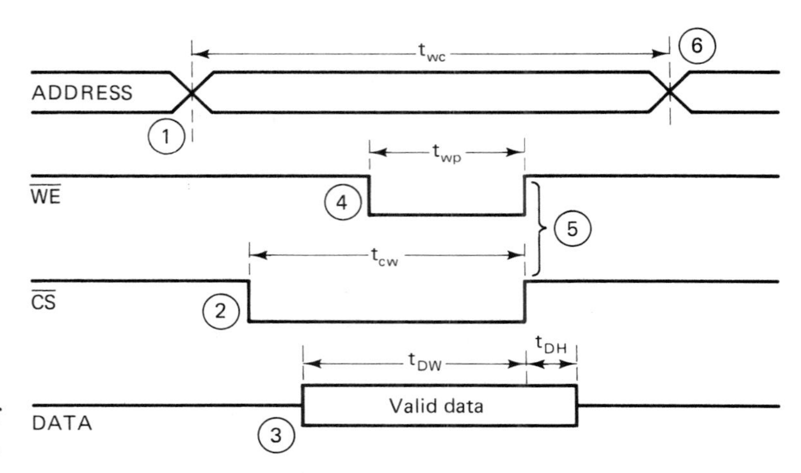

Figure 10.7 Write cycle of the 6116.

tive chip select and write enable inputs designates a write operation.

Event 1. The microprocessor outputs the address of the memory location that will be written into.

Event 2. The chip select is taken to an active-low level.

Event 3. The microprocessor places the byte of information to be stored onto the data bus.

Event 4. The write enable input on the 6116 is driven by the $\overline{WR}$ output of the microprocessor. When chip select and write enable are both low, the input buffers on the 6116 are enabled to receive data. Notice that the microprocessor placed valid data on the data bus before write enable was taken active.

Event 5. $\overline{WE}$ and $\overline{CS}$ are taken to inactive levels.

Event 6. The microprocessor removes the address and data from the buses. This completes the write operation.

Let's consider next the major specifications describing the memory write operation.

Write-cycle time (t_{WC}) This is the write equivalent of the read-cycle time. Read- and write-cycle times are equal. A 200-ns memory IC has read- and write-cycle times of 200 ns.

Write pulse width (t_{WP}) This specifies the minimum active pulse width of $\overline{WE}$.

Chip select width (t_{CW}) This is the minimum active pulse width of the chip select signal.

Data to write time overlap (t_{DW}) This is the minimum length of time that the data must be stable before the write operation takes place. If the data are on the bus for a shorter length of time than is called for by this specification, an incorrect data byte may be stored into memory.

Data hold from write time (t_{DH}) This is the length of time that the data must be held stable after the specified minimum write times have been met. Remember that we examined the hold times associated with flip-flops. The data hold time in a memory device is similar to the hold times in flip-flops.

This timing diagrams and specifications may seem complex. In the majority of cases, troubleshooting memory boards does not require the inspection of these specifications, but there will be times when an in-depth knowledge of these specifications is invaluable.

10.3.3 Constructing Memory Systems with 6116 RAMs

Several memory ICs can be connected together to form large blocks of memory (Figure 10.8). Because so many connections are involved, schematics of memory boards can get complex to the point of being unreadable. To avoid this confusion the signals that are common to all the RAMs are only shown on the first RAM. The four memory ICs in Figure 10.8 share these common lines: data bus (D0 through D7), first 11 bits of the address bus (A0 through A10), and the microprocessor's read and write control lines.

The only microprocessor signals that are not connected directly to the RAMs are $\overline{\text{MEMRQ}}$ and address bits A11 through A15. These are the signals from which the independent chip selects are derived. Notice that address lines A11 and A12 and memory request are connected to the 74LS138. With the C select input grounded, the 74LS138 performs the function of a 2-line-to-4-line decoder. Each of the decoded outputs drives a 6116's chip select. From your knowledge of decoders, you know that it is possible for only one decoded output to be active low at any time. This assures that it is impossible for more than one RAM to be selected simultaneously.

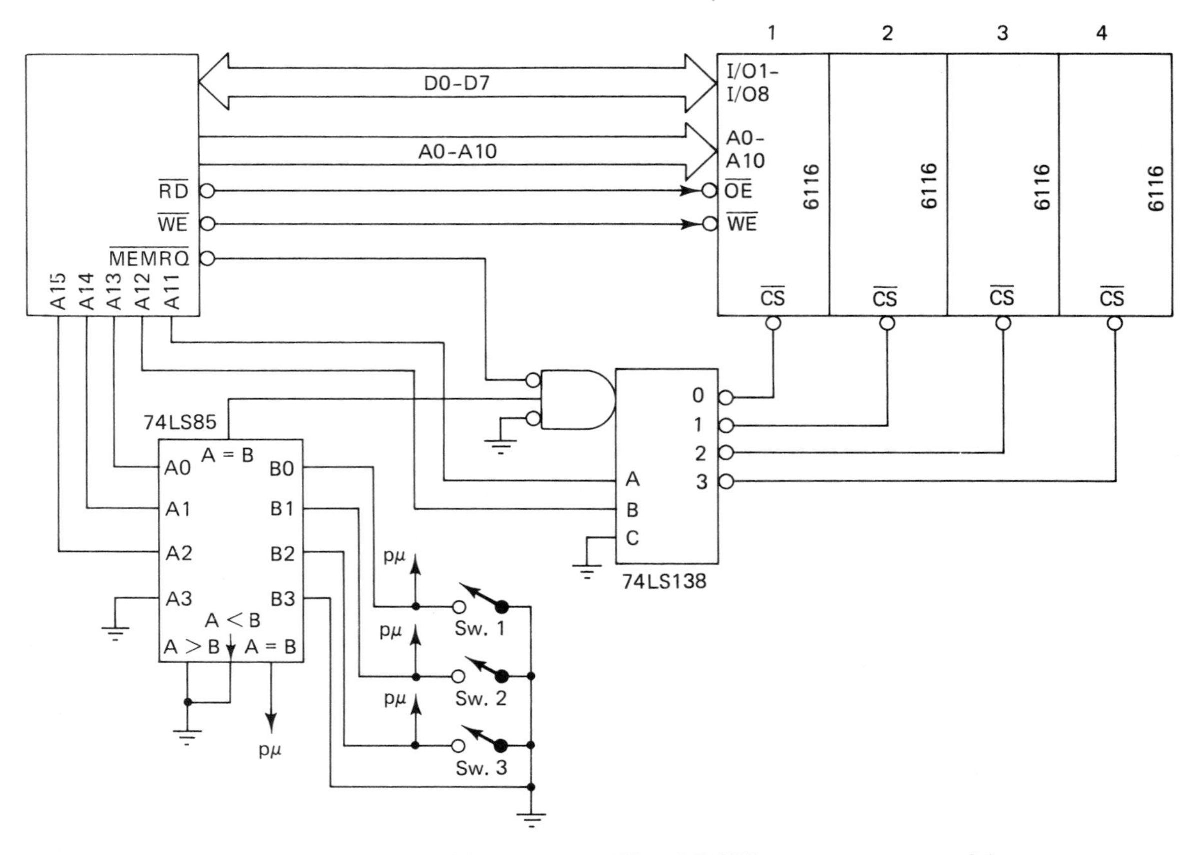

Figure 10.8 8K × 8 RAM system constructed from 6116s.

The most significant address bits (A13 through A15) are connected to the A inputs of the 74LS85 4-bit magnitude comparator. Because inputs A3 and B3 both grounded, the 74LS85 functions as a 3-bit magnitude comparator.

Now that we have examined each device in Figure 10.8, we must consider how they interact to form an 8K memory system. Another important question concerns the actual address of each RAM: With 16 address bits, the microprocessor can address 64K of memory. Where in this 64K space will the 8K of 6116 RAM appear? To solve this problem, you must create a memory map. Memory maps are diagrams that illustrate the actual address of each memory IC within the 64K memory space.

To understand how the system functions, we must work backward from the chip selects. An individual RAM will perform a read or write function only when it is chip selected. The $\overline{CS}$ inputs are driven by the output of the decoder. The decoded outputs can become active only if the decoder is enabled. One of the decoder's active low enable inputs is tied to ground. Another of the active-low enable inputs is driven by the $\overline{MEMRQ}$ line from the microprocessor. This means that the decoder will be enabled only when the microprocessor is requesting a memory access. This makes perfect sense. The RAMs should never be selected unless the microprocessor is requesting memory.

The active-high enable input is driven by the A = B output of the 74LS85. Address bits A13 through A15 must match the settings of switches 1 through 3 before the decoder is enabled. These switches will give us the ability to move our 8K block of RAM to different locations within the 64K addressing space of the microprocessor.

10.3.4 The Memory Map

Now that we understand how the hardware functions, a memory map must be created to illustrate the location of each RAM in the 64K memory space of the microprocessor. The easiest way to approach this is to make an initial assumption about the settings of switches 1 through 3. Let's assume that the three switches are closed. That makes four B inputs all equal to logic 0's. The only time that the A = B output will go active high is when address lines A13, A14, and A15 are all at logic 0 levels.

Assume that switches 1 through 3 in Figure 10.9 are closed. What happens when the microprocessor requests memory at address 0000H? The A = B output of the comparator will go active high, and the memory request line will go active low. The decoder is now enabled. The select inputs are driven by addresses A11 and A12, which are currently logic 0's. This causes decoded output 0 to go active low. RAM 1 is now chip selected. It will be chip selected starting at memory location 0000H. The 6116 is a 2K RAM. 2K is equal to 800H. Therefore, RAM 1 will be chip

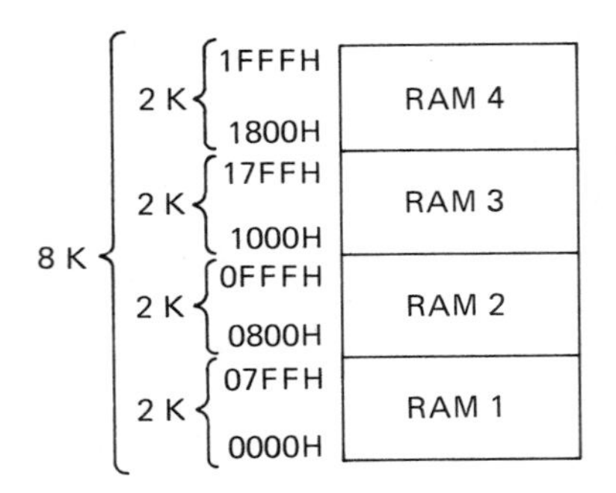

Figure 10.9 Memory map of 8K system constructed from 6116s.

selected for addresses 0000H through 07FFH, a total of 2K (800H) locations.

Address line A11 will go high at address 800H. This will cause decoded output 1 to go active low, chip selecting RAM 2. RAM 2 will be selected from memory location 800H through 0FFFH: a range of 800H. At address 1000H, A11 will return low and A12 will go high. This will cause decoded output 2 to go active low, chip selecting RAM 3. RAM 3 will be active from addresses 1000H through 17FFH: again a range of 800H. At address 1800H, both A11 and A12 are high. This will cause decoded output 2 to go active low, chip selecting RAM 3. RAM 4 will be active from addresses 1800H through 17FFH: again a range of 800H. At address 1800H, both A11 and A12 are high. This will cause decoded output 3 to go active low, chip selecting RAM 4. RAM 4 will be selected from addresses 1800H to 1FFFH, a range of 800H.

At address 2000H, A13 will go high. This will cause the A = B output to go inactive low. The 74LS138 will be disabled for addresses 2000H through FFFFH. Notice that although the data, addresses, and read and write signals are applied to the 6116s for the full range of 64K, they will not be visible except for the 8K space (0000H through 1FFFH), where they are chip selected.

What happens if we open switch 1? Instead of being ena-

Address		Switch settings
	Base 8	111
E000H		
	Base 7	110
C000H		
	Base 6	101
A000H		
	Base 5	100
8000		
	Base 4	011
6000H		
	Base 3	010
4000H		
	Base 2	001
2000H		
	Base 1	000
0000H		

Figure 10.10 Memory map indicating base addresses.

bled at addresses 0000H through 1FFFH, the decoder will now be enabled from 2000H through 3FFFH; the 8K of RAM will then appear to occupy that memory space. Refer to Figure 10.10

Let's draw some general conclusions concerning the 16 address lines and their affect on the actual address of the 8K bank of RAM.

Address Lines A0 through A10. These address lines are applied to each memory device. They have no influence on the status of the decoder. $2^{11} = 2K$: The function of these address lines is to choose the particular memory cell within the presently chip-selected RAM.

Address Lines A11 and A12 ($2^2 = 4$). These address lines select one of four RAMs to chip select. $2^{13} = 8K$: This indicates that the 13 address lines (A0 through A12) can control 2000H of memory space. That is the size of our memory system.

Address Lines A13 through A15 ($2^3 = 8$). These three address lines select the base address of the 8K of memory. The base address refers to the first physical address where RAM 1 is chip selected. As Figure 10.10 illustrates, we have the choice of eight different base addresses in which to place our 8K of memory.

These switches provide our system with an extremely important flexibility. Assume that you have a microprocessor system that already has 24K of memory and you would like to add another 8K of memory, for a total of 32K. The switches allow you to install the new memory board on any 8K boundary in your system. This is a great advantage over memory boards with fixed addresses.

10.4 OTHER STATIC RAMs

The 6116 is a byte-oriented memory IC. This means that at each unique memory location resides a group of eight memory cells. There are many RAMs that are not byte oriented. A 4-bit quantity (half a byte) is called a *nibble.* Many popular RAMs are nibble oriented, having four memory cells at each unique address. Other RAMs are bit oriented. The RAM in Figure 10.1 is an example of a bit-oriented memory. These memories have only one memory cell at each unique location. Let's examine a nibble-oriented memory device and discover how it must be connected to interface with a byte-oriented data bus.

10.4.1 The 2114: 1K × 4 Static RAM

The 2114 has 1K unique memory locations. At each unique address, there are four memory cells. Because the 2114 is a static RAM, the memory cells are flip-flops. A 1K × 4 RAM should have 10 address lines ($2^{10} = 1K$) and four data input/output lines. The block in Figure 10.11 supports this conclusion. The

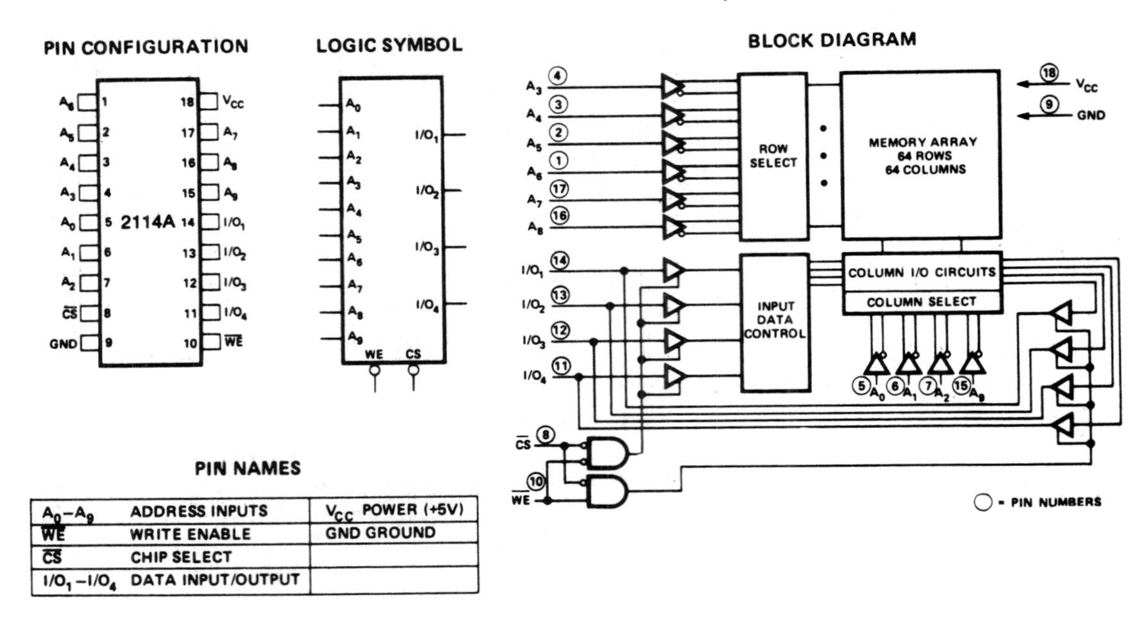

Figure 10.11 Block diagram, logic symbol, and pinout of 2114.

block diagrams of the 6116 and 2114 are extremely similar, but the 2114 does not have an output enable pin.

Refer to the two gates at the bottom of the block diagram. When $\overline{CS}$ and $\overline{WE}$ are both active low, the input data buffers are enabled and a write operation occurs. When $\overline{CS}$ is active low and $\overline{WE}$ is inactive high, the output buffers are enabled. This designates a read operation. This is the type of read operation that the 6116 would use if the $\overline{OE}$ pin were hardwired to ground. When $\overline{CS}$ is inactive, the output of both control gates will be inactive logic 0's.

The read and write cycles are similar to those of the 6116. Other than different specifications, all static RAMs will have similar timing diagrams.

Our major concern is: How do we interface a nibble-oriented memory IC with a byte-oriented microprocessor data bus? The answer is simple—we must use two 2114s to create 1K bytes of storage. The first 2114 will be connected to the least significant nibble (D0 through D3) and the second 2114 will be connected to the most significant nibble (D4 through D7). We will want both memory ICs to be active at the same time. Therefore, the chip selects for each pair of 2114s must be connected together. The major part of the circuit in Figure 10.12 is constructed from 2114 1K × 4 bit static RAMs. Together they form an 8K × 8 block of memory. Notice that we accomplished the same storage capcity using 6116 2K × 8 RAMs with a total 4 instead of 16 memory ICs.

The top row of RAMs will handle the least significant nibble of the data byte. The second row of RAMs will handle the most significant nibble of the data bus. Address lines A0 through A9 and WE- are applied to all 16 memory ICs. Each

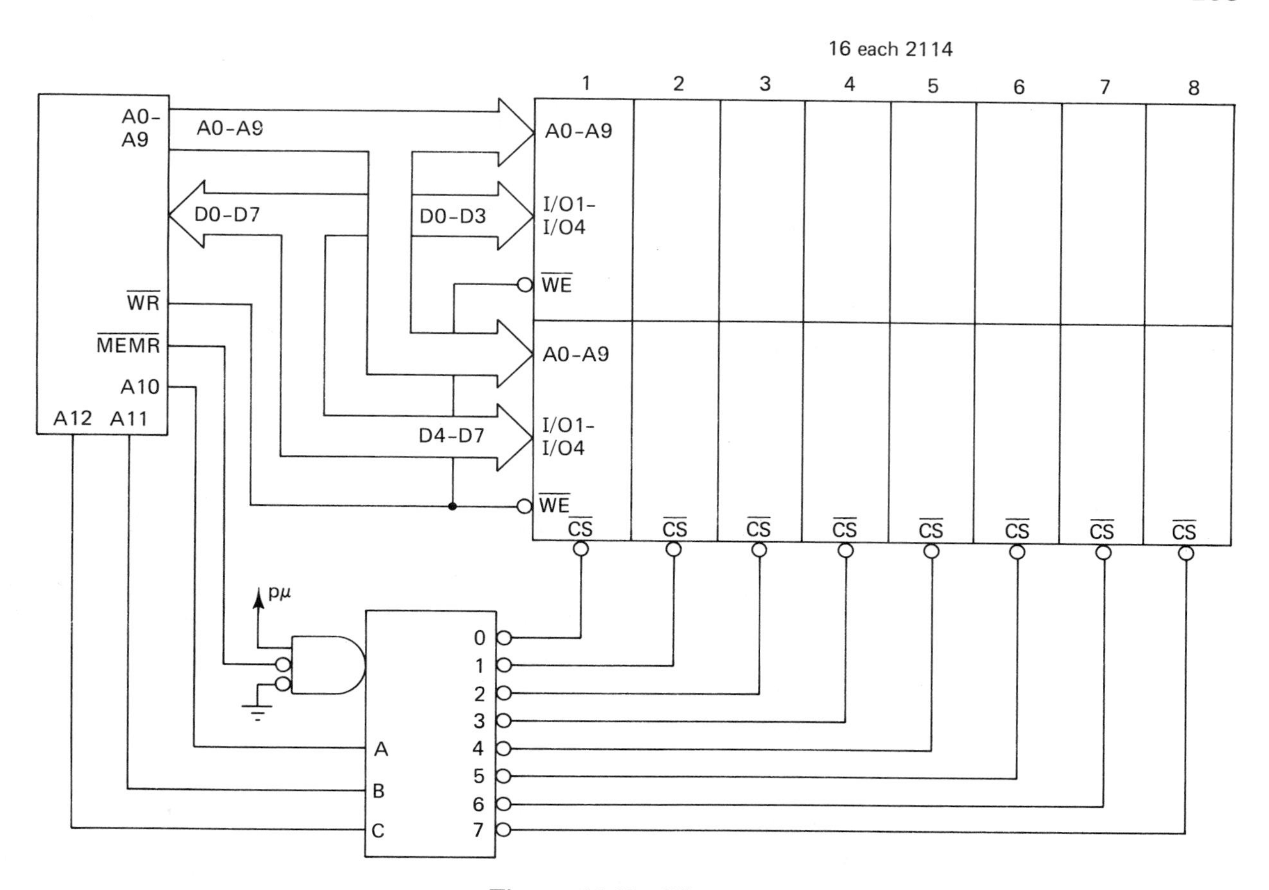

Figure 10.12 8K memory system constructed from 1K × 4 memory ICs.

column of 2114s has its chip select inputs tied together. You can think of each pair of 2114s as one 1K × 8 memory IC.

The chip selects are driven by a 74LS138. Because each chip select represents 1K of memory (instead of 2K as in Figure 10.8) all eight decoded outputs must be used to access the 8K of memory. Address lines A10 through A12 are used to drive the select inputs of the decoder. We have chosen not to use address bits A13 through A15. In this application they can be considered as don't-care values. A magnitude comparator could easily be added to the circuit for greater addressing flexibility.

The decoder will be enabled whenever the memory request is active (Figure 10.13). Because address bits A13, A14, and A15 are not used, this system is said to be *partially decoded.* No expansion memory can be added to the system until these upper-order address bits are decoded in some manner.

10.4.2 Special-Purpose RAMs

There are many types of special-purpose RAMs in common usage. Many systems employ battery-backed-up CMOS RAMs. When V_{cc} is turned off, the information in static RAM is lost. There are many applications when information vital to the sys-

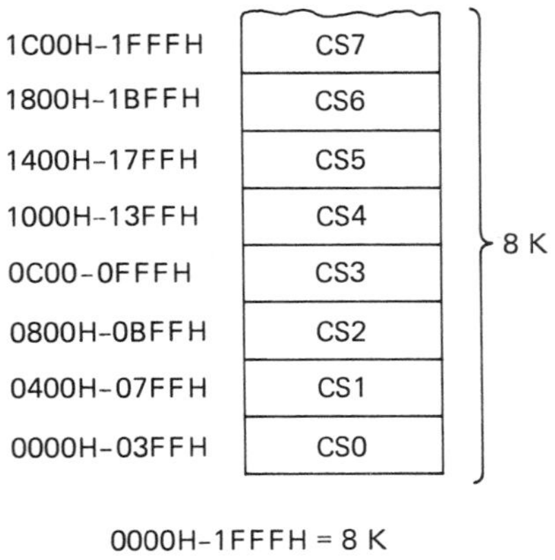

Address range	Chip select
1C00H–1FFFH	CS7
1800H–1BFFH	CS6
1400H–17FFH	CS5
1000H–13FFH	CS4
0C00–0FFFH	CS3
0800H–0BFFH	CS2
0400H–07FFH	CS1
0000H–03FFH	CS0

8 K

0000H–1FFFH = 8 K

Figure 10.13 Memory map of Figure 10.12.

tem's operation must not be lost. In those situations CMOS RAMs and some type of battery backup is used. Because CMOS devices are ultra-low power, they can be operated by batteries for extended periods of time.

Two different types of batteries are commonly used. Lithium batteries have a shelf life of over five years and can supply current for great lengths of time. Their only drawback is that they cannot be recharged. They must be replaced at periodic intervals.

Ni-Cad (nickel–cadmium) batteries are also used. Ni-Cads are rechargeable batteries. When V_{cc} is applied, the batteries are constantly being recharged. When V_{cc} is turned off, the Ni-Cad batteries supply V_{cc} to the CMOS RAMs. Ni-Cad batteries require more support circuitry than do lithium batteries.

There are also memory devices called *nonvolatile RAMs.* These CMOS RAMs have small batteries built into them. When the battery wears out, the whole package must be replaced.

10.5 DYNAMIC RAMs

Instead of using a flip-flop as the memory cell, dynamic RAMs use a MOS capacitor. Dynamic RAMs can pack much more storage capacity in a smaller space using less power than a standard static RAM. Dynamic RAMs are extremely popular.

10.5.1 A Dynamic RAM Cell

Figure 10.14 illustrates a possible way to implement a dynamic memory cell. Notice that the data-in and data-out lines are separate. The static RAMs that we examined before had common data input/output lines. The input buffer is enabled when the write input goes active low. If the input data is a logic 1, the capacitor will charge to a positive voltage; if the input data is a logic 0, the capacitor will discharge to 0 V.

The sense amplifier is a voltage comparator with an extremely high input impedance. (A high input impedance will ensure that the sense amplifier does not load down the storage capacitor.) If the voltage on the storage capacitor is greater than

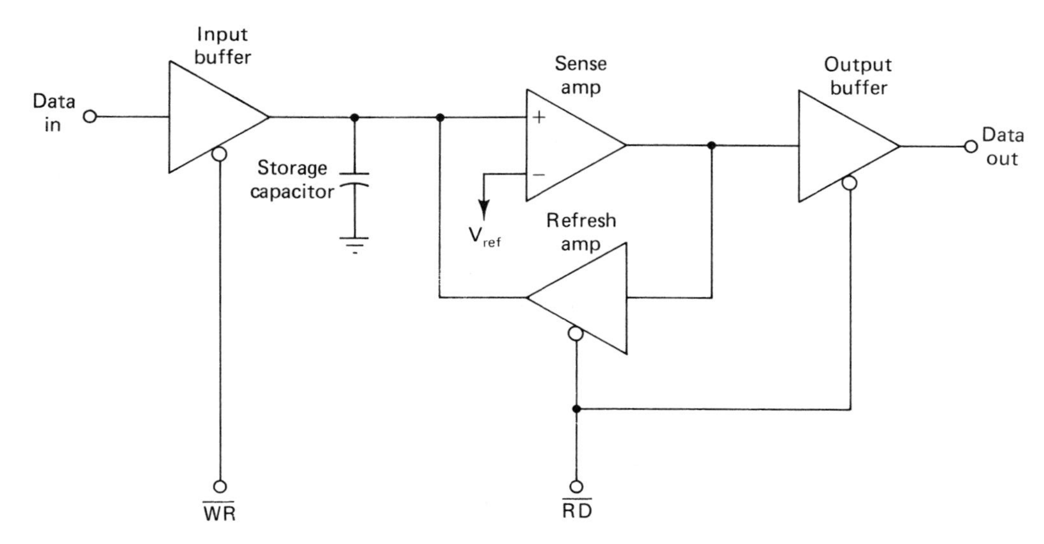

Figure 10.14 Dynamic memory cell.

the reference voltage, the output of the sense amplifier will be a logic 1; if the voltage on the capacitor is less than the reference voltage, the output of the sense amplifier will be a logic 0.

The output of the sense amplifier drives the output buffer and the refresh amplifier. Three things occur each time the read input is taken to its active level: (1) the output buffer is enabled, (2) the output of the sense amplifier passes onto the data output pin, and (3) the refresh amplifier rewrites the output of the sense amplifier onto the capacitor. All capacitors suffer from leakage current. After a few milliseconds, a logic 1 on the storage capacitor could leak down to a logic 0 level. By reading the memory cell, the voltage on the storage capacitor is refreshed.

10.5.2 The 4116: 16K × 1 Dynamic RAM

Figure 10.15 depicts the logic symbol of a 16K dynamic RAM. Dynamic RAMs are bit-oriented memory ICs. At each unique address is only one memory cell. We have discovered that two 1K × 4 RAM's must be used to create 1K bytes of memory. In the same manner, eight 16K × 1 RAM's must be used to con-

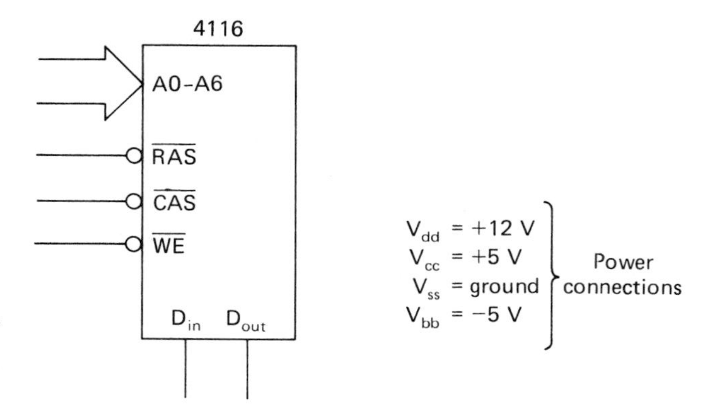

Figure 10.15 Logic symbol of the 4116 16K × 1 dynamic RAM.

struct 16K bytes of memory. Each RAM in the eight-IC array will handle 1 bit of the data bus. Notice that the 4116 has separate data-in and data-out pins. In some applications they can be connected together to form a common I/O line, as we saw in static RAMs. In other applications, an 8-bit bidirectional bus driver will be required to interface the 4116s with the bidirectional data bus.

2^{14} = 16K We would expect to find 14 address lines on the 6116; instead, only seven are indicated. The 14-bit address must be multiplexed onto the 7 address bits of the 4116. This will save seven pins, which will allow the 4116 to be manufactured in a standard 16-pin DIP package. You will remember that memory cells are constructed in arrays. Each memory cell is uniquely identified by a row and a column address.

Two control signals are used to multiplex the row and column addresses: $\overline{\text{RAS}}$ (row address strobe) and $\overline{\text{CAS}}$ (column address strobe). The row address is the least significant 7 bits of address (A0 through A6); the column address is the most significant 7 bits of address (A7 through A13). The row address will first be presented to the address inputs of the 4116. The $\overline{\text{RAS}}$ input will be taken low to latch the row address into an internal 7-bit latch. The column address must then be presented to the address inputs of the 4116. The $\overline{\text{CAS}}$ input will be taken low to latch the column address into another internal 7-bit latch. The 14-bit address has now been reconstructed inside the 4116. An active low on the $\overline{\text{WE}}$ input indicates a write operation.

The 4116 requires three voltages: +12 V, +5 V, and −5 V. At a higher price, improved versions of the 4116 (such as the Intel 2118) are available that require only a single +5-V power supply.

10.5.3 A Typical Dynamic RAM Read Cycle

Figure 10.16 illustrates a typical read-cycle timing diagram. Assume that the $\overline{\text{WE}}$ input is held at a high level throughout the read cycle. The signal labeled MUX is the multiplexer select input. When MUX is low, the row address will appear on A0 through A6 of the 4116; when MUX is high, the column address will appear on A0 through A6. The MUX signal and multiplexing of the address will be generated by additional MSI devices. The multiplexing circuitry will be examined in a few pages.

Let's examine each event in Figure 10.16.

Event 1. The MUX signal is low, so the row address is on A0 through A6 of the 4116. The falling edge of $\overline{\text{RAS}}$ will latch the row address into the 4116.

Event 2. The multiplex signal goes high. The column address will now appear on A0 through A7 of the 4116.

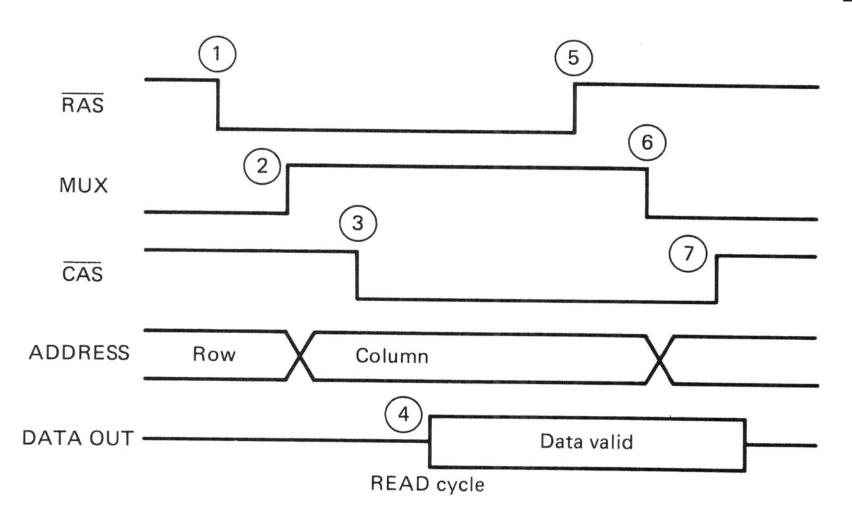

Figure 10.16 Read-cycle timing diagram.

Event 3. The falling edge of $\overline{\text{CAS}}$ will latch the column address into the 4116. The 4116 now has all the information required to access the addressed memory location.

Event 4. The valid data are now on the data-out pin.

Events 5, 6, and 7. The read operation will end by all control signals returning to their inactive levels.

Dynamic RAMs function under many critical timing constraints. A 4116 has approximately 15 timing specifications that must be met to accomplish a read or write cycle. At this point in your digital education, these complex timing parameters would probably cause more confusion than enlightenment.

10.5.4 A Typical Dynamic RAM Write Cycle

There are many ways to write information into a dynamic RAM. Figure 10.17 illustrates a *late write cycle.* A late write cycle uses the $\overline{\text{WR}}$ control signal from the microprocessor. An *early write cycle* uses the inverted $\overline{\text{RD}}$ signal to drive the $\overline{\text{WE}}$ pin on the dynamic RAM. The late write cycle is actually a read and write access. The output data of the addressed location will also be available on the data-out pin (although it is not normally used). To avoid indeterminate levels on the data bus, a microprocessor using a late write cycle must employ a bidirectional bus driver to isolate the data input and data output pins. The circuitry required for such a system will be illustrated later.

Let's examine each event in Figure 10.17.

Event 1. The row address is latched into the 4116 on the falling edge of $\overline{\text{RAS}}$.

Event 2. The multiplexer select goes high and the column address is steered onto the A0 through A6 inputs of the 4116.

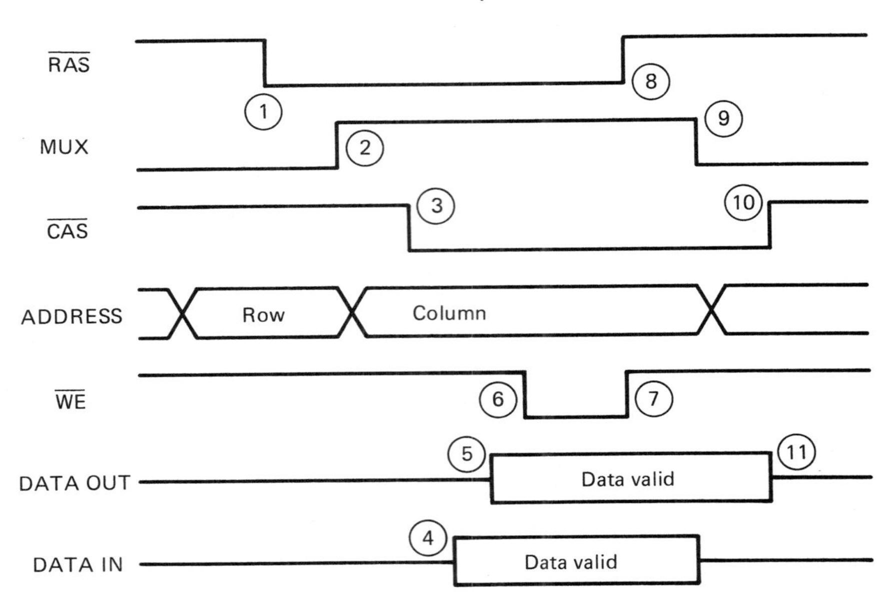

Figure 10.17 Late write cycle.

Event 3. The falling edge of $\overline{\text{CAS}}$ latches the column address into the 4116.

The first three events of the read and write cycles are exactly the same.

Event 4. The data to be written into the memory must be given a short period to stabilize before the $\overline{\text{WE}}$ goes active.

Event 5. Although we are not going to use it, the data that happen to be residing in the addressed memory location (before the write occurs) will appear on the data out pin.

Event 6. The write enable goes active. The data bit on the data in pin will be written into the addressed memory location.

Event 7. The $\overline{\text{WE}}$ pin is returned to its inactive level.

Events 8 through 11. The control signals and data in/out return to their inactive levels.

It is important to realize that the data in and data out are valid at the same time. Any system that uses the late write cycle must isolate the data-in and data-out pins.

10.5.5 Refresh Operation

Each time a dynamic RAM is read, all the memory cells in the addressed row are refreshed automatically. If we were sure that a memory location in each row would be read at least once every millisecond, a special refresh operation would not be required. A memory that is used to refresh a high-speed computer display is an example of a dynamic RAM system that does not require refresh.

Most systems have to employ some form of dynamic RAM refresh. The memory cells in a 4116 are arranged in 128 rows of 128 columns (128 × 128 = 16K). There are many special ICs that are used to supply refresh operations for dynamic RAMs. At periodic intervals they execute the following steps:

1. The dynamic RAM controller must first indicate to the microprocessor that a refresh operation is in progress. During the refresh operation the microprocessor cannot access the memory.
2. The dynamic RAM controller will connect the outputs of a 7-bit counter onto the address bus. The counter will start at the count of 0. This would correspond to row 0 of the 4116. When the $\overline{\text{RAS}}$ line is pulsed low, the 128 memory cells in row 0 will be refreshed.
3. Step 2 is repeated for each row (0 through 127) in the 4116.
4. After all 128 rows have been refreshed, the dynamic RAM controller indicates to the microprocessor that it may once again access the RAM.

Figure 10.18 illustrates the simple refresh-only cycle. While the MUX line is held low, the row address is output from a 7-bit counter in a dynamic RAM controller. The $\overline{\text{RAS}}$ line is pulsed low. The entire 128 memory cells in the addressed row have been refreshed. Approximately 1 to 3% of the system's time is spent refreshing the dynamic RAM. This is a small price to pay for extremely dense, low-cost, low-power memories.

The Z80, manufactured by Zilog Corporation, is an extremely popular microprocessor. The capacity to refresh dynamic RAM is built into the Z80. When the microprocessor is involved with internal operations, it uses the few spare microseconds to refresh a row of dynamic RAM. The refresh process is transparent to the operation of the Z80, and no extra hardware is required.

10.5.6 A Multiplexing Scheme for Dynamic RAMs

We require a circuit that receives 14 address bits from the microprocessor and outputs the 7-bit row and column addresses at the correct moment, as indicated by the read- and write-cycle timing diagrams.

The circuit shown in Figure 10.19 is constructed around the 74LS157 quad 2-line-to-1 line multiplexer. Examine the

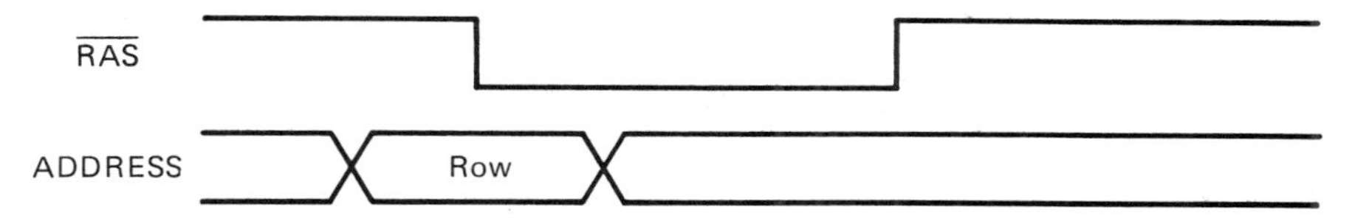

Figure 10.18 Refresh-only cycle.

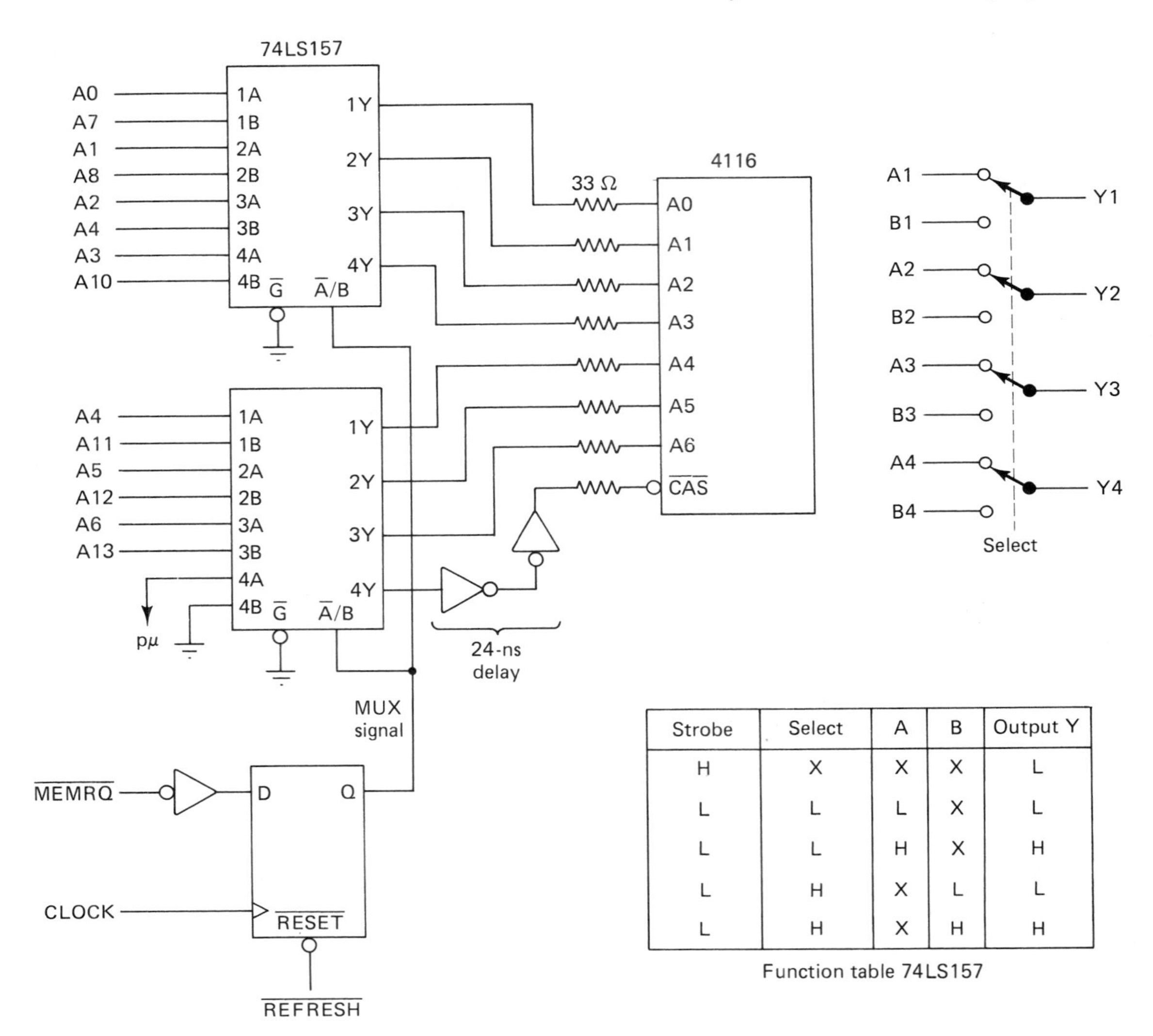

Strobe	Select	A	B	Output Y
H	X	X	X	L
L	L	L	X	L
L	L	H	X	H
L	H	X	L	L
L	H	X	H	H

Function table 74LS157

Figure 10.19 Row/column multiplexing circuit for the 4116 dynamic RAM.

function table for the 74LS157. Each 74LS157 contains four 2-line-to-1-line multiplexers with common enable and select. The data inputs are divided into two groups of four: A1 through A4 and B1 through B4. When the select input is low, the A inputs are passed onto the Y outputs; when the select input is high, the B inputs are passed onto the Y outputs. This is indicated on the schematic by labeling the select input as $\overline{\text{A}}$/B. The 74LS157 operates as the digital equivalent of a quad SPDT switch. The function table describes the operation of each multiplexer.

Line 1. If the strobe input (G) is at an inactive-high level, the select and data inputs become don't cares and the Y output is forced to a logic 0.

The strobe is held at an active level for lines 2 through 5.

Lines 2 and 3. The select input is low, indicating that the data on the A input will be passed onto the Y output. When the select is low, the level on the B input is in a don't-care condition.

Lines 3 and 4. The select input is high, indicating that the data on the B input will be passed onto the Y output. When the select is high, the level on the A input is in a don't-care condition.

It will require two 74LS157s to handle the 14 address lines and the $\overline{\text{CAS}}$ signal. The row address (A0 through A6) will be connected to the A inputs and the column address (A7 through A13) will be connected to the B inputs. The next question is: How is the MUX signal (which drives the select inputs of the 74LS157s) generated?

Remember that the MUX signal stays low until after $\overline{\text{RAS}}$ has a falling edge. The MUX signal will then go high and pass the column address (A7 through A13) onto the address inputs of the 4116s. After a short delay the $\overline{\text{CAS}}$ input will be taken low.

Every microprocessor has a clock input. The microprocessor uses this clock input as a timing reference. A memory-read or memory-write cycle may take four microprocessors clock pulses. Consider the D flip-flop in Figure 10.19. The D input is driven by the inverted memory request signal. Normally, the $\overline{\text{MEMRQ}}$ signal is high; this will place a low on the D input of the flip-flop. Therefore, the MUX signal is normally low. That fact agrees with the read- and write-cycle timing diagram. The microprocessor will initiate a memory access on a rising edge of the clock by placing the required address onto the bus and taking $\overline{\text{MEMRQ}}$ low. For the first clock cycle the MUX signal will be low. On the next rising edge of the clock, the inverted $\overline{\text{MEMRQ}}$ signal will cause the MUX signal to go high. This will place the column address onto A0 through A6 of the 4116.

The $\overline{\text{CAS}}$ signal is generated from the second 74LS157. Input 4A is pulled up to V_{cc} and input 4B is tied to ground. When the MUX signal is low, the $\overline{\text{CAS}}$ signal (4Y) is high. This agrees with our timing diagrams. When the MUX signal goes high, output 4Y also goes high. But the timing diagram indicates that there must be a short delay before the $\overline{\text{CAS}}$ signal should have a falling edge. This delay is provided by the two inverters. This 24-ns delay gives the column address a chance to settle down before it is latched into the 4116.

When the memory read or write is completed, the microprocessor will return the $\overline{\text{MEMRQ}}$ signal to its inactive-high level. On the next rising edge of the clock, the MUX signal will return to a logic 0 level. It appears that this circuit will satisfy all the demands of the read and write timing diagrams. Notice that an active-low signal called "refresh" will asynchronously reset the MUX line. This will provide us with the proper "refresh-only" timing parameters. Shortly, we will see how $\overline{\text{RAS}}$ is generated.

The 33-Ω resistors are commonly used to help impedance match the outputs of the multiplexers with the inputs of the dynamic RAM. This will ensure that the address lines settle down to valid levels in a minimal length of time.

10.5.7 A 16K DRAM System

The term *DRAM* is commonly used in place of *dynamic RAM.* We now need to create a 16K-byte memory system using eight 4116s and a bidirectional bus driver.

Figure 10.20 illustrates a 16K × 8 DRAM memory system. Each 4116 handles 1 bit of the 8-bit data byte. A0 through A7, $\overline{RAS}$, $\overline{CAS}$, and $\overline{WE}$ are common to all eight 4116s in the 16K bank. Remember that during the late write operation, both data-in and data-out pins are active. A method of interfacing the data-in and data-out pins with the microprocessor's bidirectional data bus is required.

The 8216 4-bit bidirectional bus driver was designed specifically for that purpose. Refer to the block diagram of the 8216 in Figure 10.20. On the right side of the block diagram, the input and output of each pair of drivers/receivers are connected. This side of the 8216 is interfaced with the bidirectional data bus. The left side of the 8216 is used to interface with memory ICs that have independent data-in and data-out pins. The pins labeled Out 0 through Out 3 will be connected to the data output pins of four 4116s. The pins labeled In 0 through In 3 are connected to the data input pins of four 4116s. Two 4116s are required to interface to an 8-bit data bus.

Consider the control gates at the bottom of the block diagram. If $\overline{CS}$ (chip select) is at an inactive-high level, all outputs of the 8216 will be at high-Z. The control pin labeled "$\overline{DIEN}$" is the data input enable control. When $\overline{CS}$ and $\overline{DIEN}$ are both low, the logic levels on the data-out pins of the 4116s are routed onto the data bus; this would happen during a memory-read operation. When $\overline{CS}$ is low and $\overline{DIEN}$ is high, the logic levels on the data bus are placed onto the data input pins of the 4116s; this would occur during a memory-write operation.

Once again, in an attempt to avoid confusion, all the data-in and data-out lines of the 4116s are not illustrated. Notice that the D_{in} lines are drawn together onto one line. This is also true for the D_{out} lines. This does not mean that the eight lines are electrically connected. It is just a convenient way of drawing the D_{in} and D_{out} buses of the 4116s without complicating the schematic with 16 separate lines. Each D_{in} of the 4116s is connected to an input pin on the 8216; each D_{out} pin of the 4116s is connected to a output pin on the 8216. In this manner the D_{in} and D_{out} pins of the 4116s are isolated; no conflict will happen when D_{in} is being driven by the microprocessor and the 4116 is placing data onto the D_{out} pin, as occurs during the late write operation.

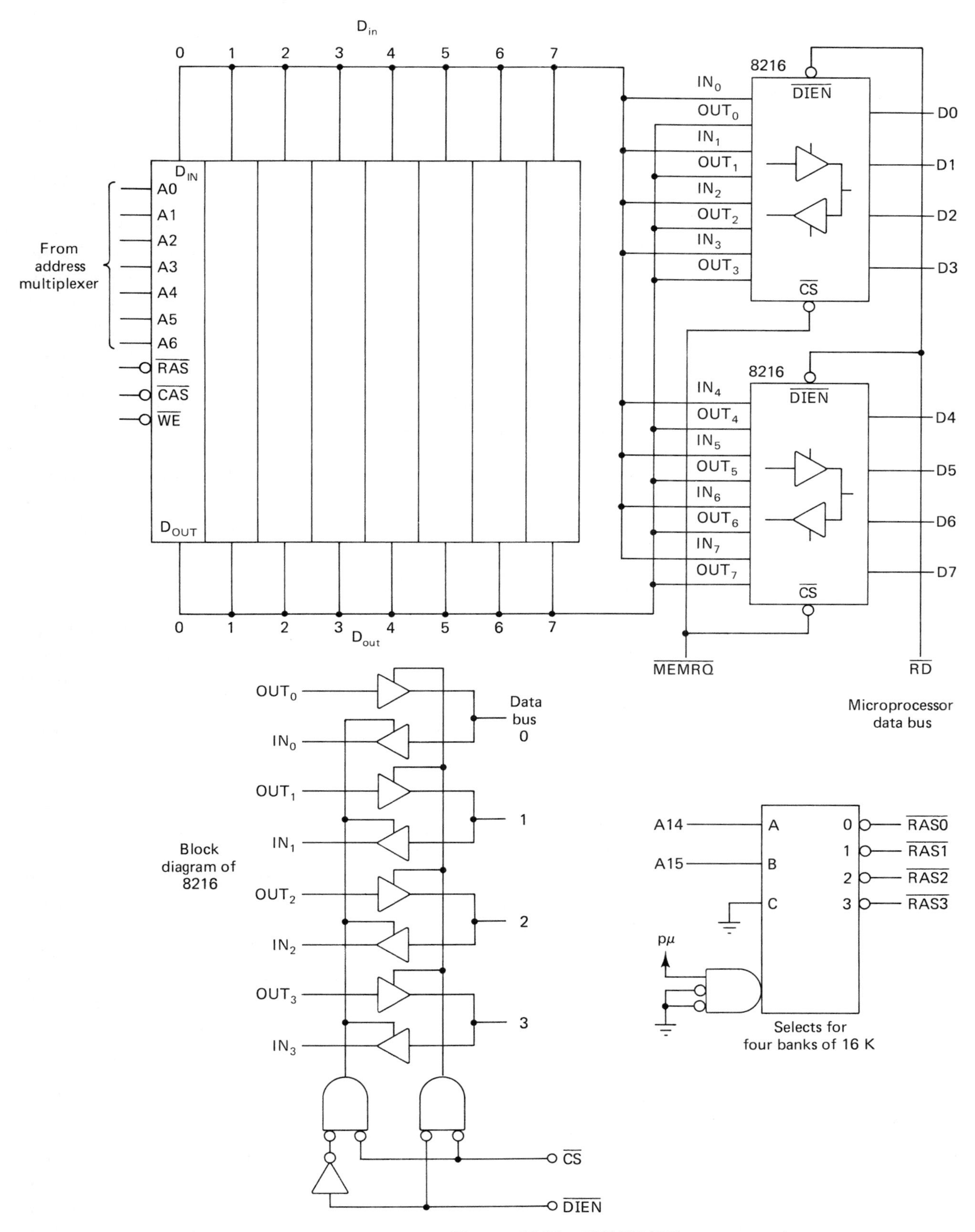

Figure 10.20 16K DRAM memory system.

The $\overline{\text{CS}}$ inputs of the two 8216s are driven by $\overline{\text{MEMRQ}}$; unless the microprocessor is requesting memory, the buffers of the 8216 will be held at high-Z outputs. The data-in enable ($\overline{\text{DIEN}}$) is driven by the $\overline{\text{RD}}$ output of the microprocessor. When the microprocessor is performing a read operation, the 8216s will pass the data out values of the 4116s onto the data bus.

Now that we have created one bank of 16K bytes of memory, how can we expand the capacity to all 64K of the microprocessor's address space? We have seen that static RAM ICs are chip selected by an active-low output of a decoder. DRAMs do not have $\overline{\text{CS}}$ inputs, but $\overline{\text{RAS}}$ provides almost the same function. To construct 64K bytes of memory using four 16K banks of 4116s, A0 through A6 (from the address multiplexer), $\overline{\text{CAS}}$, and $\overline{\text{WE}}$ will be connected to each 4116. We will use the 74LS138 illustrated in Figure 10.20 to generate four $\overline{\text{RAS}}$ signals: $\overline{\text{RAS}}$ 0 through $\overline{\text{RAS}}$ 3. $\overline{\text{RAS}}$ 0 will be applied to the eight 4116s in bank 0; $\overline{\text{RAS}}$ 1 will be connected to the eight 4116s in bank 1, with similar connections for banks 2 and 3. The $\overline{\text{RAS}}$ actually functions as a modified chip select. If a 4116 does not receive the falling edge of a $\overline{\text{RAS}}$ signal, it will ignore all activity on the A0 through A6, $\overline{\text{CAS}}$, and $\overline{\text{WE}}$ inputs.

10.5.8 Address and Data Buffers

The DRAM memory system in Figure 10.20 used the 8216 to interface the separate data inputs and outputs of the DRAMs with the bidirectional data bus. Address and data bus buffers are employed for many reasons. They are designed to drive the many inputs that occur in bit-oriented DRAM memory systems. This prevents the microprocessor's address and data bus from being excessively loaded.

Another important reason for using these buffers is to isolate banks of memory from the main buses in the microprocessor system. This aids the technician in troubleshooting memories. If banks of RAM are not buffered, one short can drag down the whole bus. Finding this problem would be extremely difficult because of the many inputs sharing each trace on the bus. Data and address buffers easily allow the technician to isolate the memory problem to a specific bank of RAM. Address bus buffers are normal drivers or three-state drivers, depending on the application. Data bus buffers will always be bidirectional, allowing both memory-read and memory-write operations.

10.5.9 Advanced DRAMs

DRAMs are now available in 64K $\times$ 1 and 256K $\times$ 1 capacities. The 4164 (64K DRAM) is pin-for-pin compatible with the 4116. One address line is added to the 4164. Because the addresses are multiplexed, the addition of one address line increases the addressing capacity by a factor of 4, instead of a factor of 2 as would occur in a nonmultiplexed RAM. An important improvement is that it requires only a +5 V V_{cc}. This creates an interesting problem; to keep compatibility with the 4116, V_{cc} is pin 8 and ground is pin 15. This is backward from normal digital ICs. Beware of this strange standard pinout when you are troubleshooting memory systems.

4164s have displaced the 4116 in all new designs. Within 18 months of the date when this book was completed, the 41256 will probably have replaced the 4164. The newer-generation microprocessors have address buses up to 24 bits wide. This would enable the microprocessor to address directly 16M (megabytes) of memory. The demand for larger memory systems will be answered by even higher density RAMs.

10.6 THE READ-ONLY MEMORY

All the memory devices that we encountered have been read/write devices. Another type of memory IC that is extremely important is the ROM (read-only memory). ROMs are nonvolatile memory ICs that store permanent digital information. This information may be programs or other forms of data. How is the information initially stored in a ROM? The process of writing information into a ROM is called *programming* the ROM. Certain ROMs are programmed by the manufacturer; others can be programmed by the user.

10.6.1 Applications of ROMs

ROMs are used for many applications. Every microcomputer must have some basic program stored in ROM. When the power is first turned on, the microprocessor will access this *startup* or *boot ROM.* The boot ROM will contain simple programs that enable the computer system to read further instructions into standard RAM from a mass memory device (typically a floppy disk). Some inexpensive microcomputers actually contain all the required programs entirely in ROM. Video games are an example of such a ROM-based system. The game cartridge that is inserted into the video game is simply a small circuit board containing a game program in ROM.

Another application of ROMs is in creating custom decoders. Consider the 74LS47. A BCD-to-seven-segment decoder can be implemented as a ROM. The BCD input could be considered as a 4-bit address, and the seven-segment output as the data outputs. Each of the 16 memory locations would contain a group of seven read-only memory cells. ROMs are often created to provide unique or custom decoder functions.

ROMs are also used as look-up tables. Instead of having a microprocessor calculate the square root of a number, it could look it up at a certain memory location. The microprocessor would use the look-up ROM in the same way that we use standard tables in textbooks. Having the miroprocessor look up the value of a particular function is much faster than calculating it with a program.

Character generators are ROM-based devices that are used to create the alphanumerics and graphic symbols on CRT displays. The ROM holds the information concerning which dots in a matrix must be illuminated to create a particular character or graphic symbol.

10.6.2 A Typical Pinout of a ROM

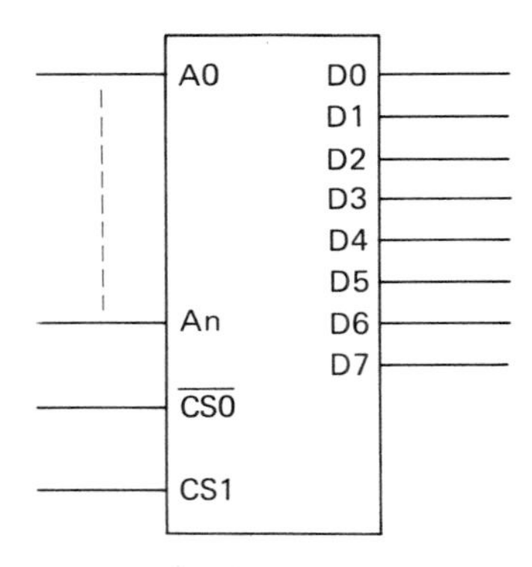

Figure 10.21 Logic symbol of a typical RAM.

The pinout of a ROM resembles the pinout of a static RAM, minus the $\overline{WE}$ input (Figure 10.21). The number of address lines will indicate how many unique memory locations the ROM contains. ROMs are available with capacities of up to 32K bytes. The eight data outputs indicate that the ROM is a byte-oriented memory device. Notice that this particular ROM has more than one chip select. It is common for ROMs, and even some RAMs, to have up to four chip selects, both active low and high. The chip selects are internally ANDed together. All chip selects must be at their active levels before the device is internally chip selected.

The read-cycle timing diagram of a ROM is almost identical to the read-cycle timing diagram of a static RAM. Each ROM will have an associated read-cycle time that describes its speed.

10.6.3 The Mask-Programmed ROM

Mask-programmed ROMs must be ordered from the manufacturer. A design engineer will create a truth table that describes the required data output for each address. The manufacturer will use this truth table to create the ROM. This is an extremely expensive process. ROMs are used only for super-high-volume products that have been thoroughly debugged.

Technicians must consider these ROMs as black boxes that output a byte of information for each address. ROMs are usually placed in DIP sockets. If a ROM is suspect, it is swapped with a known good ROM.

10.6.4 The PROM (Programable ROM)

For many applications, the initial cost of manufacturing mask-programmed ROMs is cost prohibitive. An alternative is the PROM. The advantage of PROMs is that they can be programmed by the user. PROMs contain memory cells that have fusible links. When a PROM is programmed, selected fuses are blown out, creating the required memory pattern. Because of this, the process of programming PROMs is often called *burning* PROMs.

PROMs are programmed with a device called a *PROM programmer.* The memory pattern will be burned into the PROM is down-loaded from a computer or read into the PROM programmer by a known good PROM. A new PROM is inserted into a socket on the PROM programmer. A pushbutton will initiate the programming process. A typical PROM takes 60 to 90 s to program.

Each PROM has a *signature*. This signature is a unique four- to six-place hexadecimal number. All PROMs that have been programmed with the same information will have the same signature. If a PROM is suspect, it is placed into the PROM programmer and its signature is checked. If the PROM programmer displays the wrong signature, the PROM is bad and should be discarded.

After the PROM is programmed, a new part number must be attached to it. Because a blank PROM can be programmed an almost infinite number of different ways, a new part number must be associated with each uniquely programmed PROM.

Blank PROMs are available in many different capacities and speeds. Most PROM programmers have the ability to program all the popular types of blank PROMs.

Although PROMs are more expensive than their ROM counterparts, they have important applications in small-scale production or in areas where the design is continually changing. The bit pattern programmed into the PROM can easily be modified to create new revisions.

10.6.5 The EPROM (Erasable PROM)

PROMs can be programmed only once. If the design of the circuit is modified, all the previously burned PROMs must be thrown away and new PROMs programmed. In an R&D environment, design changes are everyday occurrences. The use of PROMs in such an environment would be extremely expensive. The EPROM is a type of ROM that can be erased and reprogrammed hundreds of times. This is the perfect solution in an environment where design changes happen on a monthly or even weekly basis.

In the center of a ROM is a clear, round window. When exposed to UV (ultraviolet) light, the previously programmed information will be erased. To be erased, EPROMs are placed into devices, called PROM erasers, that expose the PROM to concentrated UV. (UV is dangerous to the eyes.) A typical EPROM takes 30 minutes to be erased.

After the EPROM is erased, it can be reprogrammed. Standard PROM programers will also program EPROMs. After the EPROM is programmed, it must also have a unique part number attached to it. This part number is written on a label that is placed over the clear window of the EPROM. If the window is not covered, stray UV, from fluorescent lights or sunlight, can erase the contents of the PROM. Direct sunlight will erase the contents of a PROM within a several weeks; fluorescent light, within a few years.

EPROMs, like their PROM counterparts, have an associated signature. If an EPROM is suspect, its signature must be verified with the PROM programmer.

10.6.6 The 2700 Family of EPROMs

The 2700 family of EPROMs is designed to be upwardly compatible. That means that as memory requirements increase, users can easily upgrade the system with the next-larger-capacity memory of the 2700 family.

Part number	*Capacity*
2708	1K × 8
2716	2K × 8
2732	4K × 8
2764	8K × 8
27128	16K × 8

Notice that the part number indicates the total number of memory locations. Divide this number by 8 to calculate the number of kilobytes that the device can store. For example, 2716 → 16/8 = 2K bytes.

Figure 10.22 illustrates the pinout of the 2716. Refer to the input of the 2716 in Figure 10.22. The 11 address inputs (A0 through A10) prove that the 2716 has 2K unique locations. There are also eight output data lines (O0 through O7). V_{cc} and ground indicate that this device is powered from a standard +5-V power supply.

The pin labeled V_{pp} is the programming voltage pin. When +25 V is applied to this pin, the 2716 can be programmed. This

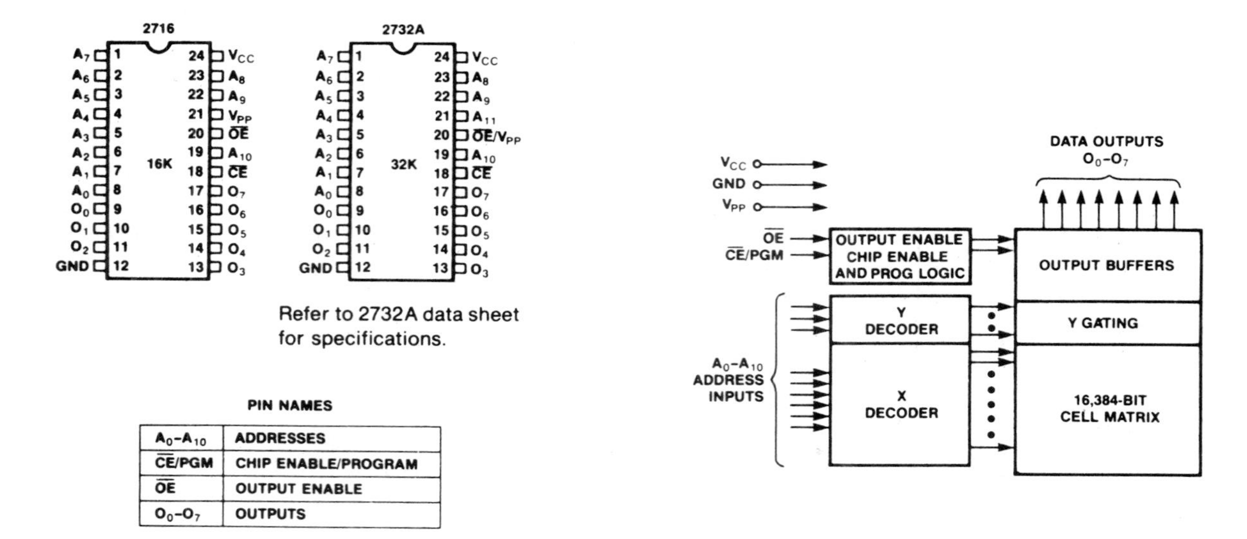

A_0–A_{10}	ADDRESSES
$\overline{CE}$/PGM	CHIP ENABLE/PROGRAM
$\overline{OE}$	OUTPUT ENABLE
O_0–O_7	OUTPUTS

Figure 10.22 Pinout and block diagram of 2716.

voltage will be required only during the programming process; it will be furnished by the PROM programmer.

There are two control signals: $\overline{OE}$ and $\overline{CE}$. Output enable functions just like the $\overline{OE}$ pins on static RAMs. Chip enable ($\overline{CE}$) is equivalent to chip select. Both $\overline{OE}$ and $\overline{CE}$ must be at logic 0 levels during the 2716 read cycle. The timing diagram describing the 2716 read cycle is similar to the static RAM read-cycle timing diagrams. Both $\overline{OE}$ and $\overline{CE}$ are furnished to simplify the expansion of ROM systems. All the output enables are connected to the $\overline{RD}$ output of the microprocessor. The $\overline{CE}$s are generated using the upper address bits and a decoder.

Notice that the 2716 is pin-for-pin compatible with the 2732. The only difference is that the V_{pp} pin of the 2716 will become the extra address line required for the 2732. The $\overline{OE}$ on the 2732 will double as the V_{pp} pin.

The 2716 is also pin-for-pin compatible with the 6116 2K × 8 static RAM. Notice the similarity between the two part numbers: 6116 and 2716. This enables a manufacturer to design a memory system that can be either ROM, RAM, or a combination of both.

10.6.7 A 16K ROM/RAM Memory System

Figure 10.23 illustrates a 16K system that can be populated with either 6116 RAMs or 2716 EPROMs. Notice the two-position jumper labeled J1. If the system is populated with RAM, J1 should be in the A-B position. This will allow the $\overline{WR}$ signal from the microprocessor to reach the $\overline{WE}$ (pin 21) of the 6116. If

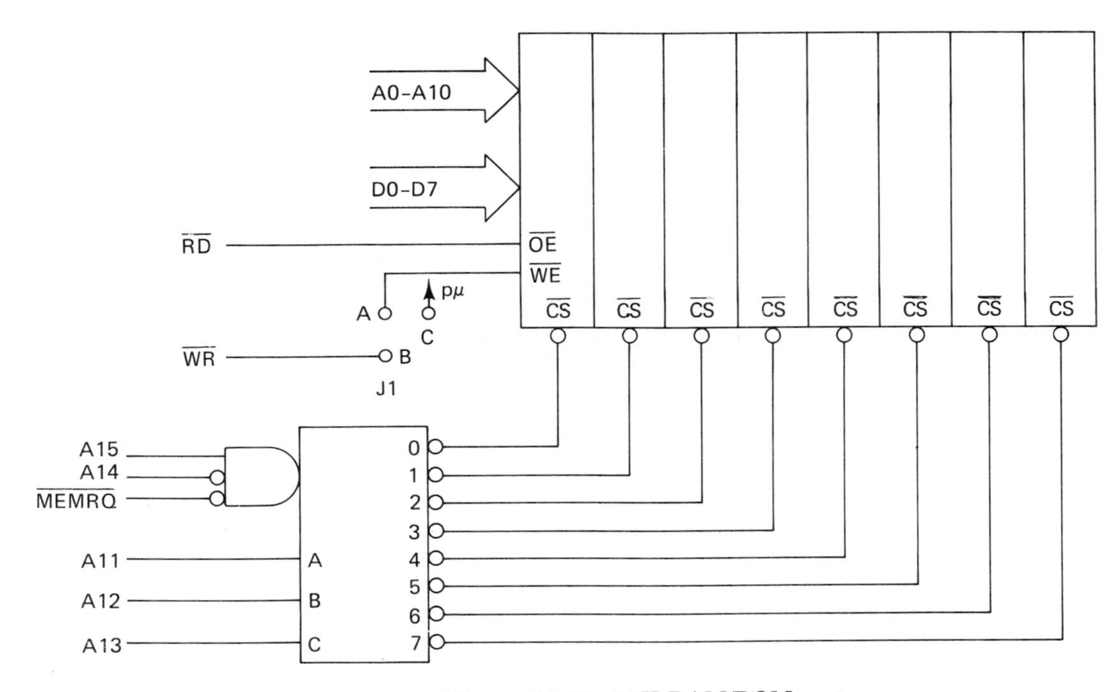

Figure 10.23 16K RAM/ROM system.

the system is stuffed with 2716s, J1 should be in the A-C position. Pin 21 in the 2716 is the V_{pp} pin and should be pulled up to +5 V when the device is not being programmed. To add even more flexibility to the system, eight separate jumpers could be used on pin 21 of each memory device. This would let us mix RAMs and ROMs in the same 16K bank of memory.

Address lines A14 and A15 drive the enable inputs of the 3-line-to-8-line decoder. The memory map of this system will be included as an exercise at the end of the chapter.

10.6.8 The E²PROM (Electrically Erasable PROM)

The E²PROM has characteristics of both ROMs and RAMs. It is nonvolatile like a ROM, but it can also be written into like a RAM. The write cycle of the E²PROM is extremely long. This device is an alternative to CMOS battery-backed-up RAMs; it is not intended to be used as a general-purpose read/write memory device. The E²PROM is gaining much greater popularity as the technology improves.

10.7 PROGRAMMABLE LOGIC DEVICES

As IC manufacturing technology improves, LSI devices can accomplish more complex functions. We have stated that SSI/MSI devices are the "glue" that bonds the LSI devices. There are many times when a design engineer wants to reduce the IC count in a new circuit design. This will result in space savings, lower manufacturing costs, and improved reliability.

Custom ICs can be manufactured that provide the equivalent of many SSI/MSI devices in a single IC. But custom devices suffer from the same cost and inflexibility faults as mask-programmed ROMs. A programmable combinational/sequential equivalent of the PROM is needed.

PALs (programmable array logic) and IFLs (integrated fuse logic) fulfill these needs. The simple PROM illustrated in Figure 10.24 is an 8 × 1 memory device. Think back to Chapter 3, where we introduced the sum-of-products method of creating combinational circuits. The PROM architecture is a standard sum of products with programmable fuses on the OR input lines. The 3-bit address creates eight different combinations. Each product of these combinations is applied to the input of the OR gate, via fusible links. A blown fuse will result in a nondynamic level being placed on that OR input. If a fuse is blown during the programming process, the input combination that creates that product will output a logic 0. If a fuse is left intact, the combination that creates that product will output a logic 1.

The PAL architecture is also the standard sum of products. But instead of fusing the inputs to the OR gates, the inputs to the AND gates are fused. A blown fuse will reduce by one the number of inputs on the AND gate. The PAL has six times as many fuses as the PROM. This gives the increased flexibility

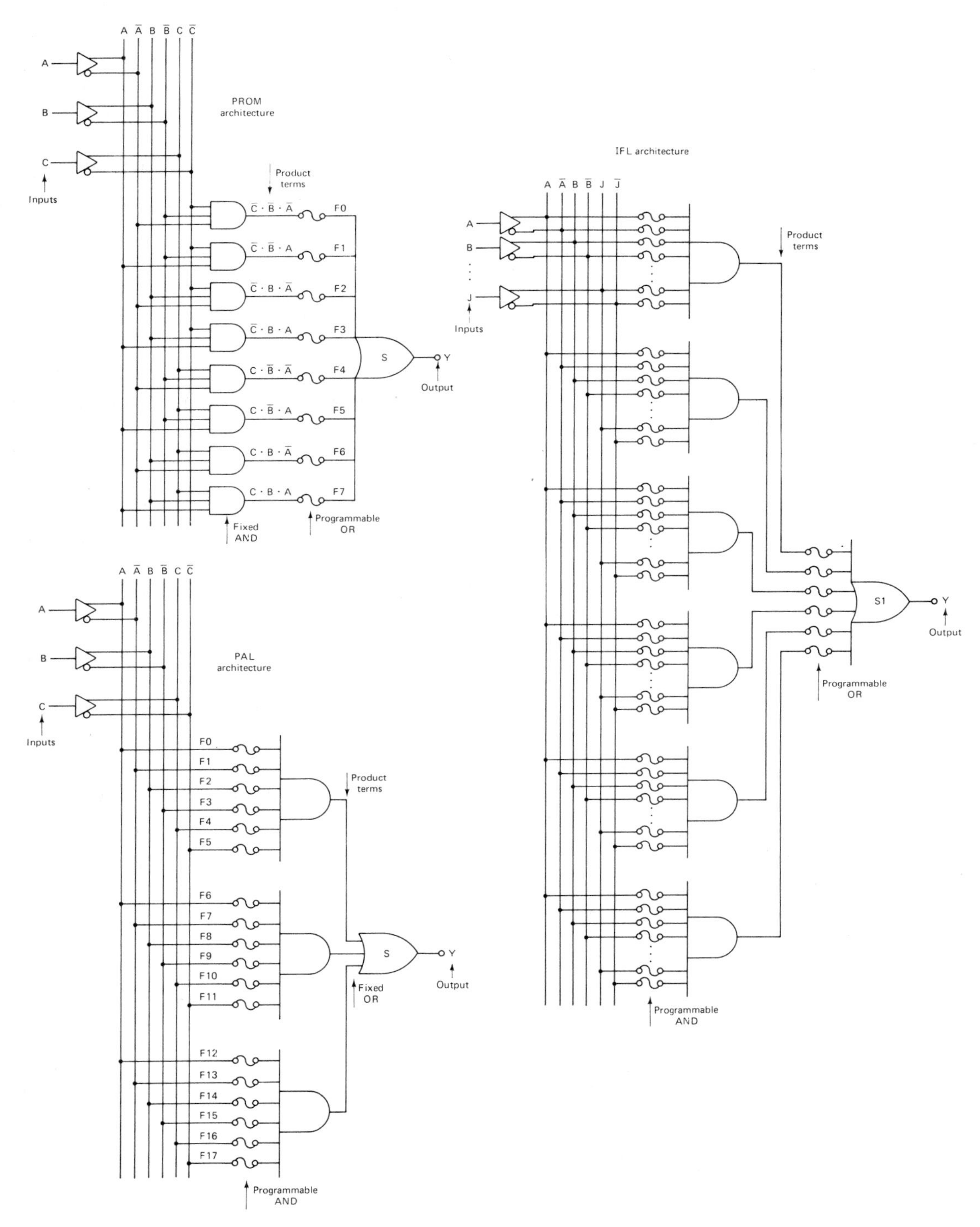

Figure 10.24 PROM, PAL, and IFL architecture.

needed to create an almost unlimited number of 3-bit combinational functions.

The IFL has fuses in both the AND-gate inputs and the OR-gate input. Obviously, the IFL device is the most sophisticated. But the sophistication does not come without a price. The development of IFL devices requires extensive computer-aided-design (CAD) programs.

The PAL is a good compromise between SSI/MSI devices and custom ICs. PALs are designed using standard Boolean algebra and truth tables. Computer programs are used to translate the Boolean equations or truth tables into bit maps that are used to program the PALs.

A typical PAL may replace 5 to 10 SSI/MSI ICs. PALs are also available with sequential devices, such as D flip-flops. These PALs are called *logic sequencers.*

The PAL can be programmed on most new PROM programmers. You will treat any programmable logic IC as a semicustom device; If a programmable logic device is suspect, its signature should be verified.

A PAL will appear as a "black box" in a circuit schematic. Separate documentation describing the circuitry that the PAL is emulating should be provided by the design engineer.

QUESTIONS AND PROBLEMS

Refer to Figure 10.1 for Problems 10.1 to 10.4.

10.1. The enable input on G2 is shorted to ground. How does this affect circuit operation?

10.2. The decoder output 2 of U1 is open. What symptoms will the circuit display?

10.3. The enable input of U1 is shorted to the data line. How does this affect circuit operation?

10.4. The ground pin of U6 is bent underneath the IC. How does this affect circuit operation?

10.5. What would happen if the $\overline{\text{WR}}$ line in Figure 10.2 went low before the data were given a chance to stabilize?

Refer to Figure 10.8 for Problems 10.6 to 10.10.

10.6. If SW 1 and 3 were closed and SW 2 open, for what range of addresses would RAM 3 be chip selected?

10.7. If all the switches were open, for what range of addresses would RAM 4 be chip selected?

10.8. Decoded chip selects for RAMs 1 and 2 are shorted together. How does this affect circuit operation?

10.9. Input C of the decoder is floating. How does this affect circuit operation?

10.10. The A > B input on the comparator is floating. How does this affect circuit operation?

Refer to Figure 10.12 for Questions 10.11 to 10.14.

10.11. Add circuitry to Figure 10.12 so that the 8K bank of memory starts at A000H.

10.12. RAM pair 5 fails to respond. What is the most likely problem?

10.13. Decoded output 7 is shorted to ground with a solder bridge. How does this affect circuit operation?

10.14. The $\overline{\text{WR}}$ line is shorted to $\overline{\text{MEMRQ}}$. How might this affect circuit operation?

Refer to Figure 10.19 for Questions 10.15 to 10.17.

10.15. How would the circuit be affected if the delay provided by the inverters were only 10 ns?

10.16. The Y outputs of the top 74LS157 appear to be stuck at logic 0 levels. What problem could create this symptom?

10.17. The reset input of the D flip-flop is shorted to ground. How does this affect circuit operation?

Refer to Figure 10.20 for Questions 10.18 to 10.20.

10.18. The $\overline{RD}$ line is shorted to ground. How does this affect circuit operation?

10.19. $\overline{MEMRQ}$ is floating. What device would appear to be malfunctioning?

10.20. Why is the 8216 required in the circuit?

10.21. How does the $\overline{RAS}$ of a DRAM compare with the $\overline{CS}$ of a static RAM?

10.22. A ROM has 14 address lines. What is its storage capacity in bytes?

10.23. Memory map the system in Figure 10.23.

PART II MICROPROCESSOR ELECTRONICS

11 The Elements of a Computer System

We have already examined the basic elements in a computer system. In this chapter we examine the different types of computers and peripherals that are in common usage.

11.1 COMPUTER SYSTEMS

Mainframes, Minis, and Microcomputers

Computers are usually classifed in three groups: mainframe, mini, or micro. The factors used to classify a computer are the sheer physical size of the machine, how many bits of data it can handle in one read or write operation, how much physical and logical memory it can address, and how fast it can process data. As technology has progressed, the distinction between the three groups has become much more subtle. Mainframes can handle multiple word lengths of data between 32 and 64 bits; minis have word lengths of data between 16 and 32 bits; finally, micros have word lengths between 8 and 32 bits.

Mainframe computers are found in environments where tremendous quantities of data must be stored, retrieved, and manipulated, and high-level mathematic and statistical applications are required. Banks, insurance companies, and manufacturing concerns all rely on mainframe computers to do heavy data and number-crunching (processing). Mainframe computer systems are really constructed from many different independent computers. The central computer in the system does all the actual data processing. Other computers in the systems are used to do front-end processing (the preprocessing of data) and communications tasks. Mainframes can support 100 to 1000 simultaneous users.

Mainframes gain their great speed by employing ECL (emitter coupled logic) circuits. ECL is a logic family that is appreciably faster than TTL or CMOS because it employs nonsaturated logic levels. Instead switching a transistor on and off between saturation and cutoff, ECL switches between two voltage levels in the linear portion of a transistor's characteristic curve. This eliminates the need to overcome the base–collector stray capacitance, greatly reducing the propagation delay. ECL uses a -5.2 V power supply and typical output voltages are -0.9 V for a logic 1 and -1.75 V for a logic 0. Although it is unquestionably the fastest logic family, ECL has many drawbacks: It consumes great amounts of power, has a low noise threshold, and cannot be highly integrated.

Minis are used in small business enviroments that require only the manipulation of moderate amounts of data. Perhaps the largest users of minis are colleges and universities. In an academic environment, minis are used to support students in programming classes, for text formatting and electronic mail, and in small-laboratory research. Other applications of minis include control computers used in factories and loading docks, and automated electronic test systems. A typical mini can support 24 to 100 simultaneous users.

The microcomputer revolution is one of the most far-reaching technical breakthoughs in history. Micros are found literally everywhere. They are used in business, education, research, and communications. Small businesses that could not justify the price of a mainframe or mini can easily afford several micros. Businesses use micros for inventory control, accounts receivables, personnel management, data base management (i.e., the storage, manipulation, and retrieval of large pools of data), word processing, and forecasting. Individuals often use micros for the same applications as those used by small businesses.

Large business that own many mainframes and minis use micros for personal workstations. These micros are connected to the mainframes and minis to appear as normal terminals. A person with a micro who has such a mainframe or mini connection can share vast amounts of data with the other users in the system. Thus a micro user has the best of all worlds: an independent personal workstation capable of running sophisticated applications and a channel to transmit and receive important data communications from all the users connected to the mainframe. Users in such a network can send and receive electronic mail, share peripherals such as printers, and have electronic bulletin boards that display company news and events. The microcomputer is presently the most dominant force in the computer world.

11.2 BASIC ARCHITECTURE OF A COMPUTER

As we examined in Chapter 8, all computer systems, from mammoth mainframes to lap-sized micros, share the same basic architecture. This architecture entails four major blocks: the central processing unit (CPU), memory, input circuitry, and output circuitry. Figure 11.1 depicts the classic block diagram of a computer system. It is important to understand that the devices and circuitry used to construct the blocks in Figure 11.1 vary greatly between different computer systems. The intent of this chapter is to describe general concepts that apply to all computers, regardless of age, size, or capability.

11.2.1 The CPU

The circuitry needed to implement a CPU in mainframes requires many circuit boards of components, whereas microcomputers contain a CPU in one LSI IC that we have already described as a microprocessor. The CPUs in mainframes and microcomputers perform the same function; the difference is that the CPU in a mainframe computer is thousands of times faster and more sophisticated than its microcomputer counterpart. The CPU can be thought of as containing the heart and brain of the computer system. CPUs typically contain the following logical blocks: arithmetic/logic unit (ALU), registers, control and timing circuitry, and microprogrammed instructions.

The ALU The arithmetic/logic unit performs arithmetic and logical functions. Addition, subtraction, and multiplication are typical arithmetic functions that an ALU can perform. AND, OR, XOR, and complement are typical logic functions that an ALU can perform.

Registers Registers are simply memory locations that reside within the CPU. CPUs contain 4 to 128 registers. These registers are classified by the type of information that they store. Some registers are used only to store addresses, others store

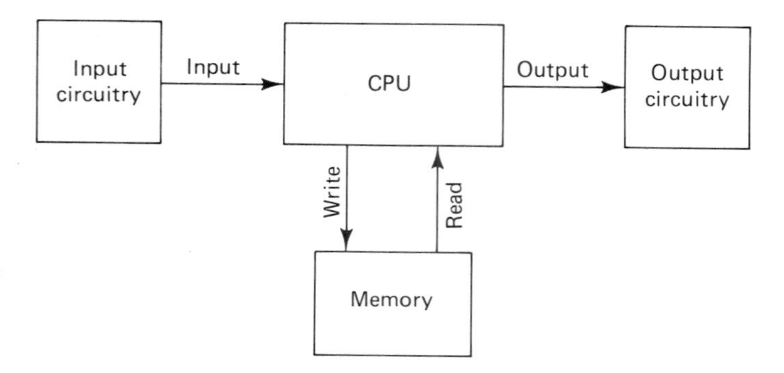

Figure 11.1 Block diagram of a generalized computer system.

specific kinds of data, and still others are general purpose in nature and can be used to store any type of information. The most important fact to remember about CPU registers is that they are used only to store data and addresses for a short period of time. They are not used like external RAM—loaded with programs or data that will be constant throughout a particular application. The information in these registers is always changing to provide the current executing instruction with the data it needs to complete its task.

When we studied memory devices we learned that to read from or write into a RAM always requires an *access time.* The microprocessor has to place the address onto the address bus, signals on the control bus have to be asserted, and decoders in the circuit and internal in the RAMs all have propagation delays. Because registers are internal to the CPU, none of the signals associated with a RAM read or write operation are required during a CPU register read or write operation. This means that the access time of registers is extremely fast! A typical CPU register has a read/write access time of less than 50 ns.

11.2.2 Control and Timing

When we discussed the microprocessor as the master of the system bus, we learned that it generated control signals during memory read and write operations. An important function of CPUs is to generate control and timing signals. All CPUs have a clock input; this clock will be used as a time standard to generate the precise signals that are required to manipulate the system bus. Complex timing diagrams are provided by the CPU manufacturer that illustrate the relationships betwen the address, data, and control signals generated by the CPU.

Microprogram We have discovered that the CPU is highly intelligent. It is the master of the system bus, generating timing and control signals, it has internal memory locations called registers, and it can perform basic arithmetic and logic functions. How can it perform all these tasks? From whence does it derive its innate intelligence? The answer is quite simple: an internal program is designed by the manufacturer of the CPU and cannot be modified by the user.

It is critical to distinguish between a microprogram and a program that is written by a CPU user and stored in external memory. As CPU users you will write programs that reside in external RAM or ROM. After the CPU is powered-on, it will fetch the first instruction written in external memory and start to execute your program. All the while that your program is executing another program is also executing—the program that resides in ROM within the CPU: the microprogram. It is this microprogram that knows how to generate the address and control signals to read the first step in your external program. It is the microprogram that actually adds two numbers together when your program requests an addition. It is also the microprogram that reads and writes information to/from the CPU registers. The microprogram is the lowest possible level of programming. Each microinstruction talks directly to the thousands of transistors that constitute a CPU. The microprogram cannot be changed; it is an element of the electronic hardware. The technician or engineer does not have to understand how the microprogram functions. Each CPU has a set of instructions that it can perform. The fact that this instruction set is actually executed by the microprogram is completely transparent to the CPU programmer.

11.3 MEMORY

The second block in Figure 11.1 is memory. In addition to RAM and ROM, computer systems also have many forms of nonvolatile mass memory. This mass memory is used to store huge amounts of programs and data. The memory devices in computers are often the largest, most expensive, most complex part of the system.

11.3.1 Main Memory

The main memory of computer is used to store the programs and data that are currently being executed by the CPU. If CPUs are to run at full speed, the main memory must have a fast access time. The only modern type of memory that meets this requirement is semiconductor memory. Before semiconductor memory was invented or cost-efficient, small ferrite doughnuts called *magnetic cores* were used to implement main memory. Because of this, the term *core memory* is still often used to refer to main memory, regardless of the fact that main memory is now constructed exclusively from semiconductor RAM.

Bootstrap Memory When a computer is powered-on, how does it know what to do? Every computer must have a low-level program in nonvolatile memory (ROM) called the *boot* or *bootstrap program*. (The term "bootstrap" is derived from the old saying about "pulling oneself up by one's own boot straps.") When the CPU is first powered up, it automatically reads the memory location that contains the first instruction of the boot program. The boot program contains instructions that initialize the system hardware. The boot program than reads into main memory a complex program called an *operating system* from one of the mass memory devices. At that point the computer starts executing the operating system program and it is up and running.

System ROM Every computer contains system ROM. We have just seen that the boot program is contained in ROM. ROM is also used to hold computer languages (such as BASIC), look-up tables for math functions, programs that interface the computer with input/output devices, and self-test programs. Inexpensive home computers, microprocessor trainers used in education, and computers that are dedicated to perform a single control function (such as monitoring a security system) keep their operating system and applications programs in system ROM.

System RAM is used for many different purposes. It holds the operating system when it is read from the mass storage device. RAM is also used to hold the current applications program that the computer may be executing and the data that the program may require. It is used as a buffer to hold data temporarily for I/O devices. It is also used for an important data storage function called a ***stack.*** A stack is a place in RAM where temporary data are kept that can be effortlessly accessed by the CPU.

On microcomputers the image of the video display is kept in fast-access RAM called *video RAM*. The contents of this video RAM are used to refresh and update the characters and graphic symbols on the microcomputer's display. Video RAM is not considered a part of main memory because it is dedicated to refreshing the display and cannot be used for any other purpose.

11.3.2 Mass Memory

The preceding description of main memory had many references to mass memory. The capacity of a computer system is often judged in terms of the computer's mass memory devices. How fast are they? What is their total storage capacity? Can they be expanded?

Just like cassette and video tape, mass memories store data in the form of magnetized regions on an oxide-coated medium. Because data are stored in this form, mass memories are nonvolatile; they retain data without the requirement of an external voltage. The data bits are written onto and read from the magnetic media in the same manner that music is recorded and played back on casssette tape. That process uses a magnetic read/write head.

Consider a simplified read operation. The head is placed (by some mechanical means) over the location where the data have been stored. Magnetic currents are induced into the head as it moves across the tape. These currents are then amplified and decoded. This results in the retrieval of the original digital information.

Semiconductor memories have no moving parts; the only delays experienced in accessing a RAM or ROM are minor propagation delays. Mass memories, on the other hand, are hybrids of electronics and mechanics. Because of simple inertia, it takes an appreciable length of time to position a read/write head over the required location on the magnetic media. This results in mass memories having access times which are thousands of times slower than those of electronic memories.

We now see that mass memories are nonvolatile and can store enormous amounts of data, but they also have great speed limitations. RAMs are extremely fast but are relatively expensive per bit of storage and lose their contents when power is lost. Computer systems require the complementary attributes of both RAM and mass memory.

Floppy Disk In microcomputer systems the most popular form of mass memory is the floppy disk. The floppy disk is a flat, circular piece of plastic coated with a magnetic oxide material. The floppy disk is placed into a protective square cardboard-like case. As the name "floppy" implies, it is flexible and easily bent. For a computer to access a floppy disk for a read or write operation, the floppy must be inserted into an appropriate floppy disk drive. Floppy disk drives are analogous to cassette decks that play and record audio cassettes. Floppy disks have an elongated access hole cut into the case that enables the read/write head of the floppy disk drive to engage the surface of the floppy disk.

Floppy disks are an extremely low cost mass storage media. They are available in three standard sizes: 8-, $5\frac{1}{4}$-, and $3\frac{1}{2}$-in. diameters. The 8-in. standard is the oldest, and was originally used as low-cost mass storage for minicomputers. Floppy disk drives are often found in pairs. A pair of 8-in. floppy disk drives are inserted into a large chassis that contains a +5-V and +12-V power supply. Together the pair of drives, power supply, and chassis weigh almost 40 lb. Typically, 8-in. floppy disks use both sides for data storage, which gives them a capacity of 1.2 MB (megabytes). Data are transferred between a computer and an 8-in. floppy in a serial stream at the rate of 500 kilobits/s.

The $5\frac{1}{4}$-in. standard, developed as a smaller, low-power alternative to the 8-in. floppy, is usually called a *mini-floppy.* Until recently, $5\frac{1}{4}$-in. floppies were the exclusive mass memory storage device for microcomputers. A $5\frac{1}{4}$-in. floppy disk drive is about one-fourth the size and weight of an 8-in. floppy disk drive. Their power requirements are low enough to use the +5 V and +12 V from the microcomputer's power supply. The $5\frac{1}{4}$-in. floppy disk drive is small enough to reside inside a microcomputer, adding little weight or bulk. A typical $5\frac{1}{4}$-in. floppy disk also uses both sides, has a storage capacity of 320 kB (kilobytes), and has a data transfer rate of only 250 kilobits/s. The newest generation of $5\frac{1}{4}$-in. floppies can store 1.2 MB and have a data transfer rate of 500 kilobits/s. They appear to be the electrical equivalent of the older 8-in. floppy.

After much competition, a new standard floppy has emerged: the $3\frac{1}{2}$-in. floppy, developed as an alternative to the $5\frac{1}{4}$-in. floppy. Because of its small size, the $3\frac{1}{2}$-in. floppy is called a *micro-floppy.* Unlike the 8- and $5\frac{1}{4}$-in. floppies, the $3\frac{1}{2}$-in. floppy is not really floppy at all. It is housed in a protective, hard plastic shell. When 8- and $5\frac{1}{4}$-in. floppies are not in a drive, they must be placed into paper envelopes to ensure that no foreign particle will contaminate the oxide surface. The $3\frac{1}{2}$-in. floppy does not require this protective envelope. The access hole on these floppies is covered by a thin piece of metal. The first $3\frac{1}{2}$-in. floppies stored information on only one side; they have a storage capacity of 400 kB. Newer $3\frac{1}{2}$-in. floppies use both sides and have therefore doubled their storage capacity to 800 kB. It appears that the new $3\frac{1}{2}$-in. floppy has been extremely well received because of its small size and rugged package.

A new floppy disk is nothing more than a piece of plastic coated with magnetic oxide. Before it can hold any data, the floppy must be formatted with tracks and sectors. All operating systems contain a formatting program. This program writes special format data onto a new floppy disk. These data define concentric rings on the surface of the floppy. These rings are called *tracks*. Each track is then broken into sections called *sectors*. A sector is the smallest block of storage on a floppy disk. The actual formatting process and the formatting data that is written onto the floppy is extremely complex. Most companies follow precise formatting standards, which include information that identifies each track and sector; a CRC (cyclic redundancy check byte), which is used to verify the validity of the data; and many different lengths of gaps and bytes of data used to synchronize the floppy disk drive and the computer attempting a read/write operation.

Consider a typical $5\frac{1}{4}$-in. floppy disk. The format used is called a standard *double-sided, double-density format.* "Double-sided" means that both sides of the floppy will be used to store data. In earlier-technology drives, only one side of the floppy (the bottom side) was used to store data. "Double-density" refers to how closely the tracks are squeezed together on the floppy. The double-density standard says that the track density is 48 tracks per inch (tpi). The first $5\frac{1}{4}$-in. floppy disk drives used a track density of only 24 tpi, this standard is called *single-density.* When a floppy is double-sided, the matching tracks on the top and bottom are called a *cylinder.* The first track on the bottom is track 0; the first track on the top is track 1. Track 0 and 1 constitute cylinder 0. This numbering scheme of even on the bottom and odd on the top is repeated for the entire floppy. Each track is divided further into eight groups called *sectors.* Notice that it is a matter of tradition to refer to the first track as track 0, but the first sector is called sector 1, *not* sector 0. Figure 11.2 describes the data capacity of each logical group on the $5\frac{1}{4}$-in. floppy.

Floppy disk drives have a switch that senses the state of the read/write notch. If it covered with a piece of tape, the disk is write protected, and the floppy will become essentially a read-only device. If the notch is left uncovered, the floppy is enabled for write operations.

The floppy disk drive uses an infrared transmitter and receiver to sense the occurrence of the index hole. Each time the index hole in the floppy disk rotates past the small, round cutout, a strobe is generated to indicate the beginning of the track.

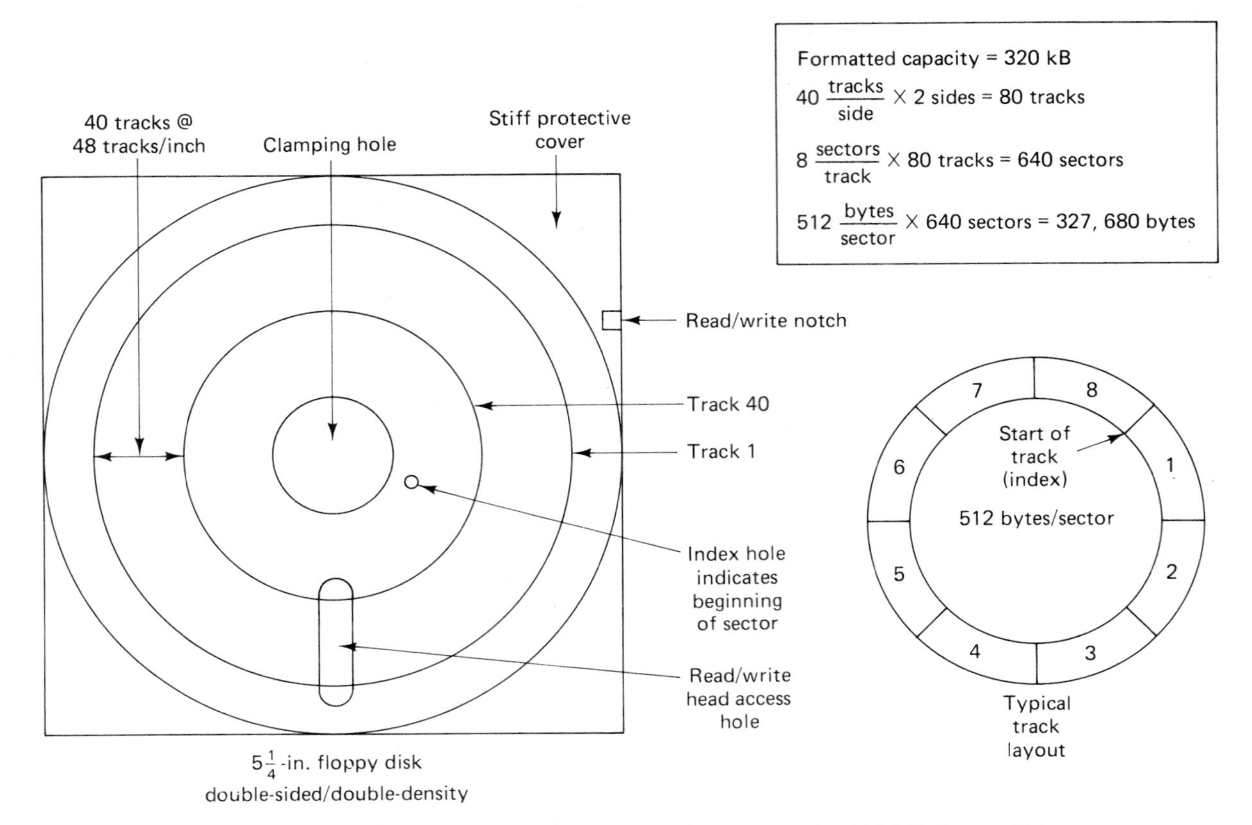

Figure 11.2 Structure of a 320-kB floppy disk.

A floppy disk that rotates at 300 rotations per minute makes 5 rotations per second. Therefore, an index pulse is generated every 200 ms.

The Hard Disk Mainframes and minis need much more mass storage and speed than floppy-disk technology can possibly offer. A hard disk is created by glazing large platters of metal with a magnetic oxide material. Floppy disks are single units. Hard disks, on the other hand, are constructed from multiple platters. A typical hard disk has 3 to 10 platters mounted on a single spindle. Both sides of each platter (except the top side of the top platter and the bottom side of the bottom platter) are used to hold data. Thus a drive that uses an eight-platter hard disk pack would require 14 read/write heads, all precisely aligned to a known reference.

The easiest way to understand the operation of hard disks is to compare and contrast them with floppy disks. The $5\frac{1}{4}$-in. floppy disk rotates at the speed of 300 rpm. When a floppy disk is accessed for a read/write operation, the magnetic head of the floppy disk drive actually comes in contact with the surface of the floppy disk. Typical hard disks rotate at 3600 rpm, and the R/W heads never touch the surface of the disk. The motors that control the movement of the read/write heads (to seek a particular track) on hard disks are thousands of times faster than those on a floppy disk. Because of these faster head-positioning motors, hard disks have access times that are much shorter than those of floppy disks. A typical floppy disk has a track-to-track access time of 75 ms, compared to 10 ms on a typical hard disk.

The track density of a floppy disk is greatly limited. This is because floppy disk drives use an open-loop system for calculating the present track position of the R/W head. When the R/W head encounters track 0, a mechanical switch is closed. Each time the floppy disk drive steps the R/W head toward the center of the floppy disk, the track count is incremented. Each time the R/W head is stepped toward the edge of the disk, the track count is decremented. There is no information recorded on the floppy disk that can be used to give positioning feedback information to the R/W head stepper mechanism.

Hard disks use a closed-loop head-positioning system. A hard disk is not only formatted with data tracks, but also extra tracks that contain servo information. These servo tracks supply feedback data to the head-positioning circuitry. The R/W head can be much more precisely positioned. This not only allows greater track density, but also decreases the track-to-track seek time. Hard disk packs have a storage capacity of 80 MB to 1 GB (gigabyte).

As we have said, hard disks come in groups of 3 to 20 platters. A group of platters is called a *disk pack*. A disk pack is removed by inserting it into a protective heavy-gauge plastic case. A disk pack weighs 5 to 20 lb. A hard disk drive with associated circuitry is often the size of a small chest of drawers.

The read/write head in a hard disk drive never touches the surface of the hard disk. It flies over the surface at a height of less than one ten-thousandths of an inch. If the read/write head touches the surface of a hard disk while the pack is spinning, this event is called a *head crash*. A head crash usually destroys both the read/write head and the hard disk platter.

The Winchester Hard Disk The hard disks that we have just examined are used in the mainframe and mini environment. They are much too bulky and expensive to be used in microcomputer applications. The newest-technology disk drives are called *Winchester drives*. IBM developed Winchester technology. The first Winchester was a dual 30-MB drive called the 30/30. The name "30/30" was soon changed to "Winchester" after the famous Winchester 30/30 rifle, "the gun that won the West." To ensure an ultraclean working environment, the hard disk platters in Winchester drives are not removable. A Winchester drive is a closed unit. For this reason Winchester drives are often called *fixed* disk drives.

Winchesters used in mini and mainframe environments are about one-fifth the physical size of conventional hard disk drives. We will focus our attention on the Winchester drives used with microcomputers.

Winchesters used in micro systems are the same physical

size as $5\frac{1}{4}$-in. floppy drives. They are available in capacities of 5 to 60 MB. Winchesters have a data transfer rate of 5 Mbps, 20 times faster than a $5\frac{1}{4}$-in. floppy. This greatly speeds up applications that require continual disk access. Because of their light weight, low cost, and small physical size, these Winchesters have become extremely popular. Most business-oriented microcomputers have both a floppy drive and a fixed disk drive. The floppy drive is used to record new programs onto the fixed drive and make backup copies of data in the event that the fixed drive experiences a failure.

Mass Memory Backup If a hard disk crashes, the data stored on the disk pack are lost. To avoid this potentially disastrous situation, data from on-line mass storage devices are perodically recorded onto a secondary form of mass storage, which is usually magnetic tape. If you have two hours of music on a cassette tape and wish to find a song near the end of the tape, you must fast-forward the tape, constantly checking to see if you have reached the song. That is an extremely slow process. Magnetic tape is also a serial form of storage; the data bits are written one after the other. It is inexpensive per bit of storage, but it is much too slow to be used as an on-line mass storage. If a hard disk crashes, a spare disk drive will be put into service and the contents of the last tape backup will be transferred to the disk pack in the new drive. The data that was stored between the time of the last backup and the time that the drive crashed will be lost. That is a good reason to make frequent tape backups of all on-line mass storage devices. A microcomputer that has one fixed disk and one floppy disk can use floppies to backup the fixed disk. A 5-MB fixed disk requires almost 20 floppies to back it up. For this reason microtape backup and removable disk cartridge are now available for microcomputer systems.

11.4 INPUT/OUTPUT DEVICES

The CPU controls the internal operation of the computer system. But computer systems must also communicate with the "outside" world. The gateways that interface the CPU with the outside world are called *I/O* (input/output) *devices*. Some I/O devices are designed to interface with human beings; other I/O devices connect the CPU with remote computer systems, transducers and sensors, or other electronic devices.

11.4.1 The Keyboard and Video Display

The keyboard is the most familiar computer input device. Using a typewriter-like keyboard, users can enter programs, commands, and data quickly and efficiently. Paired with every keyboard is a visual output device. The most common output device is the CRT display. These video displays usually have a screen of 24 lines by 80 columns. Video displays can output characters with many different attributes: high intensity, blinking, underlined, and reverse video (dark characters on a light background). Many video displays can also display colors. These display attributes and colors are used to enhance communication between the computer and the human user. Some inexpensive computers use home television sets as their video displays. Commercial-grade TV sets do not have very good resolution. (Resolution is the distance between dots on a video display. The closer the dots, the better the resolution.) Because of this poor resolution, TV sets can usually display a screen of only 16 lines by 40 columns.

Mainframe and minicomputers use keyboard/video display devices called *terminals*. Terminals are connected to computers via a serial communications link. Most computers interface with terminals using a standardized form of serial communications called RS-232. Hundreds of manufacturers supply terminals that support the RS-232 communications interface. This makes it possible for one terminal to be used with any kind of mainframe or minicomputer. (This is a great asset that you will more fully appreciate after our discussion of modems later in this chapter.) All terminals have internal memory that is used to refresh the video display. It is the job of the terminal, not the host computer, to execute this periodic refresh operation. Some terminals are called *smart terminals*. Smart terminals have an on-board CPU that enables the terminal to accomplish local processing and formatting of data. Smart terminals relieve the host computer of many time-consuming housekeeping chores. Dumb terminals have no local intelligence and are basically nothing more than a video display, keyboard, memory, and video circuitry required to support the display, and circuitry used to transmit and receive data on the serial link.

Microcomputers do not use terminals as their keyboard/video display device. A detachable keyboard communicates with the input circuitry of the microcomputer via a serial cable. The video display (called a *monitor*) usually sits on the top of the computer. Unlike the terminal, the video display in a microcomputer does not contain memory or communications electronics. It contains only the basic CRT drive circuitry. The video memory and CRT control circuitry are housed within the computer unit. This integral keyboard/video display is designed to be a cost-efficient alternative to the terminal. We will explore the difference between terminals and video monitors in a later chapter.

The greatest limitation of conventional CRTs is that they are bulky and heavy. In the quest to develop truly portable computers, the flat screen display was invented. Unlike CRTs, which require long electron guns, flat screens are solid-state devices that use LCDs or EL (electroluminescent) displays. During transportation, the flat screen is folded down on top of the computer, creating an extemely light, (12 to 14 lb) low-profile package. When the computer is to be used, the flat screen is unfolded to a comfortable viewing angle. Although flat-screen technology still has many technical problems to overcome (including the lack of color), it is sure to be the dominant video display in the very near future.

11.4.2 Printers

Other than the video display, the most common computer output device is the printer. A printer provides a *hard copy* of computer data, program runs, or program listings. The printer, like the disk drive, is constructed from a combination of electronic and mechanical components. Printers suffer a much higher percentage of mechanical failures than electronic failures.

Large computer facilities have expensive printers called *line printers*. As the name implies line printers appear to print a whole line of characters simultaneously. Line printers use many different methods to accomplish high-speed printing. Some have a chain that rotates horizontally embedded with multiple sets of print hammers. Other designs use rotating drums or vertical bars containing a set of hammers for each print column. Using high-speed mechanics, these printers can output 400 to 3600 lines per minute. Line printers are used in environments where massive amounts of data must be printed in short periods of time.

In the microcomputer world there are two major types of printers: daisy wheel and dot matrix. They are called *serial printers* because each character is printed successively, one after the other. The three major considerations used in judging printers are print quality, speed, and price.

Daisy wheel printers employ a circular printer element with 96 *spokes*. At the end of each spoke is a print block embossed with a letter or symbol. The printer derives its name from the print wheel's resemblance to a daisy, with its many petals. The daisy wheel is easily removed. This allows the use of many print types and character sets. To print a letter or character, the appropriate petal is rotated into position and struck by a hammer, impacting the ink ribbon and paper. The print head is then moved horizontally to the next column and the process repeats. Notice that unlike a conventional typewriter, the print head moves, not the carriage.

Like a good-quality typewriter, a daisy wheel printer produces *letter quality* print. Letter-quality print is the easiest type or print to read and is preferred in office environments. Because of the rotating petals, daisy wheel printers are extremely slow. Low-cost daisy wheel printers print 8 to 16 characters per second (cps), while the more expensive daisy wheel printers print 60 to 80 cps. You will find daisy wheel printers wherever letter-quality printer output is required.

The most popular printers used in the microcomputer environment is the *dot-matrix printer*. Dot-matrix printers do not print a solid character. Instead, each character is constructed from many small dots. Usually, a 7×5 matrix of overlapping dots is used to create a character. The print head contains a vertical column of pins. These pins are used to strike the appropriate dots to create the required character. The pins of the print head can quickly bang out a pattern of dots; unlike the daisy wheel printer, no rotating head is required. This greatly increases the print rate, to 140 to 160 cps. But because the representation of each letter is composed of many small dots, the print quality of most dot-matrix printers is considered quite low. A method used to greatly improve the appearance of dot-matrix characters is to print each character twice, offset by a minute distance. This improves the appearance of the characters, but only at the sacrifice of print speed.

Another advantage of dot-matrix printers is that they are not limited to the standard letters and numbers. Unlike the daisy wheel printer, which must have multiple print elements to create different sets of characters, an infinite number of patterns can be created with a dot-matrix printer. Therefore, dot-matrix printers can output letters, numerals, standard symbols, and graphics characters. The same dot-matrix printer can print in many different languages. The final advantage of dot-matrix printers is price—a top-of-the-line dot-matrix printer costs less than half as much as a comparable daisy wheel printer.

The printers that we have just examined are called *impact printers* because the character is created by impacting a head or pin against an ink ribbon and onto paper. There are many different methods of nonimpact printing. The laser printer is used in environments where letter quality and speed are required. A complex process uses a laser to produce high-quality characters up to the rate of 20,000 liner per minute. Thermographic printers, an older technology, use a head implanted with a heating element and special heat-sensitive paper. A character is essentially burned onto the paper. Ink-jet printing is a new technology that is in competition with dot-matrix printers. Charged drops of ink are fired at the paper, creating a dot-matrix character. Ink-jet printers ar extremely quick, and the print quality is consistent because there is no ribbon to wear out.

11.4.3 The Mouse and Digitizer Pad

The mouse and digitizer pad are popular forms of computer input devices. A *mouse* is a small box the size of a deck of cards that is used to control the movement of a cursor. The mouse sits on the desk top next to the micro. As the mouse is moved, the physical movement is translated into an electronic signal that moves the cursor on the computer's display. The mouse is equipped with one or more switches that can be used to issue commands to the computer. Mice are useful in word processing, selecting options from computer menus, and for using computers as electronic paint screens. Mice come in many different forms, but mechanical mice are the most popular. On the bottom of the mechanical mouse is a round ball that rests on the table top. As the mouse is moved, the ball also moves. This mechanical movement is used to drive vertical and horizontal gears. The output of these gears are converted into horizontal and vertical electronics signals that move the cursor.

Mice are too clumsy to produce high-quality artwork. *Digitizer pads* are used to enter detailed, high-resolution graphics images into a computer. These pads appear to resemble a normal pen and tablet. A typical digitizing pad is constructed with a grid of hundreds of wires. A stylus is used that has magnetic coil in its tip. As the stylus is pulled across the pad, the motion is sensed and converted into an electronical signal that positions the cursor. Digitizer pads have great potential in the computer graphics industry.

11.5 COMPUTERS AND COMMUNICATIONS

Computers do not exist as isolated entities. It is important that computers can communicate with terminals and other computers. Consider the following example. A salesperson for a small company is on a sales trip. The current inventory, which is stored in the mass memory of the computer in the home office, must be checked and an order placed.

The odds are that wherever this salesperson is, there is also a telephone. Communicating directly with a computer can not be done over an ordinary telephone line. Telephone lines have too much capacitance and resistance to send high-speed digital signals for any appreciable distance. A device called a *modem* is used to solve this problem. The modem (MOdulator–DEModulator) takes the digital signal from the terminal's communications I/O port and modulates it onto a carrier. This newly created analog signal can easily be sent along the telephone line. At the other end of the phone line, another modem is used to demodulate the analog signal and recapture the original digital information. In this manner terminals and computers can communicate at great distances. Notice that the modem transmits data from the terminal to the computer and receives data from the remote computer. A modem is both an input and an output device. With a modem and a two-wire telephone line, inventories can be checked, orders placed, applications programs written and run, remote devices controlled, and information exchanged.

Communication programs are available that enable a microcomputer to emulate an RS-232 terminal. With a microcomputer or terminal and a modem, one can access information networks that have stock reports, news stories from hundreds of sources, on-line research libraries, technical computer users' groups, and general-purpose communications bulletin boards, all of which are available at a reasonable hourly rate. The teleprocessing field is one of the most exciting and fastest growing of all computer disciplines.

11.6 BLOCK DIAGRAM OF A TYPICAL MICROCOMPUTER SYSTEM

Now that you understand the components of a generalized computer system, consider Figure 11.3, which illustrates a typical microcomputer personal work-station block diagram. The large rectangle that contains the CPU, boot ROM, system RAM, and I/O circuitry is often called the *system unit.* Sometimes this circuitry is contained on a single PCB; in other applications the system unit may consist of one large *mother board* and many specialized PCBs which are connected to the mother board with edge connectors.

The floppy disk drive and the fixed disk may be inserted into the system unit or they may be stand-alone units. Each external I/O device is interfaced to the system unit with special I/O circuitry. Notice that some devices are output only (printer

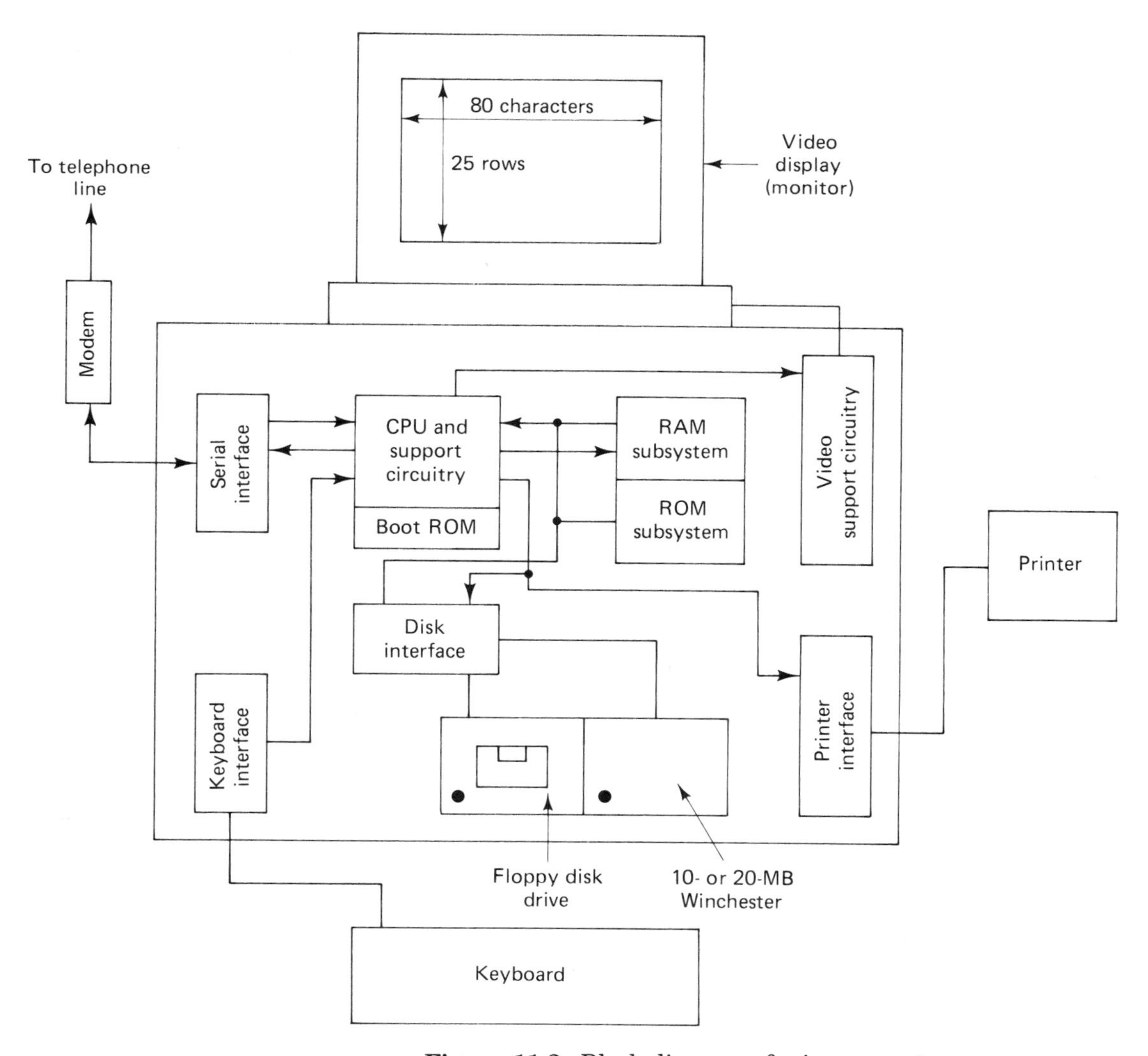

Figure 11.3 Block diagram of microcomputer.

and video display), input only (keyboard and ROM), or bidirectional (modem, disk interface, and RAM).

QUESTIONS AND PROBLEMS

11.1. What are the three goups of computers?

11.2. Where do you expect to find minicomputers?

11.3. What is a "personal workstation"?

11.4. How do the CPUs in mainframes, minis, and micros differ?

11.5. What basic function does bootstrap memory provide?

11.6. What are the uses of system RAM?

11.7. What are the two major types of mass memory?

11.8. Why are semiconductor RAMs much faster than hard disk memory?

11.9. What are some of the differences between floppy disk and hard disk storage?

11.10. What are the two most common forms of mass storage found in microcomputers?

11.11. What does it mean to "format" a floppy disk?

11.12. What is the major difference between older-technology hard disk drives and Winchester disk drives?

11.13. What is a disk "crash?"

11.14. What are the most common input and output devices?

11.15. What does the term "hard copy" mean?

11.16. What are the two types of printers used in microcomputer systems?

11.17. What is a modem?

12 The Microprocessor

In Chapter 11 we learned that the CPU is the brains and heart of a computer system. The CPUs in mainframes are constructed from many PCBs. The CPUs in minis are usually constructed from only one PCB of circuitry. The CPU in a micro resides in one LSI (or VLSI) device called a microprocessor. In its most basic sense, a microprocessor is a CPU on a chip. Because a microprocessor embodies a CPU in a single IC, it requires much less physical space and power than a multicomponent or multiboard CPU, and the cost of a microprocessor is trivial compared to that of a mini or mainframe CPU. The chances of a microprocessor failing (because it is a single IC) are thousands of times less than they are for a conventional CPU. Even when a failure occurs, the inherent simplicity of a microprocessor-based system is conducive to fast and efficient repair. Of course, a microprocessor is much less powerful than the CPU in a mini or mainframe, but for single-user personal computers, the microprocessor provides all the required fire-power.

There are many different microprocessors available commercially. They range in price from $1.50 to $250, which indicates that there is a great range in performance. The simplest and cheapest microprocessors are used as the CPUs in inexpensive home computers, appliances, and other small computer-controlled systems. The most powerful microprocessors rival minicomputer CPUs in their speed and complexity. But from the simplest to the most complex, microprocessors all share a common architecture. In this chapter we consider the generalized architecture of a microprocessor in much the same manner as we considered the generalized architecture of a computer system in Chapter 11. Once you fully understand the structure of a generalized microprocessor, that knowledge can easily be applied to understanding any commercially available microprocessor.

12.1 GENERALIZED INTERNAL ARCHITECTURE

We know that a CPU contains an ALU, registers, and control and timing circuitry. Let's examine these three groups as they appear within a microprocessor.

12.1.1 The Accumulator

Every CPU contains a special register called an *accumulator.* The accumulator is the location where the result of arithmetic and logical operations are automatically stored. The accumulator is always the center of action. Some microprocessors have two accumulators. With these microprocessors it is the programmer's responsibility to designate which accumulator will be used.

12.1.2 General-Purpose Registers

General-purposes registers are not dedicated to a specific function (as is the accumulator). They are used to temporarily hold operands and other data that the CPU currently requires. Some microprocessors have many general-purpose registers, whereas others do not have any and must rely on RAM to hold temporary operands and data. The major advantage of having on-board general-purpose registers is that they have a much faster access time than external RAM.

12.1.3 Dedicated CPU Registers

CPUs also have many registers which are not used to hold temporary operands or general-purpose data. These registers are dedicated to store specific types of information that the microprocessor may require during the execution of an instruction. Consider the following dedicated registers that appear in every microprocessor.

Flag Register This register is really a handful of independent flip-flops. Each flip-flop is a flag that reflects the current state of an operational parameter of the microprocessor. A typical flag is the *carry flag.* Assume that all the registers in a microprocessor are byte-wide (8 bits). If the addition of two 8-bit binary numbers results in a sum that is greater than 8 bits, the result of the addition placed into the accumulator will be 256 less than the true value.

```
     1111 0000    (240)
   + 0001 0011    ( 19)
  (1)0000 0011    (259)
```

The carry generated in the preceding addition cannot fit into the accumulator. The accumulator indicates that the result of the addition is 3 when the actual sum is really 259. The carry flag is set to indicate an operation that results in a carry, and reset to indicate an operation that does not generate a carry. The status of the carry flag can be tested by the program; if the carrry flag is set, the program may branch to an error routine.

That is an example of how the flip-flops in the flag register are used to indicate the operational status of the microprocessor after the completion of each instruction. For that reason, flag registers are often called *status* or *condition code registers.* It is important to remember that the microprocessor automatically sets or clears the flags, giving the program a means of making intelligent decisions.

Other typical flags indicate whether the result of an operation is 0 (the zero flag), the arithmetic sign of the number in the accumulator (sign flag), and if the number of logical 1's in the accumulator is an odd or even number (the parity flag). All these flags (and others) have important meanings that will be examined in later chapters.

Stack Pointer Often, microprocessors will be in the middle of performing one job when they are suddenly required to suspend execution of the current task to service a second, higher-priority task. At that time the CPU registers will contain information that is important to the first task. This information must be saved in RAM before the microprocessor can began servicing the second task. After execution of the second task is complete, the microprocessor will read the information from RAM back into the proper CPU registers, and execution of the first task can continue. A dedicated register is used to simplify the operation of "preserving the CPU environment." This dedicated register is called the *stack pointer.*

A portion of RAM in main memory is always reserved to store CPU registers temporarily. This area of memory is called the *stack.* The stack register holds the address of the next available memory location in the stack. Using the contents of the stack pointer as an address, the microprocessor can easily access the stack. The address stored in the stack pointer is automatically updated each time the stack is accessed for a read or write operation. One of the first steps in any program is to load the stack pointer with the first address in the RAM stack. We will learn more about the stack later in this chapter.

Program Counter The program counter, like the stack pointer, is used to hold an address. The contents of the program counter always point to the memory location where the next instruction or data byte will be "fetched." The program counter is incremented automatically after each instruction or data byte is fetched from memory.

Instruction Register When a microprocessor fetches a byte of data from memory that is to be interpreted as an instruction, it places this *operation code* into a temporary register called the instruction register.

Temporary Registers The CPU also contains many registers that are used by the microprogram to store data, addresses, and operands for short periods of time. These registers are not available to the programmer.

12.1.4 Address and Data Buffers

The microprocessor also contains a unidirectional address bus buffer and a bidirectional data bus buffer. These buffers isolate the internal bus system of the microprocessor from the external computer system bus and provide enough current gain to drive a few LS TTL inputs. Almost all microprocessor systems also use external TTL buffers. As you are aware, TTL buffers are designed to drive the large number of inputs that are connected to the system bus.

12.1.5 Control, Status, and Timing

As we learned in Chapter 8, the microprocessor is the master of the bus system. Microprocessors have control signal outputs that are used to access data in memories and I/O devices. Typically, microprocessors have 5 to 10 control signals that are used to manipulate memory, I/O devices, and other external logic.

Microprocessors also have inputs that are used to monitor the status of the devices on the external bus. Typically, a microprocessor has 5 to 10 status inputs. These inputs are used to monitor the operational status of memories, I/O devices, and other information that may be useful to the microprocessor.

Microprocessors require an extremely accurate and stable clock. The clock is the timing reference that synchronizes the address and data buses and the control signals. Some microprocessors have an internal oscillator that requires only the connection of an external crystal. Other microprocessors require that the clock input be generated with external logic. The clock must meet precise rise- and fall-time specifications. All internal activity within the microprocessor and external memory and I/O accesses are synchronized to the clock.

Microprocessors have a reset input. When power is first applied to a system, the reset input of the microprocessor is held at an active level for 100 to 500 ms. This allows time for the external logic to be powered-up and ready before the microprocessor begins to execute the boot program and initialize the system.

In most systems the microprocessor is not the only device capable of taking control of the system bus. When a large block of data needs to be moved between a disk drive and main memory, the microprocessor is not always the best device for the job. Another LSI device, called a *direct-memory-access controller* (DMAC) is used for this purpose. The DMAC and microprocessor communicate by means of a status line called the DMARQ (direct-memory-access request) and a DMAK (direct-memory-access acknowledge). When the block of data is ready to be moved, the DMAC signals to the microprocessor, via a DMARQ input, that it wants to take control of the system bus. When the microprocessor has finished its current instruction, it responds with an active level of a DMAK output and surrenders the system bus. After the block move is complete, the DMAC surrenders the system bus to the microprocessor and normal processing resumes. Every microprocessor has the ability to support DMA data transfers. In Chapter 19 we examine a DMAC chip.

12.1.6 Servicing I/O Devices: Polling and Interrupts

Consider a microprocessor system that is used to monitor many important parameters in an automobile: oil pressure, radiator temperature, and brake and clutch fluid levels. A pressure sensor will be used to monitor the oil pressure; if the oil pressure falls below a certain level, the output of the pressure sensor will go to an active logic 0 level. The temperature sensor monitoring the radiator and the float sensors monitoring the brake and clutch fluid levels will operate in a similar manner. Devices in the "outside" world are interfaced to the CPU via I/O ports. Assume that the output of each sensor drives an input port connected to the CPU.

How is the CPU going to know when the output of a particular sensor has switched to an active level? There are two methods that a microprocessor can use to monitor input ports. The first method uses a program that periodically checks the value on each of the four input ports. If the microprocessor finds an input port that is at an active level, it sets the appropriate alarm LED on the car's dashboard. Periodically, this program may first check the value on the pressure sensor input and determine if the oil sensor LED should be illuminated. It will then check the status of the remaining input ports in a similar manner. Notice that the program does all the work; the input ports do not indicate to the microprocessor that a sensor output has gone to an active level. The method of using a program for periodic interrogation to check the status of input devices is called *polling.* (The work "poll," in its most common usage is a process of questioning people to obtain information. Therefore, a program that periodically samples the status of input devices is called a polling program.)

The action of polling input devices is like a telephone with no bell; every minute or so, you have to lift up the receiver and ask if there is anyone on the line. To ask a person to poll a telephone line is unreasonable. But if a computer is not otherwise loaded down and the device requiring service can afford to wait, polling is a good, simple method of monitoring input devices.

If polling is like a telephone with no bell, is there a method of monitoring inputs devices analogous to the bell on a telephone? Yes. That method is called an *interrupt.* Let's continue with the telephone analogy. You are busy performing an important task, such as reading a book or doing your math homework, when the telephone rings. The first thing you must decide is whether you want to bother answering the telephone. (You could always ignore it, and risk missing an important call.) If you decide to answer the phone, you must stop what you are doing and mark your place in the book or write down the intermediate result of a math problem, and then answer the phone. After you have finished the conversation, you must hang up the phone and resume your reading or math homework. (Remember that you placed a book marker or wrote down an intermediate answer so that you can continue at the place where you were interrupted.)

The major difference between a telephone without a bell (polling) and one with a bell (interrupts) is that the telephone with a bell indicates to you when a caller is on the line. In the same manner, if an input device signals to the microprocessor whenever it has an active input level, it is "ringing a bell" that will gain the microprocessor's attention. Like a person, the microprocessor can choose to ignore the input device or stop what it is currently doing (and, of course, place book markers and record intermediate results) and service the input device.

Microprocessor systems use both polling and interrupts to interface with input devices. The most important input devices will be connected to interrupt inputs on the microprocessor. They will immediately gain the microprocessor's attention. Other devices that are less critical will wait their turn to be polled by the microprocessor.

Sometimes a combination of polling and interrupts is used. The interrupt request outputs of many devices can be ORed together and drive a single interrupt request input on a microprocessor. When the microprocessor senses the interrupt, it can then poll each device to establish which one requested service.

12.2 GENERALIZED INSTRUCTION SET

As you learned in the first chapter, digital circuitry "thinks" in binary: logic 0 and logic 1 levels. When a CPU fetches an instruction from memory, the instruction is in the form of binary data. After the CPU decodes the byte in its internal microinstruction ROM, the instruction can be executed. It is extremely difficult for human beings to memorize and recognize the patterns of 1's and 0's that constitute an instruction.

For the reason, each instruction is available in two forms. The first form is a sequence of 1's and 0's that the microprocessor can understand. It is called *machine language.* The second form is a three- to five-letter instruction *mnemonic.* A mnemonic is a symbolic name that is assigned to each microprocessor instruction. It helps us understand what the instruction actually does. Mnemonics make much more sense than do a string of 1's and 0's. Engineers and technicians write microprocessor programs using symbolic mnemonics to represent each microprocessor command. This symbolic form of instructions is called *assembly language.* A program called an *assembler* is invoked to translate human-level mnemonics into machine-understandable 1's and 0's.

All microprocessors have the same basic instruction set, but each uses a different set of mnemonics to represent its instruction. Once you understand one instruction set, others are easy to learn. This section describes basic instructions that are common to all microprocessors.

12.2.1 Data Transfer Instructions

The most basic group of microprocessor instructions is called *data transfer instructions.* Data is stored in both CPU registers and RAM. Data transfer instructions accommodate the movement of data from register to register, register to memory, or memory to register. The mnemonics for these instructions are often MOV (move) or LD (load). The MOV or LD instruction copies the contents of one register or memory location (called the *source*) to another register or memory location (called the *destination*).

There are MOV and LD operations that involve *immediate data* as opposed to data already stored in a register or memory. If you want to load a register with an explicit byte of data (for example, load the accumulator with C4H), the data is called immediate because the source of the data is not another register or memory location. It immediately follows the instruction in memory. An instruction that uses immediate data must specify the actual data and the destination of the data.

12.2.2 Arithmetic Instructions

You have learned that the ALU in a CPU can perform basic arithmetic and logic functions. All microprocessors can perform simple binary addition and subtraction. If an addition results in a carry or a subtraction in a borrow, the carry flag is set to indicate a potential program error. There are special addition and subtraction instructions that are used to add or subtract strings of numbers that must manipulate carries or borrows.

The result of an addition or subtraction is stored automatically in the accumulator.

The operation of adding or subtracting 1 is so common that increment and decrement instructions are included in every instruction set. This enables registers to be used as event counters.

Only advanced microprocessors have explicit multiplication and division instructions. Simple microprocessors perform multiplication and division indirectly. What does it mean to multiply a binary number by 2? Consider the following example:

(a) 0001 0000 → (b) 0010 0000 → (c) 0100 0000
(16) (32) (64)

The byte in (a) is equivalent to decimal 16. If we shift the byte once to the left and fill in the least significant bit (LSB) with 0, the result is decimal 32. If we shift the byte in (b) once to the left and fill the LSB with 0, the result is 64. We can now derive the simple rule that links logical shifts with multiplication:

To multiply a binary number by a factor of 2, simply shift it once to the left and insert a 0 in the LSB. This operation is called an arithmetic shift left.

Because division is the inverse of multiplication we can state:

To divide a binary number by a factor of 2, shift it once to the right, inserting a 0 into the MSB. This operation is called an arithmetic shift right.

What is the process of multiplying a number by 6? Two consecutive arithmetic-shift-left operations will multiply the number by 4. Another arithmetic shift left would multiply the number by 8. The question is: How do we handle multiplications that are not a power of 2? Consecutive addition is the basic definition of multiplication. Applying that thought to the problem of multiplying a number by 6, we can see that after two arithmetic shift-left operations, we should perform two consecutive additions. This results in a times-6 operation.

For example, multiply the number 8 (0000 1000) by 6 (0000 0110).

(a) Arithmetic shift left: 0000 1000 → 0001 0000 (16)
(b) Arithmetic shift left: 0001 0000 → 0010 0000 (32)
(c) Add 8: 0010 0000 + 0010 1000 → 0010 1000 (40)
(d) Add 8: 0010 1000 + 0000 1000 → 0011 0000 (48)

0000 1000 × 0000 0110 = 0011 0000

Of course, the actual assembly language program would be much more complicated than in the preceding example. The product would have to be checked after each operation to ensure that the accumulator does not overflow to create a carry. Also, the problem of negative numbers has to be addressed. What is important now is that you understand the concept of arithmetic-shift-left and shift-right operations.

Shift and Rotate Operations The preceding example illustrated the arithmetic-shift-left and shift-right instructions. Notice that in an arithmetic-shift-left instruction the MSB is lost; it is essentially shifted off the end of the register. The same is true for the LSB in an arithemetic-shift-right operation. The hardware analogy of an arithmetic-shift-left instruction is illustrated in Figure 12.1 as an 8-bit serial in/parallel out shift register. The data input of the LSB is tied to a logic 0 level. After the clock pulse all the data have shifted to the left by 1 bit. The data that was in the MSB is lost.

Microprocessor instruction sets also contain rotate instructions. The difference between a shift and rotate instruction is quite simple: If the output of the MSB flip flop (data bit 7) is tied to the D input of the LSB flip-flop, the hardware configuration will execute a rotate-left instruction on each positive edge of the clock. No bits are lost as a result of a rotate instruction. They are merely moved to a different flip-flop in the storage register. Rotate instructions are used in programs for two major puposes: as a method of moving a data bit to affect a flag in the status register, and to control hardware devices, such as the four-phase digital stepper motors that control the position of the R/W heads in a disk drive.

12.2.3 Logical Instructions

The Boolean functions AND, OR, XOR, and NOT can also be performed under program control. Refer to Figure 12.2. Consider an instruction that tells the ALU to AND the contents of two registers: the accumulator and a register that we will call register B. The result will be placed in the accumulator, overwriting the original operand, and the contents of the B register

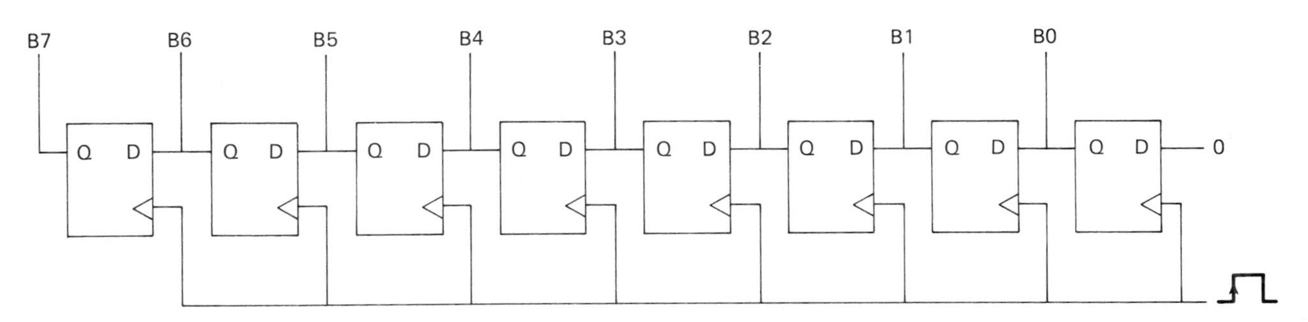

Figure 12.1 Hardware equivalent of arithmetic-shift-left instruction.

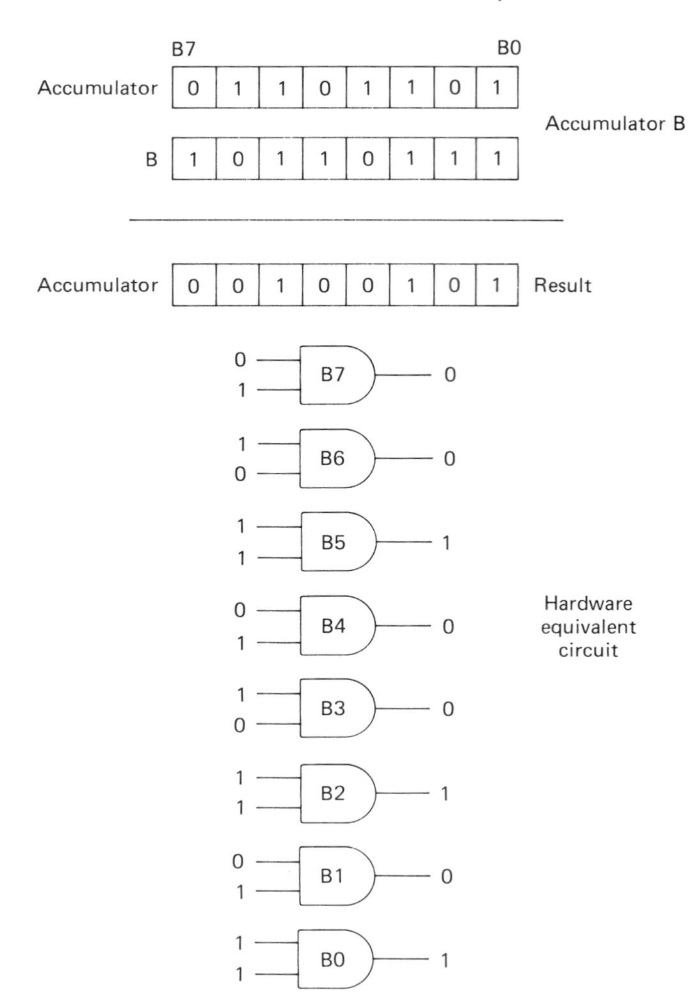

Figure 12.2 Hardware equivalent of AND instruction.

will not be affected. Each bit in the accumulator is ANDed with the same data bit in the B register. This instruction is the software equivalent of the eight two-input AND gates pictured in Figure 12.2. The same method is used to perform the OR and XOR functions.

The invert function requires only a single operand. Suppose that we wanted to NAND together the contents of the accumulator and B register in Figure 12.1. After the result of the AND operation is placed into the accumulator, the contents of the accumulator can then be complemented and the NAND function is complete.

The Boolean operations in microprocessor instruction sets are used in the same manner that we have used hardware AND, OR, and XOR gates and inverters. A microprocessor can be made to accomplish any function with a program that can be performed with digital hardware.

12.2.4 Jump and Branch Instructions

Consider the general flow chart of a microrocessor system that is used to control the heating and air-conditioning units in a

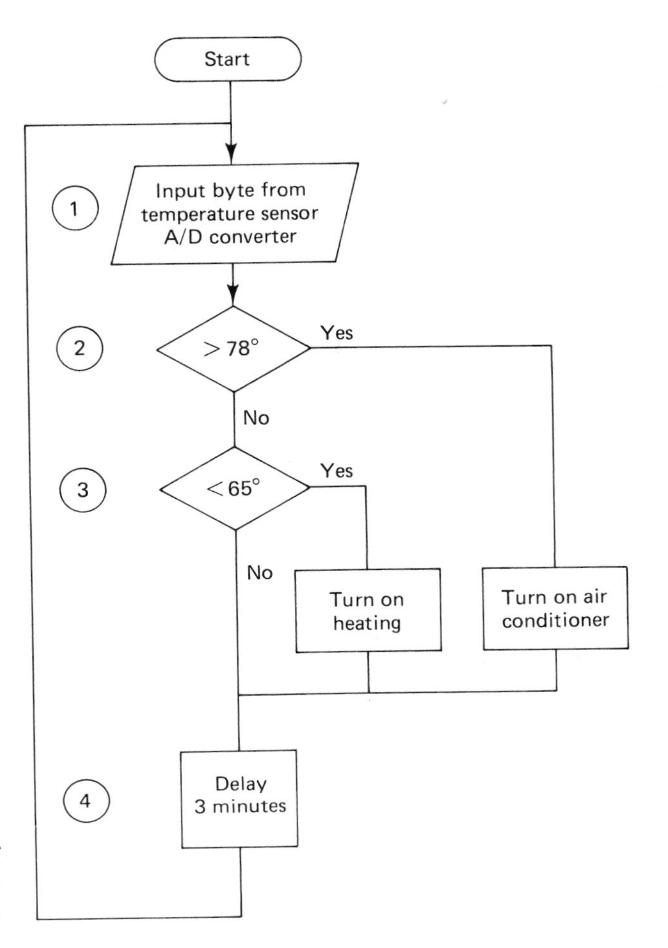

Figure 12.3 Flowchart of heater/air conditioner program.

house (Figure 12.3). The control sequence is initiated with the microprocessor reading the output of the temperature A/D converter and storing this value in a CPU register. If the temperature is greater then 78°F, the microprocessor jumps to a subroutine (a small program) that engages the air conditioner. If the temperature is less then or equal to 78°F, the microprocessor executes the next sequential instruction. This instruction checks to see if the temperature is less then 65°F. If the condi tion is true, the microprocessor jumps to a subroutine that engages the heater.

The different paths of the flowchart intersect at the box labeled as a 3-minute delay. After 3 minutes enough time has passed for the heater or air conditioner to affect the temperature. The program execution resumes with another temperature sample.

The microprocessor is considered to be an intelligent device because it can respond to as many different conditions as the program allows. Normally, a microprocessor executes a program in a sequential manner. Figure 12.3 demonstrates a situation where the execution of a program will be based on the condition of certain parameters. The variable parameter in this example is temperature.

Jump and branch instructions are used to change the normal sequential execution of a program. They are often used in conjunction with the flag register to test for certain conditions. Jump instructions place the address of the next instruction to be executed into the program counter of the microprocessor. Jump instructions are said to use *absolute addressing,* because they provide a completely new address.

Branch instructions are similar to jump instructions except that they add a value to the contents of the program counter. This value, called an *offset,* will also change the sequence of a program's execution. Because a positive or negative offset is added to the current contents of the program counter, the operation is said to be *relative* to the program counter.

12.2.5 Calls and Returns

Microprocessor programs are usually constructed from many small subprograms. These subprograms are called *subroutines.* Each subroutine accomplishes a small, discrete task. Programs written as a collection of subroutines are much easier to debug and modify. Subroutines can be tested as individual units. After each subroutine is debugged and running correctly, they can be strung together to perform a complex operation. This method of programming is called *structured* or *modular programming.* Each subroutine is a separate module (a building block) that is used to construct a whole program.

The *call instruction* is used to invoke a subroutine. When a microprocessor executes a call instruction, two important things happen: (1) the current contents of the program counter are automatically saved. (remember that this is the address of the next instruction that would be normally executed); and (2) the beginning address of the subroutine is placed into the program counter. The microprocessor jumps to the first instruction of the subroutine, but unlike the jump instruction, a return address has been saved.

The return instruction is the last instruction in a subroutine. When the microprocessor sees a return instruction, it places the return address into the program counter. The program then effectively jumps back to the next instruction after the original cell. For every call instruction there must be a matching return instruction.

Breaking large programs into many simple subroutines not only eases program writing and testing, but also saves memory because once a subroutine is in a program, it can be called an unlimited number of times. Figure 12.4 illustrates a block diagram of a program that calls three subroutines. Each subroutine ends with a return statement. Notice that the second subroutine calls yet another subroutine. This practice is called *nesting* subroutines. When the second-level subroutine reaches the return statement, it returns to the subroutine that called it. It is common practice to have many levels of nested subroutines.

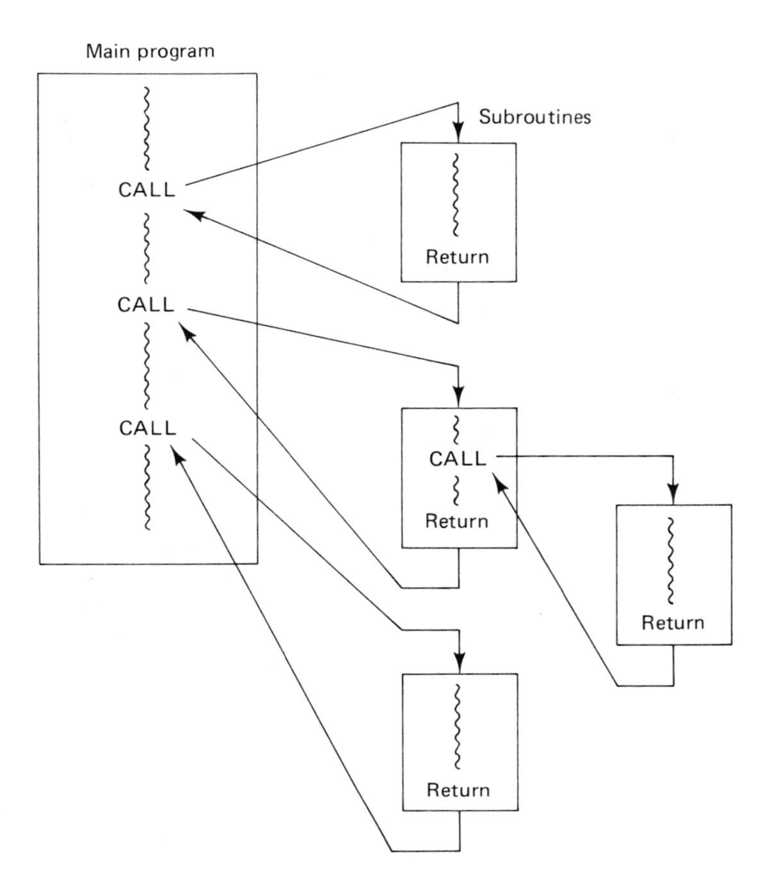

Figure 12.4 Calling subroutines from a main program.

To experienced programmers who have built large libraries of subroutines, creating a new program is little more than picking out the required subroutines and tying them together.

In Chapter 11 we mentioned the concept of an operating system. We said that an operating system was a large program that is used to help the programmer efficiently interface with the computer hardware. An operating system is really a large collection of subroutines. Each subroutine accomplishes a small task such as writing a character to the video display or inputting a character from the keyboard. The programmer is allowed to use any of the operating system's subroutines. This is like having a built-in library that one can use at will.

12.2.6 Maintaining the CPU Working Environment: Stack Operations

We have just stated that microprocessor programs are constructed from many subroutines. If the main program is using the registers to hold important information and a subroutine is called, it may also require the use of the registers. This would result in important data being overwritten and destroyed.

To ensure that subroutines can have free use of the CPU registers and still not destroy information, a portion of RAM is set aside to store data temporarily. This block of RAM is called the *stack*. You already know that every CPU has a register called the stack pointer. The stack pointer holds an address that points to the next free location in the stack. When data is saved on the stack, it is said to be *pushed* onto the stack. When data is read from the stack it is said to be *popped* off the stack.

The easiest way to understand the stack is to follow an example that pushes data onto the stack and then pops it off the stack, restoring the original CPU environment. Figure 12.5a illustrates the initial contents of the accumuator (Acc), flag register, B register, C register, and SP (stack pointer). The B and C registers can be thought of as general-purpose CPU registers. The SP is initialized to 0FFF. The memory location 0FFF is said to be the *top of the stack.* It is the highest memory location in the RAM stack. We are interested in the first five bytes of the stack. The "XX" denotes an unknown value in a memory.

The subroutine segment in Figure 12.5a illustrates saving the contents of the four registers on the stack, and then proceeding with the subroutine execution. The last instructions before the return must restore the CPU environment, previous to the subroutine call. That is the function of the pop instructions. Refer to Figure 12.5b.

Push Acc

1. Decrement the contents of the SP, 0FFF − 1 = 0FFE.
2. Write the contents of Acc (F7) into the memory location pointed to by the contents of the SP: F7 → 0FFE.

Push Flag Register

1. Decrement the contents of the SP, 0FFE − 1 = 0FFD.
2. Write the contents of the flag register into the memory location being pointed to by the contents of the SP: 04 → 0FFD.

Push B Register

1. Decrement the contents of the SP, 0FFC − 1 = 0FFB.
2. Write the contents of the B register into the memory location pointed to by the contents of the SP, FF → 0FFC.

Push C Register

1. Decrement the contents of the SP, 0FFC-1 = 0FFB.
2. Write the contents of the C register into the memory location pointed to by the contents of the SP, 4C → 0FFB.

The subroutine can now execute normal instructions. The original contents of the Acc, flags, B, and C registers are being held in the stack. Before the return instruction is executed, the contents of the Acc, flags, B, and C registers must be restored. Refer to Figure 12.5c.

Pop C Register

1. Read the contents of the memory location (4C) pointed to by the stack pointer (0FFB), and place it in the C register.

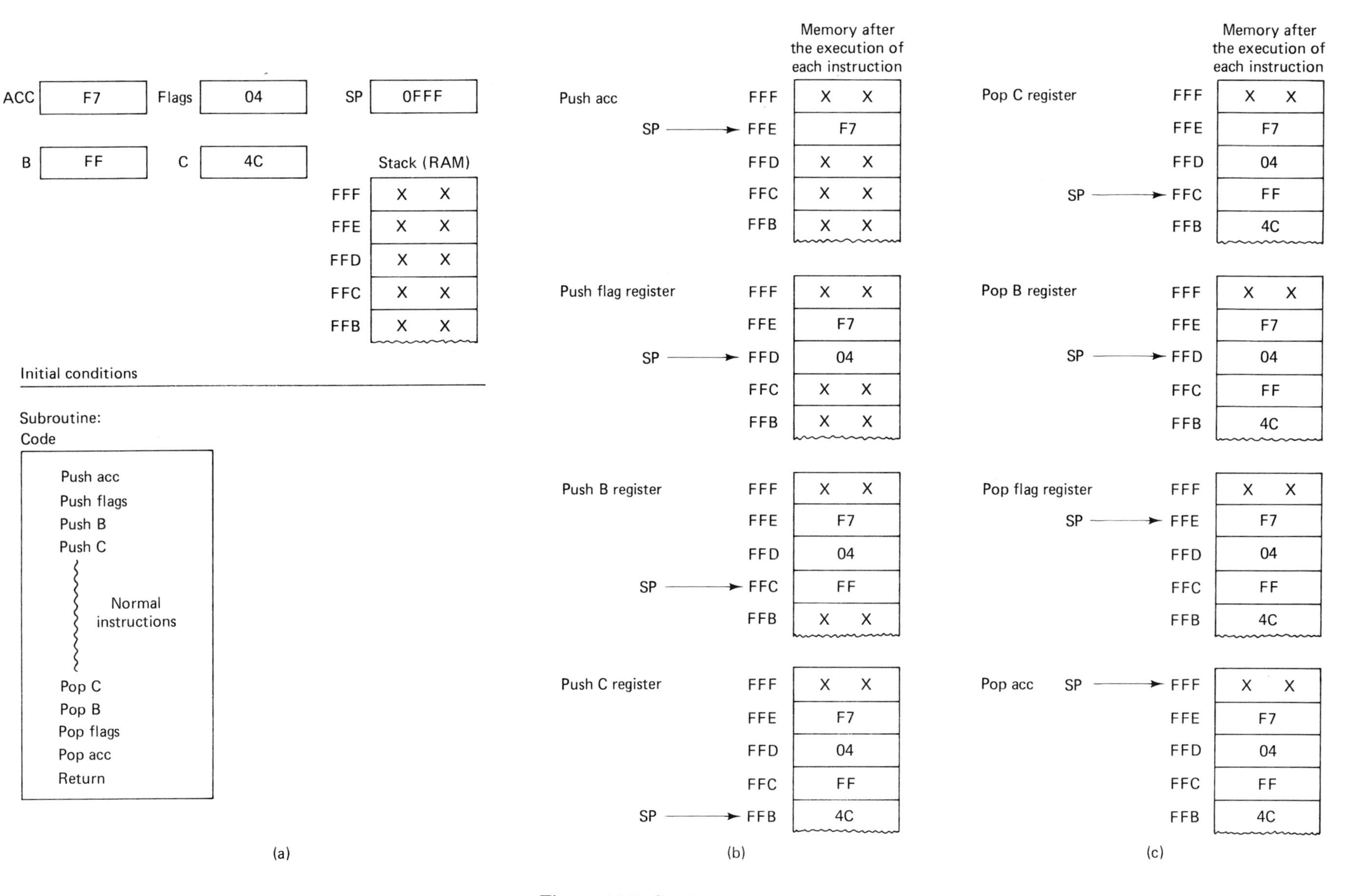

Figure 12.5 Stack operations.

2. Increment the contents of the SP, 0FFB + 1 = 0FFC.

Pop B Register

1. Read the contents of the memory location (FF) pointed to by the stack pointer (0FFC), and place it in the B register.
2. Increment the contents of the SP, 0FFC + 1 = 0FFD.

Pop Flag Register

1. Read the contents of the memory location (04) pointed to by the stack pointer (0FFD), and place it in the flag register.
2. Increment the contents of the SP, 0FFD + 1 = 0FFE.

Pop Acc

1. Read the contents of the memory location (F7) pointed to by the stack pointer (0FFE), and place it in the Acc.
2. Increment the contents of the SP, 0FFE + 1 = 0FFF.

There are many important observations that we can make from the push/pop example. Notice that the stack grows downward in memory. Each byte that is pushed onto the stack resides at a lower address than the previous byte. Also notice that the registers must be popped off the stack in the opposite order from that in which they were pushed onto the stack. The stack pointer always points to the last byte that was pushed onto the stack. Even after a byte is popped off the stack, it still resides in the stack—a pop instruction is a normal nondestructive memory-read operation. The contents of the stack memory will not change until a new push instruction overwrites a previous byte or power is lost.

The push instruction decremented the SP pointer first and then wrote the register to memory. The pop instruction did exactly the opposite; it read the byte from memory first and then it incremented the SP. The most powerful aspect of the push and pop instructions are that the programmer need not keep track of the stack addresses where the registers are stored. This is done automatically by the microprocessor. After the SP is loaded with the top of stack address, all further stack operations are referenced to the contents of the SP. This makes saving registers on the stack and restoring them from the stack a simple, painless operation.

Although push and pop instructions are not complex, the concept of the stack is still widely misunderstood. Carefully follow the sequence of operations in Figure 12.5 until you thoroughly understand the basic operation of the stack.

12.2.7 Input and Output Instructions

Microprocessors can treat I/O devices in one of two ways. An I/O device can be addressed using an I/O read or I/O write instruction. The I/O read and I/O write instructions are analogous to memory-read and memory-write instructions. The I/O read or write control signal can be used to enable an I/O decoder, just as we have seen memory-read and memory-write signals enable memory address decoders.

When the microprocessor treats I/O devices separately from memory devices, the system is said to be *I/O mapped.* The second way to handle I/O devices is to treat them as if they were just another memory location. An input device would be accessed with a memory-read operation, and an output device would be accessed with a memory-write operation. In essence, the microprocessor would not differentiate between an I/O access or a memory access. Such a system is called *memory-mapped I/O.*

Both I/O-mapped and memory-mapped I/O have advantages. Some microprocessors do not have I/O control signals and all I/O must be memory mapped. Other microprocessors have I/O control signals. This gives the circuit designer the option to choose I/O-mapped or memory-mapped I/O. We will study the advantages of each I/O treatment in a later chapter.

12.3 OPERAND ADDRESSING MODES

We have just studied the general types of instuctions that one will find in every microprocessor instruction set. We will now consider the general methods used to provide the required data for each instruction. For example: If we want to add two numbers together, an address or location must be provided to help the microprocessor locate each operand. Understanding addressing modes is as important as understanding the actual microprocessor instructions.

Implied Addressing Implied addressing is the simplest form of addressing an operand. Some instructions have the address of the operand implied as part of the instruction. Consider an instruction whose only purpose is to reset the carry flag. This instruction need not call out an explicit address, because the carry flag is the address and is implied in the instruction.

Register Addressing Many times, the operands of an instruction will be in the CPU registers. If this is the case, an instruction need only call out the register(s) that are involved in the instruction. A typical instruction to copy the contents of the B register into the C register is

```
LD C, B
```

Where LD is the mnemonic that represents the load or copy instruction, B is the source register (the place where the data will be found), and C is the destination register (the location to place the copy of the B register).

Immediate Addressing This addressing mode provides the data as part of the instruction. Reconsider the LD instruction. We may want to load the C register with the byte of data 7FH.

We must include this data as part of the instruction:

```
LD C, 7FH
```

The source of the data is the instruction itself. The destination is the C register. The C register will be loaded with the byte of data, 7FH. The "H" tells the microprocessor that the byte 7F is in hex.

Direct Addressing Many instructions need to supply the address of a memory location or I/O port. The jump instruction is used to force a new value into the program counter. As you remember, the contents of the program counter are used as an address to find the next instruction to execute.

```
JMP FF00H
```

This instruction loads the program counter with the address FF00. The microprocessor will then fetch its next instruction from memory location FF00. This is called *direct addressing* (or *absolute addressing*) because the instruction provides two bytes of data which represent the address where the microprocessor will jump.

Register Indirect Addressing The push and pop instructions use the contents of the stack pointer as an address to access the stack.

```
PUSH B
```

This instruction says: Write the contents of the B register into the memory location pointed to by the SP. This is an example of register indirect addressing. Unlike the jump instruction that provided the address, the push instruction uses an indirect method of finding the address—the contents of the SP. You will find that register indirect addressing is one of the most useful addressing modes.

There are many more advanced addressing modes that are available on sophisticated microprocessors. But once you understand these basic addressing modes, the others will be much easier to understand.

QUESTIONS AND PROBLEMS

12.1. What is the simplest definition of a microprocessor?

12.2. Name the three types of registers found in a microprocessor.

12.3. What is the SP (stack pointer)?

12.4. What is the PC (program counter)?

12.5. Why do microprocessors require a clock input?

12.6. What does it mean to "poll" an I/O device?

12.7. Compare polling with interrupts.

12.8. What does it mean when it is said that computers "think" in binary?

12.9. What is the difference between assembly language and machine language?

12.10. What are data transfer instructions?

12.11. How can microprocessors perform tasks that are equivalent to logic gates?

12.12. Do microprocessor programs always execute in a sequential manner?

12.13. What is the concept of "structured" or "modular" programming?

12.14. How does the SP keep track of all the elements on the stack?

12.15. Why are registers popped off the stack in the opposite order from that in which they were pushed onto the stack?

12.16. What is the difference between I/O-mapped and memory-mapped I/O?

13 Microcomputer Communication Techniques and Interfacing

Computers communicate with the outside world through I/O ports. In this chapter we investigate the methods used to connect computers with common peripherals such as printers and modems. The most critical aspect of any form of communications is that all parties understand and follow the same set of rules.

If two people are trying to converse, and one is speaking English while the other is speaking Japanese, very little actual information will be exchanged. The same is true of electronic devices; both the computer and peripheral must be speaking the same language. This concept of "language" involves not only the data being exchanged, but also the manner in which it is presented. You will discover that communications between electronic devices is defined by strict and upcompromising rules, formally called *communications protocols.* As we proceed through this chapter, remember to continually compare microcomputer communications techniques with normal conversation; you should see many similarities.

13.1 PARALLEL AND SERIAL CONNECTIONS

There are two ways that we can transfer data between computers and peripherals. *Serial interfaces* transfer data as a stream of logic 1's and 0's (refer to Figure 13.1). In a basic serial interface one device will act as the transmitter and the other

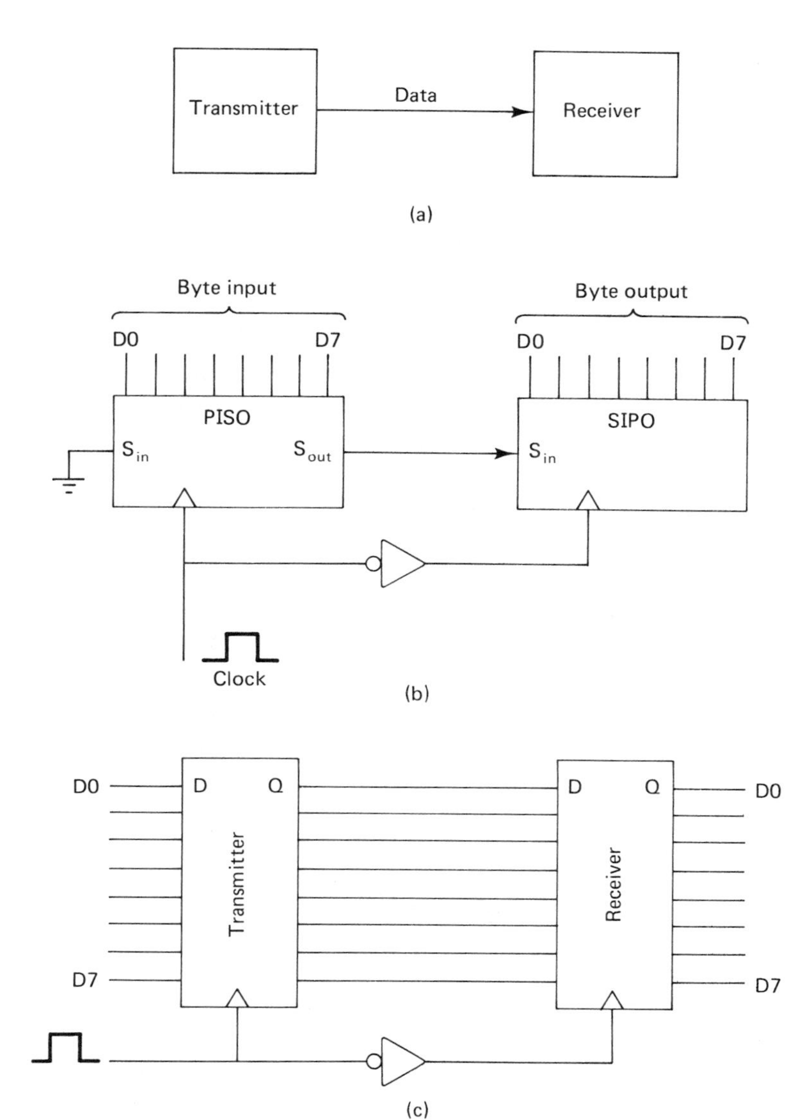

Figure 13.1 Serial and parallel interfaces.

device as the receiver. The transmitter performs a parallel-to-serial conversion (parallel in/serial out shift register) on the byte of data, and the receiver reconstructs the serial bit stream with a serial-to-parallel conversion (serial in/parallel out shift register).

The transmitter and receiver share a common clock in the serial communication system in Figure 13.1. The transmitter clocks out a bit on the rising edge and the receiver clocks in this bit on the falling edge. In practical serial interfaces, a common clock is not used. The idea is to communicate over a single conductor. This constraint is what leads to the inherent complexity of serial communications. If a common clock is not used, how are the transmitter and receiver synchronized? The concept of synchronization entails two ideas:

1. How does the receiver know when the transmitter has started sending data?

2. How does the receiver know how fast the data is being sent?

In this chapter we answer these two significant questions in great detail.

Parallel communication uses one conductor for each bit in the data byte. You can think of a parallel interface as a small bus system between two devices. Figure 13.1c illustrates a simple parallel communication system. Because eight lines are already needed for transmitting data, adding an extra line for a common clock is trivial.

Serial interfaces are used when data must be sent over one line. This is what a terminal uses to send data over phone lines via a modem. The trade-off for saving lines is the complex synchronization scheme. Parallel interfaces are very simple but require a line for each data bit and at least one more line for a handshaking signal.

Serial communication can take three forms: asynchronous, block mode, and synchronous. *Asynchronous serial communication* is performed on a character-by-character basis. The term "asynchronous" implies that no clock information is exchanged, only data. Asynchronous communications require special start and stop bits to mark the beginning and end of a character. These start/stop bits form a character *frame.*

Block mode is a modified form of asynchronous serial communications. Each character is framed with start and stop bits, but instead of sending characters as single units, a whole block of characters is assembled and sent with one command.

Synchronous serial communication is used in high-speed systems. Instead of framing each character with start and stop bits, special sync characters are sent when the computer is not sending standard data. These special sync characters serve as clock pulses to synchronize the transmitter and receiver. Data are sent in large blocks that include three distinct parts: the header, body, and trailer. The header contains information regarding the size of the data block. The body contains the actual data. The trailer has a block check character (BCC) that is used to ensure the integrity of the data. The technical details of asynchronous serial communications are discussed in great detail in this chapter. Block and synchronous communications are advanced topics that will not be pursued in this book.

13.2 THE ASCII CODE

Think of the keyboard on a conventional typewriter. It has approximately 48 keys. Each key defines two letters (upper- and lowercase) or two symbols (such as "!" and "1"). Each of these letters, numerals, and symbols is an important element in written communications.

Computers, like all digital circuits, recognize only logic 0 and logic 1 levels. When the lowercase "a" is depressed on a keyboard, how does the computer sort out the byte of 1's and 0's? How does it distinguish between an "a" and the "#" sign or the numeral "8"? Each letter or symbol must have a unique combination of 1's and 0's that distinguishes if from all others. When the computer receives a byte of input data from the keyboard, it can decode the byte of 1's and 0's from a look-up table. It seems obvious that it would benefit the whole computer community if a standard code was developed that all computers, from mainframes to micros, would recognize. There is such a code, the ASCII code. (There are other standard codes, but ASCII is the most widely employed and accepted. In this book we examine only ASCII.)

The ASCII (American Society for Computer Information Interchange) code is used in almost all computers. It is a 7-bit code that represents a total of 128 characters, letters, numerals, and control characters. The combination of 1's and 0's that represent each symbol was not chosen at random. On the contrary, there are many important reasons for the assignment of each symbol or group of symbols to particular bit patterns. Figure 13.2 is a chart that represents all 128 symbols in the ASCII code. The ASCII table in Figure 13.2 is formed of 16 rows by 8 columns. A binary and hexadecimal value is shown for each row and column. The 16 rows represent the 4 least significant bits

				Bit 7	0	0	0	0	1	1	1	1
				Bit 6	0	0	1	1	0	0	1	1
				Bit 5	0	1	0	1	0	1	0	1
Bit 4	Bit 3	Bit 2	Bit 1	Row \ Col.	0	1	2	3	4	5	6	7
0	0	0	0	0	NUL	DLE	SP	0	@	P		p
0	0	0	1	1	SOH	DC1	!	1	A	Q	a	q
0	0	1	0	2	STX	DC2	"	2	B	R	b	r
0	0	1	1	3	ETX	DC3	#	3	C	S	c	s
0	1	0	0	4	EOT	DC4	$	4	D	T	d	t
0	1	0	1	5	ENQ	NAK	%	5	E	U	e	u
0	1	1	0	6	ACK	SYN	&	6	F	V	f	v
0	1	1	1	7	BEL	ETB	'	7	G	W	g	w
1	0	0	0	8	BS	CAN	(	8	H	X	h	x
1	0	0	1	9	HT	EM	)	9	I	Y	i	y
1	0	1	0	A	LF	SUB	*	:	J	Z	j	z
1	0	1	1	B	VT	ESC	+	;	K	[	k	{
1	1	0	0	C	FF	FS	,	<	L	\	l	¦
1	1	0	1	D	CR	GS	—	=	M	]	m	}
1	1	1	0	E	SO	RS	.	>	N	∧	n	~
1	1	1	1	F	SI	US	/	?	O	–	o	DEL

Figure 13.2 ASCII table.

of the ASCII code, and the eight columns represent the 3 most significant bits. Take a moment to examine the structure of the table in Figure 13.2.

The ASCII code can be logically divided into two distinct groups: the nonprintable control characters (codes 0 through 31) and the printable characters (codes 32 through 126). ASCII F7H is a special code that will be discussed with the control characters.

13.2.1 ASCII Control Codes

Let's compare the familiar operation of a typewriter with that of a computer keyboard. The typewriter has a standard keyboard as the input device and 48 hammer mechanisms that print the characters onto paper. The computer uses a keyboard that is very similar to the typewriter keyboard. Instead of paper, the default output medium of the computer is the video display.

Here is a typical session at a typewriter: Normal text is entered, such as upper- and lowercase letters and common symbols (., ", ?, $, #, @, !, etc..), by striking a key labeled with the desired character. That character is transferred to the paper via the hammer and ink ribbon. It is extremely important to note that on a typewriter keyboard there are many keys that do not produce a character, but rather a mechanical action. The carriage return, back space, space bar, and tab bar are examples of typewriter "control" keys.

The first 32 ASCII codes do not produce printable characters, but are control codes, similar to the space bar, back space, carriage return, and tab bar on a typewriter. Many of these control codes are used only in complex block mode and synchronous communications. We will center our discussion on understanding the control characters commonly used in asynchronous communications.

ASCII 00H NULL character This character is used as a "filler" when two devices are synchronized but not exchanging any actual information.

ASCII 07H: BEL character When the computer receives a BEL character it will cause the terminal or keyboard to beep, the equivalent of ringing a bell. The BEL character is an important method of attaining a user's attention.

ASCII 08H: Back space (BS) Just like a typewriter, a computer must have a mechanism to back-space the cursor over the previous character.

ASCII 09H: Horizontal tab (HT) This control character will move the cursor to the next tab position on the present line. This operates like the standard tab function of a typewriter.

ASCII 0AH: Line feed (LF) Every typewriter has a knob attached to the paper-roller mechanism. When this knob is rotated, the sheet of paper will roll up one line. The line-feed control charcter produces the equivalent result on the video display—the cursor will move down one line.

ASCII 0BH: Vertical tab (VT) This is the line equivalent of the horizontal tab. The vertical tab character will move the cursor down a specified number of lines, just as the horizontal tab moves the cursor over a specified number of columns. Vertical tabs are usually used to control hard-copy outputs of printers.

ASCII 0CH: Form feed (FF) Like the vertical tab, form feed is also used in printer control applications. Feeding single sheets of paper into a printer is a time-consuming task. Computer printer paper is a continuous roll folded into standard-length sheets. Each sheet is called a *form.* The form-feed command instructs a printer to advance the paper to the first line of the next form.

ASCII 0DH: Carriage return (CR) There is an important difference between the carriage return on a typewriter and the carriage return of a computer. A typewriter return is a combination carriage return and line feed. It not only returns the carriage to the beginning of the line, but also feeds to the next line. A computer carriage return only returns the cursor or the print head to the beginning of the line; it does not additionally execute a line feed. Almost every command entered into a computer is terminated with a carriage return. The carriage return without a line feed enables printers to write more than once on the same line. This technique is used to improve the quality of dot-matrix characters, and to provide underlining and overstrike capability.

ASCII 11H through 14H: Device control 1 through 4 (DC1 through DC4) These control characters are most often used to control the flow of communications between a computer and a video display or a computer and a printer. Of special interest are DC1 and DC3. DC1 is called XON and DC3 is called XOFF.

Computers are fast and printers are slow. A computer can output information so quickly that a printer cannot keep up. The XON–XOFF protocol enables devices such as computers, printers, and terminals, which process data at different rates, to communicate in a simple and effective manner. Printers have a small buffer memory where the codes of characters to be printed are stored. When this buffer fills up, the printer sends an XOFF command to the computer. This causes the computer to stop sending data to the printer. When the printer buffer is empty, the printer issues an XON command. This tells the computer to start sending data again.

The use of XON and XOFF is called *flow control* or *software handshaking.* We will discover that computers and peripherals can also communicate using hardware handshaking. For now, remember that XON and XOFF are control codes that can be used to enable or disable the flow of data between a computer and a peripheral.

ASCII 1BH: Escape (ESC) The escape character is used as a preface for one or more printable characters. When the computer receives an escape character, it knows to interpret the next character or series of characters (terminated by a carriage return) as a control sequence, not as data. Escape sequences are used to exchange complex commands between the computer, display, and keyboard.

ASCII 7FH: Delete (DEL) This is the highest ASCII code. Delete will erase the character that is presently under the cursor.

You may have noticed that DEL does not follow in sequence with the rest of the ASCII control characters. The reason for DEL consisting of all logic 1's can be traced back to the days of *paper tape.* Paper tape was used as a means of mass storage media for data and programs. An output device called a *paper tape punch* punched a pattern of holes along a paper tape. Each ASCII character had a unique pattern of punched holes. The punched paper tape was read by an input device called a *paper tape reader.* The paper tape reader would translate the pattern of holes back into an ASCII character.

No hole represented a logic 0, and a hole represented a logic 1. The only way to delete a character was to punch out all seven holes. Because of that, DEL was required to be the ASCII code of all logic 1's 7FH.

Control Key Sequences You may wonder how these control codes can be executed from a standard keyboard. Computer keyboards have many extra keys. The key that enables operators to send control codes is the *control key.* The control key operates in a similar manner to the shift key. The control key must be held down while pressing another key to generate a control code. To generate an ASCII code in column 0 (codes 0 through 15), press the control key and the matching row character in column 4 of the ASCII chart. For example, a control-@ will produce an ASCII NUL; a control-G will produce the BEL character. To generate the second column of ASCII codes (16 through 31), the control is used with the characters in column 5. Control-S generates a DC3-XOFF character.

The "^" character is used to symbolize the word "control." A control-G sequence is written as ^G. This is the standard nomenclature used throughout the computer industry.

Control and escape sequences are used to control the many complexities of creating cursor-addressable video displays. The study of control and escape sequences is better left to an advanced programming class. But it is these key combinations that give every terminal a slightly unique personality.

13.2.2 Printable ASCII Characters

The ASCII codes 20H through 7EH represent printable characters. Notice the codes for uppercase "A" (41H) and lowercase "a" (61H). They differ only by the state of bit 6. Converting between upper- and lowercase letters is accomplished by toggling 1 bit.

ASCII 20H is the space. If a computer display has blank spaces, the screen refresh memory for those locations contains the ASCII code 20H. (It is important to distinguish between the ASCII NUL and the ASCII space.)

Consider the following example. When a computer is first powered up, its display is usually blank. Instead, the screen is full of "$"s. What is the problem?

Refer to the ASCII table in Figure 13.2. We know that the display memory should be full of ASCII spaces (010 0000), but instead the display is indicating ASCII "$"s (010 0100). The only difference between the two codes is the third bit. In a space it is low, but with the dollar sign it is high. It appears that bit 2 is stuck high coming out of the display memory. If that is the case, an "!" would be displayed as a "%" and the number 0 would appear as the number 4.

Notice that the digits (0 through 9) are continuous in ASCII code (30H through 39H). The last hex digit in the ASCII code is the same as the BCD code. That makes it easy to test for digits in a program. If the ASCII code is between 30H and 39H, it represents a digit. By forcing the most significant hex digit of the code to 0's, the actual number is easily derived.

The set of uppercase letters (41H through 5AH) and lowercase letters (61H through 7AH) are also continuous, starting from the first letter "A" with the lowest ASCII code to the last letter "z" with the highest ASCII code. In the BASIC language, programs use conditional expressions to test letters in string variables. The ASCII code of these letters is what is really being compared. Therefore, it can be said that B > A, because the ASCII code that represents the letter B is greater than the code that represents "A." Also notice that "Z" < "a."

13.2.3 Extended ASCII for 8 Bits and Block Graphics

You may be concerned that we normally work with bytes (8 bits) and the ASCII code is only 7 bits. What about the eighth bit? Microcomputers use an 8-bit ASCII code. The first 128 codes match standard ASCII and the additional 128 codes are used to create graphics symbols, foreign letter sets, Greek letters, and special mathematical and scientific symbols. But beware! Only the first 128 characters of the ASCII code are standard. The extended ASCII characters are different in every microcomputer.

13.3 SERIAL COMMUNICATIONS AND THE RS-232 INTERFACE

There are hundreds of companies that manufacture modems, terminals, printers, and computers. How can different combinations of all these devices easily be integrated to create a working computer system? The answer is the *RS-232 serial interface standard.* (The letter "RS" stands for "Recommended Standard.") This standard defines the mechanical and electrical connections of computers and peripherals that communicate via a serial interface. Theoretically, all RS-232 devices should be 100% compatible. In reality, this is not always the case. Nonetheless, it is a widely accepted standard that is used throughout the world.

The EIA (Electronics Industry Association) has defined RS-232 in a lengthy, complex technical document. In this section we examine RS-232 in a practical light that will enable you to interface and troubleshoot serial communication systems.

13.3.1 Half-Duplex (HDX) and Full-Duplex (FDX) Connections

When a key is pressed on a terminal and the symbol of that key appears on the video display, did the ASCII code for the depressed key get transmitted directly from the keyboard to display, or did it get transmitted from the keyboard to the computer and then echoed back to the display by the computer? The answer to that question depends on whether the terminal is operating in half-duplex or full-duplex mode. In *half-duplex* operation, the ASCII code of any key pressed is transmitted directly to the display, even if a computer is not connected to the terminal. In *full-duplex* the computer receives an ASCII code from the keyboard and retransmits (echoes) it to the display. This ensures the user that the computer actually received the correct key code. We will assume that all the connections in this chapter are full-duplex. Half-duplex is an old technique that is used only in rare situations. Notice that a full-duplex connection requires two wires—one to transmit data (from the keyboard) and another one to receive data (at the display). This concept is completely compatible with standard two-wire phone lines.

13.3.2 Data Terminal Equipment and Data Communications Equipment

RS-232 was defined before the advent of microcomputers. Its original purpose was to set a standard for terminals and computers communicating remotely over telephone lines.

In Figure 13.3, the terminal and computer are labeled with

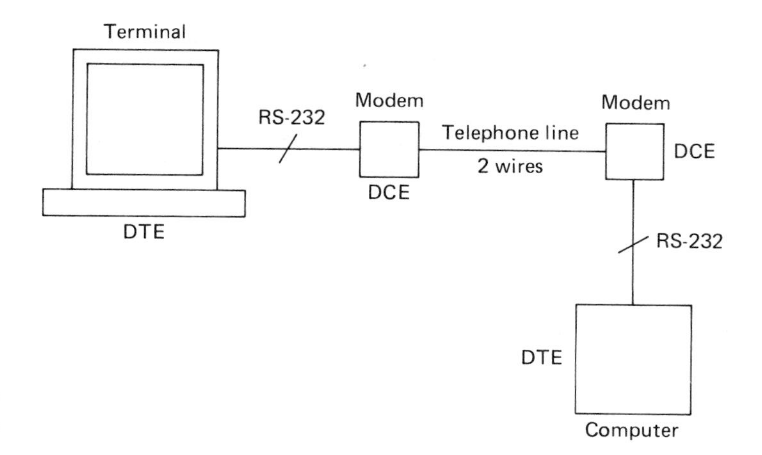

Figure 13.3 Remote terminal–computer connection.

the letters "DTE," which stands for "data terminal equipment." The modems are both labeled "DCE," for "data communications equipment." (DCE is sometimes called "data circuit terminating equipment," although "data communications equipment" is the preferred interpretation of DCE.) In every RS-232 connection, one device must play the role of DTE and the other DCE. In Figure 13.3 it is easy to see that the modems are strictly communications equipment, and the terminal and computer are data terminal equipment. The word "terminal" implies the end of a data line. The terminal is at one end of the data line and the computer is at the other end of the data line. The DTE and DCE each has specific responsibilities in the RS-232 communications scheme.

13.3.3 The RS-232 Connector Pinout

RS-232 states that all devices will be interconnected with a 25-pin D-type connector. This connector has 13 pins on the top row

Pin	*Name*	*Abbreviation*	*Function*
1	Frame ground	FG	Connected to frame or earth ground
2	Transmit data	TD	Data transmitted from DTE to DCE
3	Receive data	RD	Data received by DTE from DCE
4	Request to send	RTS	DTE-to-DCE handshake signal
5	Clear to send	CTS	DCE-to-DTE handshake signal
6	Data set ready	DSR	DCE-to-DTE handshake signal
7	Signal ground	SG	TD and RD are referenced to this ground
8	Data carrier detect	DCD	DCE-to-DTE handshake signal
20	Data terminal ready	DTR	DTE-to-DCE handshake signal

and 12 pins on the bottom row. Each of the 25 pins performs a specific function. In the great majority of applications only nine pins are used, pins 1 through 8 and pin 20. The other 16 pins are used only in exotic and specialized applications. We will concentrate our efforts on understanding the nine pins that are widely used.

13.3.4 Grounds, Data, and Handshaking

Pins 1 and 7 are grounds. Pin 1 is *frame* or *earth ground.* Pin 7 is called *signal ground.* The transmit and receive data are referenced to signal ground. Usually, pins 1 and 7 are tied together to create a common frame and signal ground..

Pins 2 and 3 handle the full-duplex data that we just covered. Pin 2 is "transmit data" and pin 3 is "receive data." But you must remember that the transmit data leaving the DTE is the receive data for the DCE. That implies that if both the DTE and DCE are transmitting data from pin 2 on the RS-232 connection, we will have two outputs connected together. The concepts of transmit and receive data are always referenced to the DTE. If the DTE is transmitting data on pin 2, the DCE will be receiving data on pin 2. Similarly, if the DTE is listening for receive data on pin 3, the DCE will transmit data on pin 3. That is why, in every RS-232 connection, one device must act as DTE and the other device as DCE.

The simple concept of DTE and DCE can be so confusing that entire books are written just to explain how to connect RS-232-compatible devices. As we delve deeper in RS-232, remember that simple relationship between the DTE and DCE and you will avoid a great deal of confusion.

Handshaking Signals We have already seen the flow control signals: XON and XOFF. XON and XOFF are ASCII codes that, through software, can enable and disable the flow of data between RS-232 devices. To complement XON and XOFF, RS-232 offers several hardware handshaking signals. These signals are used to exhange control and status information between the DTE and DCE. We will cover the basic uses of these handshake signals as they relate to Figure 13.3 and the handshaking between a typical terminal (DTE) and a modem, (DCE).

RTS Request-to-send indicates to the DCE that the DTE wants to transmit data.

CTS The clear-to-send line is used by the DCE to indicate to the DTE that the line is clear and the DTE can send data. The clear-to-send line must be active before the DTE can transmit.

DTR Data-terminal-ready is used by the DTE to indicate that it is connected on the RS-232 interface. Before the DCE tries to communicate with the DTE, it looks at DTR to see if a "live" terminal is on the interface.

DSR The original modems were called *data sets.* You should think of the "data set" and "modem" as interchangeable terms. Data-set-ready indicates to the DTE that a DCE is connected to the RS-232 interface. It is the modem's equivalent of DTR.

DCD Data-carrier-detect is used by the DCE to indicate to the DTE that a connection with another modem has been made. That means that DCE can sense the carrier signal of the other DCE on the remote end of the connection.

These five handshake signals are used in many different ways to facilitate coordination between DTEs and DCEs on the RS-232 connection. There are dozens of ways to use these five handshake signals. An RS-232 interface can have as few as two lines or use all 25. Technicians who interface serial devices must be creative, patient, and prepared to dig deep to find handshaking information in the manufacturers' cryptic device specifications.

13.3.5 RS-232 Logic Levels

The logic levels in RS-232 are:

Logic 0	*Logic 1*
+3 to +25 V	−3 to −25 V

In TTL, CMOS, and ECL circuits, the more positive voltage represents a logic 1 and the more negative voltage a logic 0. In RS-232, the more positive voltage represents a logic 0 and the more negative voltage a logic 1. Notice that an RS-232 level between −3 V and +3 V is undefined, just as is a TTL level between 0.8 and 2.0 V. An undefined RS-232 level indicates the same type of problem as an undefined TTL or CMOS level—two outputs are shorted together or the line is excessively loaded. Most RS-232 circuits are driven by ±12-V power supplies. A typical RS-232 logic 0 is +9 to +12 V and a logic 1 is −9 to −12 V.

An RS-232 logic 0 has many different names: +12 V, a space, active level, or control on. All RS-232 handshaking or control signals are active at the logic 0 level.

An RS-232 logic 1 also has many different names: −12 V, a mark, an idle level, inactive level, or control off. An RS-232 line that is at a steady logic 1 is said to be at a *marking level.*

13.3.6 The Break-out Box

A basic piece of required RS-232 test equipment is the break-out box. A break-out box has male and female RS-232 connections that enable it to be inserted in-line between the DTE and

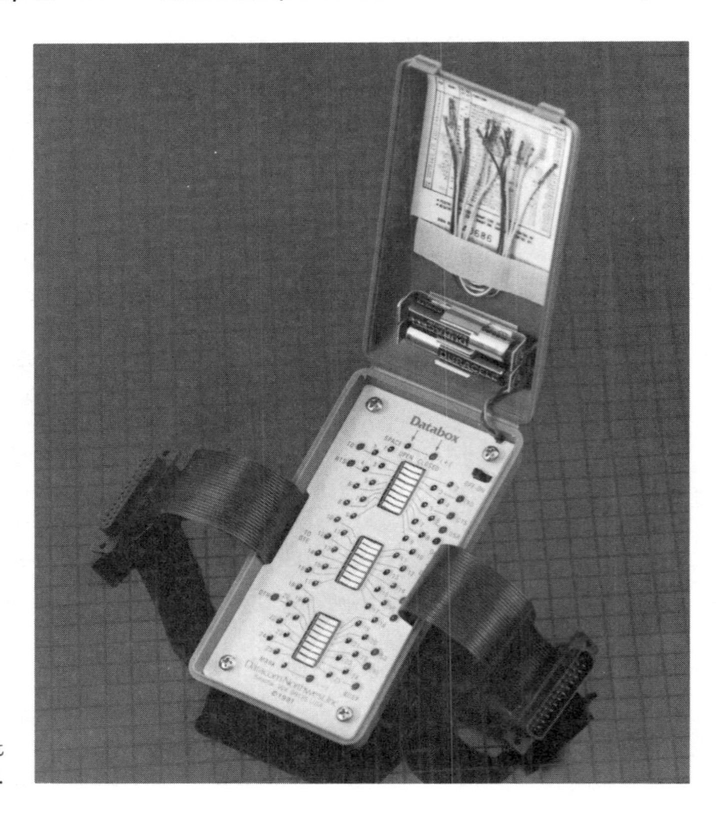

Figure 13.4 RS-232 breakout box.

the DCE. Figure 13.4 illustrates a typical break-out box. Each of the 25 RS-232 lines is routed through a switch, monitored by an LED, and has an exposed jumper pin. Using a combination of switches and jumper wires, the interface can be modified to perform any type of handshaking. The TD and RD lines can be crossed to enable direct conversation between two DTEs. (This process is known as a modem *eliminator* or *crossover connection.*) Often, DTR is looped back to DSR and DCD, to automatically generate the enable handshake signals.

The LEDs are used to monitor the transmit and receive data lines and the handshake signals. On standard break-out boxes the LED will illuminate to indicate a spacing level. Newer break-out boxes use dual, side-by-side LEDs to monitor each line. One LED illuminates to display a space, and the other to display a mark. The latest break-out boxes use *tri-state LEDs.* A tri-state LED (like a three-state digital device) has three possible states: glow red to indicate a mark, glow green to indicate a space, do not light to indicate an illegal RS-232 level (−3 to +3 V). Typical break-out boxes cost between $100 and $500.

13.3.7 The Data Line Monitor

The *data line monitor* (also known as a *protocol analyzer*) is a sophisticated piece of test equipment used to monitor serial communications. The conversation between the DTE and DCE is displayed on a CRT and written into a memory buffer where many screens of data can be saved and reviewed. The nonprintable ASCII codes (control codes) have special representations on the video display of the data line monitor. In this manner, all the ASCII characters exchanged between the DTE and DCE can be analyzed for malfunctions in the data link. Data line monitors can be purchased for $3000 to $20,000.

13.4 THE UART: UNIVERSAL ASYNCHRONOUS RECEIVER AND TRANSMITTER

The UART is a complex, programmable LSI device that performs the parallel-to-serial transmission and the serial-to-parallel reception processes, manages all the DTE/DCE handshaking lines, and controls the data speed and format of the serial bit stream. The UART was one of the first LSI ICs. Its development revolutionized the terminal manufacturing industry. One UART replaces dozens of SSI and MSI ICs and adds the flexibility of a programmable, intelligent device. UARTs are complex ICs from both a hardware and software point of view. We will examine a generalized UART in much the same manner that we examined a generalized microprocessor.

Figure 13.5 is a block diagram of a typical UART. Let's examine the function of each block.

Data Bus Buffer This bidirectional buffer isolates the microprocessor data bus from the internal UART data bus.

Programmable Operations Controller Notice the six input lines connected to this block. The clock input is used to control the speed with which the data is transmitted and received. The

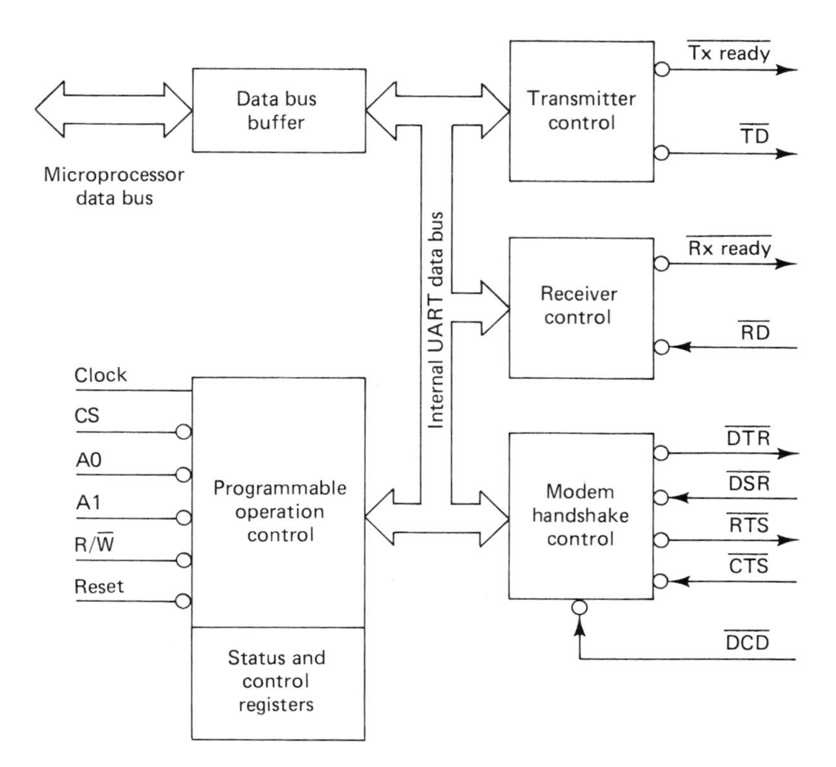

Figure 13.5 Block diagram of a UART.

frequency of the clock is usually 16 times the speed of the serial bit stream. This fast clock will enable the UART to sample the receive input in the exact center of the data window. (In Section 13.4.1 we expand on that idea.)

Like memory devices, UARTs have a chip select input. Unless this input is at an active-low level, the UART will ignore all commands from the microprocessor. The read/$\overline{\text{write}}$ line is used to indicate whether a data or control byte is being written to the UART, or receive data or a status word is being read from the UART. Control words are used to program the UART, and status words indicate to the microprocessor the current status of the UART. The two address lines are used to select a particular register within the UART. These registers contain data to be transmitted, data that has been received, control words, and status words. The UART needs a reset input to initialize its operational state to default conditions.

Transmitter Control This block performs parallel-to-serial conversion on the transmit data, inserts special framing start and stop bits, and has a status output called transmitter ready. An active level on the TxReady line indicates that the transmit buffer is empty and the UART is ready to receive a byte of transmit data.

Receiver Control This block is analogous to the transmitter control block. It performs the serial-to-parallel conversion of receive data, strips off the special framing start and stop bits, and has a line called receiver ready that is used to notify the microprocessor that a byte of data has been received and is waiting to be read.

Modem Handshake Control This block is used to assert the control handshaking lines and to monitor the status input handshaking lines. On most UARTs the receiver is enabled when DCD is sensed at an active-low level and the transmitter is enabled when CTS is sensed at an active-low level. The other handshake lines are manipulated and monitored through software control.

As you can now appreciate, the UART is a complex device that greatly simplifies the overhead associated with the RS-232 interface.

13.4.1 Structure of the Bit Stream

We still have to answer the questions that were posed at the beginning of the chapter concerning asynchronous serial communications. How does the receiver know:

1. When the transmitter has started sending data?
2. How fast the data is being sent?

The speed with which the data is transmitted on a serial communications line is called the *baud rate.* Baud rate is expressed in bits per second (bps). An RS-232 line running at 9600 baud is capable of transmitting 9600 bits of data and framing information in 1 second. In asynchronous communications, characters are transmitted on an irregular basis, independent of an absolute clock reference. On an asynchronous data line you would see bursts of data bits, not a steady, consistent stream.

How wide is each data bit traveling on a 9600-baud line? The inverse of the baud rate is equal to the period of one data bit.

$$\frac{1}{\text{baud rate}} = \text{period of 1 bit} = \frac{1}{9600} = 104\ \mu\text{s}$$

Each data bit in a 9600-baud line is 104 μs wide. Assume that the transmitter and receiver agree to use a particular baud rate. On a 9600-baud line the receiver will interpret the data bits with a length of 104 μs, and the transmitter will transmit the data stream in units of 104-μs bit cells. The critical concept here is that although the transmitter will send data at irregular (asynchronous) intervals, the width of a particular data bit will always be the inverse of the baud rate.

How does the receiver know that the transmitter has started sending data? The answer is—the start bit. The first data bit sent by the transmitter will be a space (logic 0). This space will alert the receiver that the transmission of data has started. On a 9600-baud line, precisely 104 μs after the edge of the start bit, data bit 0 will be sent; 104 μs after bit 0 is sent, bit 1 is sent. The bit stream will arrive at the transmitter in small, adjacent packets that are 104 μs wide.

In Section 13.4 we noticed that the clock input was 16 times the speed of the data stream, enabling the UART to sample in the center of the data window. Sampling the center of the data bit will reduce noise problems induced on the leading and trailing edges. After the receiver senses the initial edge of the start bit, it waits eight clock pulses and then samples the line. If it is still a logic 0, indicating a true start bit, it then starts sampling the rest of the data stream every 104 μs, in the center of each bit cell.

Another factor that must be agreed on is the number of data bits sent for each character. Normal ASCII uses only 7 data bits, whereas extended ASCII requires 8 data bits.

Added onto the end of each bit stream is a stop bit. This stop bit is a mark (logic 1) that lasts the length of a normal data bit. The start and stop bits are said to *frame* the data bits. The receiver first senses a start bit and then samples the agreed number of data bits. The bit following the last data bit is the stop bit, a marking level. If it isn't, a *framing error* has occurred, and the data is considered in error. The UART will re-

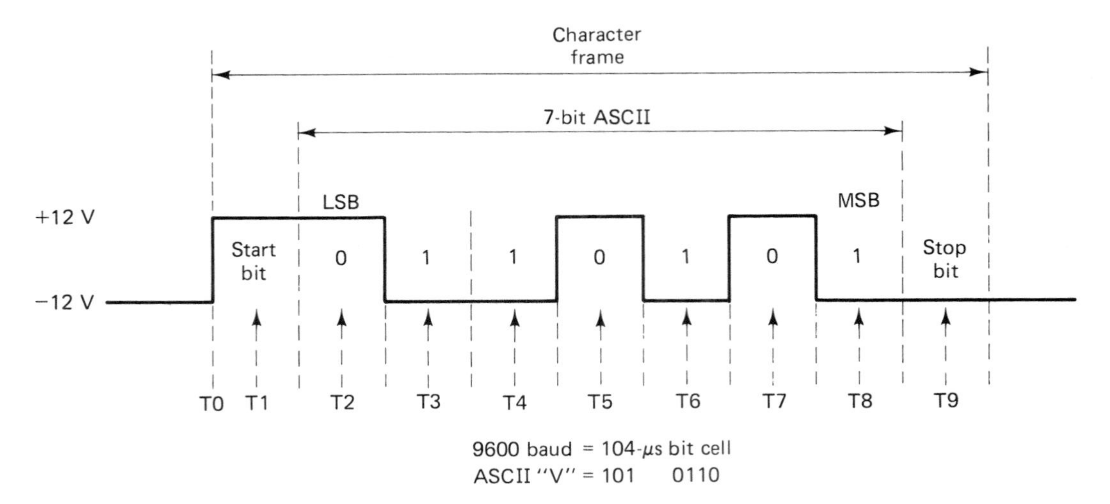

Figure 13.6 ASCII "V" at RS-232 levels.

port the framing error to the microprocessor, which may request a retransmission of the data.

Let's examine a "snapshot" of the letter "V" (ASCII code 101 0110) as it is transmitted on a serial line. We will assume that the character will have 1 start bit, 7 data bits, and 1 stop bit.

Figure 13.6 illustrates the snapshot of an ASCII "V" (101 0110) with 1 start bit and 1 stop bit, transmitted at 9600 baud, as it would be captured by an oscilloscope. UARTs operate at TTL voltage levels. Transmit data from the UART drives the input of a TTL to RS-232, inverting-line driver (typically a MC1488). TTL logic 1 levels will be converted to −12 V (mark) and logic 0 levels (space) to +12 V. An RS-232-to-TTL line receiver (tyically an MC1489) reconverts the bit stream to TTL levels.

At T0 the RS-232 line has a mark-to-space rising edge. This alerts the receiving UART that the transmission of a character has begun. Half a bit cell later (52 μs) at T1, the receiver will resample the transmit data. If it is still at a logic 0 level, the UART will prepare to receive the next 7 data bits in the stream. If it is a logic 1 level, the UART will assume that the initial edge was a noise glitch, and the reception process will be aborted.

At times T2 through T8, the receiver UART will sample the line to establish the logic level of each bit. (Although it may be confusing, keep in mind that a logic 1 = −12 V and a logic 0 = +12V.)

At T9 the receiver expects to see a marking level. This indicates the stop bit and the end of the data frame. If the sample at T9 produced a space, the UART would report a framing error to the local microprocessor, and the character received would be considered invalid. The stop bit gives the UART time to complete the serial-to-parallel conversion and signal the microprocessor that a byte of receive data is waiting to be read.

Figure 13.6 indicates that the least significant bit is sent first, and the others follow in ascending order.

13.4.2 Parity

If two things are equal, they are said to be in *parity.* To increase the reliability of serial communications, another bit can be added to the stream. This bit is called the *parity bit.* It is added to ensure the integrity of the *data received.* The parity bit is sent after the last data bit and before the stop bit. If *even parity* is selected, the UART will set or reset the parity bit to make the total number of logic 1s in the bit stream an even number. IF *odd parity* is selected, the UART will set or reset the parity bit to make the number of logic 1's in the bit stream odd. The use of a parity bit enables the receiving UART to sense single bit errors. The receiving UART will generate its own parity bit, based on the receive data stream. If the received and calculated parity bits agree, the character received is valid. If the parity check of the receiver UART and the parity bit in the bit stream disagree, the receiving UART will generate a parity error to the local microprocessor.

Figure 13.7 illustrates the first (NUL) and last (DEL) ASCII codes as they are sent with odd and even parity. Figure 13.7a illustrates the DEL character sent with odd parity. The 7 data bits are logic 1's. The parity bit is a logic 0 to maintain the odd number of logic 1's. In Figure 13.7c, DEL is sent with even parity. The parity bit is a logic 1 to make the number of logic data bits an even number. In Figure 13.7b, the ASCII "NUL" has no logic 1 bits. The parity bit is taken to a logic 1 level to maintain odd parity. In Figure 13.7d, the parity bit is taken to a logic 0 level, because 0 is considered an even number. A typical 7-bit ASCII RS-232 frame has 10 bits: 1 start bit, 7 data

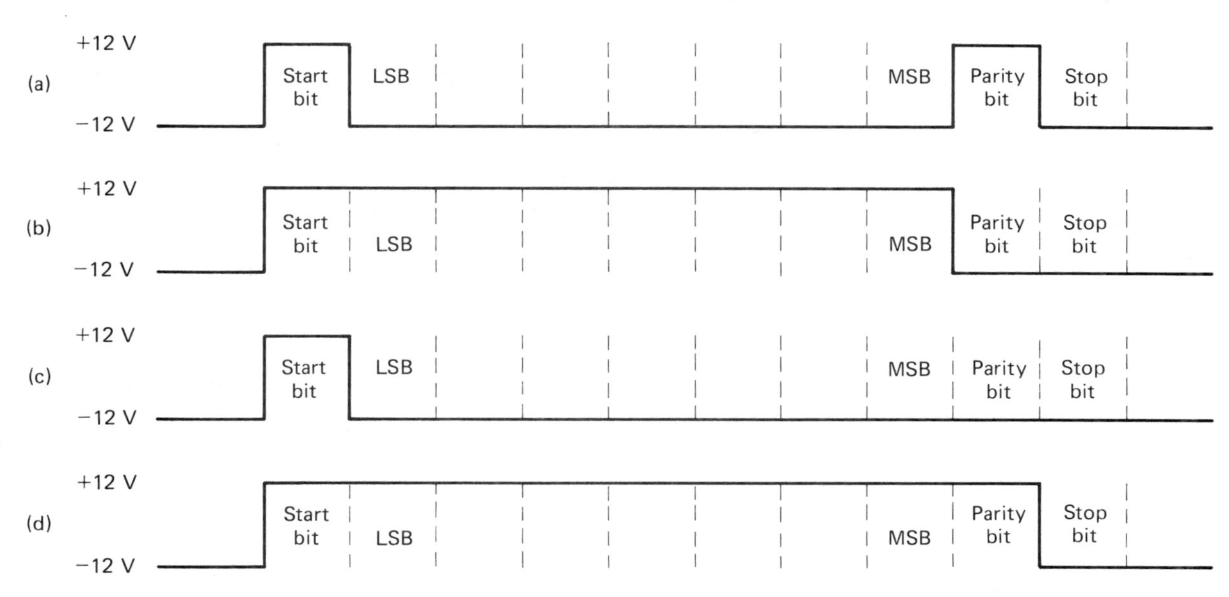

Figure 13.7 Examples of even and odd parity.

bits, 1 parity bit, and 1 stop bit, which means that a 9600-baud line can communicate at a maximum of 960 characters per second.

13.4.3 RS-232 Serial Parameters

Before two devices can communicate over an RS-232 line, they must both agree to the following parameters: baud rate, number of start and stop bits, number of data bits, and parity scheme.

Baud Rate Typical baud rates are

110, 300, 600, 1200, 2400, 4800, 9600, 19,200

The lower baud rates (110 througth 2400) are used in printers. Some printers operate as high as 9600 baud, by quickly filling up a printer buffer and then sending an XOFF character to the computer or terminal. When the buffer is empty, the printer sends an XON character to receive another buffer of printer data.

Asynchronous modems communicate at 300, 1200, and 2400 baud. (1200 baud is the most common.) Terminals operate up to 19,200 baud. Phone lines and twisted pair have appreciable amounts of capacitance. The maximum speed of any serial link is a function of the two devices communicating and the limitations of the transmission medium.

Number of Start Bits UARTs require only one start bit to synchronize to the bit stream.

Number of Data Bits 7 or 8. Both UARTS must be expecting the same number of data bits. Seven data bits is most common.

Parity Even, Odd, or None. Even, odd, and no parity are all widely accepted, with even parity being the most common.

Number of Stop Bits 1, $1\frac{1}{2}$, or 2. Stop bits give the receiving UART time to reconstruct the serial data and inform the microprocessor of an available data byte. One or $1\frac{1}{2}$ stop bits is common. (The length of $1\frac{1}{2}$ stop bits is simply 50% more than that of 1 stop bit. The idea here is the length of recovery time required by the receiving device.)

13.4.4 A Quick Word on Synchronous Communication

In asynchronous communications each ASCII character must be framed by a start and stop bit. Include another bit for parity, and the 7-bit ASCII code has an additional 3 bits of overhead, an increase of over 40%. Synchronous serial communication does not send data on a character-by-character basis. Instead, data is sent in a block or group. When no data is being sent, a special "sync" character is transmitted to maintain synchronization between the transmitter and receiver.

Synchronous protocols are extremely technical and complex. Some synchronous protocols are character-oriented (such as IBM's Bi-Sync) and others are bit-oriented (such as IBM's SNA/SDLC). An examination of synchronous communication is better left to an advanced text on data communications.

13.5 PARALLEL COMMUNICATIONS

Serial communications is inherently complicated: baud rates, stop bits, number of data bits, parity, stop bits, and complex hardware handshaking. Its real strength is communicating, short or even great distances (via a modem), over a single pair of wires.

Parallel communications is simple because an ASCII character is sent as a parallel byte, at standard TTL levels; no bit stream conversion or voltage translation is required. Eight data lines are needed, plus several handshaking and status signals. Because standard TTL drivers and voltage thresholds are used, the length of a parallel interface cable is limited to under 20 ft.

13.5.1 The Centronics-Type Parallel Interface

Just as RS-232 is the serial communications standard, the Centronics interface has emerged as the standard parallel printer interface. A typical microcomputer/parallel printer interface cable has two different types of connectors. On the microcomputer side, a standard 25-pin D-type connector is used; on the printer side of the cable, a 36-pin Centronics connector is used. Figure 13.8 illustrates a typical Centronics microcomputer/parallel printer interface.

Let's examine each pin or group of pins:

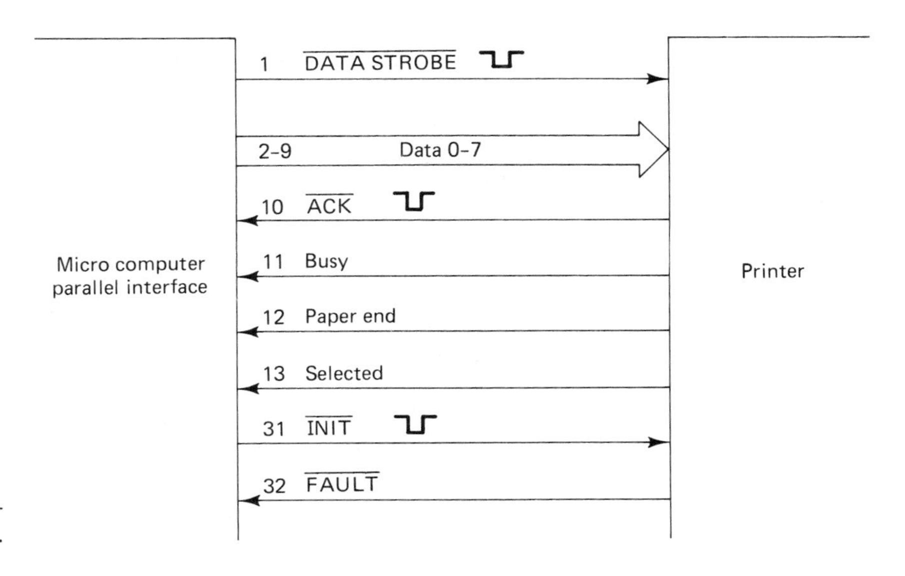

Figure 13.8 Centronics parallel interface.

Pins 2–9: Print Data The microcomputer places the 7-bit (or extended 8-bit) ASCII code on these eight lines. Notice that the Centronics interface is unidirectional; the microcomputer is the transmitter and the printer is the receiver. Unlike the RS-232 connection, the printer in a parallel interface cannot send XOFF-XON characters back to the computer. All handshaking must take place via hardware lines. The computer can still send the printer XON–XOFF characters to effectively enable or disable it.

Pin 1: $\overline{\text{Data Strobe}}$ This is an active-low short-duration pulse (typically 500 ns to 5 μs). The computer pulses $\overline{\text{Data Strobe}}$ to indicate to the printer that an ASCII character is latched onto the data outputs.

Pin 10: $\overline{\text{Acknowledge}}$ This active-low short-duration pulse (typically 5 μs) is issued by the printer. It indicates to the computer that the character has been received.

Pin 11: Busy This active-high signal indicates to the computer that the printer cannot presently receive data. Busy will go high during the printing process, when the printer is in an off-line state, or when a print error occurs.

Pin 12: Paper End (PE) Parallel printers have a microswitch that senses the presence of printer paper as it enters along the back of the platen (paper roller). If this switch opens, indicating the absence of paper, PE will go active high. The computer driving the printer should sense this occurrence and halt the printing process.

Pin 13: Selected Parallel printers have a pushbutton switch which is used to enable or disable the printer. If you are in the process of tearing off a recently printed form, or setting the paper to the beginning of a page, you do not want the computer to start the printing process. The select switch lets you take the printer off-line and place it back on-line. The select line will indicate to the computer the present state of the select switch.

Pin 31: $\overline{\text{Init}}$ The computer will pulse pin 31 low to initialize the printer. This entails clearing the print buffer and returning programmable printers to normal print mode.

Pin 32: Fault An active-low level on this pin indicates that the printer cannot receive data. This can be due to a paper end state, off-line state (disselected), or a printer error state. The fault line can be monitored by the computer's printing program or drive an interrupt line on the microcomputer. When fault goes active, the microprocessor will read the other printer status lines to find the cause of the fault.

The Centronics interface also defines many ground pins and unused pins. Manufacturers implement the unused pins to provide advanced status signals between the computer and the printer.

13.5.2 Hardware Handshaking on the Centronics Interface

Figure 13.9 illustrates a character-by-character hardware handshaking timing diagram. The timing diagram illustrates the transfer of two characters from the computer to the printer.

Event 1. The Busy line is low, indicating that the printer is ready. The computer outputs the ASCII character code onto data 0 through data 7.

Event 2. After waiting for the logic levels on the eight data lines to stabilize, the computer asserts the $\overline{\text{strobe}}$ line for a short, active-low pulse.

Event 3. Reacting to the falling edge of strobe, the printer will bring Busy to an inactive-high level. This indicates that the printer is busy processing the ASCII character.

Event 4. After a short delay, the computer may remove the character from data 0 through data 7. In most systems the character is actually latched, and it is not removed until the next character is output by the computer.

Event 5. After a fairly long period which symbolizes the time required to actually print the character, the printer will pulse $\overline{\text{Acknowledge}}$ low.

Event 6. The rising edge of $\overline{\text{Acknowledge}}$ will bring Busy to an inactive-low level.

Event 7. The computer senses an inactive Busy line and initiates the process of outputting the second character.

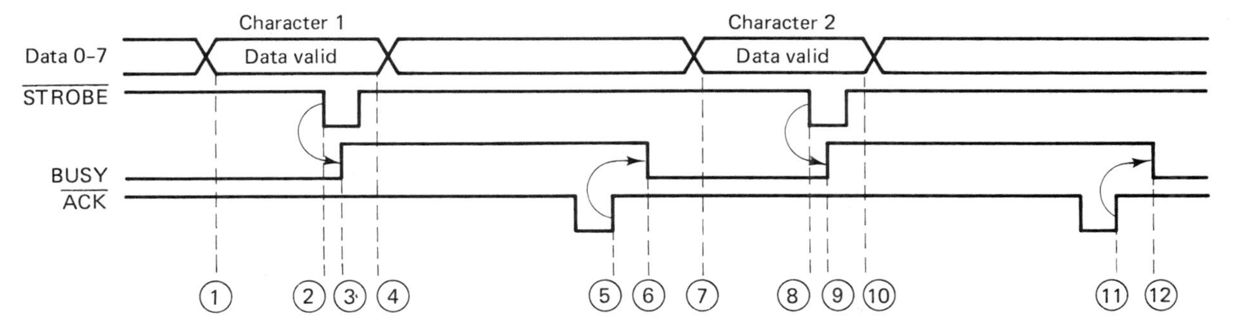

Figure 13.9 Centronics parallel hardware handshaking.

Events 8 through 12. The same character processing and handshaking that occurred during the first six events is repeated.

Notice how the $\overline{\text{strobe}}$, busy, and $\overline{\text{acknowledge}}$ lines work together to provide coordination between the computer and the printer. Take a few moments to reread the handshaking process that describes Figure 13.9. It is critical to your understanding of the parallel interface.

13.5.3 The PPI: Programmable Peripheral Interface

Just as serial communications has a dedicated LSI support device, the UART, the parallel interface also has a complex, LSI support device, the PPI. Figure 13.10, which illustrates the block diagram of a popular PPI, the 8255, should remind you of Figure 13.5, the block diagram of the UART. Take a moment to refer to Figure 13.5 and refresh your memory as to the functions of its major blocks.

The data bus buffer and read/write control logic in Figure 13.10 provide the same functions as the data bus buffer and programmable operations control in Figure 13.5. Group A and group B blocks in Figure 13.10 contain control and status registers just like the status and control register block in Figure

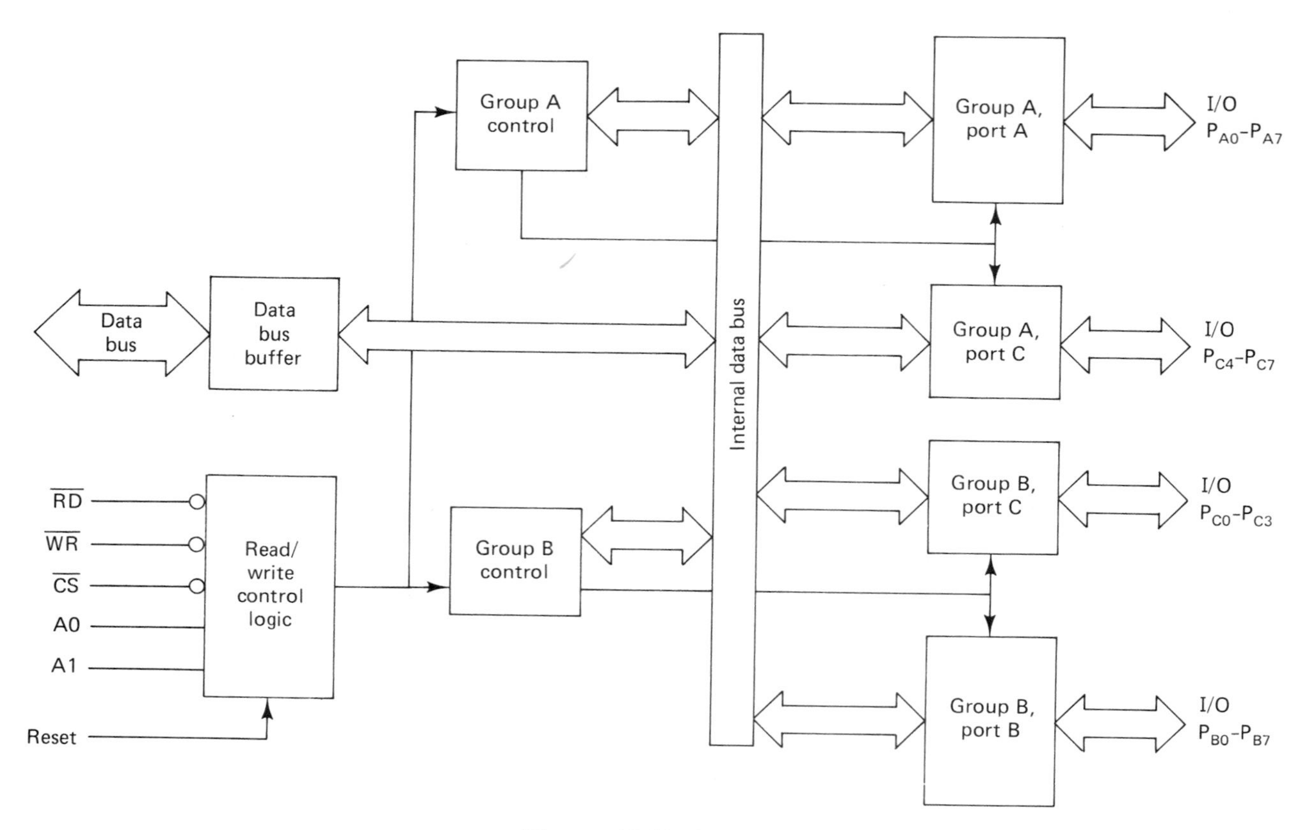

Figure 13.10 8255 programmable peripheral interface.

13.5. The only real difference between the block diagram of the PPI and the block diagram of the UART is the interface with the I/O devices. The PPI provides parallel interfaces (based on flip-flops) and the UART provides serial interfaces (based on shift registers). With all these similarities in mind, let's analyze the operation of the 8255 PPI.

Because the 8255 is programmable, it provides many modes of operation. A control word is written to configure the 8255's operational parameters. The 8255 has three operational modes: mode 0, mode 1, and mode 2.

In mode 0 the 8255 functions with three independent 8-bit I/O ports. Each port (port A, port B, and port C) can be programmed as an input port or an output port. Notice that port C is subdivided into two 4-bit ports, Pc0 through Pc3 and Pc4 through Pc7. Each of these two 4-bit ports can be independently programmed as a 4-bit input or 4-bit output port.

Mode 0 is called the basic I/O mode because during mode 0 operation, the 8255 is incapable of hardware handshaking. Mode 1 is used to support the full hardware handshaking required for the Centronics parallel interface. During mode 1 operation, port A and port B can function as 8-bit input or output ports. Port C upper nibble (Pc4 through Pc7) is used to support the standard $\overline{\text{strobe}}$, busy, and $\overline{\text{acknowledge}}$ handshaking signals as illustrated in Figures 13.8 and 13.9. The port C lower nibble is used to support hardware handshaking for port B. The 8255 can fully support two 8-bit parallel hardware handshaking I/O ports. This greatly simplifies interfacing a microcomputer with a Centronics parallel interface.

Mode 2 supports a bidirectional I/O port. This is used in applications where a parallel port must transmit and receive data. (Remember that the Centronics interface was unidirectional.) Mode 2 is used in advanced interfacing applications.

We have just briefly covered the capabilities of the 8255. This PPI provides extremely simple, yet powerful and flexible support of parallel interfaces. In Chapter 19 we will see how the 8255 is used in a microcomputer system.

QUESTIONS AND PROBLEMS

13.1. What is the basic circuit used to perform serial I/O? Parallel I/O?

13.2. How do computers differentiate between commands and characters?

13.3. What is the difference between a carriage return on a typewriter and a carriage return on a computer?

13.4. What keys must you depress to make a terminal or computer "beep"?

13.5. What control character is used to feed printer paper to the first line on the next page?

13.6. Explain the concept of flow control using XON and XOFF.

13.7. What is RS-232?

13.8. What do the terms DTE and DCE designate?

13.9. How do two DTEs communicate on RS-232?

13.10. You have just purchased a personal computer and a modem and decide to try your new system by dialing up a local electronic bulletin board.

(a) What serial parameters must be set before you can communicate?

(b) Once you connect with the bulletin board, a strange phenomenon occurs. Each time you type a key it appears twice on the video display. What is the problem?

13.11. What are the logic levels of an RS-232 signal?

13.12. What is the most popular RS-232 troubleshooting tool?

13.13. Give a concise description of a UART.

13.14. Define "baud rate."

13.15. Define "frame" as it pertains to serial communications.

13.16. Define "parity" as it pertains to serial communications.

13.17. You are given a terminal, breakout box, and an oscilloscope and told to derive the baud rate of the terminal. How would you accomplish this?

13.18. What are the three handshake signals used in parallel interfaces?

13.19. Busy is shorted to +5 V. How does this affect the interface?

13.20. Why is the distance that a parallel interface can communicate so limited?

13.21. What is the advantage of using a PPI instead of standard octal latches?

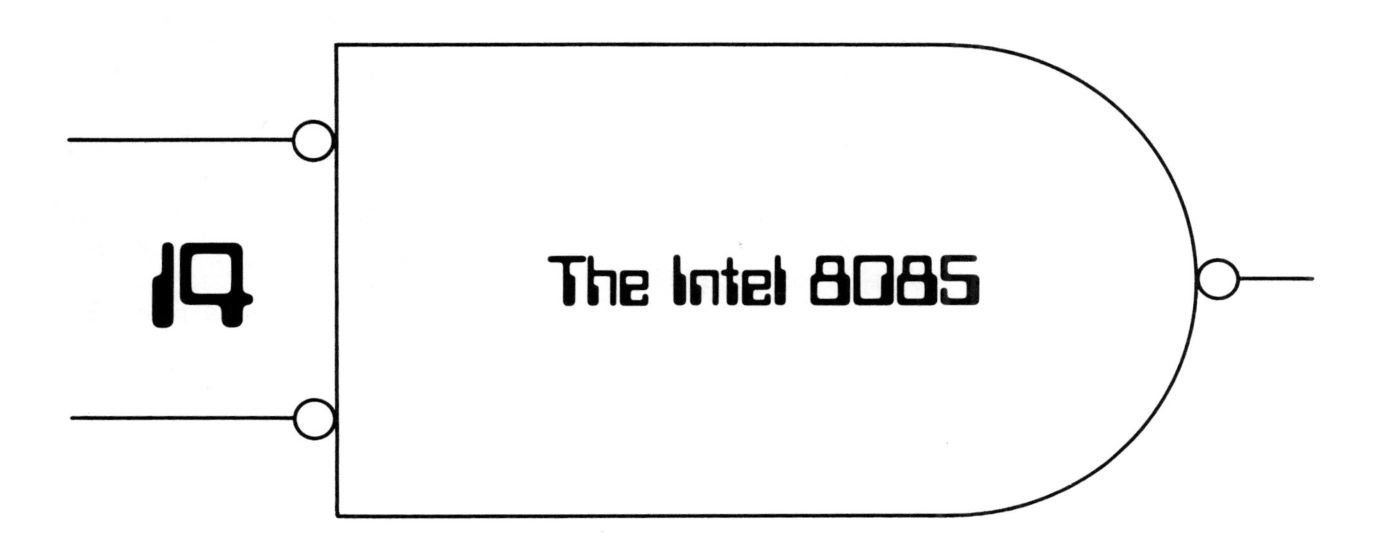

Intel Corporation of Santa Clara, California, was one of the pioneers in microprocessor research and development. They developed the 4004 microprocessor in 1971. Although the 4004 was the first commercially produced microprocessor, it was really intended to be used as a calculator CPU, not as a general-purpose microprocessor. It could only read or write 4 bits of data (a nibble) at a time.

Intel then developed the 8008 microprocessor. The 8008 had a data bus that was a full byte wide. It could read or write 8 bits at a time. Like the 4004, the 8008 was not intended as a general-purpose CPU. It was developed to be used as a CRT controller in terminals.

In 1973 the first general-purpose microprocessor was introduced—the Intel 8080. The 8080 was an improved version of the 8008. The 8080 was the first of many 8-bit microprocessors. After the success of the 8080, many IC manufacturers developed their own microprocessors. The largest forces in the market are Intel and Motorola. In this chapter we examine the hardware aspects of the 8080's successor, the 8085. In Chapter 17 we study Motorola's most popular 8-bit microprocessor, the 6800.

Consider a company that designed a product incorporating

the 8080. That company would have a considerable investment not only in hardware design, but also in software development. Much time is invested in writing the programs that drive the microprocessor system. When the successor to the 8080 was developed (the 8085), an important factor was software compatibility. Could all the programs written for the 8080 also run on the 8085? The answer to that question is "yes." The 8085 is said to be *upwardly compatible* with 8080 software. That means that the 8085 could run all the programs written for the 8080, but the reverse is not true—the 8080 cannot run all the programs written for the 8085.

Upward compatibility does not come without a price. To maintain this compatibility, many of the original design flaws of the 8080 have been propagated through the third- and fourth-generation microprocessors.

14.1 THE 8080

The 8080 is obsolete; it has long since been replaced by third- and fourth-generation microprocessors. Nonetheless, a quick overview of the 8080 will help give you insight into the advanced microprocessors. Figure 14.1 illustrates a block diagram of the basic 8080 system.

The 8080 requires two support chips: the 8228 system controller/bus driver and the 8224 clock generator and driver. The technology to integrate all three devices into the same IC was

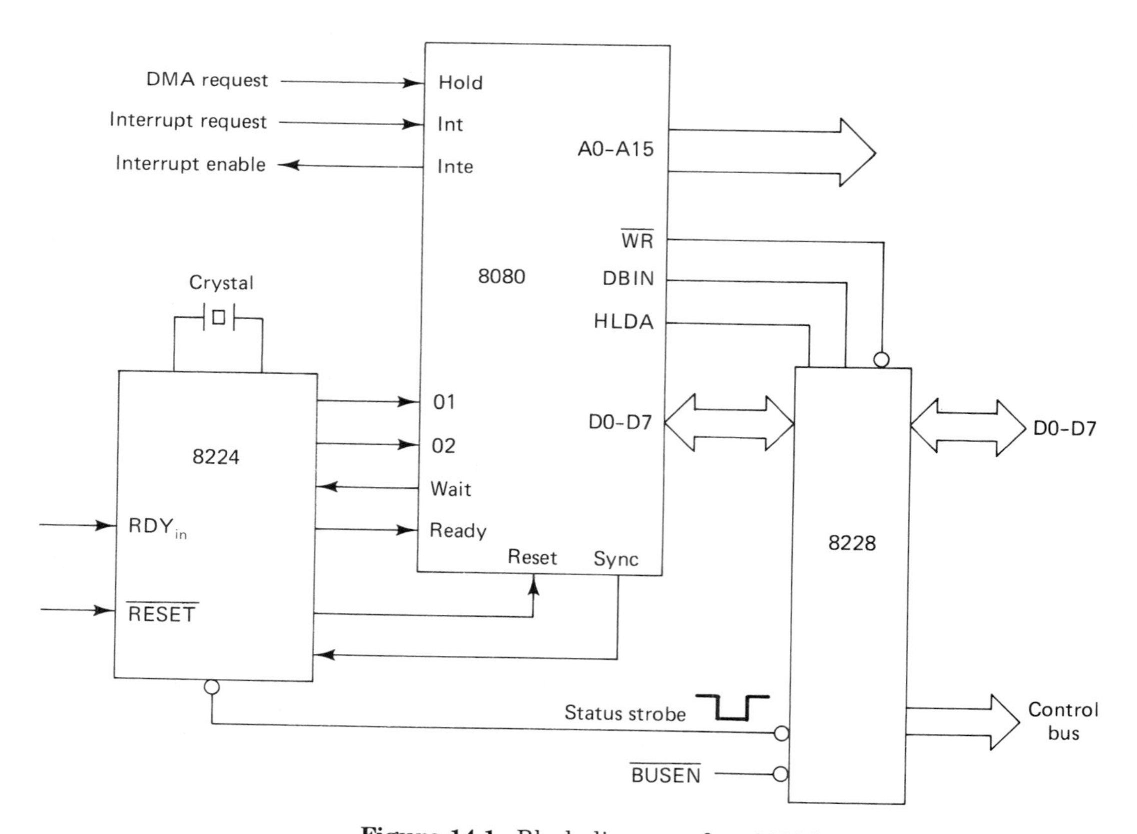

Figure 14.1 Block diagram of an 8080-based system.

not available in 1973. These three ICs constitute the kernel of the 8080 system.

The 8080 has 16 address lines and 8 data lines. Microprocessors are classified as *n-bit microprocessors*, where *n* is equal to the number of lines on the data bus. That is why the 8080 is called an 8-bit microprocessor. With 16 address lines the 8080 can directly address 64K of memory. This memory can be a mixture of RAM and ROM, or even I/O devices that are memory mapped. The control bus output of the 8228 is the memory and I/O read and write outputs, and interrupt acknowledge.

Notice that the 8080 required three power supply voltages: +5 V, −5 V, and +12 V. All third-generation microprocessors require only +5 V, which makes them compatible with TTL power supplies.

14.2 THE 8085

The 8085 was developed to overcome the problems of the 8080. All programs written for the 8080 run without modification on the 8085. In fact, the 8085 added only two instructions to the 8080 instruction set. The major advances of the 8085 were hardware in nature: +5 V powered, no clock generator or system controller required, and a much more sophisticated hardware interrupt structure.

Like any intelligent device, the 8085 can be considered from either a hardware or a software point of view. In this chapter we will first investigate the hardware aspects of the 8085. In the next chapter we analyze the 8085's instruction set and addressing modes. To thoroughly understand a microprocessor, one must not only know the hardware and software, but also the intimate interaction between the two.

14.2.1 Analysis of the 8085's Functional Pinout

You should have a good idea of what to expect in the 8085 pinout. You already understand the concepts of three-bus architecture and how microprocessors interface with memory. Remember to apply your present knowledge of microprocessors while we are analyzing the 8085. Figure 14.2 indicates the functional pinout of the 8085; all signals are grouped according to common functions.

Power Group To be compatible with TTL power supplies, the 8085 requires only a V_{cc} of +5 V and ground. Ground is called V_{ss} (substrate voltage) because the 8085 is a MOS device.

Clock Group You know that every microprocessor requires a clock. All of the microprocessor's activity is referenced to this clock. The 8085 does not require an external clock generator; a crystal is simply placed between the X1 and X2 inputs of the 8085. The input frequency of the crystal is divided by 2 to produce the internal reference frequency. The clock output is a copy

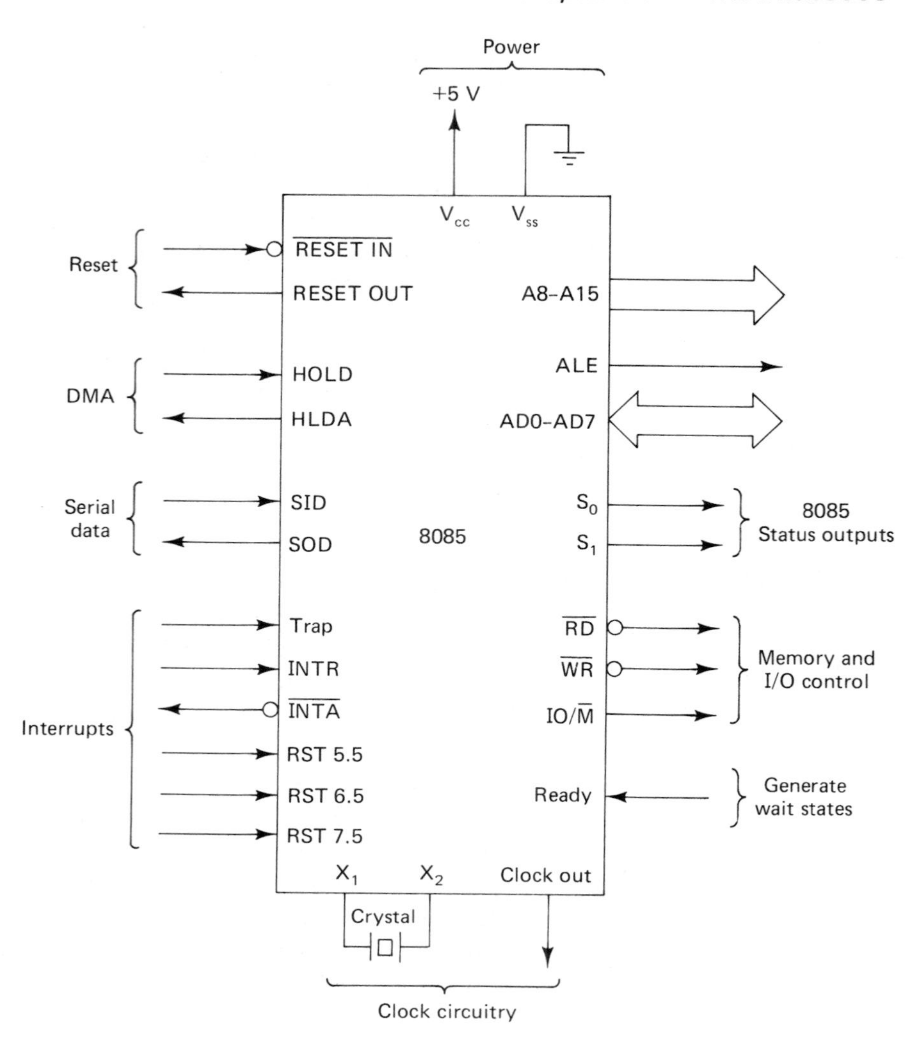

Figure 14.2 Pinout of 8085—grouped according to function.

of the 8085's internal clock. It is used to synchronize the rest of the system to the 8085.

Reset Group When the reset input is at an active-low level, the internal operation of the microprocessor stops. During reset the program counter is set to an address of 0H. When the reset input goes to an inactive-high level, the 8085 will fetch an instruction from memory location 0H. Memory location 0H usually contains a jump instruction to a boot program in ROM. Reset out is active-high and is used to reset other devices in the system.

Address/Data Bus Group Until recently, 40-pin packages for LSI devices were considered standard. With the advent of VLSI devices the 40-pin standard has been dropped. New VLSI devices are opting to use 64- to 86-pin packages. Eight-bit microprocessors were designed to be implemented in 40-pin packages. Because of this limitation, the 8085 multiplexes the lower byte of the address bus (A0 through A7) and the data bus (D0 through D7). Here the word "multiplexed" refers to the time division of the address/data bus.

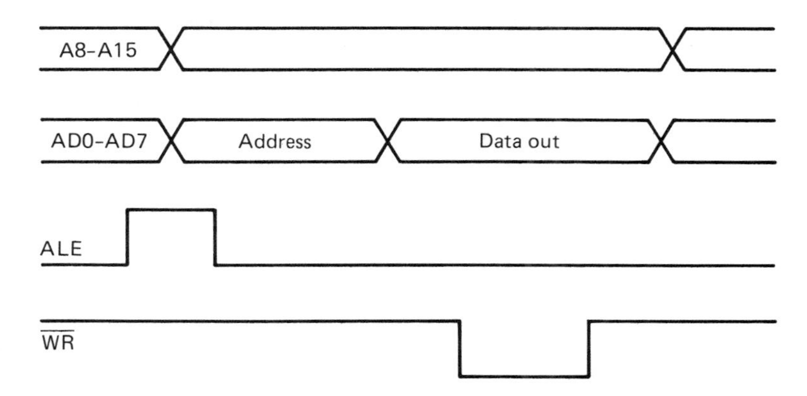

Figure 14.3 8085 memory write using ALE control signal.

Think back to a microprocessor read or write operation as it was described in Chapter 10. The first action of the microprocessor is to provide a 16-bit memory address. During this period the 8085 uses AD0 through AD7 as the lower part of the address bus. The system hardware will provide an 8-bit latch to demultiplex the address/data bus. The 8085 uses the ALE (address latch enable) signal clock to operate the external address latch. After the address on AD0 through AD7 is latched, AD0 through AD7 can be used as a standard bidirectional data bus. Figure 14.3 illustrates an 8085 memory-write operation. Notice that address lines A8 through A15 contain valid address information throughout the entire write operation. For the first part of the write operation AD0 through AD7 contains the lower 8 bits of the 16-bit address. On the falling edge of the ALE, the lower 8 bits of data will be latched into an external latch. This will free AD0 through AD7 to be used as a data bus for the remainder of the write operation.

The use of a multiplexed address/data bus saves a total of seven pins for other functions. (The eighth pin that would be saved is used for the ALE control line.) Contrary to popular belief, multiplexing microprocessor buses does not add any appreciable complication to the hardware support system of the 8085. Just remember that on the falling edge of ALE, the lower 8 bits of the address must be externally latched. It is as simple as that.

Memory and I/O Control Group The 8085 provides three control lines for the management of memory and I/O accesses: $\overline{RD}$, $\overline{WR}$, and IO/$\overline{M}$. Figure 14.4 depicts a truth table and simple combinational circuit that decodes these control signals into $\overline{\text{memory read}}$, $\overline{\text{memory write}}$, $\overline{\text{I/O read}}$, and $\overline{\text{I/O write}}$ signals. The first four lines of the truth table have already been examined thoroughly in Chapter 10. The fifth line indicates that $\overline{RD}$ and $\overline{WR}$ should never be active simultaneously. In the last line neither $\overline{RD}$ or $\overline{WR}$ is active. This indicates that the microprocessor is not currently accessing memory or I/O.

You are already aware that during a memory access the 8085 will place a 16-bit address onto AD0 through AD7 and A8 through A15. During an I/O access the 8085 will place an 8-bit address on AD0 through AD7. With an 8-bit address, the 8085

$IO/\overline{M}$	$\overline{WR}$	$\overline{RD}$	Action
0	1	0	Memory read
0	0	1	Memory write
1	1	0	I/O read
1	0	1	I/O write
X	0	0	Illegal
X	1	1	No memory or I/O access

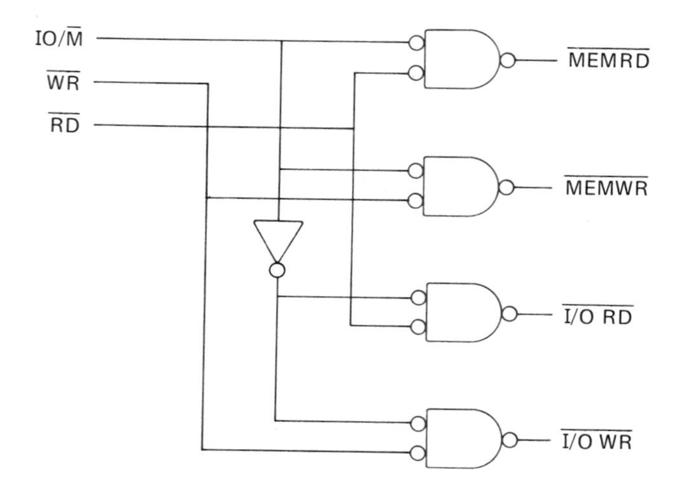

Figure 14.4 Truth table and decoding circuit for memory and I/O control.

can access 256 unique I/O ports. The number of ROM and RAM locations in a microprocessor system is extremely large compared to the number of I/O ports. You can be assured that a capacity of 256 I/O ports is more than sufficient.

An interesting question is: What does the 8085 do with A8 through A15 during an I/O access? The port address on AD0 through AD7 is also placed on A8 through A15. The reason for this redundant address is simple: Many hardware designs use the upper address bits of the address bus for decoding chip selects. Repeating the port address on A8 through A15 allows the designer to take advantage of decoders which may be used for the dual purpose of memory and I/O port chip selection.

DMA Group We have considered the situations where another device may take control of the system bus. When this happens the microprocessor must bring its address, data, and control lines to a high-Z level, essentially removing itself from the circuit. This will ensure that the current bus master will be able to access memory and I/O devices without contending with the microprocessor's outputs.

When another device wants to take over the system buses, it must drive the Hold input active-high. The hold input on the 8085 functions as a DMA request input. After the 8085 finishes executing its current instruction, it will bring the address, data, and control bus to high-Z. The 8085 will then place an active-high level on the HLDA (hold acknowledge) output. This will inform the requesting device that the system buses are available. The 8085 will poll the hold input until the current bus master returns it to an inactive level. This indicates that the system buses have been released. The 8085 will respond by bringing HLDA low and resuming normal processing.

Ready Input We have seen timing diagrams illustrating a microprocessor performing memory read and write operations. What happens if a particular memory or I/O device has a long read or write access time? For example, a particular memory device has a read access time of 10 μs. The 8085 is using a 6-MHz clock crystal, which would give it a read or write cycle period of 1 μs. It appears that the slow memory would be incompatible with the fast microprocessor. The solution to this problem is the ready input on the 8085. Each time a slow memory or I/O device is accessed, external hardware will bring the ready input to an inactive-low level. As long as the ready input is low, the 8085 will not change the levels on the address and control lines. When the ready input returns to an active-high level, the 8085 will complete the read or write cycle. If the ready input is not used, it must be tied to an active-high level.

The process of bringing the ready input low to accommodate slow memory and I/O is called *inserting wait states*. Each wait state is equal to one period of the microprocessor clock. An external counter can be programmed to hold the ready line at a low level for the required number of wait states. The ready input on the 8085 makes it compatible with any memory or I/O device, regardless of access time. In a later section we analyze a simple wait-state generator for the 8085.

Interrupt Group The 8085 is often used as an intelligent controller. It must be able to efficiently handle interrupt requests from external devices. The 8085 has three types of interrupts: TRAP (nonmaskable interrupt), INTR (8080-type interrupt), and RST (hardware restarts). We have already discussed the concept of servicing external devices using polling routines and interrupts. Let's examine the 8085's interrupts.

The *TRAP interrupt* is known as a NMI (nonmaskable interrupt). The other types of interrupts can be enabled or disabled through microprocessor instructions. When a DI (disable interrupts) instruction is executed, the INTR and RST lines will be ignored. The EI (enable interrupts) instruction will force the 8085 to monitor the INTR and RST lines for active levels. A programmer disables interrupts during any sequence of instructions that are time dependent and must be executed without interruption.

TRAP is intended to indicate serious system problems; it cannot be masked. When TRAP goes active the microprocessor will finish executing its current instruction, the contents of the PC (program counter) will be automatically saved on the stack, and the address 0024H will be written into the PC. An interrupt service routine will reside at address 0024H. The last instruction in the service routine is a return. The contents of the program counter prior to the interrupt will be popped off the stack and the 8085 will continue normal program execution.

An active TRAP input actually invokes an automatic subroutine call at address 0024H. The TRAP is often connected to an ac power failure indicator. This will give the 8085 a few milliseconds (which is a long time in microprocessor terms) to shut the system down before the impending power failure. Consider a microprocessor-based system that controls the operation of valves in a chemical plant. During a TRAP power-failure service routine, the 8085 would close all valves in the plant to prevent flooding. It may also save important data in CMOS battery-backed-up memory, or write the data to a mass storage device.

The INTR input is know as an *8080-type interrupt*. The only interrupt facility on the 8080 was the INTR pin. This same INTR was implemented on the 8085. In a typical 8085 system, many different devices may be able to bring the INTR input to an active-high level. When the 8085 senses an active INTR input, it will complete its current instruction. The 8085 will then take the $\overline{\text{INTA}}$ (interrupt acknowledge) output to an active-low level. The external device that created the interrupt will then be required to identify itself. It will do so by placing a special one-byte instruction onto the data bus. The 8085 will read the data bus and then jump to a service routine. Figure 14.5 illustrates the hardware required to support 8080-type interrupts.

The circuit in Figure 14.5a is composed of a 74LS148 8-line-to-3-line priority encoder and the 74LS240 octal-inverting three-state buffer. This circuit can control the interrupt requests from eight devices. Each of the eight interrupt request lines is connected to an active-low input on the 74LS148. As you should remember, input seven has the highest priority and input zero the lowest priority. The encoded active-low outputs of the priority encoder drive three inputs on the inverting three-state buffer. The other five inputs to the buffer are tied to ground.

When any of the eight interrupt requests driving the in puts of the 74LS148 go active, the $\overline{\text{GS}}$ output will go low. The inverter will cause the INTR input to the 8085 to go active high. This will signal to the microprocessor that a device requires service. After the 8085 completes its current instruction, it will bring the $\overline{\text{INTA}}$ output low. The inputs of the octal three-state buffer will be inverted and output onto the microprocessor's data bus. The 8085 will treat this data byte as an instruction. Figure 14.5b illustrates the special group of 8085 one-byte restart instructions. These instructions cause the 8085 to push the contents of the PC onto the stack and jump to a particular memory location, where the proper service routine will reside. At the end of each service routine is a return instruction. This will cause the address to be popped off the stack, and the microprocessor will continue normal programming (just like the TRAP interrupt).

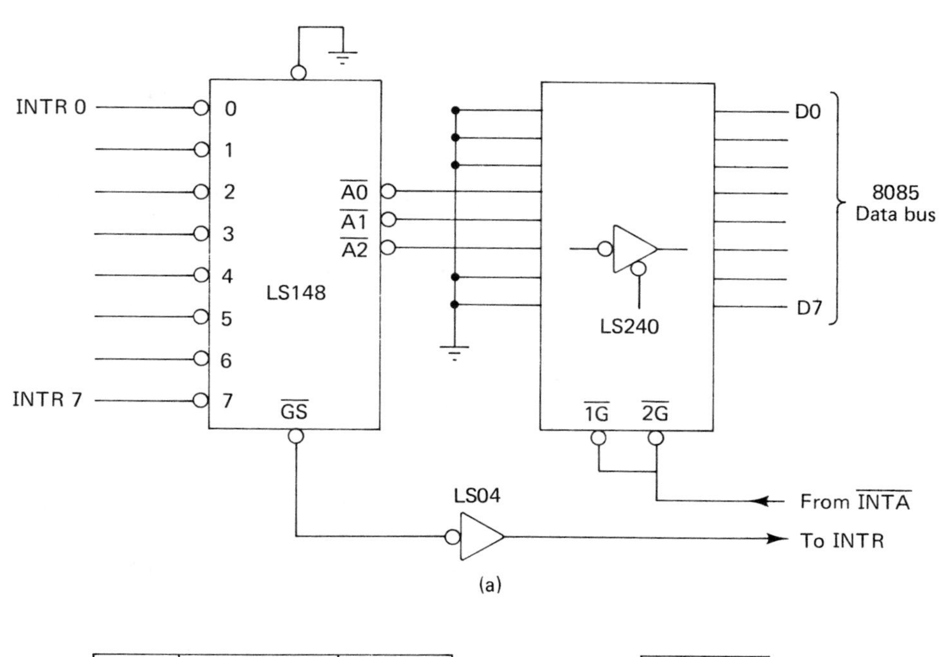

(a)

Name	Op code	Reset address
RST 0	1100 0111	0 H
RST 1	1100 1111	8 H
RST 2	1101 0111	10 H
RST 3	1101 1111	18 H
RST 4	1110 0111	20 H
RST 5	1110 1111	28 H
RST 6	1111 0111	30 H
RST 7	1111 1111	38 H

8085 restart instructions

(b)

B5	B4	B3
0	0	0
0	0	1
0	1	0
0	1	1
1	0	0
1	0	1
1	1	0
1	1	1

Bit action of B3, B4 and B5

(c)

Figure 14.5 Priority-interrupt control for 8080-type interruptions.

The table in Figure 14.5b indicates the proper bit pattern for each instruction. Notice that bits B0 through B2 and B6 and B7 are logic 1's for each RST instruction; only B3 through B5 change, and they change in a normal binary count sequence (as indicated in Figure 14.5c). Our concern here is: How do the 74LS148 and the 74LS240 generate the correct RST code? Because the bits that do not change are pulled to ground on the input side of the buffer, they will appear as logic 1 levels when $\overline{\text{INTA}}$ goes active.

Consider the case when the highest-priority interrupt (pin 7 of 74LS148) goes active. The $\overline{\text{GS}}$ will go active, indicating a pending interrupt, and the output of the 74LS148 will be 000—the inverted binary-encoded value of input 7. When the microprocessor acknowledges the interrupt by bringing $\overline{\text{INTA}}$ low, the code of 1111 1111 will be output onto the 8085's data bus—a RST7 instruction. Work through the rest of the restart instructions to ensure that the correct RST instruction is placed onto the data bus for each interrupt request.

It is important to realize that the 74LS240 is a bus talker and must be held at high-Z until the 8085 acknowledges the interrupt. The 8085 will read the data bus (the RST instruction) and then bring the $\overline{\text{INTA}}$ input to an inactive level. This will return the outputs of the octal buffer to high-Z.

The third type of 8085 interrupt is called the *hardware restarts*. We have just seen the software restart instructions—RST0 through RST7. They are called software restarts because they are actual 8085 instructions. There are three hardware restart pins on the 8085: RST5.5, RST6.5, and RST7.5. One external device can be connected to each hardware restart input. The hardware restart pins function exactly like the software restart instructions. When a hardware restart pin goes to an active-high level, the 8085 will finish its current instruction and jump to a specific location in memory. TRAP is really nothing more than a nonmaskable hardware restart input.

Name	Restart address
RST 5.5	002CH
RST 6.5	0034H
RST 7.5	003CH

Figure 14.6 Restart locations for RST5.5, 6.5, and 7.5.

Figure 14.6 indicates the restart locations for each of the three hardware restart pins. The TRAP and RST inputs provide a fast, efficient means of servicing interrupts without the requirement of external circuitry like the 8080-type interrupt. Trap and RST interrupts are often called *vectored interrupts*. The use of the term "vector" in describing interrupts implies that the interrupt has an associated address (i.e., a direction). When TRAP or a hardware RST goes active, the device requiring service need not identify itself because an automatic address vector is associated with each pin. Compare this with the 8080-type interrupt, where a device must identify itself by placing a one-byte restart instruction onto the data bus when $\overline{\text{INTA}}$ goes active low.

Status Output Group S0 and S1 are status output pins. They indicate to the outside world the current operation being performed by the 8085. Figure 14.7 illustrates the encoded status outputs.

	S_1	S_0
HALT	0	0
Write	0	1
Read	1	0
Fetch	1	1

Figure 14.7 Status information on S0 and S1.

The status information should be latched by the falling edge of ALE. Using S0, S1, and IO/$\overline{\text{M}}$, advance information of 8085 activities can be decoded. Most 8085 circuits do not make use of the status pins. In Chapter 19 we examine an advanced microprocessor that does make important use of its status outputs.

Serial Data Group SID (serial input data) and SOD (serial output data) provide the 8085 with the basic functions of a UART. The accumulator is used to buffer the byte being transmitted or received. We have mentioned that the 8085 instruction set contains only two more instructions than the 8080 instruction set. These instructions, RIM and SIM, are used for multiple purposes. One function of RIM and SIM is to provide software support for the SID and SOD pins.

In some microprocessor systems, keeping a low parts count is an important criterion of the design. The vectored interrupts of TRAP, RST5.5, RST6.5, RST7.5, and elementary serial I/O using the SID and SOD pins make the 8085 an attractive processor for low-parts-count intelligent controller applications.

14.3 INTERFACING THE 8085 WITH MEMORY AND I/O

There are two specific ways to interface the 8085 to a conventional three-bus architecture. The first method employs the exclusive use of Intel components: RAMs, ROMs, UARTs, PPIs, and other system support devices. Intel provides an on-chip 8-bit latch to demultiplex the address/data bus. Figure 14.8 illustrates a minimum 8085 system.

14.3.1 Minimum-Component 8085 System

This will be your first analysis of a microprocessor-based system. You know that every microprocessor system must have a CPU, memory (RAM and ROM), and I/O. The system in Figure 14.8 employs three Intel LSI devices: the 8085 microprocessor, the 8355/8755 ROM/PROM and parallel I/O ports, and the 8155/8156 RAM/timer/parallel I/O ports. We will examine each device and then analyze the manner in which they are interconnected.

On the 8085 we will concentrate on the lines required to interface memory and I/O: the address, data, and control buses. In Figure 14.8 the interrupt, DMA, status, and serial I/O capacities of the 8085 are not used.

The circuit connected to the $\overline{\text{RESET}}$ input is a basic *RC* power-on delay. Remember from basic dc theory that a capacitor appears to be a short at the moment when a circuit is first energized. The *RC* time constant will determine the length of time required for the capacitor to charge up to a logic 1 level. (A typical reset time is a few hundred milliseconds.) The diode will ensure that the spike generated by the capacitor when the power is turned off will be clamped to a maximum of V_{cc} + one diode drop. While $\overline{\text{reset}}$ is held at an active-low level, the 8085 will initialize the PC to 0H, and all the buses will be at a high-Z level. This will allow the rest of the system to be powered-up and ready.

When the $\overline{\text{RESET}}$ input goes to a logic 1 level, the 8085 will fetch its first instruction at memory address 0H (ROM or PROM). The 8355 contains 2K × 8 mask-programmed ROM and two fully programmable 8-bit I/O ports. The 8755 is the version of the 8355 used in short-run or R&D environments. Instead of 2 KB of ROM, the 8755 contains 2 KB of EPROM. The memory in the 8755 can be erased and reprogrammed hundreds of times.

The two chip enables ($\overline{\text{CE}}$ and CE) must both be at active levels before the 8355 can be accessed. The $\overline{\text{CE}}$ doubles as a program pin for the 8755 EPROM. You should think of it as an active-low chip enable. In Figure 14.8 CE is tied at an active-high level and $\overline{\text{CE}}$ is tied to A11.

SYSTEM OPERATION

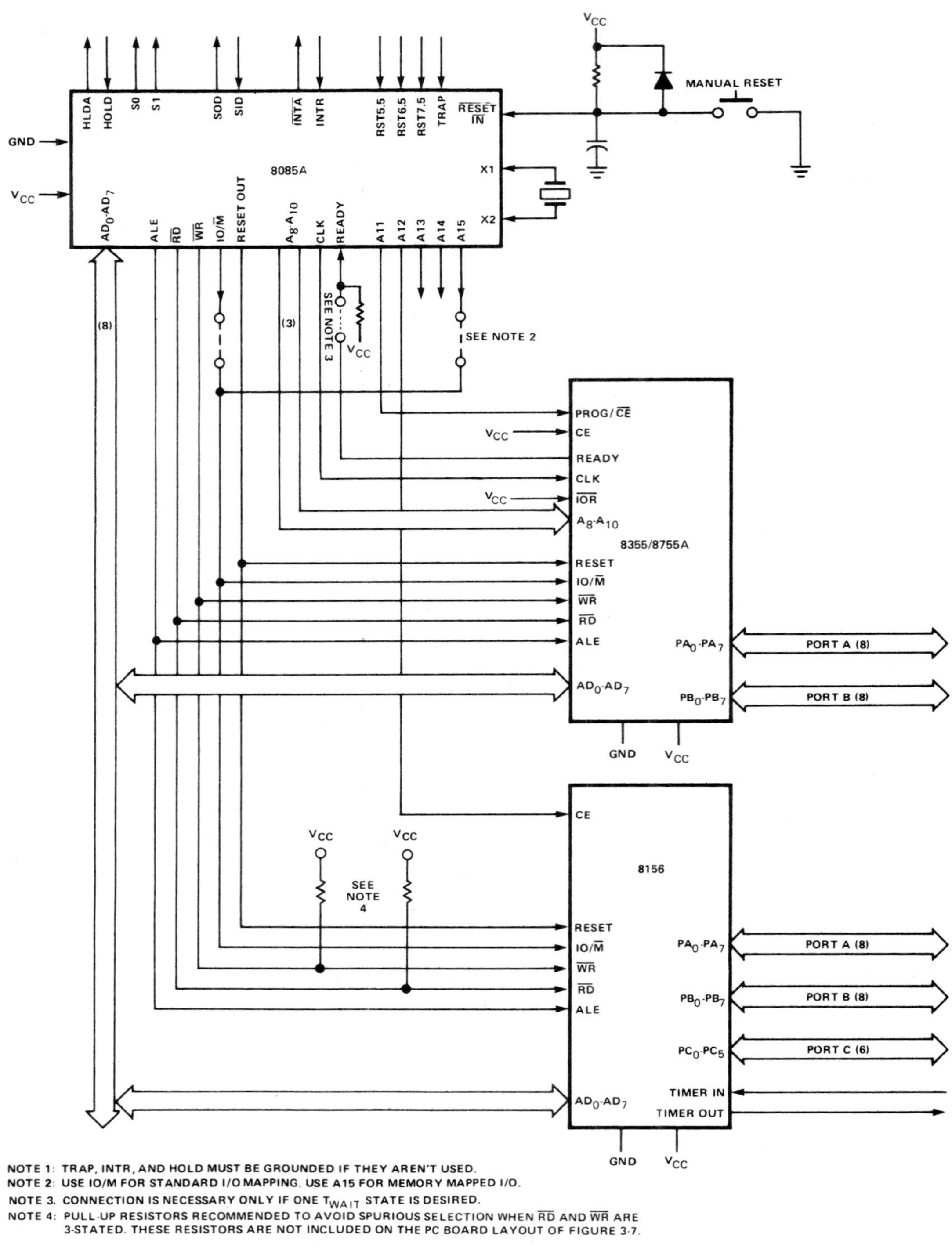

Figure 14.8 Minimum 8085 system.

The ready line on the 8755 is an output that can be used to insert one wait state onto the 8085. If the jumper on the 8085's ready line is out, the ready input is pulled active high and the 8085 runs at full speed with no wait states. If the jumper is in, the 8755 automatically inserts one wait state during each memory access.

The clock input is used to synchronize the 8355's ready output to the microprocessor's memory-read cycle.

The I/O read ($\overline{IOR}$) is used as an auxiliary method of performing an I/O read. In this application it is not used and must be tied to V_{cc}.

The 8355 has 2K of unique storage locations. Therefore, it requires 11 address lines, A0 through A10. A8 through A10 are the three most significant address lines.

Because the 8355 has programmable I/O ports, it must be reset by the 8085 on power-up. The active-high reset input will initialize the two programmable 8-bit I/O ports as inputs.

The IO/$\overline{M}$ line serves to differentiate between an 8085 memory read and I/O access. When $\overline{CS}$ is low, IO/$\overline{M}$ is high, and $\overline{WR}$ is low, the 8355 is informed that the 8085 is performing an I/O output operation. That means that the 8085 is writing data to one of the I/O ports of the 8355. Because it contains only ROM, the 8355 will never be accessed for a memory-write operation.

When $\overline{CS}$ is low, IO/$\overline{M}$ is high, and $\overline{RD}$ is low, the 8085 is performing an I/O read of one of the 8355's I/O ports. If the same situation occurs when IO/$\overline{M}$ is low, the 8085 is performing a memory-read operation of the 8355's ROM.

Because the 8355 was designed specifically to be used in an 8085-based microprocessor system, it contains an internal 8-bit latch that is used to demultiplex AD0 through AD7. A high level of the ALE input will cause the gate of the 8355's internal transparent latch to go active. The lower 8 bits of address will be latched on the falling edge of ALE. This will allow the 8085 to use AD0 through AD7 as a data bus for the remainder of the memory or I/O access. This saves the requirement of an external latch.

Finally, we see the multiplexed address/data bus, AD0 through AD7. The 8355 has four internal registers that are used to support the two parallel 8-bit I/O ports. Address bits A0 and A1 are used with the $\overline{CS}$, IO/$\overline{M}$, $\overline{RD}$, $\overline{WR}$ control lines to access the I/O circuitry.

A1	*A0*	*Selected*
0	0	Port A data register
0	1	Port B data register
1	0	Port A direction control register
1	1	Port B direction control register

Port A and port B can be programmed on a bit-by-bit basis as inputs or outputs. A port is programmed by writing the appropriate byte to its direction control register, called the DDR (data direction register) by Intel. Each bit in the direction control register will control whether the corresponding bit in the data register is an input or output. A logic 0 causes the corresponding bit in the port to be an input and a logic 1 causes it to be an output. Writing the byte—1111 0000—to the direction control register of port A will cause the lower nibble of port A to be inputs and the upper nibble of port A to be outputs.

The 8155/8156 is a 256-byte static RAM, programmable timer, and three-I/O-port LSI support device for the 8085 family. The only difference between the 8155 and 8156 is the active level of the CE inputs; the 8155 CE is active low and the 8156 CE is active high. The circuit in Figure 14.8 employs the 8156. As the 8156 is a complex, highly integrated device, we will examine its pinout in the same manner as we examined the 8355.

Unlike the 8355, the 8156 has only one chip enable. When CE is active high, the 8156 is selected. The 8156 CE is tied to A12. The active-high reset on the 8156 serves the same purpose as the reset on the 8355. When reset goes active, the I/O ports have initialized as input ports, and the timer is reset.

The control lines—$IO/\overline{M}$, $\overline{WR}$, $\overline{RD}$ and ALE—function exactly as they did with the 8355, with one exception. Because it contains RAM, the 8085 will perform memory reads and writes on the 8156. Although 256 bytes may not seem like very much RAM, this three-chip system is designed primarily as an intelligent controller. The program that runs the system is contained in the 8355's 2 kB of ROM. The 256 bytes of RAM in the 8156 will be used primarily to hold temporary data and, most important, will serve as a stack, so the 8085 can execute subroutines and interrupt service routines.

The 8156 also has an internal address demultiplexing latch, which is controlled by the ALE. Address bits A0 through A7 are used to address one of the 256 unique RAM locations and also the I/O and timer ports control and data registers.

The 8156 contains two 8-bit I/O ports and one 6-bit I/O port. The I/O capabilities of the 8155/8156 are a slightly scaled down version of the 8255 PPI. Like the 8255, the 8156 supports three modes of I/O, including full parallel hardware handshaking in mode 1. The 40-pin limitation of the 8156 reduces port C from a full 8-bit port (as in the 8255) to a 6-bit port but does not affect its ability to handshake because only three handshaking lines are required: $\overline{\text{strobe}}$, busy, and $\overline{\text{acknowledge}}$. Imagine the power of an 8255 integrated into a device that also contains RAM and a programmable timer.

The programmable timer in the 8085 is a 14-bit downcounter that has four operational modes. The system clock can be connected to the timer input. The timer can be programmed to count a certain number of *clock ticks* and then interrupt the 8085. The timer can also be configured as a baud-rate clock for a UART.

14.3.2 The 8085 in a Standard Component System

The second method used in interfacing the 8085 with the system employs standard support devices. To use standard support devices, AD0 through AD7 must be demultiplexed with an external latch. Figure 14.9 illustrates a typical method of demultiplexing the bus.

The 74LS373 is an octal-transparent latch with an active-high gate and an active-low output enable. When the gate input is high, the data on the D inputs of the eight latches pass through to the Q outputs. On the falling edge of the gate signal, the data is latched. Notice that the gate is driven by ALE from the 8085. When $\overline{OE}$ is active, the 74LS373 places the Q outputs of the eight flip-flops onto the lower byte of the address bus. When $\overline{OE}$ is inactive, the eight outputs of the 74LS373 go to high-Z. In Figure 14.9 the $\overline{OE}$ input of 74LS373 is tied to an active-low level. In a system capable of DMA operation, the $\overline{OE}$

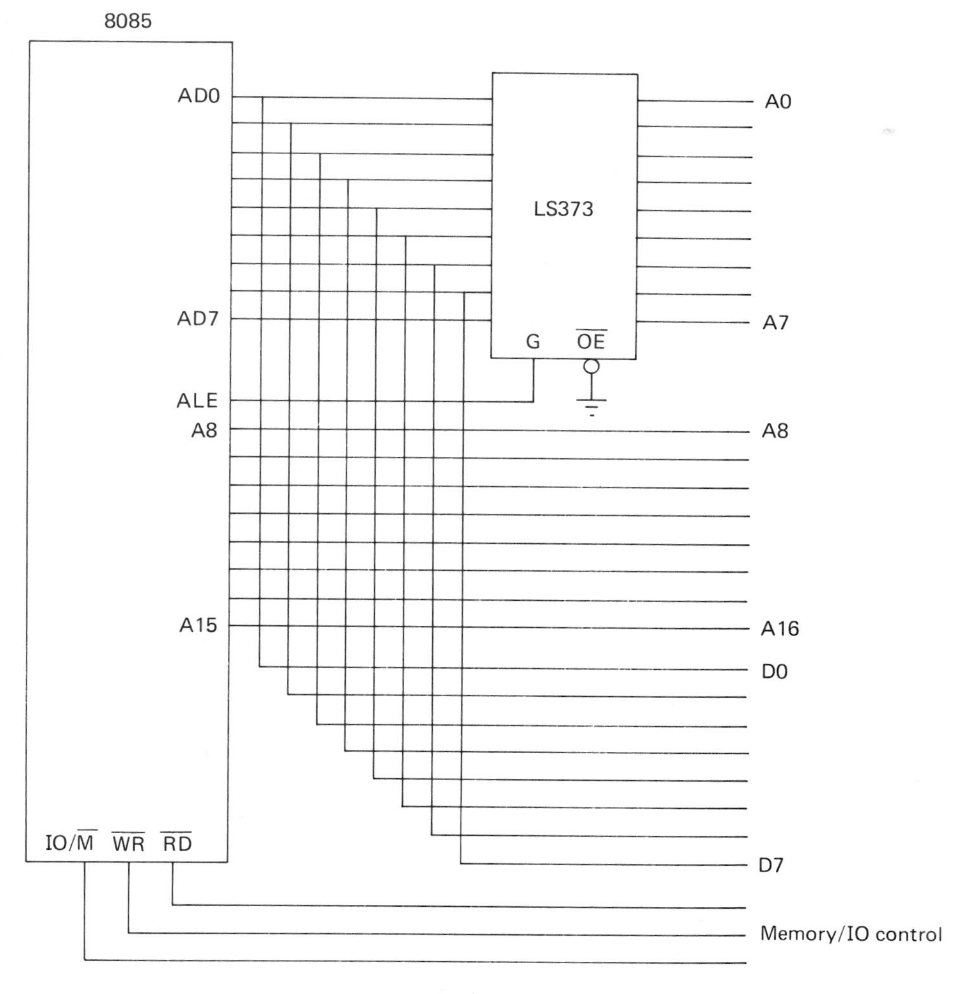

Figure 14.9 Demultiplexing AD0 through AD7 with an octal transparent latch.

would be tied to the HLDA output of the 8085. During a DMA operation the data, address, and control buses must go to high-Z to enable a second bus master to take charge. Because the 74LS373 is driving the lower byte of the address bus, it must also go to high-Z during a DMA operation.

We have examined many timing diagrams, describing simple flip-flops to complex DRAMs. Figure 14.10 illustrates the simplified read and write timing diagrams of the 8085. The read/write process of the 8085 is a simple, straightforward procedure involving the data, address, and control buses.

Figure 14.10a illustrates a memory or I/O read operation. Remember that the state of the IO/$\overline{\text{M}}$ pin will determine what type of read operation will actually occur. Notice that the operation takes three ticks (cycles) of the system clock. A tick of the 8085 clock is the time between falling edges. The three ticks that constitute the read cycle will be referenced as T1, T2, and T3. Let's examine what occurs during each tick of the clock.

T1. The falling edge of T1 will initiate the read cycle. AD0 through AD7 and A8 through A15 will be driven with the mem-

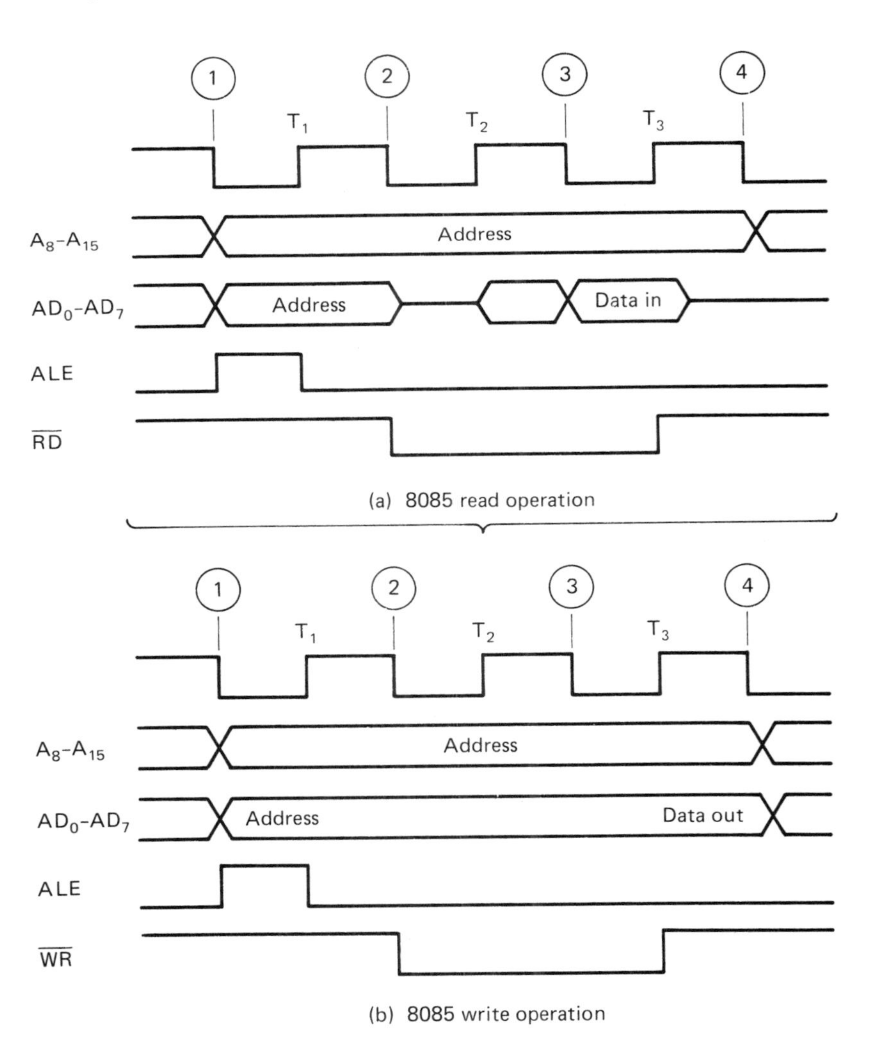

Figure 14.10 Read and write timing diagrams for 8085: (a) read operation; (b) write operation.

ory or I/O address. ALE will go to an active-high level, enabling the transparent demultiplexing latch. If the cycle is a memory read, $\text{IO}/\overline{\text{M}}$ will go to an active-low level.

On the rising edge of T1, ALE will return to a logic 0 level. At this point the stable low byte of address is latched into the internal latch on Intel LSI support chips or into the external latch, as illustrated in Figure 14.9.

You may wonder why a transparent latch (instead of a negative-edge-triggered flip-flop) is used to demultiplex the address/data bus. The answer is simple—the address actually becomes stable a short time after the falling edge of T1. A transparent latch allows the memory devices to start decoding the address long before the falling edge of ALE. This decreases the read and write cycle time. An edge-triggered flip-flop would have to wait for the falling edge of ALE to latch AD0 through AD7. One-half of a clock tick would be wasted.

T2. At the beginning of T2, AD0 through AD7 will go to high-Z, and $\overline{\text{RD}}$ will go active. The system's address decoders are now enabled. After a short delay, the proper memory or I/O device will be chip selected.

On the rising edge of the clock (during T2) the 8085 will sample the level on the ready input. If it is an inactive low, the 8085 will insert a wait state. During a wait state, the 8085 will maintain the levels on the address and control buses. This will give a slow memory or I/O port some extra time. At each subsequent rising edge of the system clock, the 8085 will resample the ready input. When it finally returns to an active level, the 8085 will execute T3 and complete the access.

T3. The 8085 assumes that the read data has been placed on AD0 through AD7. On the rising edge of the clock the $\overline{\text{RD}}$ output will go high, latching the read byte of data inside the 8085. The read cycle is complete at the falling edge of T3. The address, data, and control buses are returned to their inactive levels.

Figure 14.10b illustrates the 8085 write access. The write cycle functions exactly like the read cycle. $\overline{\text{WR}}$ goes active at the beginning of T2, the ready is sampled on the rising edge of T2, and the write data is latched in the memory on the rising edge of $\overline{\text{WR}}$.

14.3.3 Wait-State Generation

What happens if the 8085 accesses a slow memory or I/O device that needs to generate wait states? It is the responsibility of the external hardware to pull the ready input of the 8085 to an inactive-low level. At the rising edge of the clock in T2, the 8085 will sample the ready input. If it is low, a wait state will be generated. That means that the 8085 will hold the address on A8 through A15 stable and $\overline{\text{RD}}$ active low. One clock tick later, the 8085 will resample the ready input. As long as Ready is a logic 0, the 8085 will continue to generate wait states. This will continue until the external hardware that generated the wait state returns ready to an active-high level.

Figure 14.11 illustrates a simple hardware scheme for generating wait states. Assume that output 2 of the 3-line-to-8-line decoder is used to chip select a slow memory device. This memory device requires that the 8085 will have a read or write cycle

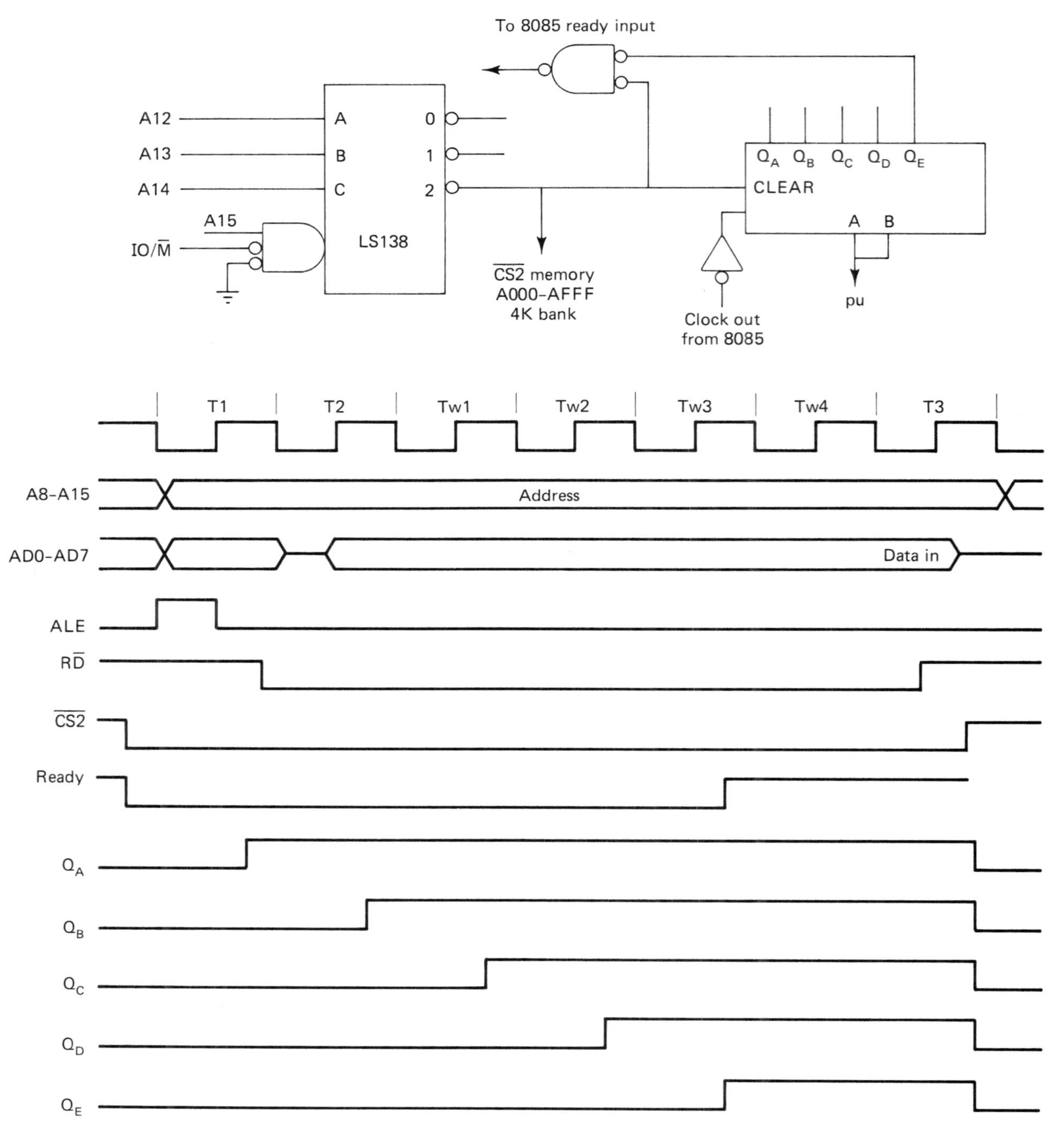

Figure 14.11 Insertion of wait states.

of seven clock ticks instead of the normal three clock ticks. We must pull the ready input to the 8085 low for four clock cycles.

In normal operation $\overline{CS2}$ is at an inactive-high level. This will force the output of the OR gate high. This high level is driving the ready input of the 8085. Notice that the shift register has an active-high clear input. The shift register is normally held in a cleared state. Also notice that the shift register is clocked by the negative edge of the system clock, via an inverter, and the serial data input is pulled up to +5 V. Refer to the timing diagram.

The read access described in the timing diagram of Figure 14.11 takes seven clock ticks; the standard T1, T2, and T3, plus four wait states, TW1 through TW4. Assume that the timing diagram illustrates a read access of a memory in the range A000H through AFFFH.

T1. On the falling edge of T1 the address is output on AD0 through AD7 and A8 through A15. After a short delay, $\overline{CS2}$ will go active low. The shift register is now enabled. With $\overline{CS2}$ low and Q_e of the shift register low, the output of the OR gate will go low, pulling the 8085's ready line to an inactive level.

T2. On the falling edge of T2, the shift register will be clocked, and a logic 1 will be shifted into Q_a. On the rising edge of the clock in T2, the 8085 samples the ready input and finds an inactive-low level. The 8085 will now enter a wait state.

TW1. This is the first wait state. On the falling edge of the clock in TW1, the shift register is clocked and a logic 1 is shifted into Q_b. On the rising edge of the clock, the 8085 resamples the ready line. Because Ready is still inactive, the 8085 will enter another wait state.

TW2 and TW3. The 8085 continues to execute wait states, and the logic 1 is shifted one bit further with each falling edge of the system clock.

TW4. The falling edge of the clock at TW4 will shift a logic 1 into Q_e. This will force the output of the OR gate to a logic 1 level. The ready line has now returned to an active level. The 8085 samples the ready line on the rising edge of the clock during TW4. Because the ready line is now high, the 8085 will complete the read access with T3.

After the read access is complete, $\overline{CS2}$ will return to an inactive-high level. This will keep the output of the OR gate at a high level, and the shift register will enter a cleared state.

Shift registers or counters are commonly used to generate wait states. All that is required is a device that can count clock ticks and pull the ready line inactive for the required number of wait states.

You have now had a comprehensive introduction into the

hardware characteristics of a typical 8-bit microprocessor. All 8-bit microprocessors share similar architectures and timing diagrams. You should now go back to Chapter 10 and insert the 8085 and 8-bit latch into each circuit. This will illustrate how the 8085 is connected in conventional memory systems.

QUESTIONS AND PROBLEMS

14.1. How can one tell how much memory a microprocessor can directly address?

14.2. Why is the 8085 considered an "8-bit" processor?

14.3. How can the 8085 use the same pins for both address and data?

14.4. What are the three 8085 control signals associated with memory and I/O access?

14.5. How does the 8085 interface with other "bus masters"?

14.6. How does the 8085 accommodate slow memories and I/O devices?

14.7. What does it mean that the TRAP is a NMI (nonmaskable interrupt)?

14.8. Describe the operation of an 8080-type interrupt.

Refer to Figure 14.5 for the next three questions.

14.9. The trace running between INTA- and the enable inputs of the LS240 is cut. How would this affect the circuit?

14.10. The input of the LS04 is shorted to the enable pins on the LS240. How might this affect the circuit?

14.11. $\overline{\text{INTA}}$ is shorted to ground. How does this affect the circuit?

14.12. What advantage is there to using Intel components in the 8085 system?

Refer to Figure 14.8 for the next three questions.

14.13. The capacitor in the reset circuit is shorted. How does this affect the circuit action?

14.14. The diode in the reset circuit is stuffed backwards. How does this affect the circuit?

14.15. When is the 8156 chip selected?

14.16. How many clock cycles are there in a standard 8085 memory or I/O access?

Refer to Figure 14.11 for the next two questions.

14.17. The inverter that drives the clock input to the shift register is internally open. How does this affect wait-state generation?

14.18. The 8085 appears to be reading bad data from the memory located in A000 through AFFF. You place a scope probe on the ready input of the 8085 and it is 3.6 V. What are some likely malfunctions that would cause this problem?

14.19. What facilities of the 8085 make it an excellent low-parts-count intelligent controller?

15 Introduction to the 8085 Instruction Set and Addressing Modes

In this chapter we introduce you to the 8085's instruction set and addressing modes. Be sure to equate each instruction to the manner in which it is executed by the 8085 and the system hardware. Remember that hardware and software do not exist independently; strive to understand this hardware/software relationship.

15.1 THE 8085 REGISTER SET

Microprocessor systems are only as good as the software that runs them. The 8085 has not been used as the CPU for many popular microcomputers. This is mainly due to its lack of sophisticated addressing modes. The 8085 is extensively used as a control processor: from video tape recorders to electronic PBX systems, the 8085 has been employed as an intelligent controller. We will briefly study the simple instruction set and addressing modes of the 8085.

Figure 15.1 illustrates the instruction set of the 8085. The instruction set is subdivided into groups of related instructions. Each instruction has a mnemonic representation and an 8-bit code. The mnemonic is the symbolic name of the instruction. A microprocessor program written in symbolic mnemonics is called an assembly language or source code program.

Because microprocessors understand only 1's and 0's, each

DATA TRANSFER GROUP

Move

Instruction	Opcode
MOV A,A	7F
MOV A,B	78
MOV A,C	79
MOV A,D	7A
MOV A,E	7B
MOV A,H	7C
MOV A,L	7D
MOV A,M	7E
MOV B,A	47
MOV B,B	40
MOV B,C	41
MOV B,D	42
MOV B,E	43
MOV B,H	44
MOV B,L	45
MOV B,M	46
MOV C,A	4F
MOV C,B	48
MOV C,C	49
MOV C,D	4A
MOV C,E	4B
MOV C,H	4C
MOV C,L	4D
MOV C,M	4E
MOV D,A	57
MOV D,B	50
MOV D,C	51
MOV D,D	52
MOV D,E	53
MOV D,H	54
MOV D,L	55
MOV D,M	56

Move (cont)

Instruction	Opcode
MOV E,A	5F
MOV E,B	58
MOV E,C	59
MOV E,D	5A
MOV E,E	5B
MOV E,H	5C
MOV E,L	5D
MOV E,M	5E
MOV H,A	67
MOV H,B	60
MOV H,C	61
MOV H,D	62
MOV H,E	63
MOV H,H	64
MOV H,L	65
MOV H,M	66
MOV L,A	6F
MOV L,B	68
MOV L,C	69
MOV L,D	6A
MOV L,E	6B
MOV L,H	6C
MOV L,L	6D
MOV L,M	6E
MOV M,A	77
MOV M,B	70
MOV M,C	71
MOV M,D	72
MOV M,E	73
MOV M,H	74
MOV M,L	75
XCHG	EB

Move Immediate

Instruction	Opcode
MVI A, byte	3E
MVI B, byte	06
MVI C, byte	0E
MVI D, byte	16
MVI E, byte	1E
MVI H, byte	26
MVI L, byte	2E
MVI M, byte	36

Load Immediate

Instruction	Opcode
LXI B, dble	01
LXI D, dble	11
LXI H, dble	21
LXI SP, dble	31

Load/Store

Instruction	Opcode
LDAX B	0A
LDAX D	1A
LHLD adr	2A
LDA adr	3A
STAX B	02
STAX D	12
SHLD adr	22
STA adr	32

byte = constant, or logical/arithmetic expression that evaluates to an 8-bit data quantity. (Second byte of 2-byte instructions).

dble = constant, or logical/arithmetic expression that evaluates to a 16-bit data quantity. (Second and Third bytes of 3-byte instructions).

adr = 16-bit address (Second and Third bytes of 3-byte instructions).

* = all flags (C, Z, S, P, AC) affected.

** = all flags except CARRY affected; (exception: INX and DCX affect no flags).

† = only CARRY affected.

ARITHMETIC AND LOGICAL GROUP

Add*

Instruction	Opcode
ADD A	87
ADD B	80
ADD C	81
ADD D	82
ADD E	83
ADD H	84
ADD L	85
ADD M	86
ADC A	8F
ADC B	88
ADC C	89
ADC D	8A
ADC E	8B
ADC H	8C
ADC L	8D
ADC M	8E

Subtract*

Instruction	Opcode
SUB A	97
SUB B	90
SUB C	91
SUB D	92
SUB E	93
SUB H	94
SUB L	95
SUB M	96
SBB A	9F
SBB B	98
SBB C	99
SBB D	9A
SBB E	9B
SBB H	9C
SBB L	9D
SBB M	9E

Double Add †

Instruction	Opcode
DAD B	09
DAD D	19
DAD H	29
DAD SP	39

Increment**

Instruction	Opcode
INR A	3C
INR B	04
INR C	0C
INR D	14
INR E	1C
INR H	24
INR L	2C
INR M	34
INX B	03
INX D	13
INX H	23
INX SP	33

Decrement**

Instruction	Opcode
DCR A	3D
DCR B	05
DCR C	0D
DCR D	15
DCR E	1D
DCR H	25
DCR L	2D
DCR M	35
DCX B	0B
DCX D	1B
DCX H	2B
DCX SP	3B

Specials

Instruction	Opcode
DAA*	27
CMA	2F
STC†	37
CMC†	3F

Rotate †

Instruction	Opcode
RLC	07
RRC	0F
RAL	17
RAR	1F

Logical*

Instruction	Opcode
ANA A	A7
ANA B	A0
ANA C	A1
ANA D	A2
ANA E	A3
ANA H	A4
ANA L	A5
ANA M	A6
XRA A	AF
XRA B	A8
XRA C	A9
XRA D	AA
XRA E	AB
XRA H	AC
XRA L	AD
XRA M	AE
ORA A	B7
ORA B	B0
ORA C	B1
ORA D	B2
ORA E	B3
ORA H	B4
ORA L	B5
ORA M	B6
CMP A	BF
CMP B	B8
CMP C	B9
CMP D	BA
CMP E	BB
CMP H	BC
CMP L	BD
CMP M	BE

Arith & Logical Immediate

Instruction	Opcode
ADI byte	C6
ACI byte	CE
SUI byte	D6
SBI byte	DE
ANI byte	E6
XRI byte	EE
ORI byte	F6
CPI byte	FE

BRANCH CONTROL GROUP

Jump

Instruction	Opcode
JMP adr	C3
JNZ adr	C2
JZ adr	CA
JNC adr	D2
JC adr	DA
JPO adr	E2
JPE adr	EA
JP adr	F2
JM adr	FA
PCHL	E9

Call

Instruction	Opcode
CALL adr	CD
CNZ adr	C4
CZ adr	CC
CNC adr	D4
CC adr	DC
CPO adr	E4
CPE adr	EC
CP adr	F4
CM adr	FC

Return

Instruction	Opcode
RET	C9
RNZ	C0
RZ	C8
RNC	D0
RC	D8
RPO	E0
RPE	E8
RP	F0
RM	F8

Restart

Instruction	Opcode
RST 0	C7
RST 1	CF
RST 2	D7
RST 3	DF
RST 4	E7
RST 5	EF
RST 6	F7
RST 7	FF

I/O AND MACHINE CONTROL

Stack Ops

Instruction	Opcode
PUSH B	C5
PUSH D	D5
PUSH H	E5
PUSH PSW	F5
POP B	C1
POP D	D1
POP H	E1
POP PSW*	F1
XTHL	E3
SPHL	F9

Input/Output

Instruction	Opcode
OUT byte	D3
IN byte	DB

Control

Instruction	Opcode
DI	F3
EI	FB
NOP	00
HLT	76

New Instructions (8085 Only)

Instruction	Opcode
RIM	20
SIM	30

ASSEMBLER REFERENCE

Operators

- ( )
- NUL
- LOW, HIGH
- *, /, MOD, SHL, SHR
- +, —
- NOT
- AND
- OR, XOR

ASSEMBLER REFERENCE (Cont.)

Pseudo Instruction

General: ORG, END, EQU, SET, DS, DB, DW

Macros: MACRO, ENDM, LOCAL, REPT, IRP, IRPC, EXITM

Relocation: ASEG, DSEG, CSEG, PUBLIC, EXTRN, NAME, STKLN, STACK, MEMORY

Conditional Assembly: IF, ELSE, ENDIF

Constant Definition

Examples	Type
0BDH, 1AH	Hex
105D, 105	Decimal
72O, 72Q	Octal
11011B, 00110B	Binary
'TEST', 'A' 'B'	ASCII

Figure 15.1 Intel programmers 8085 quick reference card.

INTEL® 8080/8085 INSTRUCTION SET REFERENCE TABLES

INTERNAL REGISTER ORGANIZATION

A Reg. (8)

B Reg. (8)	C Reg. (8)
D Reg. (8)	E Reg. (8)
H Reg. (8)	L Reg. (8)
Program Counter (16)	
Stack Pointer (16)	

FLAG BYTE (D_7 … D_0)

S	Z	X	AC	X	P	X	C

C: CARRY; P: PARITY; AC: AUX. CARRY; Z: ZERO; S: SIGN

X: UNDEFINED

REGISTER-PAIR ORGANIZATION

PSW

A (8)	FLAGS (8)

B (B/C) (16)
D (D/E) (16)
H (H/L) (16)
Prog. Ctr. (16)
Stack Ptr. (16)

NOTE: Leftmost Byte is high-order byte for arithmetic operations and addressing. Left byte is pushed on stack first. Right byte is popped first.

BRANCH CONTROL INSTRUCTIONS

Flag Condition	Jump		Call		Return	
Zero=True	JZ	CA	CZ	CC	RZ	C8
Zero=False	JNZ	C2	CNZ	C4	RNZ	C0
Carry=True	JC	DA	CC	DC	RC	D8
Carry=False	JNC	D2	CNC	D4	RNC	D0
Sign=Positive	JP	F2	CP	F4	RP	F0
Sign=Negative	JM	FA	CM	FC	RM	F8
Parity=Even	JPE	EA	CPE	EC	RPE	E8
Parity=Odd	JPO	E2	CPO	E4	RPO	E0
Unconditional	JMP	C3	CALL	CD	RET	C9

ACCUMULATOR OPERATIONS

	Code	Function
XRA A	AF	Clear A and Clear Carry
ORA A	B7	Clear Carry
CMC	3F	Complement Carry
CMA	2F	Complement Accumulator
STC	37	Set Carry
RLC	07	Rotate Left
RRC	0F	Rotate Right
RAL	17	Rotate Left Thru Carry
RAR	1F	Rotate Right Thru Carry
DAA	27	Decimal Adjust Accum.

RESTART TABLE

Name	Code	Restart Address
RST 0	C7	0000_{16}
RST 1	CF	0008_{16}
RST 2	D7	0010_{16}
RST 3	DF	0018_{16}
RST 4	E7	0020_{16}
TRAP	Hardware* Function	0024_{16}
RST 5	EF	0028_{16}
RST 5.5	Hardware* Function	$002C_{16}$
RST 6	F7	0030_{16}
RST 6.5	Hardware* Function	0034_{16}
RST 7	FF	0038_{16}
RST 7.5	Hardware* Function	$003C_{16}$

*NOTE: The hardware functions refer to the on-chip Interrupt feature of the 8085 only.

REGISTER PAIR AND STACK OPERATIONS

	PSW (A/F)	Register Pair B (B/C)	Register Pair D (D/E)	Register Pair H (H/L)	SP	PC	Function
INX		03	13	23	33		Increment Register Pair
DCX		0B	1B	2B	3B		Decrement Register Pair
LDAX		0A	1A	7E(1)			Load A Indirect (Reg. Pair holds Adrs)
STAX		02	12	77(2)			Store A Indirect (Reg. Pair holds Adrs)
LHLD				2A			Load H/L Direct (Bytes 2 and 3 hold Adrs)
SHLD				22			Store H/L Direct (Bytes 2 and 3 hold Adrs)
LXI		01	11	21	31	C3(3)	Load Reg. Pair Immediate (Bytes 2 and 3 hold immediate data)
PCHL						E9	Load PC with H/L (Branch to Adrs in H/L)
XCHG			EB	EB			Exchange Reg. Pairs D/E and H/L
DAD		09	19	29	39		Add Reg. Pair to H/L
PUSH	F5	C5	D5	E5			Push Reg. Pair on Stack
POP	F1	C1	D1	E1			Pop Reg. Pair off Stack
XTHL				E3			Exchange H/L with Top of Stack
SPHL					F9		Load SP with H/L

Notes: 1. This is MOV A,M. 2. This is MOV M,A. 3. This is JMP.

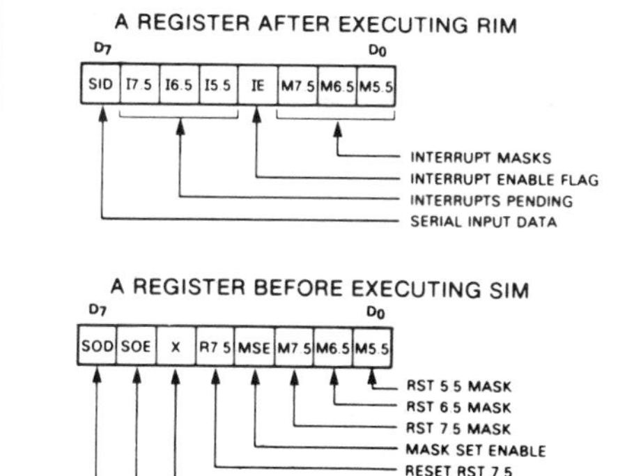

Figure 15.1 (Continued)

instruction has an 8-digit binary code. This 8-bit code (usually represented by two hexadecimal digits) is called an *op-code* (short for "operation code"). A program written as series of 1's and 0's is called a *machine language* or *object code program.* A utility program called an *assembler* is used to translate source code programs written in mnemonics into object code programs that the machine (microprocessor) can execute directly.

The first item that we should examine is the internal register and register-pair organization illustrated in Figure 15.1. Notice that the 8085 has an 8-bit accumulator called the A register. There are also six general-purpose 8-bit registers: B, C, D, E, H, and L. You should also expect to see two 16-bit registers: the PC (program counter) and SP (stack pointer). There is also a flag register consisting of five status flags.

The register-pair organization chart shows that the six general-purpose registers can be concatenated and used as three 16-bit register pairs: the B register pair (BC), the D register pair (DE), and the H register pair (HL). You will discover that the H register pair is most often used to hold a 16-bit address. The letters H and L are said to stand for "high address byte" and "low address byte." For stack operations the A register and flag register will be paired together and called the PSW (program status word). The six general-purpose registers are used as pairs for two reasons: They can be used to hold a 16-bit address, and they can be used to hold a number greater than 255.

15.2 THE 8085 FLAG REGISTER AND 2's-COMPLEMENT NOTATION

In Chapter 16 you will study the conditional jump and call instructions. These instructions use the state of a particular flag to determine whether program execution should branch to another instruction or a subroutine. Three bits in the flag register are denoted as don't-care positions. Let's examine each of the five flags.

Carry Flag The carry flag is set when the result of an operation produces a number that will not fit into the 8-bit accumulator. Arithmetic, logical, and rotate instructions affect the state of the carry flag.

Parity Flag In chapter 13 you learned how parity is used to check for single bit errors in data streams. The parity flag is set when the number of logic 1's in the accumulator is even and reset when it is odd. The parity flag is the least used of all the flags.

Auxiliary Carry (AC) The AC flag is used to simplify the performance of BCD arithmetic of the 8085. An 8-bit accumulator can hold two 4-bit BCD digits. The AC flag is used to indicate a BCD carry or an illegal BCD digit residing in the lower or upper nibble of the accumulator. After the BCD addition the 8085 instruction DAA (decimal adjust accumulator) should be executed.

If the AC is set and the DAA instruction is performed, the contents of the accumulator will be modified to reflect the true result of the BCD addition or subtraction. Consider the following example:

```
 0001 1001      19
+0000 0001     + 1
 0001 1010      20
   1    A is an illegal BCD number
```

Execute DAA and the result is:

00010 0000 20—the correct BCD answer.

In this example the number in the accumulator will be interpreted as two BCD digits: 1 and 9. Adding 19 and 1 should result in the decimal number of 20. The binary result of the addition is not a legal BCD code. Using the state of the AC flag and the DAA instruction, the accumulator is readjusted to reflect the correct decimal answer of 20. If the AC flag is not set and the DAA is executed, the result of the addition will not be modified. Because the AC flag cannot be directly tested, many people forget that it exists.

Zero Flag The zero flag is one of the most useful flags. If an operation leaves a result of 0000 0000 in the accumulator, the zero flag will be set; if the accumulator contains any number other then zero, the zero flag will be cleared. As confusing as it may first sound, just remember that the zero flag is a logic 1 when the contents of the accumulator is equal to zero, and a logic 0 when the contents of the accumulator is not equal to zero.

Why is the zero flag so useful? It is used to indicate equality between two numbers whose magnitudes are being compared, and it also indicates when the contents of registers that are being used as software counters are decremented to a value of zero.

The Sign Flag The sign flag is defined as a copy of the most significant bit of the accumulator. Although technically correct, the preceding definition gives no clue as to why this flag is called the sign flag. To understand the significance of the sign flag, we must study the manner in which negative numbers are represented in the microprocessor world.

15.2.1 2's-Complement Notation

Two's-complement notation is used to represent negative binary numbers. The process of creating a 2's-complement number is quite simple:

1. Represent the number in a normal 8-bit binary fashion.

2. Complement the number, inverting all 8 bits.
3. Add one to the number. Discard any carry.

As an example, let us represent the number 23 in 2's-complement notation.

1. 23 = 0001 0111
2. $\overline{0001\ 0111} = 1110\ 1000$
3. 1110 1000
   $+\underline{0000\ 0001}$
   1110 1001 is the 2's complement of 23

The most significant bit (B7) of a negative 2's-complement number is always equal to logic 1. That is why B7 is called the sign bit, and the sign flag is nothing more than a copy of B7 of the accumulator. Any byte of binary data can be interpreted as either an unsigned or a signed number. This interpretation depends on the context of the application. Assume that all 8-bit quantities are unsigned unless you are instructed otherwise.

Let's take a moment to consider the range of numbers that are represented in an 8-bit signed number. Remember that B7 is the sign bit—if it is a 0, the number is positive; if it is a 1, the number is negative. Because B7 is used as a sign bit, only 7 bits remain to represent quantity. Seven bits can represent 128 unique numbers. That means 128 unique positive numbers and 128 unique negative numbers. We have not lost any range using 2's-complement notation; the 256 unique numbers now exist on both sides of zero on the number line.

That brings up an important point: Is the number zero a positive or a negative number? Formally speaking, zero is a positive number. Therefore, we can make the following statement:

A signed 8-bit number can represent 256 unique numbers: −128 through +127. The 128 negative numbers are −128 through −1, and the 128 positive numbers are 0 through 127.

Figure 15.2 illustrates the number line representing the range of 2's-complement numbers. The positive numbers from 0 through 127 have a sign bit of 0 and appear in normal binary representation. The negative numbers from −127 through −1 have a sign bit of 1 and also progress in a normal binary manner. As the numbers increase from left to right, they are incremented by one. From −1 to 0 the number 1111 1111 is incremented to 1 0000 0000 and the carry is dropped, to produce the representation of 0000 0000 for the positive number zero.

To derive the magnitude of a negative 2's-complement number, execute the 2's-complement operation in reverse order.

1. Subtract one from the number.
2. Complement the number.
3. Perform a binary-to-decimal conversion.

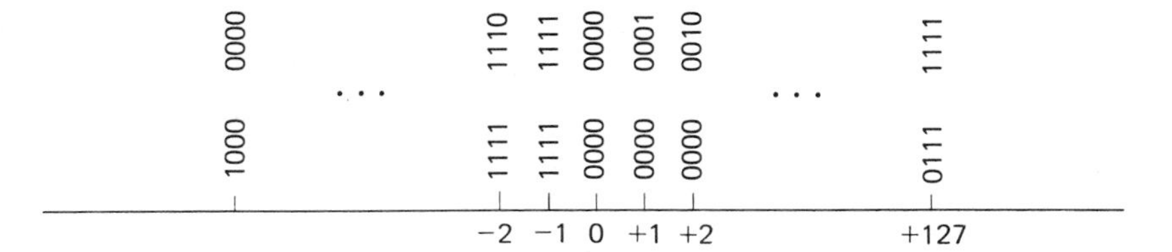

Figure 15.2 2's-complement number line.

For example, assume that 1100 1010 is a signed 8-bit number. What is its magnitude?

Because the sign bit is a logic 1, this number is negative. We must now derive its magnitude.

1. 1100 1010
   $-\underline{0000\ 0001}$
   1100 1001
2. $\underline{1100\ 1001}$ = 0011 0110
3. 0011 0110 = 54

As an interesting exercise, calculate the 2's complement of zero.

When we subtract two numbers, we are really adding the negative value of the second number to the first number: 39 − 25 is equivalent to 39 + (−25). Microprocessors subtract numbers by adding the 2's-complement representation of the second number to the first. Let's perform the preceding subtraction with 2's-complement arithmetic.

1. 39 = 0010 0111   25 = 0001 1001
2. The 2's complement of 0001 1001 is
   a. $\overline{0001\ 1001}$ = 1110 0110
   b. 1110 0110
      $+\underline{0000\ 0001}$
      1110 0111
3. Add 39 + (−25): 0010 0111
   $+\underline{1110\ 0111}$
   1 0000 1110
4. Discard carry and 0000 1110 = 14, the correct answer.

If the answer to a 2's-complement operation has a sign bit of 1, it is a negative number. The number must be converted to derive the actual magnitude.

Now you should understand the function of the sign bit and 2's-complement representation. When will you use 2's-complement numbers? The most common application of 2's-complement numbers is calculating offsets in relative jump commands. You will learn about relative jump operations when we study the Z80 microprocessor. For now just remember that the sign bit is a copy of B7 of the accumulator. Any binary number can be interpreted as a signed or unsigned number, and assume that a number is unsigned unless you are informed to the contrary.

15.3 THE DATA TRANSFER GROUP

Now that you know the internal register organization of the 8085, we can study the actual instruction set. The data transfer group contains instructions that are used to copy data from register to register, register to memory, memory to registers, register pairs to memory, and memory to register pairs. These instructions only copy data from a source to a destination; they do not manipulate data in any manner or affect the settings of any flag.

15.3.1 The MOV Instruction

The most basic 8085 instruction is MOV. MOV copies the contents of the source register into the destination register. The contents of the source register is not affected. The format of the MOV instruction is

```
MOV reg1, reg2
```

where reg2 is the source of the data and reg1 the destination. You already understand the difference between a memory location and the contents of the memory location. You must use the same idea to differentiate between a register and the contents of a register. If a register or a memory location is enclosed in parentheses, this indicates a reference to the contents of the register or memory location.

FF00H refers to a memory location; (FF00H) refers to the contents of memory location FF00H. In the same manner the MOV instruction can be symbolized as

$$(reg2) \rightarrow reg1$$

This says: "the contents of reg2 are MOVed into reg1."

Notice that the MOV instruction concerns only 8-bit quantities of data. The first group of MOV instructions illustrates the accumulator as the destination, and each of the other general-purpose registers as the source. Closely examine this first group of MOV instructions.

There are three concepts that need further examination. Following each instruction is an 8-bit number expressed in hexadecimal. In fact, each instruction in Figure 15.1 is followed by a unique 8-bit number. This number is the instruction's op-code. When an 8085 fetches an instruction it reads an 8-bit quantity from memory. It stores this byte in an internal instruction register. The 8085 then matches this byte with the op-code of one of the instructions in Figure 15.1.

The second thing that you may find confusing is the instruction MOV A,A. This instruction moves the contents of the accumulator into the accumulator, which seems to do absolutely nothing. You will find that every register has a MOV instruction that refers to itself as both the source and destination registers. These instructions are indeed, useless. They are best ignored.

The third problem in the first group of MOVs is the instruction

MOV A,M.

The 8085 does not contain a register M. Actually, register M is not a physical register, it is a symbolic register. Symbolic register M is defined as:

The memory location pointed to by the contents of the H register pair.

The idea of using a register pair to hold a 16-bit address is called *register indirect addressing.* By loading the address of a frequently accessed memory location into the H register pair, the program does not have to continually provide the address; that saves space and program overhead, and greatly speeds program execution. The symbolic representation of MOV A,M is

$((HL)) \rightarrow A$

Notice the double parentheses around HL. The inner parentheses indicate that the contents of the H register pair should be interpreted as an address. The outer parentheses indicate that the byte stored at the memory location pointed to by the contents of the H register pair will be copied into the accumulator. Consider the following example.

(HL) = 07FFH and (07FFH) = 3CH
MOV A,M will move the byte 3CH into the accumulator.

Take a moment to understand register indirect addressing. It is an extremely popular and powerful addressing mode that is supported by every microprocessor.

You know that a basic read/write operation takes three clock ticks. In addition to external memory or I/O accesses, microprocessors consume time performing instruction decoding and other internal operations. You must learn to equate each instruction with the actual sequence of discrete hardware operations that the 8085 performs. Consider Table 15.1. Each instruction will be summarized with a similar table.

Let's examine the information contained in Table 15.1. The

TABLE 15.1

Instruction:	MOV reg, reg
Bytes:	1
Total clock ticks:	4

Ticks	*8085 Operation*
3	Op-code fetch
1	Internal operations

TABLE 15.2

Instruction:	MOV reg, memory — MOV memory, reg
Bytes:	1
Total clock ticks:	7

Ticks	*8085 Operation*
3	Op-code fetch
1	Internal operation
3	Memory read or memory write

instruction mnemonic is MOV; the source and destinations are registers, it is a one-byte instruction and takes a total of four clock ticks. The 8085's activity can be further divided into discrete operations. The first three clock ticks are used to perform a memory-read operation—an op-code fetch. The last clock tick is used to perform internal operations: decoding the op-code and performing the register to register copy.

Table 15.2 summarizes the MOV instruction involving symbolic register M. The 8085 will execute a MOV instruction involving symbolic register M as follows: op-code fetch, internal operation, memory access; if the destination is a register, a memory read of symbolic register M will occur; if the destination is register M, a memory write will occur. The contents of the H register pair will be used to provide the memory address.

15.3.2 The MVI Instruction

The logical extension of the MOV instruction is the MVI instruction. MVI stands for "move immediate data" into a register. In Chapter 12 we discussed the concept of immediate data—the data immediately follows the op-code in memory. The form of an MVI instruction is

```
MVI REG, DATA
```

Notice that the data takes the place of the source register in the MOV instruction. The register is the accumulator or any general-purpose 8-bit register (including symbolic register M).

MOV is a one-byte instruction; all the information required to carry out a MOV instruction (source and destination registers) is included in the op-code. MVI is a two-byte instruction. The first byte is the op-code and the second is the 8 bits of immediate data. When the 8085 fetches a MVI instruction, it knows that the next successive byte in memory contains the data. To execute a MVI instruction, the 8085 performs two memory-read operations. The first memory read is the op-code fetch. (The execution of every instruction begins with an op-code fetch.) The second operation is a data read.

MVI C, 0BH is a typical move immediate instruction. The quantity 0BH, will be written into the C register.

TABLE 15.3

Instruction:	MVI reg, byte
Bytes:	2
Total clock ticks:	7

Ticks	*8085 Operation*
3	Op-code fetch
1	Internal operation
3	Data fetch

TABLE 15.4

Instruction:	MVI memory, byte
Bytes:	2
Total clock ticks:	10

Ticks	*8085 Operation*
3	Op-code fetch
1	Internal operation
3	Data fetch
3	Memory write

Tables 15.3 and 15.4 summarize the two forms of the MVI instruction. You must remember that after each memory access, the PC is automatically incremented. After the op-code is fetched, the PC is incremented and points to the memory location that holds the data byte. After the data byte is fetched, the PC is incremented and then points to the op-code of the next instruction. This is a critical process to understand.

Again, notice that a MVI to memory requires an extra three clocks ticks, to write the retrieved data byte to symbolic register M.

15.3.3 The LXI Instruction

The MOV and MVI instructions both manipulate byte-wide quantities of data. Because the six general-purpose registers can be used as the B, D, and H register pairs, the 8085 must provide some 16-bit instructions. LXI is the 16-bit equivalent of the MVI instruction.

```
LXI REG-PAIR, DBLE-BYTE
```

The register pair can be the B, D, H, or SP, and the data are a 16-bit quantity. LXI is a three-byte instruction—one byte for the op-code and two bytes (a double byte) for data. Let's examine the mnemonic LXI. "L" represents the load instruction; during a load instruction, the register (or register pair) is the destination—it is "loaded" with data. In 8085 mnemonics, the letter "X" always designates a register pair, and you have seen that the letter "I" designates immediate data. Therefore, LXI can be said aloud as:

"Load the register pair with immediate data."

Now is a good time to tackle the concept of what is known as "Intel format." Consider the instruction LXI H, F07A, which Figure 15.1 indicates has an op-code of 21H. The result of LXI H, F07A is that the byte F0 is placed in the H register and 7A is placed into the L register: F0 → H and 7A → L. Intel format indictates how the actual three bytes that constitute this instruction will appear in memory. Let's assume that this instruction has been stored at memory location 080A.

```
(080A) = 21 )
(080B) = 7A } → LXI H,F07A
(080C) = F0 )
```

To start the operation, the 8085 will place 080A onto the address bus and perform an op-code fetch. The byte "21" will be read and placed into the instruction register of the 8085. The 8085 will then decode "21" into the instruction LXI H, dble-byte. The two bytes of data must now be read. Because the PC was incremented after the op-code fetch, the 8085 will place 080B onto the address bus and read the byte "7A." This byte will be placed into the least significant register, register L. The 8085 will perform a third memory-read operation by placing 080C onto the address bus. The byte "F0" is read and placed into the most significant register, register H.

The important point to notice is that the order of the data in the assembly language statement—LXI H, F07A—resides in memory in a reverse order. Intel format states that all three-byte instructions will reside in memory in the following order: op-code, least significant byte, most significant byte. Remember that assembley language instructions are used to improve the ease of understanding microprocessor programs. You must realize that the order of data as it actually appears in memory will be opposite to the way it appears in an assembly language instruction.

Table 15.5 indicates how the double byte of data is stored

TABLE 15.5

Instruction:	LXI reg-pair, dble byte
Bytes:	3
Total clock ticks:	10

Ticks	*8085 Operation*
3	Op-code fetch
1	Internal operation
3	Fetch low-order data
3	Fetch high-order data

in memory. After the normal op-code fetch and internal operation, the first byte data fetched will be placed in the low-order register of the pair. The second byte of data fetched will be placed in the high-order register.

15.3.4 The Load/Store Instructions

Let's examine the last group of instructions in the data transfer group—the Load/Store instructions. We have defined "load" as the process of writing data into a register. Store is the complementary version of load. Store is defined as the process of writing data from a register into a memory. From the microprocessor's point of view, a load operation is a read from memory, and a store operation is a write to memory.

Consider the LDAX B and LDAX D instructions. LD stands for load, the letter "A" for accumulator, and "X" always refers to a register pair; in this case the register pair will be BC or DE. The LDAX instruction says:

"Load the accumulator with a copy of the byte stored at the memory location pointed to by the contents of the register pair."

LDAX B = ((BC)) → A
LDAX D = ((DE)) → A

You have already seen register indirect addressing with the H register pair and symbolic register M. The LDAX instruction also uses register indirect addressing. The address pointer can be placed in either BC or DE. The destination register is always the accumulator.

The STAX B and STAX D instructions are the store equivalent of the LDAX instructions. The contents of the accumulator are stored at the memory location pointed to by the contents of either the B or D register pairs.

STAX B = (A) → (BC)
STAX D = (A) → (DE)

The parentheses around the register pair indicate that the contents of the accumulator are copied into the memory location whose address is the contents of the register pair. Table 15.6

TABLE 15.6

Instruction:	LDAX — STAX
Bytes:	1
Total clock ticks:	7

Ticks	*8085 Operation*
3	Op-code fetch
1	Internal operation
3	Memory read (load) or memory write (store)

illustrates the manner in which the 8085 performs the load or store accumulator indirect instructions.

LHLD addr and SHLD addr are load and store instructions for the H register pair. You know that register M is the memory location pointed to by the contents of the H register pair. The H register pair can only hold one pointer at a time. A program may require the use of many different memory pointers. The LHLD addr instruction allows the programmer to load a memory pointer directly from memory. The SHLD allows the programmer to store a memory pointer directly into memory.

LHLD ADDR = (ADDR) → L AND (ADDR+1) → H

The preceding formalized statement indicates that LHLD will copy the contents of the memory location (whose address is indicated by bytes two and three of the instruction) into the L register. It will then increment the address and copy the contents of that memory location into the H register. For example, if we assume that (A080) = A4 and (A081) = 0C, executing LHLD A080 will result in (HL) = 0CA4.

SHLD ADDR = (L) → ADDR AND (H) → ADDR+1

The SHLD instruction provides the store equivalent of the LHLD instruction. For example, if we assume that (HL) = 0E4D, executing SHLD 010F will result in (010F) = 4D and (0110) = 0E.

LHLD and SHLH direct are both three-byte instructions. The first byte is the op-code and the second two bytes comprise the address of the source memory location (LHLD) or the destination memory location (SHLD).

The instruction "LHLD 7F00" appears in memory as a sequence of three bytes: 2A, 00, 7F. Notice that in accordance with Intel format, the second byte of the instruction is the least significant address byte and the third byte of the instruction is the most significant address byte. These instructions use *direct addressing,* which means the actual address follows the op-code. An instruction that uses direct addressing is three bytes long.

The LHLD and SHLD instructions may seen confusing. Why should one store memory pointers in memory? Using RAM to store indirect pointers is an advanced programming technique. At this point you should strive to understand the mechanics of the LHLD and SHLD instructions.

Table 15.7 summarizes the load and store H-register-pair direct instruction. Notice that LHLD and SHLD require five memory accesses and a total of 16 clock ticks. As you can now start to appreciate, memory accesses consume a great deal of the microprocessor's time and effort.

Once again, concentrate on understanding Intel notation. Notice that the direct address is fetched low-byte and then high-

TABLE 15.7

Instruction: LHLD — SHLD
Bytes: 3
Total clock ticks: 16

Ticks	*8085 Operation*
3	Op-code fetch
1	Internal operation
3	Read low address byte
3	Read high address byte
3	Load (memory read) or store (memory write) L reg
3	Load (memory read) or store (memory write) H reg

byte, and the actual load or store operation is accomplished L register first (least significant) and then H register (most significant).

Load or Store Accumulator Direct The LDA (load accumulator direct) and STA (store accumulator direct) are extremely simple instructions. They both use direct addressing and are therefore three-byte instructions.

LDA ADDR = (ADDR) → A

LDA loads the accumulator with the data stored in the memory location whose direct address is supplied in the second and third bytes of the instruction. The instruction LDA 0C30 would appear in memory as a sequence of three bytes: 3A, 30, 0C. Notice that the address 0C30 is stored in Intel format.

STA ADDR = (A) → ADDR

STA simply copies the contents of the accumulator into the address presented in the second and third bytes of the instruction. The sequence of 32, 40, C7 will store the contents of the accumulator at memory location C740.

LDA and STA are strong-arm instructions that are used only if a particular memory location is going be accessed infrequently. Register indirect addressing is always preferred to direct addressing because an instruction using register indirect addressing is only one byte long, whereas an instruction using direct addressing is three bytes long. Longer instructions not only take more room to store in program memory, but also execute much more slowly because the microprocessor has to do an op-code fetch and two memory reads just to assemble all the information required to complete the instruction.

Examine Table 15.8. STA and LDA require four memory accesses and therefore consume 13 clock ticks.

TABLE 15.8

Instruction:	LDA — STA
Bytes:	3
Total clock ticks:	13

Ticks	*8085 Operation*
3	Op-code fetch
1	Internal operation
3	Fetch low address byte
3	Fetch high address byte
3	Load (memory read) or store (memory write) reg-A

15.3.5 Exchange HL and DE

The last instruction under the data transfer group is XCHG. This simple instruction exchanges the contents of the H register pair with the D register pair.

XCHG = (HL) ⟨=⟩ (DE)

Try and think of a good use for XCHG. Remember that the H register pair often holds a memory pointer to symbolic register M. By using XCHG, two pointers can be held in register storage: one in HL and the other in DE. The active pointer will reside in the H register pair. XCHG is a fast way to swap pointers between the HL and DE registers.

Refer to Table 15.9. The speed at which an instruction is executed is directly proportional to the number of external (outside the microprocessor, such as memory or I/O) accesses required. You have just seen that XCHG, which swaps two 16-bit quantities, is executed in only four clock ticks, whereas LHLD or SHLD requires five memory accesses, and consequently 16 clock ticks.

TABLE 15.9

Instruction:	XCHG
Bytes:	1
Total clock ticks:	4

Ticks	*8085 Operation*
3	Op-code fetch
1	Internal operation

Knowing how many clock ticks an instruction requires is important during program routines that are time dependent. We can use this knowledge to create calibrated delay routines that can be used to debounce switches through software instead of hardware.

In Chapter 16 we will study the remaining instructions in the 8005 instruction set. All microprocessors have extremely similar instruction sets. After you feel comfortable with the 8085 instruction set, you should be able to learn the instruction set of any other 8-bit microprocessor in a minimum length of time.

QUESTIONS AND PROBLEMS

15.1. Consider the concept of register-indirect addressing. Why is it convenient that the address bus of the 8085 is exactly twice as wide as the data bus?

15.2. What are the general-purpose registers of the 8085?

15.3. What are the dedicated registers of the 8085?

15.4. Why is the flag register really considered as a group of five independent flip-flops?

15.5. When an operation results that places a value 00H in the accumulator, what flag is affected?

15.6. Convert the following decimal numbers into binary 2's-complement form.
(a) −59 **(b)** −128 **(c)** −23
(d) 0 **(e)** −1 **(f)** −37

15.7. Assume that the following are signed numbers. Convert them to decimal.
(a) 1001 1101 **(b)** 0111 1111 **(c)** 1110 0001
(d) 1111 1001 **(e)** 0010 0000 **(f)** 1111 0001

15.8. What is symbolic register M?

15.9. Why are MVI instructions two bytes long?

15.10. What action does the 8085 execute during the following instruction: MVI M, 07H?

15.11. What does the letter "I" designate in an 8085 instruction mnemonic?

15.12. What does the letter "X" designate?

15.13. What is the definition of a load instruction? A store instruction?

15.14. Describe Intel format.

15.15. What is the meaning of direct addressing?

15.16. How many clock ticks do the shortest 8085 instructions take to execute?

15.17. How do the instructions in the data transfer group affect the flag.

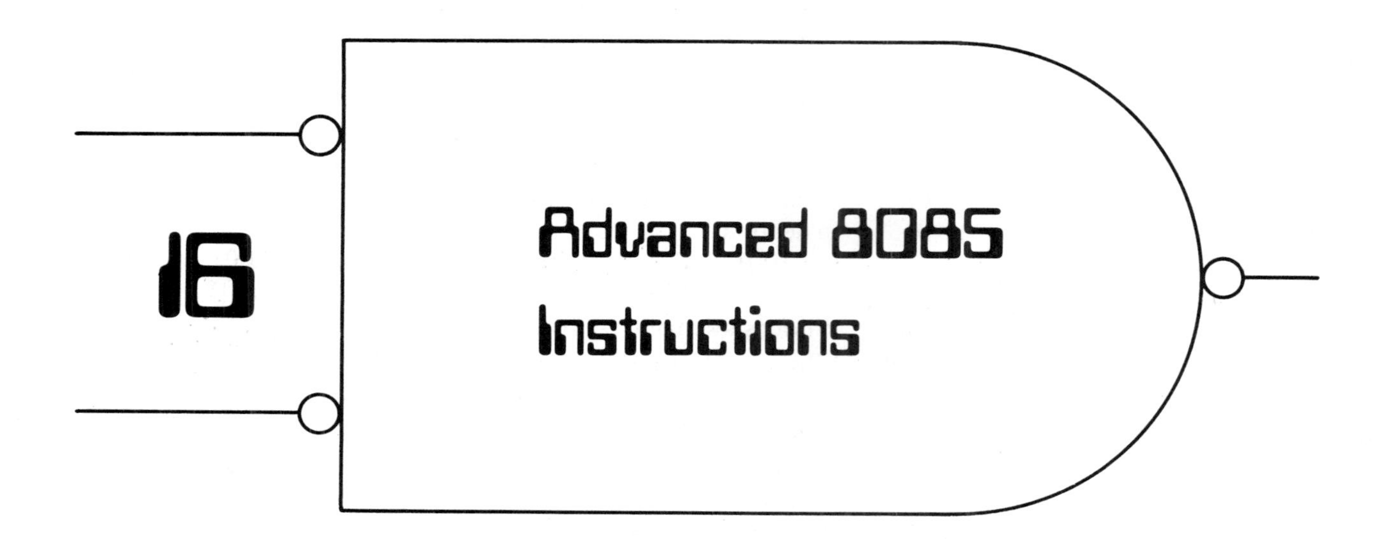

In Chapter 15 we examined the internal registers of the 8085, the flags, and the instructions in the data transfer group. In this chapter we introduce the remaining instructions in the 8085's instruction set.

16.1 ARITHMETIC AND LOGICAL GROUP

This large group of instructions provide the basic arithmetic and Boolean algebra instructions.

16.1.1 The ADD, ADC, SUB, and SBB Instructions

These four instructions give the 8085 addition and subtraction capabilities. They are all one-byte instructions with the accumulator providing one operand and a selected register providing the second operand. The result of the addition or subtraction is always placed in the accumulator.

ADD REG → (A)+(REG) → A
ADC REG → (A)+(REG)+(CARRY FLAG) → A
SUB REG → (A)−(REG) → A
SBB REG → (A)−(REG)−(CARRY FLAG) → A

where "reg" is the accumulator or any general-purpose 8-bit register.

Notice that the accumulator is always one of the operands and that the result of the operation is stored back into the accumulator, overwriting the original operand. All the flags (Carry, Zero, Sign, Parity, and Aux Carry) are affected by these instructions. The flag register is updated by the 8085 during the last clock tick of the operation.

ADD is used for normal 8-bit additions between the accumulator and a register (including register M). If the result of the addition is greater than 127, bit 8 will be high and the sign flag will be set; if the result of an addition is greater than 255, the carry flag will be set to indicate that the accumulator contains the correct answer minus 256 (the value of the carry).

ADC is used to perform multiple-byte precision addition. Assume that we want to add two 24-bit numbers. Because the 8085 is an 8-bit microprocessor, the process of adding these two 24-bit numbers must be broken down into three 8-bit additions. Refer to the following example.

	ADC	ADC	ADD
Operand 1	1000	01010	1101
Operand 2	+ 0011	10100	1100
Carry	0	1	
	1011	11111	1001

Step 1. Add the two least significant bytes using the ADD instruction. After the addition, the carry flag will be set to indicate the carry from bit 7 of the accumulator.

Step 2. Add the next two more significant bytes and the current setting of the carry flag using the ADC instruction. The carry flag is essentially added to bit 9 of the 24-bit addition. Because this addition does not generate a carry, the carry flag will be reset.

Step 3. Add the two most significant bytes and the current setting of the carry flag using the ADC instruction. Any carry that may have been generated from bit 15 is now added to bit 16.

There are many concepts introduced in the multiple-byte addition example that we must examine in greater detail. This 24-bit addition cannot be accomplished simply by performing an ADD and two consecutive ADCs. The correct operands must be moved into the accumulator and the other register used in the addition. Because it is overwritten after each ADD or ADC instruction, the intermediate results must also be saved somewhere other than the accumulator. The carry flag must be checked after the second ADC to ensure that a carry out of bit 23 did not occur. Notice how even a simple microprocessor program can generate a great deal of programming overhead.

TABLE 16.1

Instruction:	ADD reg, ADC reg, SUB reg, SBB reg
Bytes:	1
Total clock ticks:	4

Ticks	*8085 Operation*
3	Op-code fetch
1	Internal operation

The SUB and SBB (subtract with borrow) instructions operate exactly like their counterparts, ADD and ADC. SUB is used for normal 8-bit subtraction, and SBB is used for multiple-precision subtraction. Internally the 8085 uses 2's-complement subtraction. The carry flag functions as a borrow flag for the SUB and SBB instructions. If the 2's-complement result of the subtraction indicates a negative number, the carry flag will be set to indicate a borrow.

Table 16.1 summarizes the add and subtract accumulator with register operations. If symbolic register M is an operand in any of these operations, another three clock ticks must be added to the instruction cycle. This will enable the 8085 to fetch the second operand from the memory location pointed to by the H register pair.

16.1.2 The DAD Instruction

The 8085 has an elementary 16-bit addition instruction called DAD (double add). DAD helps overcome the overhead that we would encounter by accomplishing a 16-bit addition using an ADD and ADC sequence. DAD uses the H register pair as a 16-bit accumulator. Consider the general form of DAD:

DAD REG-PAIR = (HL) + (REG − PAIR) → (HL)

where reg pair is B, D, H pairs or SP. An important fact to know is that DAD affects only the carry flag. If the result of the addition is greater than 65,535, the carry flag is set. No other flags are affected.

Table 16.2 summarizes DAD. Notice that the 8085 consumes six clock ticks performing the internal 16-bit addition.

TABLE 16.2

Instruction:	DAD reg-pair
Bytes:	1
Total clock ticks:	10

Ticks	*8085 Operation*
3	Op-code fetch
1	Internal operation
6	Internal 16-bit addition

16.1.3 Increment and Decrement Instructions

Adding or subtracting the quantity of one from registers is an extremely common operation. The 8085 has four instructions that are used to perform that task.

INR REG = (REG) + 1 → REG
INX REG = (REG-PAIR) + 1 → REG-PAIR
DCR REG = (REG) − 1 → REG
DCX REG = (REG-PAIR) − 1 → REG-PAIR

Remember that whenever you see an "X" in an 8085 assembly language mnemonic, that indicates an operation on a register pair. The INR and DCR instructions increment and decrement registers; while INX and DCX increment and decrement register pairs. These four instructions affect all flags except the carry flag.

This allows us to use any general-purpose register as a counter—a value is placed into a register and the register is decremented each time a particular event occurs. When the contents of the counter register is decremented to zero, the zero flag is set and the program can respond accordingly. Implementing software counters is an extremely simple, yet powerful programming technique.

Tables 16.3 through 16.5 summarize the increment and decrement instructions. As a matter of interest, refer to Table 16.4. Notice the manner in which the content of a memory location (register M) is incremented or decremented. The content of the memory location is read into a temporary register in the 8085, where it is incremented or decremented. The new value is then written back into register M. Two extra memory cycles are required, adding six clock ticks to the operation.

TABLE 16.3

Instruction: INR reg — DCR reg
Bytes: 1
Total clock ticks: 4

Ticks	*8085 Operation*
3	Op-code fetch
1	Internal operation

TABLE 16.4

Instruction: INR memory — DCR memory
Bytes: 1
Total clock ticks: 10

Ticks	*8085 Operation*
3	Op-code fetch
1	Internal operation
3	Memory read (HL) and increment in ALU
3	Memory write (HL)

TABLE 16.5

Instruction:	INX reg-pair — DCX reg-pair
Bytes:	1
Total clock ticks:	6

Ticks	*8085 Operation*
3	Op-code fetch
1	Internal operation
2	Execute internal increment operation

16.1.4 Logical Instructions

Notice the large group of logical instructions: AND, XOR, and OR. The contents of the register selected will be ANDed, XORed, or ORed with the contents of the accumulator. The result of the operation will be placed into the accumulator, overwriting one of the original operands. The Boolean algebra operation occurs on a bit-by-bit basis as if eight separate two-input gates were used.

$$\text{ANA REG} = (A) \cdot (\text{REG}) \rightarrow A$$
$$\text{XRA REG} = (A) \oplus (\text{REG}) \rightarrow A$$
$$\text{ORA REG} = (A) + (\text{REG}) \rightarrow A$$

Assume that (A) = 1101 0101, (HL) = 40CA, and (40CA) = 0011 0110. The instruction XRA M will be executed as follows:

$$\begin{array}{ll} (A) & 1101\ 0101 \\ (40CA) \oplus & \underline{0011\ 0110} \\ & 1110\ 0011 \rightarrow A \end{array}$$

Result (A) = 1110 0011

The logical instructions affect all the flags. An important question is: How is the carry flag affected by a logical instruction? Because the logical instructions are executed on a bit-by-bit basis, a carry can never occur: A logical instruction will always reset the carry flag. This is an important detail to remember. Executing an ANA A operation will not change the value in the accumulator (remember that $X \cdot X = X$), but it will reset the carry flag. Because the 8085 instruction set does not have an explicit "reset the carry" instruction, ANA A can be used to perform that task.

Table 16.6 summarizes the logical-to-register operations. If one of the operands is contained in register M, an additional three clock ticks are required for a data-read operation.

16.1.5 Compare with Accumulator

The CMP reg instructions are performed exactly like the corresponding SUB reg instructions, with one important difference—the result of the CMP instruction is not returned to the

TABLE 16.6

Instruction:	ANA reg, XRA reg, ORA reg
Bytes:	1
Total clock ticks:	4

Ticks	*8085 Operation*
3	Op-code fetch
1	Internal operation

accumulator; it is discarded. Why perform an operation that discards the result? The CMP instruction is used to affect the condition of the zero and carry flags; it compares the contents of the accumulator with the contents of a register. The setting of the zero and carry flags will indicate if the content of the accumulator is equal to, greater than, or less than the value in the second register. This is an extremely useful instruction.

CMP REG: (A) − (REG) → DISCARDED

If zero flag = 1, (A) = (reg).

If zero flag = 0 and carry flag = 0, (A) > (reg).

If zero flag = 0 and carry flag = 1, (A) < (reg).

TABLE 16.7

Instruction:	CMP reg
Bytes:	1
Total clock ticks:	4

Ticks	*8085 Operation*
3	Op-code fetch
1	Internal operations

Why use CMP instead of SUB? CMP does not overwrite the value in the accumulator, as does SUB. One could use the SUB instruction to perform a compare by: MOV (A) to an unused register, perform the SUB, MOV the original contents of the accumulator back from the temporary storage register. You can now see why CMP was included in the 8085's instruction set.

Table 16.7 illustrates that the compare with accumulator instruction is performed exactly like the arithmetic or logical to register instructions.

16.1.6 Arithmetic and Logical with Immediate Data

We have seen how the 8085 performs addition, subtraction, AND, XOR, and OR. The result is always returned to register A (precisely why it is called the accumulator). There is also a set of instructions that uses immediate data as the second operand in the arithmetic and logical instructions.

ADI BYTE (ADD IMMEDIATE): (A) + BYTE → A
ACI (SUB IMMEDIATE): (A) + BYTE + (CARRY) → A
SUI (SUB IMMEDIATE): (A) − BYTE → A
SBI (SBB IMMEDIATE): (A) − BYTE − (CARRY) → A

ANI (ANA IMMEDIATE): (A) · BYTE → A
XRI (XRA IMMEDIATE): (A) ⊕ BYTE → A
ORI (ORA IMMEDIATE): (A) + BYTE → A
CPI (CMP IMMEDIATE): (A) − BYTE → DISCARDED

Table 16.8 summarizes the arithmetic and logical with im-

TABLE 16.8

Instruction:	Immediate Arith and Logical Instructions
Bytes:	2
Total clock ticks:	7

Ticks	*8085 Operation*
3	Op-code fetch
1	Internal operations
3	Fetch data — memory read

mediate data instructions. The second memory cycle is required to perform the data fetch.

16.1.7 Rotate Instructions

The rotate instructions perform the software equivalent of the standard 8-bit, SISO (serial in/serial out) shift register. Many people find the rotate instructions to be confusing, not because their actions are difficult to understand, but because of the poor choice of mnemonics. You will have to take a special effort to remember the mnemonics for each of the four rotate instructions.

Figure 16.1 illustrates the effect of the four rotate instructions. The rotate instructions should be considered in two groups. The first group, RAL and RAR, treat the carry flag as if it were part of the accumulator. The combination of 8-bit accumulator and carry flip-flop creates a logical 9-bit shift register, with the least significant bit of the accumulator connected to the carry flag. The mnenomic "RAL" stands for "rotate accumulator left." Most people prefer to think of RAL as "rotate accumulator left through carry." "Through carry" implies that the carry is considered to be an extension of the accumulator.

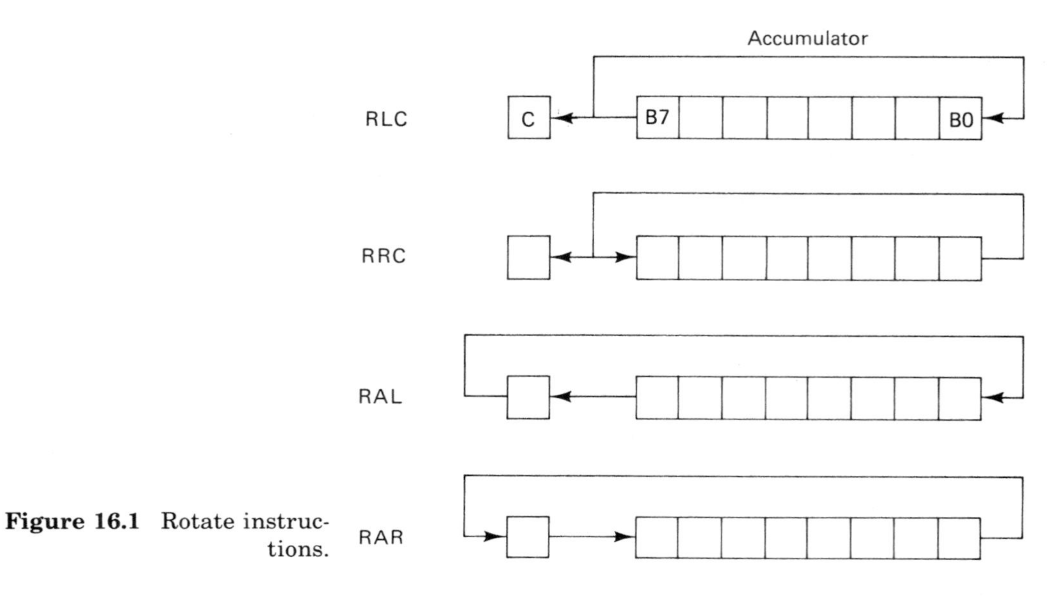

Figure 16.1 Rotate instructions.

RAR (rotate accumulator right) is also most commonly called "rotate accumulator right through carry."

Notice that RAL and RAR are nondestructive operations—no information is lost; the location of the data is only moved. RLC and RRC are rotate instructions that do not use the carry flag as part of the accumulator. They are called "rotate accumulator to carry." Notice the use of "to carry" instead of "through carry" to differentiate between using the accumulator alone (RRC and RRC) and treating the carry flip-flop as part of the accumulator (RAL and RAR).

Remember that shift instructions can be used to perform multiplication and division. To multiply the accumulator by 2, one must clear the carry flag (using the ANA A instruction) and then execute RAL. By resetting the carry flag, we will ensure that logic 0 is rotated into bit 0. After performing RAL the carry flag can be used to test for a result that is greater than 255. A divide-by-2 operation can be performed in the same manner as a RAR operation. The carry flag should be tested for a fractional result.

More common than using the rotate instructions for multiplication and division is the process of bit testing. You will often need to test a particular bit in the accumulator to establish if it is a logic 1 or logic 0. By rotating the bit into the carry flag, its logic level can be directly tested.

Table 16.9 summarizes the rotate instructions.

16.1.8 Special Instructions

These instructions are loosely grouped together because they do not fit into any specific category.

DAA: Decimal Adjust the Accumulator We have already covered the basic concepts of this instruction during our discussion of the auxiliary carry flag. After each BCD addition, DAA is executed. This ensures that any illegal BCD quantities are adjusted.

CMA: Complement the Accumulator This instruction simply complements each bit in the accumulator. It does not affect any flags.

TABLE 16.9

Instruction: RLC, RRC, RAL, and RAR
Bytes: 1
Total clock ticks: 4

Ticks	*8085 Operation*
3	Op-code fetch
1	Internal operations

STC: Set the Carry Flag STC is used with the rotate through carry (RAL and RAR) instructions to set bits in the accumulator. Remember that ANA A can be used to clear the carry flag.

CMC: Complement the Carry Flag The action of CMC is self-explanatory. The carry flag is often used to indicate the status of a particular operation. CMC is used to return the carry to its inactive state within the context of the operation.

DAA, CMA, STC, and CMC are typical one-byte four-clock-tick instructions.

16.2 BRANCH CONTROL GROUP

In Figure 12.3 we considered a program that is used to control an air conditioner/heater system. The program examined the digital output of a temperature sensor, and decided whether to enable the air conditioner, the heater, or to proceed directly to a programmed delay. We said that the program displayed intelligence because it had the ability to make decisions based on the current temperature. The branch control group of 8085 instructions enable a program to make decisions based on the current setting of a particular flag.

16.2.1 Unconditional and Conditional Jumps

This is a three-byte instruction that causes the 8085 to deviate from the normal process of sequential program execute. Bytes two and three contain the 16-bit address of the next instruction to execute. The 8085 essentially "jumps" to the new location that is placed into the PC.

JMP addr is called an *unconditional jump*; the "conditional jump" instructions are JNZ, JZ, JNC, JC, JPO, JPE, JP, and JM. Each of these instructions tests an 8085 flag for a particular level; if the condition expressed in the instruction is true, the jump occurs; otherwise, the 8085 executes the next sequential instruction in a normal manner.

Consider the following program segment:

```
WAIT:      IN    1FH     ;INPUT FROM KEYBOARD
           CPI   0FH     ;HAS A KEY BEEN CLOSED?
           JNZ   WAIT    ;IF ZERO FLAG IS NOT SET, TRY AGAIN
KEYCLOS:   MVI   C,7CH   ;SET 20-MS DELAY
```

The second line in the program segment compares the contents of the accumulator with the immediate data '0FH'; if (A) = 0FH, the zero flag is set; otherwise, the zero flag is reset. The third line indicates if (A) does not equal 0FH (the zero flag is reset), program execution should jump to the previous line labeled "WAIT." If the zero flag was set by the CPI instruction, program execution will continue with the next line.

The preceding example introduced the concept of "conditional jumps." Each conditional jump tests for a set or reset state of a particular flag. As you have seen, JNZ tests for a reset zero flag. Consider the remainder of the conditional jump instructions:

JZ Jump if zero This instruction should be thought of as meaning "jump if equal." If a compare instruction sets the zero flag, that implies that the comparison resulted in equality. The CMP or CPI instructions are not the only instructions that affect the zero flag, but they are the instructions that most commonly precede JZ and JNZ.

JNZ Jump if not zero. This is the complement of the JZ instruction. If the zero flag is reset, indicating inequality, the jump will occur. This instruction can be thought of as meaning "jump it not equal."

JC and JNC Jump if carry (is set) and Jump if no carry (carry is reset). JC and JNC are used after arithmetic or rotate instructions.

JPE and JPO Jump if parity is even (parity is set) and Jump if parity is odd (parity is reset). These two conditional jumps test the parity flag. They are the least used jump instructions.

JM and JP Jump if minus (sign is set indicating a negative 2's-complement number) and Jump if positive (sign is reset indicating a positive 2's-complement number). JM and JP are used to test the condition of the most significant bit in the accumulator. Although the mnemonics would lead you to believe that these instructions are used only in signed arithmetic, that is not true. On the contrary, JM and JP are most often employed as bit test operations for bit 7 of the accumulator.

Imagine an operation that reads a byte from an input port into the accumulator. The state of bit 7 can be used to reflect the status of a peripheral; if bit 7 is low, the peripheral is ready; if bit 7 is high, the peripheral is not ready. A JM or JP instruction can be used in the polling program. That is just one example of how the sign flag is used to reflect status.

Tables 16.10 and 16.11 summarize the unconditional and

TABLE 16.10

Instruction: JMP and Conditional Jumps with True Conditions
Bytes: 3
Total clock ticks: 10

Ticks	*8085 Operation*
3	Op-code fetch
1	Internal operations
3	Fetch low byte of jump address
3	Fetch high byte of jump address

TABLE 16.11

Instruction:	Conditional Jumps with False Condition Codes
Bytes:	3
Total clock ticks:	7

Ticks	*8085 Operation*
3	Op-code fetch
1	Internal operations—evaluate the condition
3	Read low-order address byte—then abort

conditional jump instructions. Notice that JMP and true conditional jumps require 10 clock ticks. At this point you should have been able to guess that. As always, the two data bytes following the op-code contain the jump address in Intel format.

Table 16.11 illustrates the action occurring during a conditional jump whose condition was not met. One might think that an unsuccessful jump operation would require only four clock ticks: three for the op-code fetch and one internal operations and to evaluate the condition. The 8085 burns an extra three clock ticks before aborting the instruction: a total of seven ticks for a false jump condition.

PCHL Load the contents of the H register pair into the program counter: (HL) → PC. Because the address of the next op-code fetched is always held in the PC, this instruction has the same effect as an unconditional jump. It is used in advanced applications of parameter passing values via the stack. Do not concern yourself with it at this point.

16.2.2 Unconditional and Conditional Calls

We have already discussed that assembly language programs are built from many small modules called *subroutines.* A main program will "call" a subroutine and the last instruction in the subroutine must be a "return" to the main program. Just as we discovered with the jump instructions, the 8085 provides both unconditional and conditional call instructions. Let's examine the unconditional call instruction.

The easiest way to understand the CALL instruction is to examine the manner in which it is executed by the 8085. Refer to Table 16.12. The first thing that you must understand in the 8085 is that the SP (stack pointer) always points to the last entry on the stack. (This last entry is known as "the top of the stack" even though the stack grows upside down). Before another byte can be written to the stack, the stack pointer must first be decremented to point to the next vacancy on the stack.

The first three clock ticks are consumed by the normal op-code fetch. The next phase in the CALL instruction combines the normal 8085 internal operations with decrementing the SP. The SP is now pointing to the next vacancy on the stack. The

TABLE 16.12

Instruction: CALL and True Conditional Calls
Bytes: 3
Total clock ticks: 18

Ticks	*8085 Operation*
3	Op-code fetch
2	Internal operations and SP = SP − 1
3	Fetch low address byte
3	Fetch high address byte
3	Stack write—(PC high) → (SP) and SP = SP − 1
3	Stack write—(PC low) → (SP)
1	Internal operations

next six clock ticks will fetch the 16-bit address that points to the beginning of the subroutine. This address will be held in a temporary register pair on the 8085.

The next six clock ticks are used to save the return address. Remember that the PC is automatically incremented after each op-code and data fetch. The PC is now pointing to the memory location that contains the op-code of the instruction that will be executed when the subroutine returns control to the main program. This address must be saved on the stack. Notice that PC high is first written to the stack, the SP is decremented (pointing to the next vacancy on the stack), and then the PC low is written to the stack. SP now points to the least significant byte of the return address.

During the last tick of the clock, the contents of the temporary register pair that contains the subroutine address is moved into the PC. The next instruction the 8085 executes will be the first instruction in the subroutine.

The conditional call instructions use the same criterion as the conditional jump instructions. A conditional call that does not meet the proper condition is aborted after nine clock ticks.

16.2.3 Unconditional and Conditional Returns

A return is the last instruction in a subroutine. Return must pop the return address off the stack and into the PC. The 8085 will then execute the instruction that follows the CALL instruction that initiated the subroutine. Table 16.13 summarizes the RET instruction.

Although the stack may have been used to store and retrieve data during the subroutine, you must assume that the SP is presently pointing to the least significant return address byte, as was illustrated in the analysis of the CALL instruction. After the normal four ticks (during which the op-code is fetched and internal operations are processed), the return address will be popped off of the stack. The contents of the memory location pointed to by the SP will be loaded into PC low, and the SP will

TABLE 16.13

Instruction:	RET
Bytes:	1
Total clock ticks:	10

Ticks	*8085 Operation*
3	Op-code fetch
1	Internal operations
3	Memory read ((SP)) → PC low and SP = SP + 1
3	Memory read ((SP)) → PC high and SP = SP + 1

be incremented. The contents of the memory location pointed to by the incremented contents of the SP will be loaded into PC high, and the SP is once again incremented, pointing to the last byte of unread data on the stack.

The 8085 also supports conditional returns, which are analogous to conditional jumps and conditional calls. A true conditional return requires 12 clock ticks, two more than the unconditional return. A false conditional return requires six clock ticks.

16.2.4 Restart Instructions

Notice that the CALL or CALL on condition instructions were three bytes long—op-code plus two address bytes. In Chapter 14 we studied the 8080-type interrupts. The interrupting device was required to supply the op-code of a one-byte instruction. This instruction took the form of one of the restart instructions: RST 0 through RST 7. A restart instruction is really a single-byte unconditional CALL instruction. An explicit address is not required with a RST instruction—the beginning address of the subroutine is built into the encoded op-code.

Restart instruction	*Call address*
RST 0	0H
RST 1	8H
RST 2	10H
RST 3	18H
RST 4	20H
RST 5	28H
RST 6	30H
RST 7	38H

Also remember that the hardware restarts and trap have call addresses that occur in the same general area as the RST call addresses. In actual practice, a jump instruction will be placed at each RST address. This jump instruction will direct the 8085 to a much less croweded area where the actual interrupt routine resides.

When the 8085 is interrupted by the INTR signal, it will

TABLE 16.14

Instruction:	RST
Bytes:	1
Total clock ticks:	12

Ticks	*8085 Operation*
3	Op-code fetch during interrupt operation
2	Internal operations and SP = SP − 1
3	Stack write—(PC high) → (SP) and SP = SP − 1
3	Stack write—(PC low) → (SP)
1	Internal operations

complete its current instruction and then read the RST op-code from the data bus. After the interrupt service routine is completed, the 8085 will want to return to the original location where it was interrupted. That is the precise reason why the RST instructions are one-byte calls. Refer to Table 16.14.

The current contents of the PC is saved onto the stack, exactly as is done in a standard call instruction. The last clock tick is used to insert the implied call address into the PC. Every interrupt service routine that is entered via a software restart instruction (RST 0 through RST 7) or hardware restart (5.5, 6.5, 7.5, and TRAP) will end with a RET instruction. This will return the 8085 to the program location where it was originally interrupted.

16.3 I/O AND MACHINE CONTROL

This is the last group in the 8085 instruction set. After we have covered these instructions, you will be intimately familiar with the 8085 instruction set. But most important, you will understand the manner in which the 8085 interfaces its software and hardware operations.

16.3.1 Stack Operations: PUSH and POP

One of the first (if not the first) instructions in every microprocessor program is to load the SP with the highest RAM address in the stack. On the 8085 that is accomplished with the LXI SP, dble-byte instruction. Every call, software restart, and hardware restart requires the use of the stack to save the return address. If the SP is not loaded with a legal stack address, the return instruction at the end of a subroutine will cause the 8085's PC to be loaded with a random value. The 8085 will essentially be lost, and the system will crash.

Because we have already studied call and return operations, you know how the 8085 manipulates the stack. Once program execution is directed to a subroutine, all the registers that the subroutine will use must be saved on the stack. This is called preserving the program's environment.

After the subroutine is complete, but before the return instruction is executed, the CPU registers must be restored to their original values by popping them off of the stack. The original register environment (prior to the call or interrupt) will then be restored.

To increase the speed of preserving the program's environment, the 8085 pushes and pops the registers in the form of register pairs. The accumulator and flag register are concatenated to form the PSW (program status word). Table 16.15 illustrates the operation of the PUSH reg-pair instruction. After the normal op-code fetch, the 8085 uses two clock ticks for internal operations and to decrement the SP. By decrementing the SP, it points to the next available location on the stack. Just as in the call instruction, the next six clock cycles are used to push the high byte and then the low byte of the register pair onto the stack. The final tick is used for internal operations.

As you already know, the POP reg-pair instruction is used to copy the last two bytes on the stack into the associated CPU register pair. The registers will be pushed onto the stack in the beginning of the routine and popped off the stack at the end of the routine. Table 16.16 summarizes the POP reg-pair instruction.

The last byte written on the stack is the low byte of the register pair. Therefore, the POP instruction should take the contents of the stack location pointed to by the SP and load it into the low byte of the register pair, and then increment the SP to point at the high byte. The high byte is then loaded into the high register of the pair, and the SP is incremented a second

TABLE 16.15

Instruction: PUSH reg-pair
Bytes: 1
Total clock ticks: 13

Ticks	*8085 Operation*
3	Op-code fetch
2	Internal operations and SP = SP − 1
3	Stack write—(reg-pair high) → (SP) and SP = SP − 1
3	Stack write—(reg-pair low) → (SP)
1	Internal operations

TABLE 16.16

Instruction: POP reg-pair
Bytes: 1
Total clock ticks: 10

Ticks	*8085 Operation*
3	Op-code fetch
1	Internal operations
3	Stack read—((SP))→(reg-pair low) and SP = SP + 1
3	Stack read—((SP))→(reg-pair high) and SP = SP + 1

time. The SP must always point to the last byte of unread data on the stack.

16.3.2 The XTHL Instruction

At the beginning of this chapter we mentioned that the LHLD and SHLD instructions were used to manipulate memory pointers stored in memory. It is also common practice to store memory pointers on the stack. This is actually preferred because stack operations all use register-indirect addressing (via the SP) instead of direct addressing. This cuts down on the housekeeping address chores that the program must perform.

XTHL simply exchanges the two bytes stored at the current address of SP and SP+1 with the contents of the H register pair. As we just stated, this operation is used to support register indirect addressing using the H register pair. Table 16.17 summarizes the XTHL instruction.

Let's analyze the execution of XTHL. After the op-code fetch and one tick of internal operations, the 8085 will perform two consecutive stack reads; placing these bytes into a temporary register pair in the 8085. The next two operations will entail writing the contents of the H register pair onto the stack. The contents of the temporary register pair is then moved into the H register pair. Essentially, the two bytes at the top of the stack are exchanged with (HL).

16.3.3 The SPHL Instruction

SPHL is a simple one-byte instruction that copies the contents of the H register pair into the SP: (HL) → SP. This instruction must be used with great caution because the current contents of the SP are lost. SPHL is used to support a powerful programming technique that implements many simultaneous software stacks. Thus instead of using symbolic register M for memory to CPU transfers, the contents of HL can be loaded into the SP, and the powerful push and pop instructions are used to manipulate memory.

TABLE 16.17

Instruction: XTHL
Bytes: 1
Total clock ticks: 16

Ticks	*8085 Operation*
3	Op-code fetch
1	Internal operations
3	Stack read—((SP)) → (temp reg-pair low), SP = SP + 1
3	Stack read—((SP)) → (temp reg-pair high)
3	Stack write—(H) → (SP) and SP = SP − 1
3	Stack write—(L) → (SP)

TABLE 16.18

Instruction:	IN port
Bytes:	2
Total clock ticks:	10

Ticks	*8085 Operation*
3	Op-code fetch
1	Internal operations
3	Fetch port address
3	Read contents of port into accumulator

16.3.4 Input and Output Instructions

The 8085 instruction set contains only two simple instructions for I/O accessing. OUT byte outputs the contents of the accumulator to the port whose address is the byte following the OUT op-code. IN byte is the complement to the OUT byte instruction; it loads the accumulator with the byte residing at the port pointed to by the second byte of the instruction.

OUT BYTE (A) → OUTPUT PORT
IN BYTE (INPUT PORT) → A

OUT and IN use direct addressing. Because I/O ports require only 8 address bits, these instructions are only two bytes, instead of the three bytes required to perform direct memory addressing. OUT and IN do not affect any flags. If an input byte must be tested for zero or sign (bit 7), an ANA A instruction can be performed. This will not change the value of the input byte, but will cause the flags to react to the byte in the accumulator.

The greatest advantage of I/O-mapped I/O is that the memory and I/O circuitry are completely independent, and therefore share different decoders and hardware space. The advantage of memory-mapped I/O is simple—all of the powerful register M instructions can be used to manipulate memory-mapped I/O, whereas I/O-mapped I/O has only two simple support instructions—OUT and IN. Tables 16.18 and 16.19 summarize the execution of the OUT byte and IN byte instructions. Notice that each instruction requires two fetches: an op-code fetch and a port address fetch.

TABLE 16.19

Instruction:	OUT port
Bytes:	2
Total clock ticks:	10

Ticks	*8085 Operation*
3	Op-code fetch
1	Internal operations
3	Fetch port address
3	Write (A) to port

16.3.5 Control Instructions

There are four simple control instructions. You have already seen that DI and EI are used to disable and enable the maskable interrupts: INTR, RST 5.5, RST 6.5, and RST 7.5. On power-up, the 8085 disables interrupts. The programmer must perform an EI instruction to enable the maskable interrupts. DI is executed as a preface to those segments of program code that must not be interrupted by external device requests. TRAP can never be masked through software, but it is a simple matter to gate the trap interrupt input to the 8085 via an AND gate with one input that is driven by a TRAP-enable flip-flop.

NOP (no operation) is a useful instruction that wastes four clock ticks; it does not affect the flags. NOP is used as a "filler" in delay routines and in places where one may desire to insert code at a later time, or to overwrite existing code (known as *patching*) without having to reassemble the whole program.

HLT (halt) is used to halt the 8085. When HLT is executed:

1. AD0 through AD7 goes to high-Z.
2. A8 through A15 goes to high-Z.
3. The control bus goes to high-Z.

The processor is "playing dead." Once a software halt is initiated, there are only three ways that the 8085 can exit the halted state:

1. An active-low $\overline{\text{reset}}$ input will reset the 8085 and cause it to fetch a startup instruction from memory location 0H.
2. The 8085 will allow DMA accesses through the normal HOLD and HLDA pins. But after the bus is released, the 8085 will stay in a halted state.
3. An interrupt is the only way to force the 8085 out of the halted state and to execute the next sequential instruction. The programmer must remember to execute an EI instruction before halting the 8085.

The HLT instruction was originally intended to allow the 8085 to idle until an interrupt request was generated. This would essentially synchronize the 8085's operation to the interrupt request. The halt instruction is not often employed in working systems; the 8085 is usually too busy with other processing chores to halt and wait for an interrupt.

TABLE 16.20

Instruction:	HLT
Bytes:	1
Total clock ticks:	5

Ticks	*8085 Operation*
3	Op-code fetch
1	Internal operations
1	Enter halt state

Table 16.20 illustrates the halt instruction. During the last clock tick when the 8085 is entering the halted state, ALE pulses active high to let external hardware latch the levels on the status outputs. This will be the only true indication to the outside world that the 8085 has been halted through software.

16.3.6 SIM and RIM

The 8085 instruction set contains only two additional instructions to the 8080 instruction set. These instructions were created to support the advanced hardware features of the 8085: hardware restarts and serial I/O. RIM (read interrupt mask) and SIM (set interrupt mask) are dual-purpose instructions, involving both advanced hardware features.

The SIM Instruction The instructions EI and DI enable or disable all maskable interrupts. What should a programmer do if it is desirable to disable only certain interrupts but not all of them? This brings up the concept of a *mask*. To mask an interrupt means to ignore it, even if an EI instruction has been executed. On the other hand, to say that an interrupt is unmasked means that if an EI instruction has been executed and that interrupt goes active, the 8085 will acknowledge and service it. It is important to realize that even if a particular interrupt is unmasked, if EI has not been executed, the interrupt will still be ignored.

Using SIM, a programmer can mask or unmask any of the hardware restart interrupt lines. Remember that TRAP can never be disabled (and therefore masked) through software, and INTR, the 8080 type interrupt, is not addressed in the SIM instruction.

Figure 16.2 illustrates the format of the SIM instruction. The programmer will use the MVI A, byte instruction to load the accumulator with the correct mask. SIM will then write the mask into the 8085's interrupt mask register. Because SIM is a dual-purpose instruction, there must be a means of indicating when the programmer is using SIM as a set mask instruction. Bit 3 in Figure 16.2 is called MSE (mask set enable); if this bit is set when the SIM instruction is executed, the 8085 knows that SIM is being used to set the interrupt mask. Bits 0, 1, and 2 are the mask bits for a particular hardware restart. A logic 1 in either of these bit positions will mask that restart; a logic 0 will unmask the restart.

Notice bit 4 of the SIM format. Hardware restart 7.5 is edge triggered. That means that if a rising edge occurs on the RST 7.5 pin, an internal flip-flop will be set. Even if the level on RST 7.5 returns inactive low, the internal RST 7.5 flip-flop will remain set in an active-high state until the interrupt is serviced, a reset occurs, or the flip-flop is reset by setting B4 high and executing a SIM instruction.

Consider the case when a programmer issues a DI instruction and enters a time-dependent segment of code. If RST 7.5 has an active edge, the interrupt will be ignored (because interrupts have been disabled), but the RST 7.5 flip-flop will remember that the interrupt occurred. When the programmer re-enables interrupts with an EI instruction, the 8085 will sense an active RST 7.5 interrupt and proceed to service it in a normal manner. Granted that this may take some time, but the device will be serviced, even if the RST 7.5 line has returned to an inactive level. RSTs 5.5 and 6.5 are level active and therefore do not have an associated interrupt flip-flop.

We have now covered the 5 bits in the accumulator that will be used in the mask interrupts portion of the SIM instruction.

	7	6	5	4	3	2	1	0
Accumulator	SOD	SDE	X	R7.5	MSE	M7.5	M6.5	M5.5

Figure 16.2 SIM format.

	7	6	5	4	3	2	1	0
Accumulator	SID	I7.5	I6.5	I5.5	IE	M7.5	M6.5	M5.5

Figure 16.3 RIM instruction.

Bit 5 is a don't-care condition and should be ignored. Bits 6 and 7 are used to facilitate the serial output ability of the 8085. Bit 6 is the SDE (serial data enable) bit. If this bit is high and SIM is executed, bit 7 of the accumulator will be latched onto the 8085's SOD pin. In this manner the 8085 can function as a simple UART.

The RIM Instruction RIM (read interrupt mask) is the complement to SIM. It also supports both the hardware RSTs and serial I/O. Figure 16.3 illustrates the format of the RIM instruction. Bits 0, 1, and 2 indicate the interrupt mask set by the SIM instruction. Bit 3 indicates whether EI or DI has been executed last. If bit 3 is low, the system interrupts are disabled; if it is high, the interrupts are enabled. Bits 0 through 3 help remind the programmer of how the interrupts have been masked (via SIM) and the present state of the interrupt enable.

Bits 4, 5, and 6 indicate any pending interrupts. The programmer may choose to disable interrupts and instead use the RIM command to see if any RST interrupts are active. Periodically, the program can execute RIM to look for active interrupts. This is considered a polling method of servicing interrupts. That means that the program checks for interrupts when it has a spare moment, instead of the interrupt actually breaking into the middle of an executing program sequence. A logic 1 in any of these bit positions indicates an active interrupt.

Bit 7 contains the latched value of the 8085's SID input. It is used to perform the function of the receiver in a UART.

Remember that before SIM is executed, the accumulator must be loaded with the proper bit pattern, including the MSE and SDE enable bits. In an opposite fashion, RIM is first executed and then the contents of the accumulator can be examined for interrupt information and serial input data. RIM and SIM are standard one-byte four-clock-tick instructions.

You have now been exposed to the full 8085 instruction set and all of its addressing modes. You should understand how the 8085 executes software instructions by fetching op-codes and data from memory. As a microprocessor technician, you may spend a good deal of time reading pure memory dumps. You must feel comfortable working with 16-bit quantities in Intel format.

QUESTIONS AND PROBLEMS

16.1. What is the difference between ADD and ADC?

16.2. Why is a register A designated as the "accumulator"?

16.3. What location functions as an accumulator for the DAD instruction?

16.4. What is the difference between the subtract and the compare instructions?

16.5. What type of instruction usually follows CMP or CPI?

16.6. What set of instructions is used to perform division and shift bits in the accumulator?

16.7. What is the difference between an instruction that rotates "through the carry" and one that "rotates to the carry"?

16.8. How does one set the carry flag? Reset the carry flag?

16.9. What is the difference between an "unconditional" and a "conditional" instruction?

16.10. What is the last instruction in every subroutine? Why?

16.11. Why is a restart instruction referred to as a "one-byte call"?

16.12. How does the 8085 communicate with I/O devices?

16.13. What are the advantages of memory-mapped I/O?

16.14. How do the 8085 stack operations function?

16.15. Using an oscilloscope, how could you verify that the 8085 is in a halted state?

16.16. How are the SIM instructions used to set the interrupt mask and send serial output data?

17 The Z80 and 6800 8-Bit Microprocessors

You have now examined the 8085 in great detail. You should understand the function of each of its pins and how instructions are carried out through the process of op-code fetches, data fetches, and memory-write operations.

There are many other popular 8-bit microprocessors. The Z80 microprocessor, developed by the Zilog Corporation, is also a direct descendent of the 8080. The 6800 microprocessor was developed by Motorola Corp. The 6800 differs greatly from the 8085 because it has no general-purpose registers and does not provide direct I/O capability. In 6800-based systems, all I/O ports are memory mapped.

As we analyze the Z80 and 6800 microprocessors, we will compare and contrast their functional pinouts with that of the 8085. You must always build on your past knowledge, and all microprocessors, from the first 4004 to the most advanced 32-bit VLSI microprocessor, share many similar characteristics.

17.1 THE Z80 MICROPROCESSOR HARDWARE FUNCTIONS

The 8085 is most often employed as an intelligent controller. The Z80, on the other hand, is one of the most popular 8-bit microprocessors to be used as a CPU in microcomputers and also enjoys wide use as a control processor. Figure 17.1 illustrates the Z80 functional pin configuration.

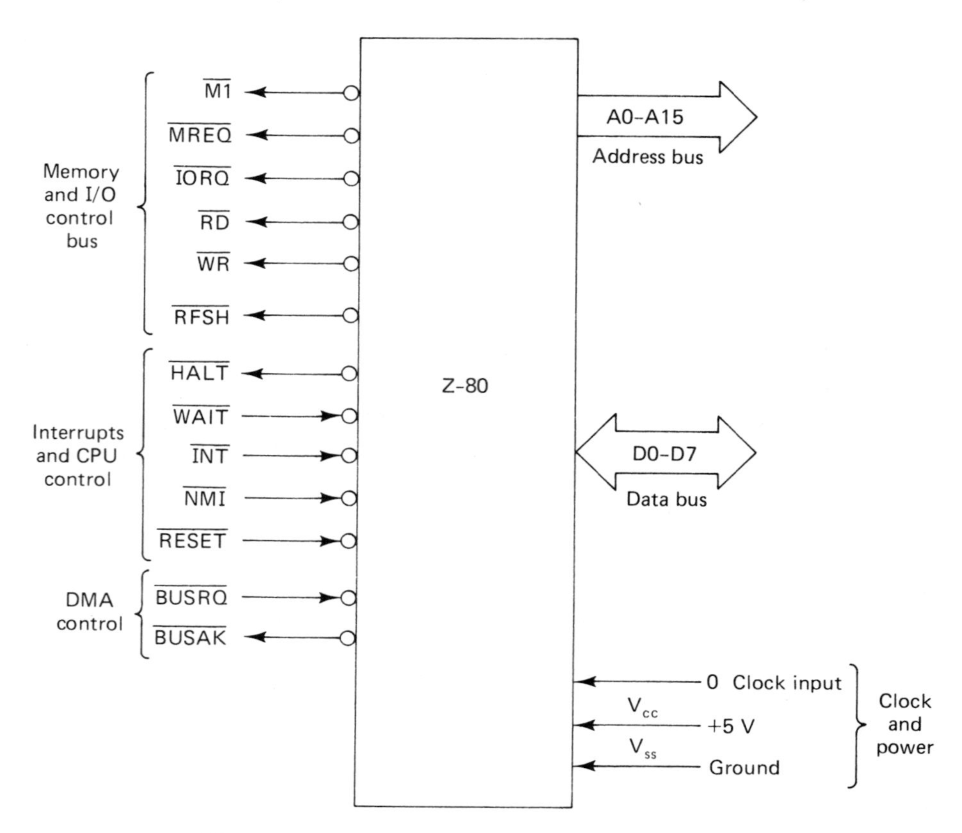

Figure 17.1 Z80 pins grouped according to function.

17.1.1 Address and Data Bus

The first thing to notice is that the Z80 does not have a multiplexed address/data bus. Although the Z80 and 8085 are both housed in 40-pin packages, the Z80 dedicates a full 24 pins for the address and data bus.

17.1.2 Clock and Power

The 8085 has an internal oscillator which is driven by an external crystal. The Z80 does not have an internal oscillator; the Z80's clock must be generated by external hardware. The power requirements for the Z80 are standard +5 V and ground.

17.1.3 Memory and I/O Control

The 8085 has IO/$\overline{\text{M}}$, $\overline{\text{RD}}$, and $\overline{\text{WR}}$ outputs to interface with external memory and I/O. These three outputs must be externally decoded to derive $\overline{\text{MEMRD}}$, $\overline{\text{MEMWR}}$, $\overline{\text{I/ORD}}$, and $\overline{\text{I/OWR}}$. The Z80 provides four control outputs to interface with external memory and I/O: $\overline{\text{MREQ}}$ (memory request), $\overline{\text{IORQ}}$ (I/O request), $\overline{\text{RD}}$, and $\overline{\text{WR}}$. Figure 17.2 illustrates the simple circuitry that translates these four Z80 control signals into the appropriate memory and I/O signals.

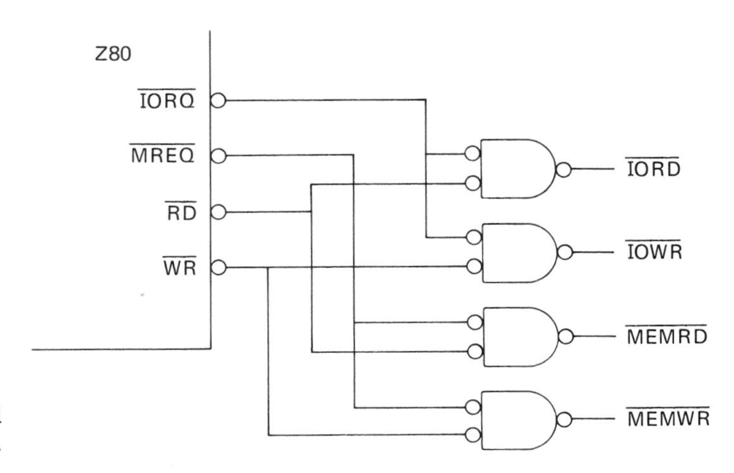

Figure 17.2 Z80 memory and I/O control lines.

The 8085 has two encoded status outputs; when S0 and S1 on the 8085 are both logic 0's, it indicates that the 8085 is performing an op-code fetch. The Z80 does not have the equivalent of the 8085's status outputs. As an alternative solution it has the $\overline{\text{M1}}$ output. $\overline{\text{M1}}$ (machine cycle 1) indicates that the Z80 is performing an op-code fetch. This is an important signal that is often used to synchronize external logic to the beginning of the Z80 instruction cycle.

The use of static RAM is usually limited to small, microprocessor-based, intelligent controllers. The 8085 microprocessor system that we examined in Chapter 14 is an example of such a system. Dynamic RAM is the only main-memory solution for general-purpose computer systems. The only drawback of DRAM is that it adds the requirement of memory refresh to the system's housekeeping chores.

Recall that the initial four clock ticks of each 8085 instruction cycle were consumed by a three-tick op-code fetch and one tick of internal processing. The Z80 functions in a similar manner except that it is synchronized to the rising edge of the clock instead of the falling edge, as is the 8085. To simplify interfacing the Z80 with DRAM, the designers incorporated an on-board refresh mechanism.

During the second two ticks of the op-code fetch/internal processing cycle, the Z80 outputs a 7-bit refresh address onto A0 through A6 and pulses $\overline{\text{MREQ}}$ low. With the aid of minimal external hardware, the Z80 refreshes a row of DRAM during each op-code fetch. The Z80 contains an internal refresh register (R register) that holds the row refresh address and is incremented after each op-code/refresh cycle. This process greatly simplifies interfacing the Z80 with DRAM.

17.1.4 Interrupts and CPU Control

One of the 8085's greatest strengths is its hardware interrupt structure: RST5.5, RST6.5, RST7.5, INTR, and TRAP. The hardware restarts and TRAP are vectored interrupts which require no external support circuitry. The Z80 also has an extremely strong interrupt-handling ability, but the Z80's strength lies in its software-interrupt processing facilities.

The Z80 has two interrupt input pins: $\overline{NMI}$ and $\overline{INT}$. $\overline{NMI}$ (nonmaskable interrupt) is the equivalent to the 8085's TRAP. When $\overline{NMI}$ goes active low, the Z80 performs a standard hardware restart operation by pushing the current contents of the PC onto the stack and vectoring to address 66H. $\overline{INT}$ (maskable interrupt request) is an enhancement of the 8080-type interrupt. Through software, the Z80 can be instructed to react to an active $\overline{INT}$ in one of three modes.

Mode 0 is identical to the 8080-type interrupt where the interrupt device (or as we have seen, the interrupt controller) must provide a one-byte restart op-code. When operating in interrupt mode 1, an active $\overline{INT}$ input will cause the Z80 to automatically execute a RST 7 instruction. As you may remember, a RST 7 instruction pushes the current setting of the PC onto the stack, and vectors program execution to memory location 38H.

Interrupt mode 2, is the most powerful interrupt processing mode. The Z80 has a dedicated 8-bit register called the I register (interrupt register). The programmer must load the I register with an 8-bit address. When the Z80 is set to interrupt mode 2, an active $\overline{INT}$ input will result in the following activity.

1. The interrupting device must provide an 8-bit address (with address bit 0 always equal to a logic 0). This step is exactly like the 8080-type interrupt, except that the device is now providing an 8-bit address, not an 8-bit op-code!
2. The Z80 will concatenate the contents of the I register and the 8-bit address provided by the interrupting device to form a 16-bit address. This address does not point to the beginning of an interrupt service routine, but rather at an entry in an interrupt vector table.
3. The Z80 then retrieves the interrupt service routine address from the interrupt table.
4. The current contents of the PC are pushed onto the stack, and the Z80 moves the address of the service routine into the PC.
5. The Z80 starts executing the service routine.

This complex indirect call to a service routine can easily be demonstrated. In Figure 17.3, the I register has been loaded with 0CH, and the programmer has directed the Z80 into mode 2 interrupts and performed an EI instruction. Starting at memory address 0C00H is an address table. Each pair of consecutive bytes in the address table (stored in Intel format) contains the starting address of an interrupt routine.

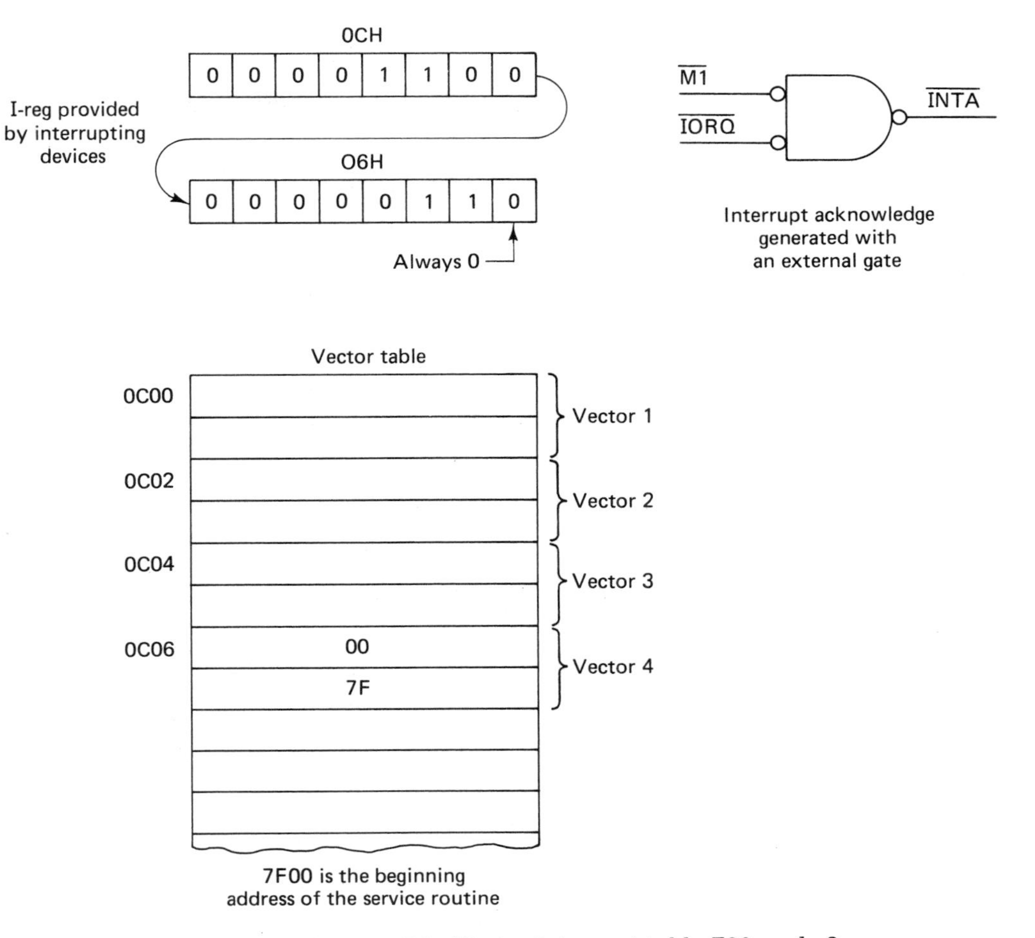

Figure 17.3 Vector interrupt table Z80 mode 2.

When $\overline{\text{INT}}$ goes active low, the Z80 will initiate the mode 2 interrupt service process. Notice that the Z80 does not have an INTA (interrupt acknowledge) output: How does the interrupting device know when to output the 8-bit op-code (in mode 0) or 8-bit address (in mode 2) onto the data bus? The Z80 INTA is not as straightforward as the 8085's interrupt acknowledge: $\overline{\text{M1}}$ and $\overline{\text{IORQ}}$ are both taken low to acknowledge an $\overline{\text{INT}}$ request. Note the clever manner in which two seemingly unrelated signals are used to create an $\overline{\text{INTA}}$ signal.

When the Z80 acknowledges the interrupt, the external device will provide an 8-bit address. The Z80 concatenates the 8-bit address with the contents of the I register. In Figure 17.3 this will create the address 0C06H. Notice that this must always be an even number to point correctly to the beginning of a 16-bit vector address.

The Z80 reads memory location 0C06 and places 00H into the low byte of a temporary register pair, and then reads memory location 0C07 and places 7FH into the high byte of the temporary register. After the PC is pushed onto the stack, the contents of the temporary register will be copied into the PC and the Z80 will execute the interrupt service routine located at address 7F00H.

It is important to realize that the original address derived from the concatenation of the I register and 8-bit address does not point to the beginning of an interrupt service routine, but rather to a memory location whose contents is the address of the routine. That is why mode 2 is called the indirect vectored interrupt mode.

You can now appreciate the powerful hardware interrupt features of the 8085 and the equally powerful software interrupt features of the Z80.

The $\overline{\text{WAIT}}$ input provides the same wait-state generation capabilities that the ready input provided for the 8085. Notice the different approach that the Z80 uses for the concept of wait. The 8085's ready input is active high; this indicates that the external devices are ready to be accessed. The Z80's wait input is active low; this indicates that the external devices are not ready to be accessed, and the Z80 must wait. The active-high ready input and the active-low wait input say the same thing in two different ways; the 8085 assumes that the device is not ready unless the ready input is taken active high; the Z80 assumes that the device is always ready unless the wait input is taken active low.

We have seen how the Z80 uses the $\overline{\text{M1}}$ output to indicate an op-code fetch, while the 8085 uses two encoded status inputs. The $\overline{\text{HALT}}$ output on the Z80 follows the same idea. When the 8085 enters the software halt state, an extra ALE signal is provided to latch the status outputs which indicate a halt state. The Z80 simply brings $\overline{\text{HALT}}$ active low whenever the Z80 enters a software halt state. Once again, the Z80 and 8085 provide the same information in two different ways.

The $\overline{\text{RESET}}$ input is an active-low reset input that is equivalent to the 8085's reset input.

17.1.5 DMA Control

When an external master wants to take control of an 8085-based system, it signals this intention by bringing the HOLD input to an active-high level.The 8085 acknowledges the DMA request by taking the HDLA output to an active-high level. $\overline{\text{BUSRQ}}$ (bus request) is the Z80's DMA request input, and $\overline{\text{BUSAK}}$ (bus request acknowledge) is the Z80's DMA acknowledge signal—different names and active levels for pins that provide the same functions.

17.2 THE Z80 INSTRUCTION SET AND ADDRESSING MODES

The Z80 is 100% software compatible with 8080 object code programs. Its instruction set includes every 8080 instruction (they even have the same op-codes) plus many new powerful instructions. To support the many advanced instructions, the Z80 uses an entirely different set of instructions mnemonics. Learning a new set of mnemonics for every microprocessor is a difficult, time-consuming task. Even with our knowledge of the 8085 instruction set to work with, a comprehensive analysis of the Z80's instruction set is beyond the scope of this text.

17.2.1 Programmer's Model of the Z80

Figure 17.4 illustrates the programmer's model of the Z80. The first thing that should strike you about Figure 17.4 is that the registers under the labels of "main" and "alternate" are exactly the same as the general-purpose registers in the 8085. This is true, with one minor difference. The parity flag in the Z80 is used for a dual purpose: during logical instructions it is used as a normal parity flag; during signed addition and subtraction it is used as an overflow flag. If an addition results in a number greater than +127 or a subtraction results in a number less than −128, the parity/overflow flag will set to indicate a 2's-complement overflow.

The DAA instruction of the 8085 works correctly only when performing BCD addition. The Z80 provides an extra nontestable flag that enables DAA to work with both BCD addition and subtraction.

What about the extra register set labeled "alternate?" This register set is used as a quick and efficient means of saving all the main registers as a preface to executing subroutines or servicing interrupting devices. Values stored in the alternate register set cannot be directly manipulated by the Z80.

The dedicated register block indicates the normal 16-bit SP and PC. We have already examined the function of the I register (interrupt) and the R register (refresh). That leaves the IX and IY registers to consider.

The Z80 supports two advanced addressing modes (in addition to the normal 8085 addressing modes). The IX and IY registers are called *index registers.* The indexed addressing mode is an extension to the register-indirect addressing mode

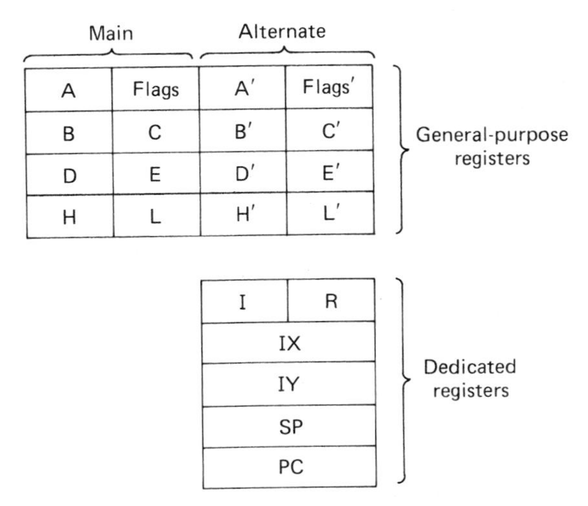

Figure 17.4 Programmer's model of Z80 registers.

that employed the H-register pair as a memory pointer to symbolic register M. Consider the vector interrupt table illustrated in Figure 17.3. Data are often kept in similar tables. An index register is used to point to the first address in such a table. To access any byte of data in the table, the base address of the table and a displacement into the table must be provided. Consider the following table of data:

Address	*Byte*	
A740	0C	A740 → IX
A741	FF	
A742	08	LD B, (IX + 3) = (A743) → B
A743	B3	
A744	00	
A745	D0	

The beginning address of the data table has been loaded into the IX register. The LD (load) instruction loads the B register with the contents of the memory location pointed to by the contents of the IX register (A740) plus the displacement byte (03). The displacement byte is an 8-bit 2's-complement number. The range of addresses that an index register can point to is IX+127 (displacement equal to 0111 1111) and IX−128 (displacement equal to 1000 0000). By using a displacement of zero, an index register functions like the H register pair in the 8085; both IX and IY are employed as index registers. Indexed addressing enables programmers to manipulate large, multidimensional tables of data.

The Z80 also provides a relative-to-the-PC addressing mode. The jump and call instructions in the 8085 are three bytes long—one byte for the op-code and two bytes for the 16-bit absolute address. The Z80 has a relative jump instruction that is two bytes long—one byte for the op-code and one byte for a relative offset. Consider the following program segment:

Address		*Instruction*
07F0		IN A7H
"		"
"		"
"		"
"		CP B
07F5	28	JR Z, F9H
07F6	F9	
07F7		

The first instruction gets a byte of data from the input port at address A7H. After some intermediate operations, the contents of the accumulator is compared to the contents of the B register. The next instuction tests the state of the zero flag. The mnemonic "JR Z" means that if the zero flag is set, jump to the location in memory that is derived by adding the second byte of the instruction to the contents of the program counter.

After the op-code for JR Z (28) and the displacement (F9) is fetched, the PC will contain the address of the next op-code (07F7). If the zero flag is set, we want to jump backward seven addresses (07F7 − 07F0 = 7). The Z80 assumes that the displacement byte is a 2's-complement number. If bit 7 of the displacement byte is low, the displacement is positive (forward); if bit 7 is high, the displacement is negative (backward). Here the displacement is F9, a negative number with a magnitude of 7. If the zero flag is set, the Z80 will subtract 7 from the PC and jump to address 07F0.

PC relative addressing is important for two reasons: The jump instruction required only two bytes, and even if the program is moved to another location in memory, the jump address will still be correct, because it is calculated relative to the PC. That is what is meant by *relocatable code.*

We have just taken a short look at the software power of the Z80. Because of its sophisticated addressing modes, extensive instruction set, alternate register set, and indirect interrupt vectoring, the Z80 is an extremely popular 8-bit microprocessor.

17.3 6800 MICROPROCESSOR HARDWARE FUNCTIONS

Because the 8085 and Z80 both evolved directly from the 8080, they share many similar hardware and software characteristics. The 6800 embodies a different approach to microprocessor design. The 40-pin package has two pins that are not used and two ground pins. Compare that to the 8085, which has so many hardware functions that it requires a multiplexed address/data bus to fit into a 40-pin package.

The 6800 is an older-technology microprocessor that is seldom used in new designs. Why bother to study the 6800? The descendants of the 6800, the 6809 and 68000, are extremely popular and powerful microprocessors. Examining the 6800 will help you understand the 6809 and 68000, whether you encounter them in an advanced microprocessor class or on the job.

The 6800 does not have any general-purpose registers; it uses a technique called *page-zero addressing to* effectively use the first 256 bytes of memory as CPU registers. The 6800 does not have explicit I/O hardware functions or instructions. All I/O devices are treated as if they are memory locations.

Comparing and contrasting the operation of the 8085, Z80, and 6800 will result in a greater understanding of microprocessors in general, helping you to develop the ability to quickly learn any new microprocessor.

17.3.1 Pins of the 6800

Figure 17.5 illustrates the pins of the 6800 grouped according to function.

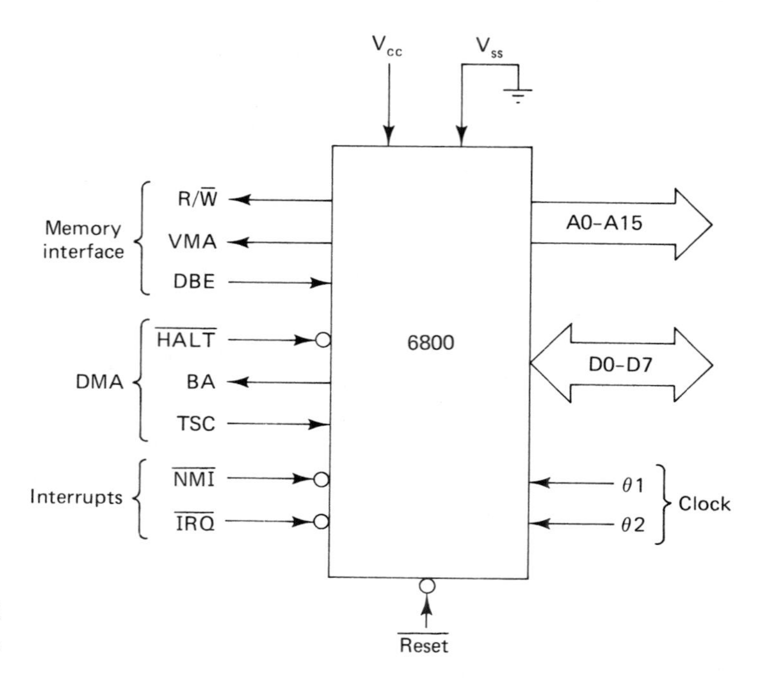

Figure 17.5 Functional pinout of the 6800.

Power inputs The 6800 requires a standard TTL-compatible power supply. It has one pin for V_{cc} and two pins for V_{ss} (ground).

Clocks The 6800 requires a two-phase, nonoverlapping clock. The frequency of both phases of the clock are equal, but both phases of the clock should never be high simultaneously. This is what is meant by "nonoverlapping." The 6800 starts a memory cycle on the rising edge of phase 1, and phase 2 is used to qualify the time that valid data are available on the bus. To meet the rigid specifications of the clock waveform, a two-phase clock generation IC is required to generate the 6800 clocks.

$\overline{\text{RES}}$ (Reset) This is the active-low reset as seen in the 8085 and Z80.

Address bus (A0 through A15) and Data bus (D0 through D7) These are the standard nonmultiplexed address and data bus, used in the Z80.

Memory/IO Interface Control Signals

R/$\overline{\text{W}}$ (read/$\overline{\text{write}}$) The level of this output indicates a read (high) or a write (low) access.

VMA (valid memory access) The VMA output goes active high to indicate that a valid address is on the bus. The 6800 does not have separate memory and I/O operations. All I/O devices in 6800 systems are memory mapped and accessed as if they were standard memory. The 8085 has IO/$\overline{\text{M}}$, $\overline{\text{RD}}$, and $\overline{\text{WR}}$ as its memory and I/O control signals, and the Z80 has $\overline{\text{MREQ}}$, $\overline{\text{IORQ}}$, $\overline{\text{RD}}$ and $\overline{\text{WR}}$ as its memory control signals. With the 8085, when $\overline{\text{RD}}$ and $\overline{\text{WR}}$ are both inactive, no memory or I/O access are occurring. The same is true with the Z80. The 6800 requires

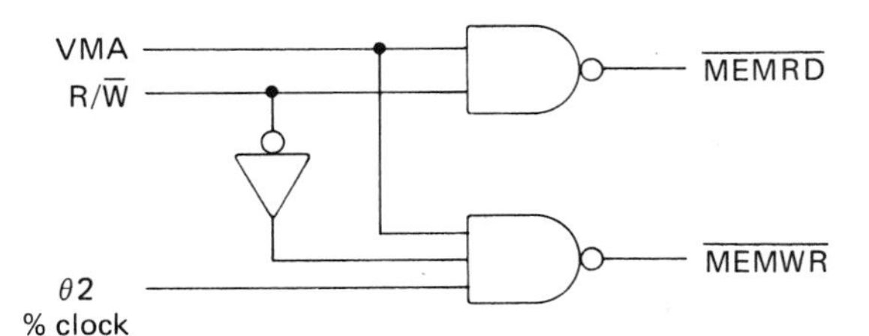

Figure 17.6 MEMRD- and MEMWR- signal generation in 6800.

the VMA signal to qualify when a memory/peripheral access is actually occurring; without VMA it would appear that the 6800 is always performing a read or write operation.

Figure 17.6 illustrates a simple circuit that generates $\overline{\text{MEMRD}}$ and $\overline{\text{MEMWR}}$ control signals. $\overline{\text{MEMRD}}$ is true when R/$\overline{\text{W}}$ and VMA are both high. $\overline{\text{MEMWR}}$ is true when VMA is high, R/$\overline{\text{W}}$ is low, and phase 2 of the clock is also high. $\overline{\text{MEMWR}}$ is qualified with phase 2 of the clock to ensure that the falling edge of the $\overline{\text{WR}}$ pulse does not occur until the 6800 has presented the data onto the data bus.

The output of an address decoder can be used to divide the 64K address space of the 6800 into memory and I/O areas. In this manner memory and I/O read/write signals can be generated that are similar to those in a 8085 system.

DBE (data bus enable) Active-high input which enables the internal data bus transceiver in the 6800. When DBE is in an inactive-high state, the data bus goes to high-Z. Because data will exit (on memory write) or enter (on memory read) the 6800 only when phase 2 of the clock is high, DBE is usually driven by the phase 2 clock input. The level on the R/$\overline{\text{W}}$ output will determine the direction of the data flow.

DMA Control

$\overline{\text{HALT}}$ This input functions in the same manner as the 8085's HOLD and the Z80's $\overline{\text{BUSRQ}}$ inputs. When $\overline{\text{HALT}}$ is driven active low, the address bus, data bus, and R/$\overline{\text{W}}$ will go to high-Z, VMA will be pulled to an inactive-low level (to ensure against spurious memory accesses), and BA (bus available) will go high to acknowledge that the memory subsystem can now be accessed by another bus master.

BA (bus available) This active-high output is the equivalent of HLDA on the 8085 and $\overline{\text{BUSAK}}$ on the Z80. BA will also go high to signal that a WAI (wait for interrupt) instruction has been executed. WAI is similar to the 8085's HLT instruction.

TSC (three-state control) TSC is an alternative DMA request input, but it suffers from one great limitation: The registers in the 8085 and Z80 are static RAM. However, the registers in the 6800 are constructed from dynamic RAM and the 6800 clock inputs are used for internal refresh timing. During TSC oper-

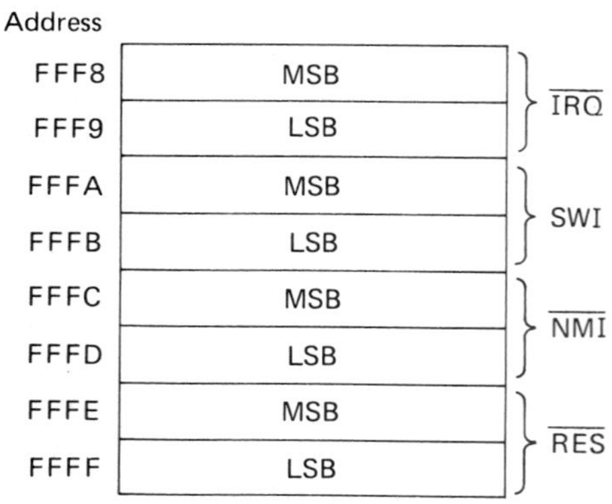

Address		
FFF8	MSB	$\overline{IRQ}$
FFF9	LSB	
FFFA	MSB	SWI
FFFB	LSB	
FFFC	MSB	$\overline{NMI}$
FFFD	LSB	
FFFE	MSB	$\overline{RES}$
FFFF	LSB	

Figure 17.7 Interrupt-Reset-vector table.

ation, these clocks must be held at a fixed level and register refresh does not occur. Employing the $\overline{HALT}$ input is the preferred manner of accomplishing DMA operations.

Interrupt Processing $\overline{NMI}$ (nonmaskable interrupt) and $\overline{IRQ}$ (interrupt request) are similar to the TRAP and INTR pins on the 8085 and $\overline{NMI}$ and $\overline{INT}$ on the Z80. When $\overline{NMI}$ or $\overline{IRQ}$ (assuming that it is unmasked) goes active low, the 6800 automatically pushes all its internal registers onto the stack. (This did not occur with the 8085 or Z80; the registers had to be saved by executing push reg-pair instructions.) The interrupt processing of the 6800 is extremely crude compared to that of the 8085 or Z80.

Figure 17.7 illustrates the 6800's interrupt/reset vector table. After the 6800 pushes the CPU registers onto the stack, it reads the appropriate memory location in the vector table to establish the starting address of the interrupt service routine. The starting address for the $\overline{IRQ}$ routine is stored at addresses FFF8 and FFF9; the address that points to the $\overline{NMI}$ routine is stored at addresses FFFC and FFFD. Figure 17.7 raises two interesting points.

1. Notice that the two-byte address is not stored in Intel format (as is employed by the 8085 and Z80). Instead, it is stored as high byte first, then low byte. The 6800, and its descendants, store all two-byte quantities in this manner.
2. The interrupt and reset vectors in the 8085 and Z80 were stored in low memory. The equivalent vectors in the 6800 are stored in high memory.

The 6800 does not need to generate an interrupt acknowledge signal because both $\overline{NMI}$ and $\overline{IRQ}$ are vectored interrupts; the interrupting device does not need to provide an instruction (as does the 8080-type interrupt) or an address (as does the Z80 mode 2 interrupt).

Figure 17.7 also indicates the starting address for the SWI (software interrupt, which is similar to a RST instruction on the 8085 and Z80) and the $\overline{RES}$ (reset), power-up routine.

17.4 THE 6800 MICROPROCESSOR INSTRUCTION SET AND ADDRESSING MODES

Figure 17.8 illustrates the 6800 programmer's model. You may be surprised by the simplicity of this model. There are no general-purpose registers as we have seen in the 8085 and Z80. Do not be confused by the appearance of two accumulators—ACC A and ACC B. Most of the 6800's instructions support both accumulators. The index register, IX, functions exactly like the index registers in the Z80. The other two 16-bit registers are the PC and the SP.

17.4.1 6800 Flags

The 6800 flag register is known as the CCR (condition code register). Although you may not recognize the name of each flag, you are already familiar with how each functions.

Carry flag The standard carry/borrow flag which indicates a carry out of bit 7 of the active accumulator.

Overflow flag Provides the same function as the Z80's overflow flag.

Zero flag The same as the zero flag in the 8085 and Z80.

N (negative) flag Called the sign flag in the 8085 and Z80.

Half-carry flag Known as the aux-carry flag in the 8085.

17.4.2 6800 Addressing Modes

Let's examine each of the 6800's addressing modes.

Immediate We have used the immediate addressing mode to supply data for an instruction as the second and third bytes in the instruction.

Extended and Direct LDA addr is an example of an 8085 instruction that uses direct addressing. The address of the operand is contained in the second and third bytes of the instruction. The 6800 calls this *extended addressing*. The term *direct ad-*

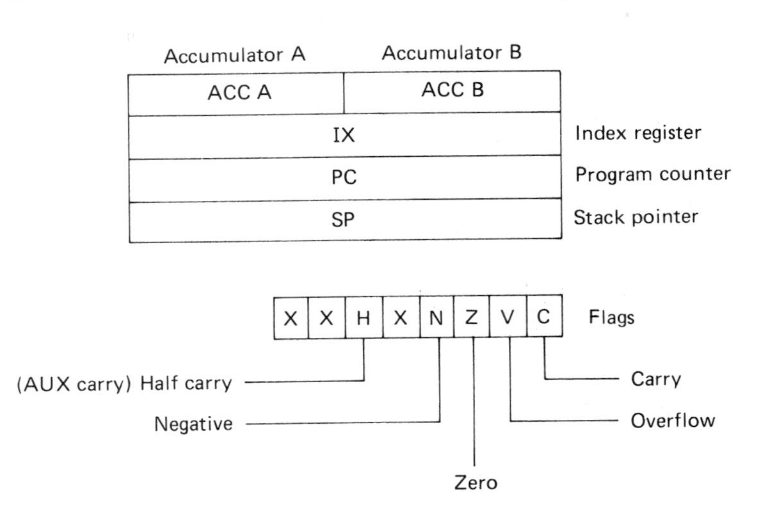

Figure 17.8 Programming model of 6800.

dressing is used to describe a slightly modified version of the 8085's direct addressing mode. A 6800 instruction that uses direct addressing supplies only one address byte. The 6800 assumes that the operand lies within the first 256 bytes of memory (00 through FF). Direct addressing requires only two bytes—an op-code and one address byte. The 6800 effectively uses the first 256 bytes of memory as its general-purpose CPU registers. Direct addressing helps facilitate this concept.

Direct addressing on the 6800 is also commonly called *page-zero addressing*. Consider the 64K memory space of an 8-bit microprocessor. This memory can be thought of as 256 pages of 256 bytes per page (256 × 256 = 64K). The least significant byte of the address is said to signify a memory location (1 of 256) within a particular page of memory. The most significant address byte is said to specify a particular page in memory. The 6800's direct addressing mode assumes that all address references will be to page zero. Therefore, only one address byte is required. Figure 17.9 illustrates the concept of page memory.

Relative Earlier in this chapter we examined the relative addressing mode of the Z80 as it applies to jump instructions. The 6800 also supports relative addressing with one 2's-complement displacement byte.

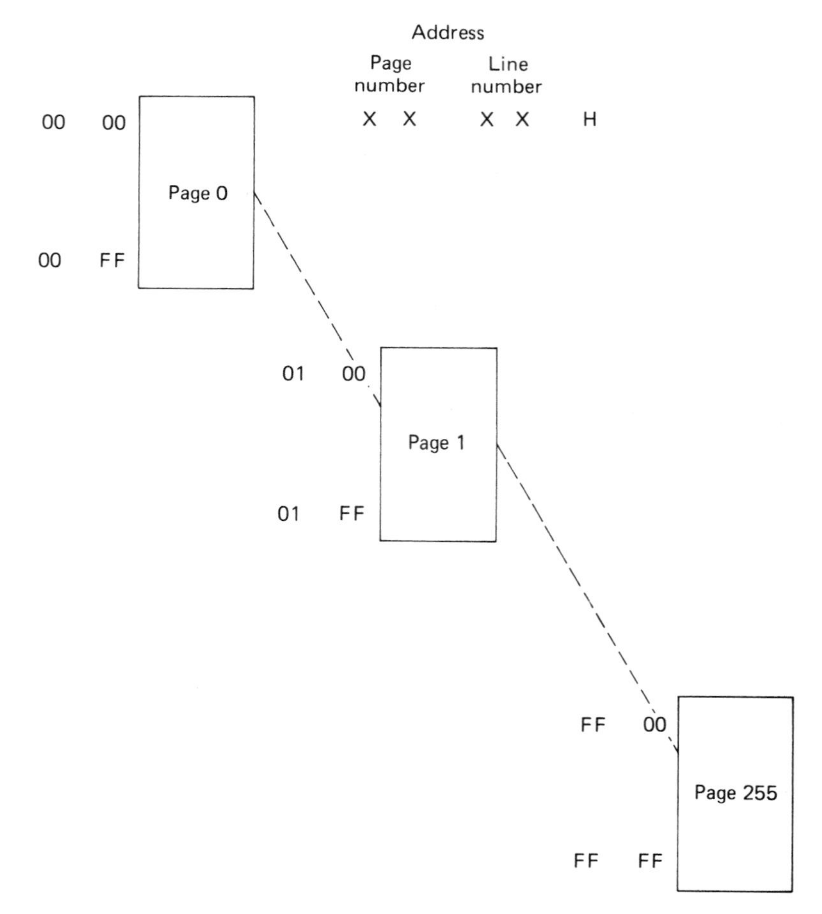

Figure 17.9 Paged memory.

Indexed The 6800 supports indexed addressing with an 8-bit unsigned offset. The Z80 also supports indexed addressing, but the offset byte is assumed to be a signed number.

17.4.3 The 6800 Instruction Set

Each instruction is supported by one or more addressing modes. Each instruction and associated addressing mode has a unique op-code. After the op-code fetch, the 6800 decodes the instruction to derive the op-code and its associated addressing mode. This will determine the number of bytes in the instruction.

As an example: ADDA adds the contents of ACC A with the second operand specified in the instruction. If immediate addressing is used, the next byte is the operand and the instruction is two bytes long. If the instruction uses extended addressing, the next two bytes contain the address of the operand, and the instruction is three bytes long. If the instruction uses direct addressing, the next byte in memory contains the address of the operand in page zero, and the length of the instruction is two bytes. The 6800 instruction mnemonics are different from the 8085 and Z80 mnemonics. The instruction set provides all the typical operations of an 8-bit microprocessor.

QUESTIONS AND PROBLEMS

17.1. How are the 8085 and Z80 related?
17.2. How do the clock inputs of the 8085 and Z80 differ?
17.3. Compare the Z80 and 8085 memory/IO control signals.
17.4. What does the Z80's $\overline{M1}$ output signify?
17.5. What are the Z80's DMA control lines?
17.6. How does the Z80 simplify interfacing with DRAM?
17.7. How do the Z80 and 8085 memory/IO accesses differ?
17.8. What are the Z80's dedicated registers?
17.9. How are index registers different from the HL register pair in the 8085?
17.10. What are the two advantages of relative addressing?
17.11. Compare the 6800 and Z80 clocks.
17.12. What does the 6800 do on a power-up reset?
17.13. How does the control signal VMA function?
17.14. What signal usually drives the DBE input of the 6800?
17.15. Name the general-purpose registers in the 6800.
17.16. What is "page-zero addressing"?
17.17. How does the 6800 interface with I/O?

18 Analysis of a Microcomputer System: The IBM-PC

In this book we have covered a great deal of material, from simple logic gates through microprocessors. All these components are designed and manufactured for one simple reason—to be integrated into an electronic system that provides a useful function. Our goal in this chapter is to examine the structure of a complex electronic system, the IBM-PC (Personal Computer).

In the late summer of 1981, IBM's Entry Systems Division announced a new product, a product that had been long awaited and anticipated: the IBM-PC. In 1981 the market was full of so-called "personal computers." Most were based on the Z80 or the 6502 microprocessor. Both the Z80 and 6502 are 8-bit processors that can directly address 64K of memory. Although many advanced technology VLSI microprocessors had been developed, none had been implemented as the CPU in a microcomputer.

IBM has always been known as a company that produces the world's most powerful mainframe computers. The concept of IBM manufacturing a desktop personal computer seemed alien to the company's mammoth image. Many people believed that personal computers would never be anything but underpowered, expensive toys.

When a large company like IBM enters a particular market, it is said to "legitimize" that market. Prior to 1981 the majority of microcomputer manufacturers were upstart companies, born in garages, basements, or similar unconventional settings. The fact that IBM planned to enter this market made many people think that the realm of the microcomputer was about to leave the home and small business environment and break into corporate America.

The IBM-PC did more than just legitimize the microcomputer market; it revolutionized and standardized the personal computer industry. Many analysts feel that the IBM-PC was a "self-fulfilling prophecy." So many corporations believed that IBM would produce a state-of-the-art, high-powered microcomputer capable of taking the demanding business world by storm that they immediately wanted the machine, sight unseen.

A computer is only as good as the software that it runs. Prior to 1981, the one rule of purchasing a personal computer was: "First locate the software that fulfills your need, then find a computer that will run it." When software manufacturers discovered that IBM was about to produce a microcomputer, they waited in line to buy the first machines on which to develop their applications programs. The software manufacturers, like the corporate world, knew that if IBM designed and marketed a personal computer, it would be a huge success.

Because of that belief, applications programs were quickly developed and launched into the marketplace with great hype and fanfare. This, in turn, fed the fires of the people who wanted an IBM-PC, which increased sales, and lured even more software developers onto the scene. So the self-fulfilling prophecy rolled on and gained more momentum with each passing month.

Looking back on the events of late 1981 and early 1982 does not lend any great insight into what really happened. Was the IBM-PC truly a revolutionary computer? In many ways yes, and in other ways no. The individual components that made up the IBM-PC were not exceptionally advanced. But the manner in which the system was integrated and its "open architecture" were inspired.

Our task in this chapter is to avoid getting caught up in the emotional issues that have surrounded the IBM-PC from its first day in the marketplace. It was chosen as an illustrative example of a microcomputer system for many reasons. Obviously, it is extremely popular, and therefore it is an important machine to know. Most important, the IBM-PC uses all the technology that we have pursued in this text. You now have the opportunity to see how the devices and concepts that we have discussed have been integrated into an exceptionally popular computer system.

The IBM-PC is such a complex machine that our analysis will be far from comprehensive. There are dozens of books available that describe the hardware and software attributes of the IBM-PC in great detail. The IBM-PC *Technical Reference Manual* is the most important document for the PC technician to possess. It describes the hardware and basic I/O system of the PC. It also contains schematics of all the printed-circuit boards used to support the PC.

18.1 HARDWARE SYSTEM OVERVIEW

The IBM-PC (for the remainder of this book called simply the PC) is housed in three units: the system unit, the keyboard, and the display.

18.1.1 The System Unit

The system unit is packaged in a rectangular metal box measuring approximately 18 in. wide, 14 in. deep, and 8 in. high. The system unit contains three major subsystems: the power supply, system board with five expansion slots, and mass-storage devices.

The power supply is a 63.5-W switching-type power supply that delivers 7 A at +5 V, 2 A at +12 V, 300 mA at −5 V, and 250 mA at −12 V. The power supply is considered to be a closed, nonrepairable unit. If it is isolated as a bad unit, it is simply swapped with a new or known-good power supply and discarded. IBM considers the power supply to be too complex and not cost-justified to repair.

The system board is the heart of the PC. It is a multilayered printed circuit board measuring approximately $8\frac{1}{2}$ in. by 11 in. Traces on the top and bottom carry signals, and the center traces are ground and power buses.

We know that a minimum computer system must have a CPU, main memory, and I/O ports. In addition, practical computers have many other LSI support devices to improve the system's processing power. The PC employs an advanced microprocessor, the Intel 8088. The 8088 is a direct descendent of the 8085 and shares many similarities with it, including a multiplexed address/data bus. The 8088 appears internally to be a 16-bit microprocessor, with 16-bit general-purpose registers and ALU, but externally the 8088 has an 8-bit data bus. It can only read or write quantities of byte-wide data during each memory access. Intel describes the 8088 as a high-performance 8-bit microprocessor; IBM describes it as a powerful 16-bit microprocessor. A more sensible description of the 8088 is a "pseudo-16-bit microprocessor" or an 8/16-bit microprocessor.

The 8088 has 20 address lines and can therefore directly address 1 MB of memory. People want complex application programs that run at fast speeds. Both these requirements entail the support of huge quantities of RAM. The PC supports up to 640 kB of DRAM. The other memory locations are reserved for system ROM (48K), display memory (128K), and a 216K undefined block.

In Chapter 14 we briefly examined the old 8080 microprocessor. It required two support chips, a clock generator/driver, and a system controller. One of the major advantages of the 8085 was that it did not require these two support chips. As if the leap forward to the 8088 was really a step backward, the 8088 requires a clock generator/driver and a system controller similar to those used in the 8080 system.

The 8284 clock generator and driver is designed specifically to be used with the 8088. It has an on-board oscillator that works with an external crystal to provide processor and peripheral clocks. It also controls two ready inputs, and the system reset lines.

The 8288 is the bus controller that was designed to function in the 8088 microprocessor system. It takes three status lines from the 8088 and derives the memory and I/O control signals, and interrupt acknowledge.

The system board also has a 40-pin socket for a co-processor. A co-processor is an auxiliary processor that is designed to assist the system microprocessor by performing specific processing chores. A typical bottleneck in programs is the processing of high-precision floating-point arithmetic and transcendental (trigonometric, exponential, and logarithmic) functions. Microprocessors are general-purpose system controllers. To perform these complex functions utilizing the basic add, subtract, multiply, and divide instructions can be dreadfully slow. Why not dedicate a special-purpose microprocessor whose only job is to perform super-high-speed math functions?

That is the function of the 8087 NDP (numeric data processor). This auxiliary processor performs floating-point arithmetic and transcendental functions 10 to 100 times faster than the equivalent 8088 emulation program. Most PCs do not employ this optional co-processor because it is not yet supported by general application programs.

The system board holds many other LSI support devices and MSI TTL buffers and latches, and a few SSI gates and flip-flops. We examine these in Chapter 19.

We have already stated that the PC supports up to 640 kB of DRAM. The system board has a capacity of 256K of DRAM. The additional 384K of DRAM is added to the system on a memory expansion board. The 256K of system board DRAM is arranged in four banks of nine chips. Each memory chip is a 64K × 1 DRAM; eight 64K DRAMS form 64 kB of storage. The ninth DRAM is used for error checking and is known as the parity bit. Each location in the parity DRAM contains a logic 0 or a logic 1 parity check bit. When the 8088 performs a memory-read operation, the parity bit that is read is checked against the calculated parity. If a mismatch occurs, the NMI of the 8088 is taken active to report the memory parity error. In this manner, memory integrity is ensured.

The system board also contains sockets for six 8K × 8 ROMs. Five of the sockets are stuffed. The last one is available for custom applications. In Chapter 12 we posed the question:

What does a microprocessor system do after a reset operation? The answer to that question established the need for system ROM bootstrap memory.

What programs or information is contained in the 40 kB of system board ROM? One ROM contains what IBM terms as its POST (power-on self-test). In Chapter 19 we examine the POST in detail. The POST ROM also contains the BIOS (Basic Input/Output System). The BIOS is the lowest-level programs that interface the system programmer with the actual PC hardware. The BIOS contains software interrupt-driven programs that enable the programmer to utilize the PC's hardware resources through high-level calls.

When programmers write application programs, they call on the BIOS functions (as if they were subroutines) to perform actual hardware chores. This isolates the programmer from the actual PC hardware environment. This also helps programs achieve a moderate level of hardware independence, by providing an extra layer between applications programs and the actual system hardware. Theoretically, if two different computers that are based on the same microprocessor provide similar BIOS functions, a high-level applications program should be able to run on both machines, with very litle modification.

The ROMs also contain the floppy-disk bootstrap program. This is a small program that loads the DOS (Disk Operating System) from a floppy disk in the default drive. Four of the ROMs also contain the BASIC programming language.

The system board has two eight-station DIP switches. An eight-station DIP switch is a plastic switch that fits into a standard 16-pin DIP socket and contains eight SPST switches. These switches are used to indicate the system's setup: how much DRAM in the system, how many floppy disk drives, if the 8087 is present, and what type of video monitor is in use. These are important parameters that the 8088 must establish on power-up. One of the simple programs in ROM reads the 16 switches (via an 8255 PPI) and establishes the system's hardware operating environment.

The most important factor contributing to the success of the PC is that of the five expansion slots on the system board. An expansion slot permits the addition of a plug-in expansion board into the system. Each expansion slot (called I/O Channels by IBM) is attached to a 62-pin edge-connector receptacle. The system address bus, data bus, control bus, and interrupt and DMA signals are run to each expansion slot. The five expansion slots are in parallel with the rest of the system. If one wants to increase the memory of the PC from the 256K to the full amount of 640K, a 384K memory expansion board can easily be plugged into any empty slot. The settings of the memory indication DIP switches must be reset (as outlined in the IBM manual) to reflect the additional memory.

In a basic PC two of the expansion slots are already stuffed with boards. The first board is called the floppy disk adapter board and is centered around the Intel 8272 floppy disk controller (FDC) LSI device. It interfaces the 8088 with up to four standard $5\frac{1}{4}$-in. double-sided, double-density floppy disk drives. Most systems have two floppy disk drives, which mount horizontally in the front of the system unit.

The other expansion slot that is always stuffed contains a display adapter board. Minicomputer systems use RS-232 terminals as displays. Terminals usually contain three PCBs: the logic board, monitor board, and power supply. The logic board contains the UART (for serial communications) video memory, character generator ROM, CRT controller LSI support chip, and a microprocessor. The monitor board interfaces the logic board outputs with the CRT. The power supply supports the dc requirements of the logic board and the video board.

Microcomputer systems have taken a slightly different approach to the display requirements. Instead of using a standard (and somewhat expensive) RS-232 terminal as a display, microcomputers employ a bare-bones video monitor. This video monitor is essentially a monitor board, power supply, and CRT. The equivalent of the display board in the RS-232 terminal is a printed circuit board that is plugged into one of the PC's expansion slots. This results in lower cost with much greater flexibility.

Many different display boards are available for the PC. The standard display adapter board has 4K of static video refresh RAM, a character generator ROM, and a 6845 CRT controller. It is designed to drive the standard PC green phosphor display. Other boards are available that support high-resolution monochrome and high-resolution full-color graphics on the appropriate displays.

That leaves three empty expansion slots that can be used in many different ways. A discussion of some types of boards available to plug into the expansion slots follows.

Parallel ports interface the PC with standard Centronics interfaced printers. The PC supports two parallel ports, called LPT1: and LPT2: (LPT is short for "line printer port.")

Serial boards (called *asynchronous adapters*) interface the PC with any standard RS-232 device. The two possible RS-232 ports on the PC are designated by the names COM1: and COM2: (COM is short for "communication port"). Typical RS-232 devices are printers, modems, mice, graphics tablets, and analog I/O converters. There are literally thousands of RS-232-compatible devices that can be directly interfaced with the PC via an RS-232 adapter board. The PC must be able to converse with each device connected to its RS-232 ports. This is accomplished by a program called a *device driver,* which is usually provided by the manufacturer of the peripheral device.

The PC has an internal time-of-day (TOD) clock that must be set by the operator every time the PC is powered up. Many people prefer to purchase and install a real-time clock that is backed up by a nickel–cadmium or lithium battery. When the PC is powered up, a special program called an *auto-execute batch file* can easily be instructed to read the current time of day and date from the battery-backed-up clock, minimizing operator effort.

We have already discussed how RS-232 terminals and modems are used to communicate with distant computers over telephone lines. Many manufacturers offer 1200-baud modems contained on a single PCB that plug into a PC expansion slot. If a standard external modem is used, it can be connected to the PC via an RS-232 expansion board. Programs are available that make the PC emulate many popular different RS-232 terminals. That enables a person to communicate with distant computers and data bases while maintaining the stand-alone ability of a microcomputer, thus deriving the best of both worlds.

If two double-density double-sided floppy disk drives do not provide enough storage or if faster disk access is required, a Winchester disk drive adapter and hard disk drive can be easily installed in the PC. The hard disk controller board plugs into any expansion slot. It contains all the circuitry required to control two Winchester-type drives. The drives are available in typical capacities of 10 MB, 20 MB, and 30 MB. The hard disk can be inserted into the system unit in place of the second floppy disk drive, or reside in an external chassis. IBM manufactures an upgraded PC called the PC XT that has a 10-MB hard disk as a standard item.

As you can now see, the extra three expansion slots in the PC can quickly be filled. To help alleviate this shortage of expansion slots, one of the most popular types of boards for the PC is the *multifunction board.* Multifunction boards provide many different functions on one board, saving precious expansion slots. A typical multifunction board provides sockets for 256K to 384K of expansion DRAM, a serial RS-232 port, a Centronics-type parallel port, a battery-backed-up clock, and a few software support functions.

There are hundreds of manufacturers that market dozens of different types of expansion boards for the PC. The preceding examples have been of some of the most popular boards.

18.1.2 The Keyboard

The PC's keyboard is a capacitive-type keyboard that interfaces with the system unit via a 6-ft coiled expansion cable. The cable has four lines: +5 V, ground, keyboard data, and data clock (keyboard clock). The keyboard communicates with the system unit in a serial fashion.

Local intelligence in the keyboard is provided by an 8048 microcomputer chip. Note the term "microcomputer chip," not "microprocessor." These two terms are often used interchangeably, but they designate two very different devices. Our initial definition of a microprocessor was "a CPU on a chip." A computer requires not only a CPU, but memory (both RAM and ROM) and I/O ports. From that definition we can assume that a mirocomputer chip not only contains a CPU, but also RAM, ROM, and I/O ports. Microcomputer chips are used as low-cost, low-parts-count alternatives to microprocessors in applications where a general-purpose microprocessor is not required.

A keyboard scanner/encoder chip is a perfect application for a microcomputer chip. The 8048 contains an 8-bit CPU (with a full instruction set), 1K × 8 ROM, 64 bytes of RAM, three 8-bit parallel ports, 8-bit counter/timer, and an internal clock oscillator. The whole computer system is contained in one 40-pin package. (For development purposes a special 8048 is available with 1K of EPROM. This enables designers to debug their firmware before the 8048 with the actual custom mask programmed ROM is produced.) The ROM in the keyboard's 8048 contains a program that scans the keyboard looking for a closed key. When a closed-key indication occurs, the 8084 interrupts the 8088 in the system unit and transmits the key code. The key code goes through a serial-to-parallel conversion at the system unit and is stored in a small buffer memory waiting to be processed by the 8088.

The 8048 frees the 8088 from the routine and time-consuming process of scanning the keyboard via software, waiting for a closed key, and then debouncing the key with a software delay. The idea of providing many local processing devices in a computer system is called *distributed processing.* This technique increases the efficiency of the system by relieving the CPU of simple chores, through the use of dedicated local intelligence.

The mechanical portion of the keyboard, like the switching power supply in the system unit, is considered to be a nonrepairable item. If a key sticks open or closed, the whole keyboard must be swapped for a new unit.

18.1.3 Video Display and Scanning Techniques

Before you can understand how the display for the PC functions, a short course in television technology is in order. Figure 18.1 illustrates the manner in which a picture is generated on a TV set.

Broadcast television in the United States follows the NTSC (National Television Standards Committee) standard. It states that a television picture is composed of 525 horizontal lines. The complete picture, called a *frame,* is painted in two consecutive sweeps, called *fields.* The first field paints the even-numbered lines and the second field paints the odd-numbered lines. This method of painting a frame is called *interlaced scanning.*

Why use such a complicated method of painting a frame of

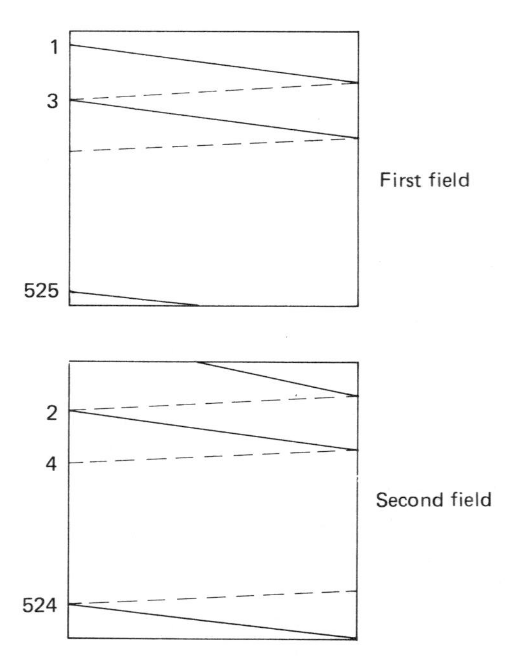

Figure 18.1 Interlaced raster scan.

video information? Why not just paint consecutive lines? The limited bandwidth of television circuitry imposes the specification that a maximum of 30 frames can be painted each second. The quality of the television picture depends largely on the persistence of the human eye; the ability to maintain the image of the screen as a new frame is being painted. If noninterlaced scanning was used and a frame was painted in one complete sweep, the television picture would appear to flicker. Interlaced scanning is a "trick" that effectively doubles the frequency that the picture is painted (each field is 16.5 ms long), but does not require expensive circuitry with a higher bandwidth. Interlacing odd and even fields "tricks" the eye into seeing the screen being painted 60 times a second, thus correcting the flicker problem.

Three signals are required to display video information on a CRT: horizontal sync, vertical sync, and video information. Horizontal sync returns the electron beam from the right side of the screen to the beginning of the next line. Vertical sync returns the electron beam from the lower right corner of the display to the beginning of the next field in the upper left corner of the screen. The video information controls the number of electrons fired at the screen. The inside of the screen is coated with a phosphor that glows when struck by electrons; if no electrons are fired at the screen, it will remain dark. If many electronics are fired at the screen, a bright dot will appear. Between the black and white levels, many different shades of gray can be generated. Figure 18.2 illustrates these signals.

Figure 18.2a illustrates a typical horizontal sync signal. The falling edge of H-sync starts the process of returning the beam to the beginning of the next line. The width of the nega-

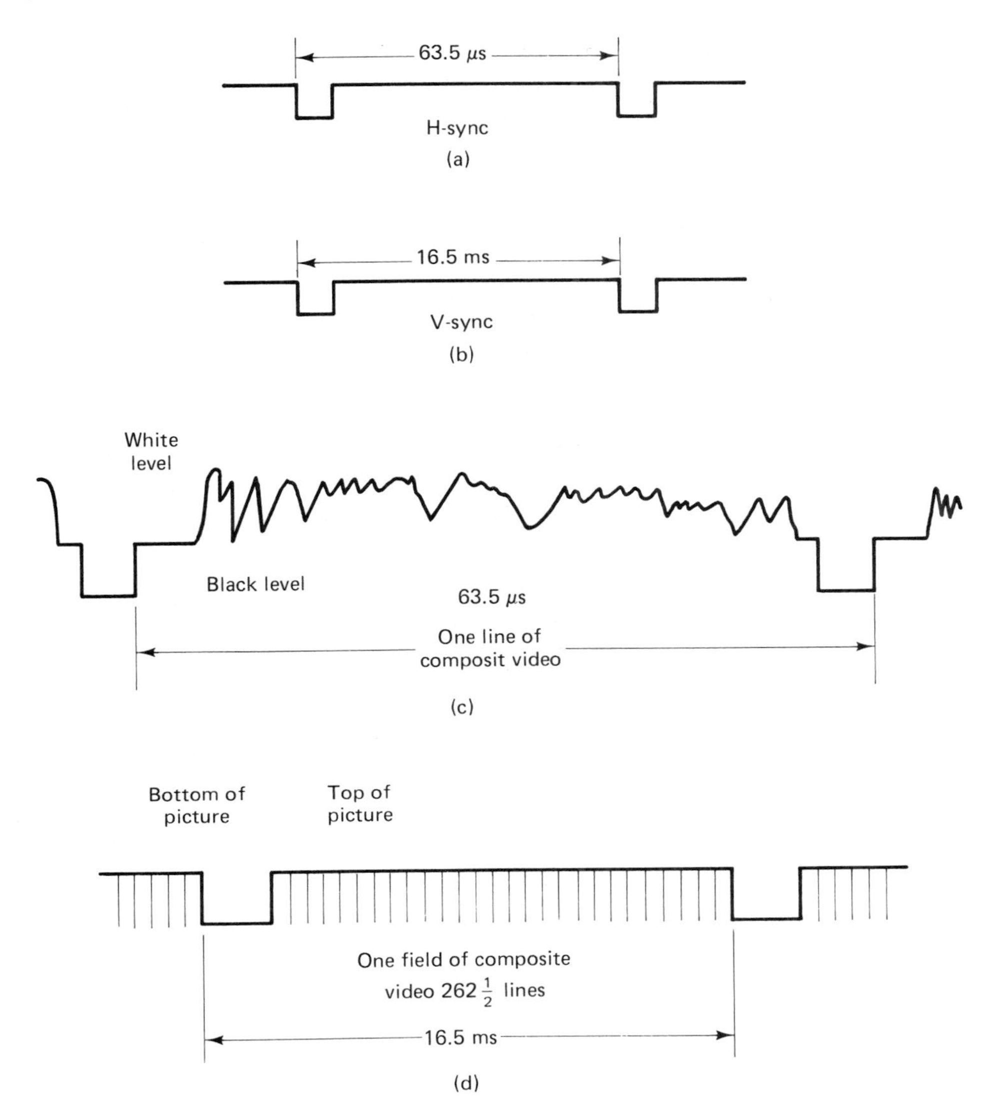

Figure 18.2 Sync pulses and video information.

tive pulse indicates the *flyback time* (the time required to move the beam to the next line). Flyback is extremely short compared to the time between the edges of the H-sync. This indicates that it takes much longer to paint a line of information than to return the beam to the beginning of the next line.

Figure 18.2b illustrates the V-sync signal. In television technology the vertical sync interval is actually a complex series of pulses. In computers it will appear as a single negative pulse, as shown in Figure 18.2b It takes much longer to return the beam from the end of the field (V-sync period) than it does to return the beam from the right side of the screen to the beginning of the next line (H-sync period).

The video signal can be sent as three separate signals—H-sync, V-sync, and video information—or it can be sent as a composite signal. Composite video contains the sync pulses and video information integrated into one signal. Figure 18.2c and d illustrate the horizontal and vertical portions of a composite video signal.

How are computer displays different from standard televisions? Televisions conform to NTSC standards so that everyone can receive signals from the commercial and public airways. Because computer displays do not receive broadcast signals, they do not need to follow any standard. The picture quality produced on a particular display is proportional to the number of horizontal and vertical lines that it supports. The number of lines dictate the rate of the sync signals. We will limit our investigation of displays to the standard IBM monochrome display that is used on the majority of PCs.

The standard PC monitor has a CRT which employs long-persistence, P-39 green phosphor. The term "long persistence" indicates that the CRT will "glow" for a long period after being struck by an electron. Interlaced scanning is not required because the increased persistence of the display overcomes the problem of flicker. Most terminals and computer displays use noninterlaced scanning. This greatly simplifies the video processing circuitry.

The PC's monochrome display is capable of displaying 25 lines of characters with 80 characters per line. Each character is constructed from a 7 × 9 dot matrix which is enclosed in a 9 × 14 character box. Figure 18.3 illustrates the character box matrix used on the PC.

A character is created by illuminating selected dots in the 7 × 9 matrix; the use of 63 dots results in extremely readable, high-quality characters. If each character box is constructed from 14 rows of dots and the display supports a total of 25 rows of characters, a total of 350 horizontal lines are displayed. Em-

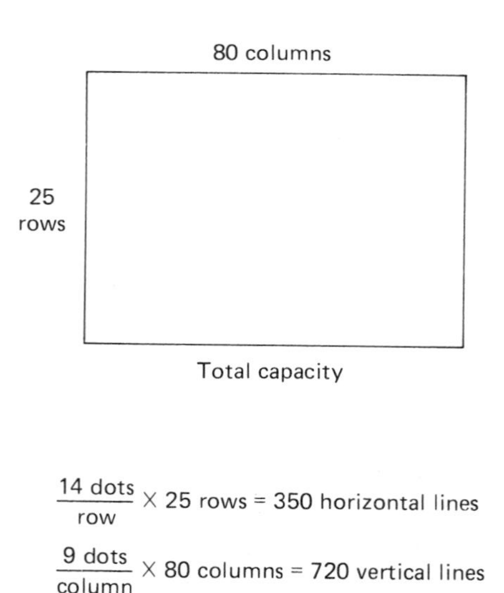

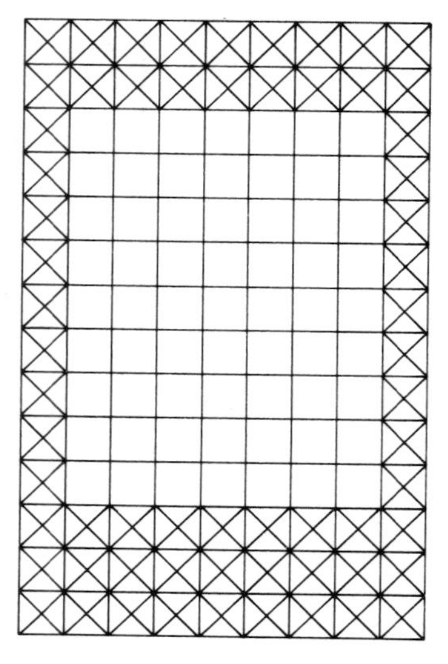

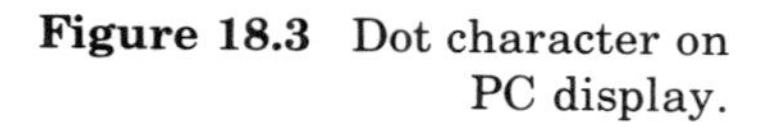
Figure 18.3 Dot character on PC display.

ploying the same reasoning, a character box that is 9 dots wide and 80 per row results in a display of 720 vertical lines.

The monochrome adapter expansion card in the PC has two banks of 2K × 8 static RAM. One bank holds the ASCII character code for each of the 2000 characters per screen (25 rows × 80 columns = 2000). Associated with each ASCII character code is an attribute byte that is held in an adjacent 2K × 8 bank of static RAM. A character attribute refers to how a particular character is displayed. In normal operation a character is displayed as a light character on a dark background. The attribute byte of a character enables it to be displayed as a dark character on a light background (reverse video), invisible (dark character on dark background), underlined, blinking, or intensified. A character that is intensified appears much brighter than a typical character.

These two banks of character and attribute RAM are collectively called the *display refresh memory.* Do not confuse the term "refresh memory" with a DRAM that requires refresh. Display memory is called refresh memory because the video display is "refreshed" each time a new frame of video information is painted. Because the display requires a total of only 4K of memory, the use of static RAM is attractive and cost-effective.

How is a character actually painted onto the screen? We have seen that a television paints one line at a time. The computer monitor also paints one line at a time, but the process is complicated because it takes 14 horizontal lines to complete the painting of one character. Figure 18.4 illustrates a simplified block diagram of a video display support circuit.

To simplify Figure 18.4, the address, data, and control bus

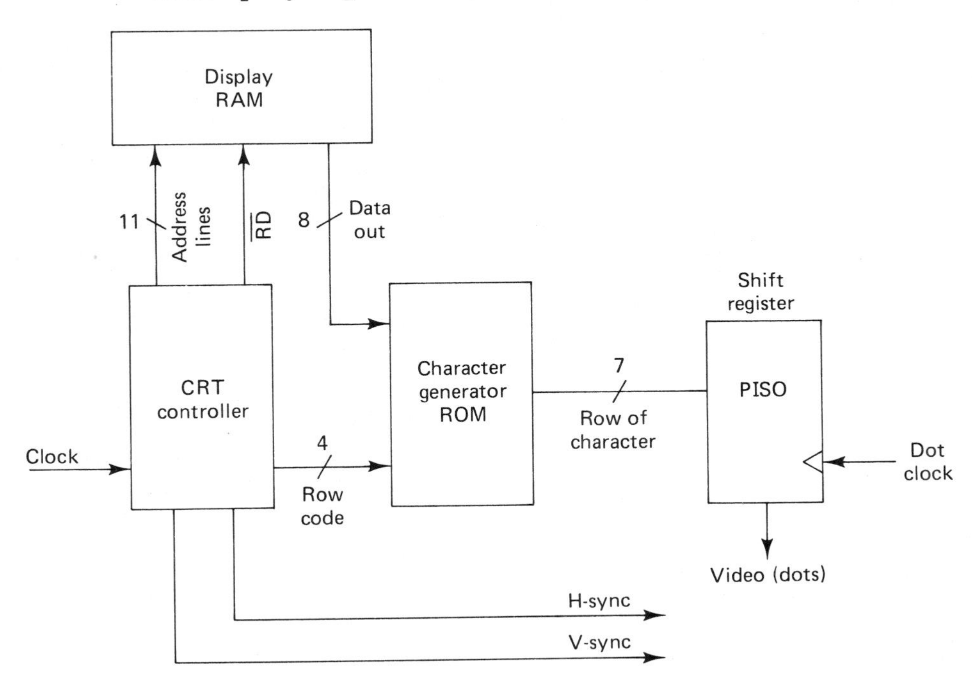

Figure 18.4 CRT controller and support circuitry.

connections between the system microprocessor and the CRT controller are not shown. The CRT controller is a complex programmable LSI device. Programming of the CRT controller is the process that defines the type of display, modes of operation, number of vertical and horizontal lines, and size of the character matrix.

The display RAM is *dual-ported.* The microprocessor writes the ASCII code of the characters to be displayed in the appropriate display RAM location. The CRT controller has 11 address output lines and a $\overline{RD}$ control line. This enables the CRT controller to read the contents of the display RAM.

The CRT controller also has four row code outputs. Remember that the character is nine rows high. A character will be painted on the CRT one row at a time. The four row code outputs indicate which of the nine rows is presently being painted.

The charcter generator is simply a ROM that contains the dot patterns for each of the nine rows of the 256 unique characters that the PC is capable of displaying. The 8 bits of the character code and the 4 bits of row code form a 12-bit select address for the character generator ROM.

The character generator outputs a 7-bit code that describes a row of dots for a particular character. This parallel 7-bit quantity drives the inputs of a parallel in/serial out shift register. The output of the shift register is digital video information.

Assume that the CRT controller has just asserted V-sync and the electron beam of the CRT has returned to the beginning of the first line of the display. The CRT controller will then read the first character code from the display RAM and output a row code of 0. This 12-bit address accesses a 7-bit row dot pattern that appears on the output of the character generator ROM. This input is shifted out of the PISO shift register. A logic 1 turns on the electronic stream of the CRT and leaves a bright spot on the display. A logic zero turns off the electron stream of the CRT and results in a dark spot on the display. After the 7 bits for the first character are shifted out, the CRT controller repeats the same process for the next 79 characters on line 1.

The first row of 80 characters has now been painted on the display. The CRT controller asserts H-sync and the electron beam returns to the beginning of line 2. The previous process is repeated with the row code set to row 1, and the second row of 80 characters is painted.

This process is repeated for the remaining rows of characters in line 1. Notice that each character is read from display RAM nine times before the complete character has been painted. The remaining 24 lines are painted in a similar row-by-row manner. After each row is painted, the CRT controller issues an H-sync pulse to return the electronic beam to the beginning of the next row.

The CRT controller greatly simplifies the process of refreshing the display. After it is programmed by the system microprocessor, the CRT controller operates independently of any further processor intervention, with one small exception. Because the display RAM is dual-ported, the CPU and CRT controller must never attempt to access it simultaneously. The CPU will ensure that this condition never occurs.

The output of the monochrome adapter board is connected to the input of the PC's monochrome display. The monochrome display contains two circuit boards, the monitor, and power supply. The monitor board translates the digital level H-sync, V-sync, and video signals into analog signals that can drive the CRT. The power supply provides dc power to the transistors and ICs on the monitor board.

18.2 TERMINALS AND VIDEO MONITORS

Figure 18.5 illustrates the difference between a standard RS-232 terminal and the display adapter expansion board/video monitor combination used in the PC. An RS-232 terminal holds the logic board, monitor board, power supply, CRT, and keyboard

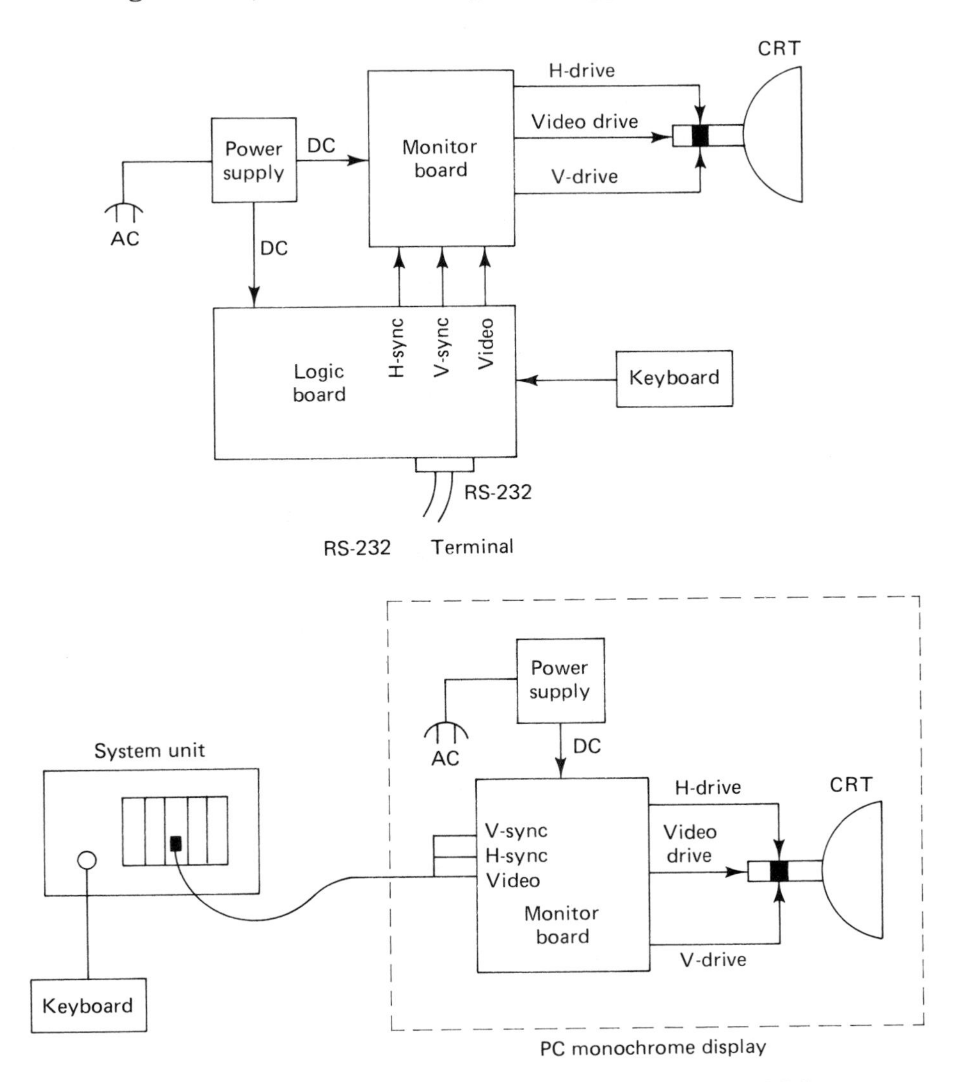

Figure 18.5 Block diagram of RS-232 terminal and PC display.

in one unit. All communication between the computer and terminal is via the RS-232 interface. The components of the traditional terminal are broken into three separate units of the PC. We have already examined how the keyboard interfaces with the system board via a connector on the back of the PC. The PC keyboard is completely independent from the display unit.

The equivalent of the logic board in the RS-232 terminal is the display adapter expansion board plugged into one of the system board's five expansion slots. This gives PC users the option of choosing from among many display adapter boards that will function with the standard monochrome display.

The output of the display adapter expansion board is connected to the PC monochrome display via a cable and a nine-pin "D" connector. The monochrome display contains the monitor board, power supply, and CRT.

We have now examined the major components that constitute the IBM-PC microcomputer system. In Chapter 19 we examine the relationship of the components on the system board and various PC troubleshooting techniques.

QUESTIONS AND PROBLEMS

18.1. What are the three units that constitute the IBM-PC?

18.2. What is a "co-processor"?

18.3. How is RAM parity checked?

18.4. What does the ROM on the PC contain?

18.5. What function do the DIP switches on the system board perform?

18.6. What are the expansion slots on the system board? What are they used for?

18.7. Name the major subassemblies in a standard PC system unit.

18.8. If floppy disk storage is not sufficient, what method may be used to increase the mass storage capacity of the PC?

18.9. What is the difference between a microprocessor and a microcomputer chip?

18.10. What are the two different methods of "painting" a CRT?

18.11. What are the three signals required to create a video signal?

18.12. Describe the video "refresh" memory in the monochrome adapter board.

18.13. What is the difference between the video monitor used on microcomputers and a standard ASCII terminal?

18.14. Which two devices in the PC system can access the video refresh memory?

18.15. What intelligent device does the keyboard contain?

19 Troubleshooting Microcomputer Systems

In Part 1 of this book you learned to analyze and troubleshoot circuits containing SSI and MSI logic elements. Troubleshooting a computer system requires many additional skills. A system such as the IBM-PC contains a minimum of three PCBs, a floppy disk drive, a power supply, a keyboard, and a video display. Most computer dealers and repair services do not troubleshoot to the component level on microcomputer systems. They troubleshoot only to the bad subassembly, which is simply swapped with a known-good unit. The bad unit is returned to the manufacturer or to a repair depot, where it is quickly and economically repaired with the aid of sophisticated test equipment. In this chapter we are going to analyze the system board of the IBM-PC in greater detail and investigate methods of troubleshooting microcomputer systems to the component level.

19.1 ANALYSIS OF THE SYSTEM BOARD

As you learned in Chapter 18, the system board is the heart of the complex PC. We will first examine the functional pinouts of the major LSI devices that reside on the system board and then consider how the devices are interconnected to create a computer system.

19.1.1 The CPU and Direct Support Circuitry

Figure 19.1 illustrates the three-chip system that performs the CPU function in the PC. In this book we examine only the hardware aspects of the 8088. The internal registers, instruction set, and addressing modes are not integral to the discussion at hand. The 8088 is the 8-bit version of the 16-bit 8086. By studying the 8088 you should have very little trouble learning 16-bit microprocessors.

The 8088 is an advanced, complex microprocessor that has two distinct modes of operation: the minimum mode and the maximum mode. When the 8088 is used in *minimum mode,* it functions like an enhanced version of the 8085. Its memory and I/O interface, DMA, ready, and interrupt lines operate in an almost identical manner to the 8085. The 8088 is used in the minimum mode when it is employed as a dedicated intelligent controller. It is a natural, high-performance extension of the 8085. Because it is so similar to the 8085, you should have little trouble understanding minimum-mode 8088-based systems.

When the 8088 is used in the *maximum mode,* many of the familiar control lines radically change functions. The maximum mode allows the 8088 to be used in multiprocessor environments. In maximum-mode systems the 8088 is often teamed with the 8087 numeric data co-processor and other co-processors to create a high-throughput, maximum-performance microcomputer system.

In minimum mode, the 8088 requires only the 8284 clock generator as a support device. The memory, I/O, DMA, and interrupt lines are handled directly by the 8088. In the maximum mode, the 8088 requires the 8284 clock generator and the 8288 system controller. The 8288 system controller becomes an extension of the 8088 CPU. It is used to decode three status outputs from the 8088 into memory, I/O, interrupt acknowledge, and buffer control signals.

The 8284 is the 8088 system clock generator and driver. It is used to generate clocks and to manage the reset and ready inputs. The crystal works with an internal oscillator to create three clock outputs: CLK is the output that drives the 8088 and 8288 clock inputs. Its frequency is one-third of the crystal frequency. OSC is the TTL-level output of the internal oscillator. OSC is routed to pin B30 on the five expansion connectors. PCLK is the peripheral clock. Its frequency is one-sixth of the crystal frequency. PCLK drives the clock input of the 8253 PIT (programmable interval timer).

The 8284 is also capable of controlling two ready input lines. RDY1 is an active-high input and $\overline{\text{AEN1}}$ is the active-low qualifier for RDY1. When RDY1 is high and $\overline{\text{AEN1}}$ is low, the ready output of the 8284 will be an active-high level. Two other inputs, RDY2 and $\overline{\text{AEN2}}$, are not used in the PC circuit. They are not illustrated in Figure 19.1; assume that they are tied to active levels.

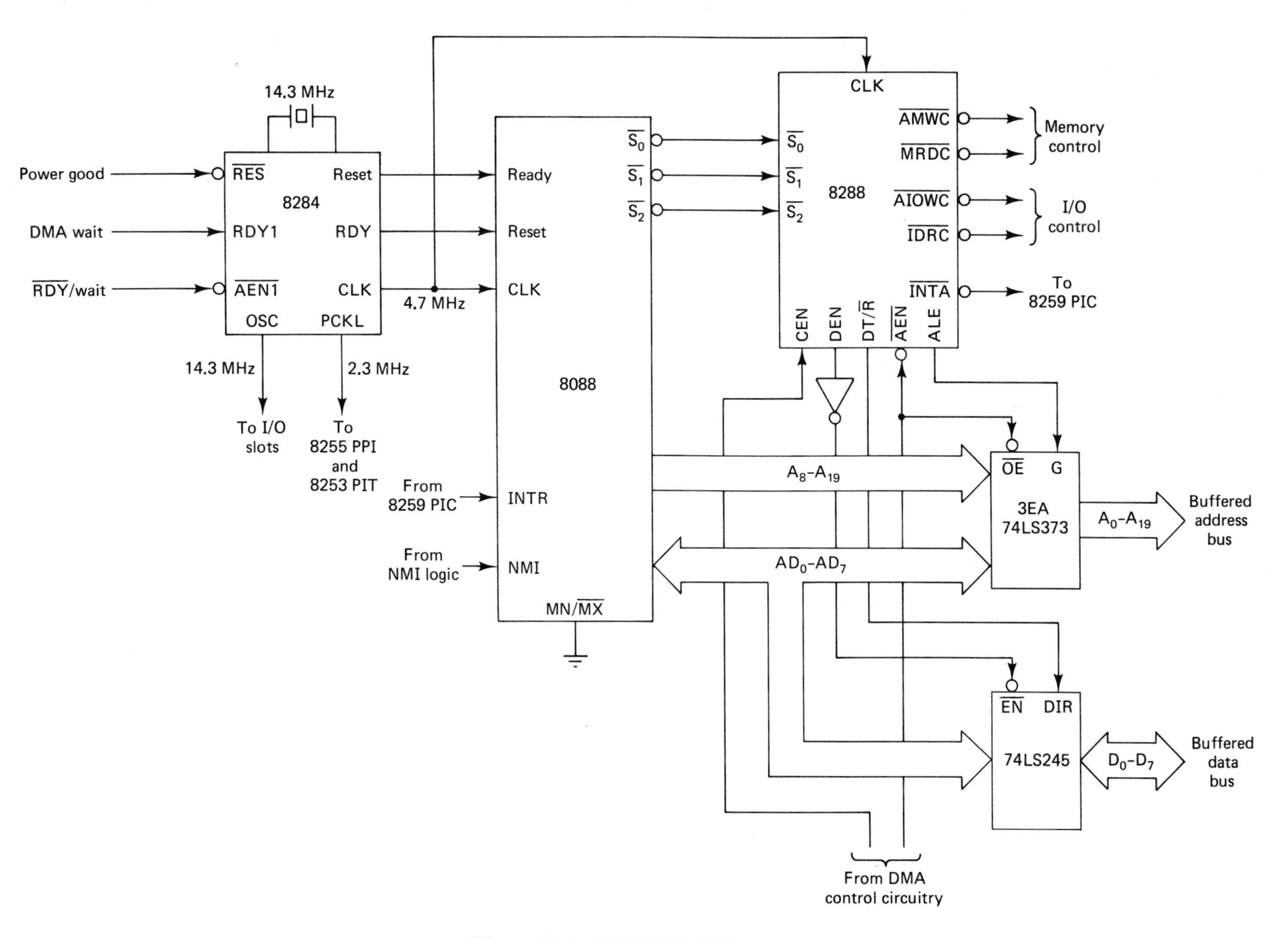

Figure 19.1 8088, 8284, 8288 system.

The $\overline{RES}$ is the system reset input. It is driven by a signal from the power supply called *power-good.* When power-good goes active high, the power supply is up and running and the 8284 brings the reset input on the 8088 to an inactive-low level. The power-good output of the PC's switching power supply negates the requirement of the classic RC reset delay that you will see in most microprocessor systems.

We will now examine the 8088 as it operates in maximum mode, as designated by the grounded MN/$\overline{MX}$ (minimum/maximum) input. The 8288 system controller is used to effectively expand the 40-pin limitation of the 8088. Its basic function is to input three encoded status outputs of the 8088 and generate memory and I/O control signals, and interrupt acknowledge pulses.

The 8088 has a 20-bit address bus. The first 8 bits are multiplexed with the 8-bit data bus. You may be wondering why the 8088 does not have an ALE output. In minimum mode the 8088 has a standard ALE output, but in maximum mode ALE is generated by the 8288 system controller. The multiplexed address/data bus on the 8088 functions exactly like the address/data bus of the 8085.

Notice the two familiar interrupts inputs, INTR and NMI. Pin limitations do not allow the 8088 to have any hardware restart lines. Recall that the mode 2 interrupt of the Z80 requires that the interrupting device provide an 8-bit address. This address is concatenated with 8 bits in the I register to form a 16-bit indirect pointer to a location in a look-up table that contains the effective address of the interrupt service routine.

The 8088 INTR input works in a similar manner to the mode 2 interrupt of the Z80. When a INTR occurs, the 8088 finishes its current instruction and pushes the contents of its instruction pointer (the 8088's equivalent of the program counter) onto the stack. The 8288 acknowledges the interrupt with an active level on $\overline{INTA}$. The interrupting device responds by driving the data bus with an 8-bit quantity. The 8088 then multiplies this byte by 4 to derive the indirect pointer to a location in the first 1K of memory that contains the effective address of the service routine.

When NMI interrupt goes active, the 8088 automatically vectors to a reserved location in the look-up table that contains the address of the NMI service routine.

In the miminum mode the 8088 employs the standard HOLD and HLDA pins of the 8085 to facilitate DMA operations. In maximum mode the 8088 uses pins called $\overline{RQ/GT}$ (bus request/grant) to provide local DMA. $\overline{RQ/GT0}$ and $\overline{RQ/GT1}$ are two bidirectional DMA request and acknowledge pins. The bus request is initiated by pulsing one of the $\overline{RQ/GT}$ pins low. The 8088 will provide the DMA ackowledge by pulsing the same pin low in a few clock cycles. In the PC system, $\overline{RQ/GT1}$ is used by the 8087 NDP for DMA requests, and $\overline{RQ/GT0}$ is tied inactive high.

Let's examine the inputs and outputs of the 8288 system controller. Advanced memory write command ($\overline{\text{AMWC}}$) and advanced I/O write command ($\overline{\text{AIOWC}}$) are the memory and I/O write commands. These write commands are derived from advanced status information. The standard control outputs of $\overline{\text{MWTC}}$ and $\overline{\text{IOWC}}$ are also available on the 8288. They more closely resemble the timing of the memory and I/O write commands that we examined with the 8085. Memory read command ($\overline{\text{MRDC}}$ and I/O read command ($\overline{\text{IORC}}$) are the same memory and I/O read commands that are generated by the 8085.

As we have already noted, the 8288 provides the interrupt acknowledge signal ($\overline{\text{INTA}}$). This signal differs from the 8085 because it consists of two active-low pulses. The first acknowledges the interrupt request and the second commands the interrupting device to place the 8-bit interrupt number onto the data bus.

CEN (command enable) is a control signal that is used to enable the memory and I/O commands outputs. When CEN is at an inactive-low level, all memory and I/O command outputs and DEN (data bus enable) go to their inactive levels.

$\overline{\text{AEN}}$ (address enable) is used in a similar manner to CEN. When $\overline{\text{AEN}}$ goes inactive-high, all of the command outputs go to high-Z. In the PC, CEN and $\overline{\text{AEN}}$ are driven by DMA circuitry. When a DMA request occurs, the DMA wait input to the 8284 clock generator goes to an inactive level, forcing the 8088 to generate wait states and the CEN and $\overline{\text{AEN}}$ lines go inactive, causing the memory and I/O control lines and the data and address bus buffers to float their outputs to high-Z. Thus the system is ready for an external bus master to take charge.

In addition to the standard memory and I/O read and write control signals, the 8288 provides three additional lines that simplify interfacing the 8088 with address and data bus buffers. Figure 19.1 indicates that the 20 address lines are buffered by three 74LS373 octal, three-state transparent latches. When $\overline{\text{AEN}}$ is active low, the Q outputs of the 74LS373s are enabled. ALE from the 8288 functions in the exact manner of the ALE signal of the 8085. On the falling edge of ALE, the 20-bit address is latched into the three 74LS373s. The 74LS373s provide five useful functions: They increase the drive capability of the 8088 address outputs, demultiplex the addresss/data bus, and latch the 20-bit address; they can be disabled to accommodate DMA accesses; and finally, they isolate any bus faults the unbuffered 8088 bus, which simplifies finding opens and shorts in the bus system.

The 74LS245 is an octal, three-state bidirectional buffer/driver. It accomplishes similar functions for the data bus that the 74LS373s accomplish for the address bus. Although bidirectional bus drivers do not need an ALE signal, they must be driven by a signal that indicates the direction of data flow. When the 8088 is reading memory or I/O devices, the 74LS245 must pass data from the buffered data bus into the microprocessor;

when the 8088 is writing data to a memory or I/O device, the 74LS245 must pass data from the microprocessor onto the buffered data bus. The direction control input of the 74LS245 is driven by DT/$\overline{R}$ (data transmit/$\overline{\text{receive}}$) of the 8288. When the 8088 is performing a read operation, DT/$\overline{R}$ is low; a write operation takes DT/$\overline{R}$ high.

DEN (data enable) from the 8288 is used to enable data bus drivers. When DEN is active high, its inverted level will enable the 74LS245. When it is inactive low, the 74LS245 will go to high-Z. Once again, notice that the 74LS245, like the three 74LS373s, can be driven into a high-impedance state to accommodate DMA accesses. When CEN is taken inactive low, the 8288 will bring DEN inactive low, causing the outputs of the 74LS245 to go to high-Z.

We have now covered the major aspects of the 8088, 8284, and 8288 three-chip CPU system. Like all microprocessor systems that we have seen in this book, the PC system board revolves around the simple concept of a three-bus architecture providing memory and I/O access, wait states, interrupt structures, and DMA access. No matter how complex a microcomputer system appears to be, those basic hardware functions must be implemented in some form.

In addition to the three-chip CPU system, the system board employs four other LSI support devices. We will briefly cover each device and the role it plays in the PC system.

19.1.2 The 8255 PPI (Programmable Peripheral Interface)

We studied the 8255 in Chapter 13. It contains three 8-bit programmable parallel ports and associated registers and control circuitry. The PC uses the 8255 to read the two sets of system configuration DIP switches, and the parallel 8-bit key code from the keyboard serial-to-parallel converter circuit; it drives the input of a device (8253 PIT) that controls the tone of the system board's small audio speaker; and it is also used to enable or disable various control clocks used throughout the system.

19.1.3 The 8259 PIC (Programmable Interrupt Controller)

In Chapter 14 we examined a classic priority-interrupt controller circuit based on the 74LS148 8-line-to-3-line priority encoder. Its usefulness is limited to generating software RST instructions as employed by the 8080-type interrupt. The 8088 requires that the interrupting device provide an 8-bit interrupt number, not a restart op-code. This interrupt service routine number can range from 00H to FFH. We need a programmable interrupt controller, a device that can be programmed to supply a particular interrupt number for each prioritized interrupt input. The 8259 PIC is such a device. It was originally developed to function in 8080/8085 systems. In those systems, instead of generating a simple one-byte RST op-code, it generated a programmed three-byte CALL instruction that greatly increased the interrupt-handling abilities of the 8080/8085.

In an 8088 system it is programmed to provide an 8-bit interrupt number for each of its eight prioritized interrupt inputs. Figure 19.2 is a functional pinout of the 8259 PIC as it is used on the PC system board.

The connection to the microprocessor data bus enables the 8259 to be programmed with command words, and provide status information and interrupt vectors to the 8088. All LSI devices that we will study in this chapter are I/O mapped. The $\overline{\text{CS}}$, $\overline{\text{WR}}$, and $\overline{\text{RD}}$ inputs are used to program the 8259 and read its status registers. A0 is used as an address bit to choose particular command and status registers in the 8259.

Interrupt request inputs IR0 through IR7 are the eight prioritized interrupt inputs. IR0 has the highest priority and IR7 the lowest priority. The 8259 also has four pins (which are not illustrated) that are used to cascade up to eight 8259s, providing 64 levels of prioritized interrupts.

INT is used to drive the INTR request input of the 8088. The 8288's $\overline{\text{INTA}}$ output drives the $\overline{\text{INTA}}$ input of the 8259. $\overline{\text{EN}}$ is usually an active-low signal that is used to enable the data bus buffers to allow the interrupt vector to reach the data inputs of the microprocessor. In the PC the 8259 PIC is connected to the nonbuffered address/data bus; therefore, $\overline{\text{EN}}$ is used to disable the 74LS245 data bus transceiver. This ensures that no bus contention will occur between the 8259 and memory when the interrupt vector is placed onto the 8088's data bus.

From a hardware point of view, the 8259 PIC is a simple device that does not greatly differ from the interrupt controller

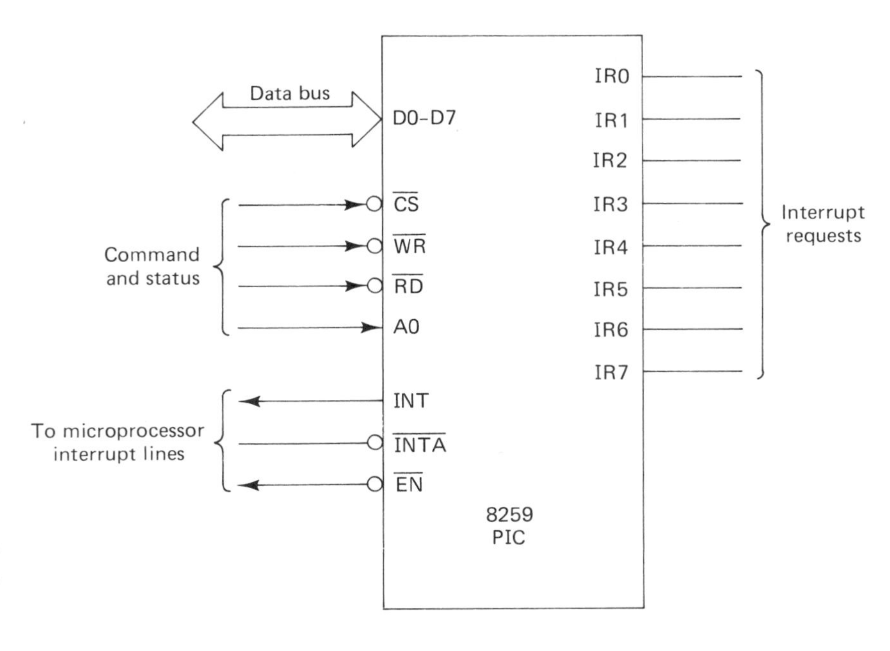

Figure 19.2 8259 PIC—programmable interrupt controller.

in Chapter 14. From a software point of view, the 8259 can be a complicated device, offering many methods and modes of interrupt-handling services.

The PC employs the 8259 prioritized interrupts in the following manner. IR0 is driven by a time-of-day (TOD) clock that is used to update an internal time-keeping memory location. IR1 is driven by the keyboard interrupt circuitry. Each time a key is depressed, IR1 goes active to alert the 8088 to read the keyboard input buffer. IR2 through IR7 are bused to the five expansion slots. Because the usage of the six extra interrupts are not explicitly defined, the technician must be sure that no more than one expansion board is trying to use the same interrupt. The interrupts that each expansion board uses are documented in manufacturers' user manuals.

19.1.4 The 8253 PIT (Programmable Interval Timer)

We dedicated a chapter to the study of MSI counters. MSI counters are internally constructed from J-K flip-flops that are configured to toggle on each active edge of the clock. These counters have many Q outputs that are decoded to derive the present count state.

In the context of MSI counters the term "programmable" implied that a starting count could be jammed into the Q outputs on the active level of the load input. In that usage, the terms "programmable" and "presettable" are equivalent. We have seen LSI support devices that are also described as "programmable." As it applies to LSI devices, "programmable" implies much greater power and flexibility then their MSI counterparts. Flexibility and power are the key words. Flexibility means that a particular device can be programmed to function in many different modes, satisfying semicustom requirements. Powerful implies that the device is capable of performing moderately complex tasks without the constant attention and intervention of the system processor.

The 8253 is a programmable interval timer. A PIT can be used to perform many useful tasks. The 8253 is used to generate precise timing delays, count external events, generate baud rate clocks for UARTs, and function as a one-shot. After we examine the functional pinout of the 8253 in Figure 19.3, we will see how the PC utilizes this flexible device.

You should expect to see the familiar data bus connection and control inputs. A0 and A1 are used to select from four internal programmable control, mode, and status registers. The other lines in Figure 19.3 form the clock interface inputs and outputs. The 8253 contains three identical 16-bit negative-edge-triggered, down counters. Each counter is associated with three pins. The clock is the standard negative-edge-triggered clock.

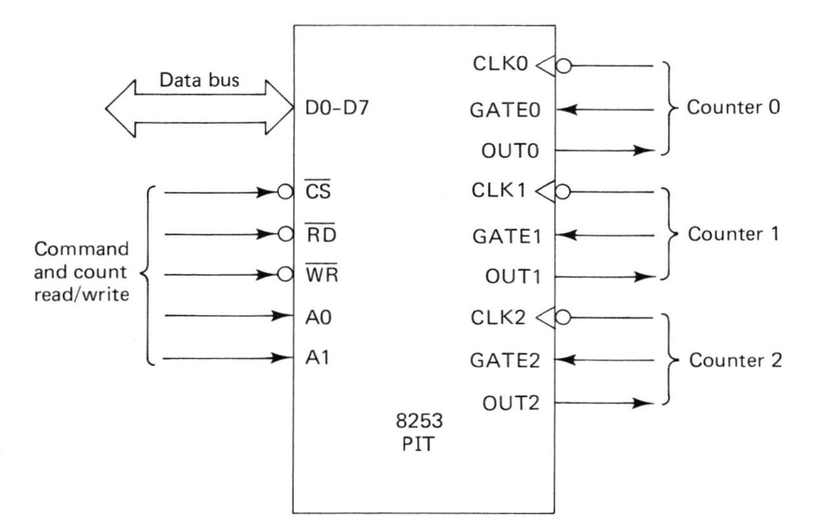

Figure 19.3 8253—programmable interval timer.

On the falling edge of each clock pulse the programmed contents of the counter is decremented. The gate input acts as a counter enable signal. If the gate is active high, the counter is enabled. When the gate is taken to a logic 0, the counter is disabled and will not react to any falling edges on the clock input. Finally, there is one output for each counter.

It may seem strange that a 16-bit counter has only one output. If we constructed a 16-bit counter from MSI devices, each Q output would be accessible. The outputs could then be decoded with external logic to produce the desired output signal. When devices become programmable, hardware chores are handled quite easily by software. Essentially that means that the 8253 can be programmed to produce many different types of output signals. In this manner, external decoding hardware is displaced by software. Each counter can be programmed to perform in one of six modes. These output modes entail active-low terminal pulses, one-shots, symmetrical square waves, and software- or hardware-triggered strobes.

The three counter clock inputs on the 8253 are driven by the output of a D-type flip-flop that divides the 8284's 2.38-MHz PCLK by 2. Because this 1.19-MHz signal is derived from a 14.318-MHz crystal, it can be used to provide an extremely stable and precise time reference.

How does the PC employ these three channels of programmable timers? The output of timer 0 drives the IR0 input of the 8257 PIC. This highest-priority input is used to update the 8088's time-of-day (TOD) clock. The output of timer 1 starts a DRAM refresh cycle which involves the 8237 DMA controller. The output of timer 2 controls the speaker driver circuitry. By placing the appropriate values into counter 2's count register and using one output of the 8255 PPI to drive the gate input, many different tones and sound effects can be generated under program control.

19.1.5 The 8237 DMAC (Direct Memory Access Controller) and Programed I/O

In many places in this book we have discussed the concept of DMA. Every microprocessor that we have examined has had DMA request and DMA acknowledge pins. To fully understand the need for DMA, we will examine the transfer of a block of data from a floppy disk drive into the system's main memory exclusively under software control.

Refer to the programmed I/O (PIO) block diagram illustrated in Figure 19.4a. The FDC (floppy disk controller) is an LSI device that interfaces a maximum of four floppy disk drives with the microcomputer system. In the PC, the FDC is contained on an expansion board called the *floppy disk drive adapter.* The microprocessor programs the FDC by writing command and mode bytes into its internal registers. Like all peripheral interface devices that we have examined, the FDC is I/O mapped and responds to $\overline{\text{IOR}}$ and $\overline{\text{IOW}}$ commands.

The flowchart is Figure 19.4b indicates the program steps that a microprocessor must perform to move a sector of data from a floppy disk into RAM. Let's examine each block in Figure 19.4b.

Block 1. The HL register is loaded with the base address of RAM where the sector of data is to be stored.

Block 2. The BC register is loaded with the number of data bytes that will be transferred. Recall from Chapter 1 that a $5\frac{1}{4}$-in. double-density floppy disk contains 512 bytes of data per sector.

Block 3. The microprocessor must command the FDC to read a sector of data. This command typically specifies the track number, sector number, and the side of the floppy to be read.

Block 4. The microprocessor may poll the FDC's status register or wait for an interrupt that indicates that a byte of data is ready to be read.

Block 5. The microprocessor will read the FDC's data register with a standard I/O read (IN port) command.

Block 6. The byte must now be transferred from the accumulator of the microprocessor into the appropriate RAM location. This can easily be accomplished by using register indirect addressing via the HL register pair.

Block 7. The destination RAM pointer must be incremented to point to the memory location that will receive the next byte of floppy disk data.

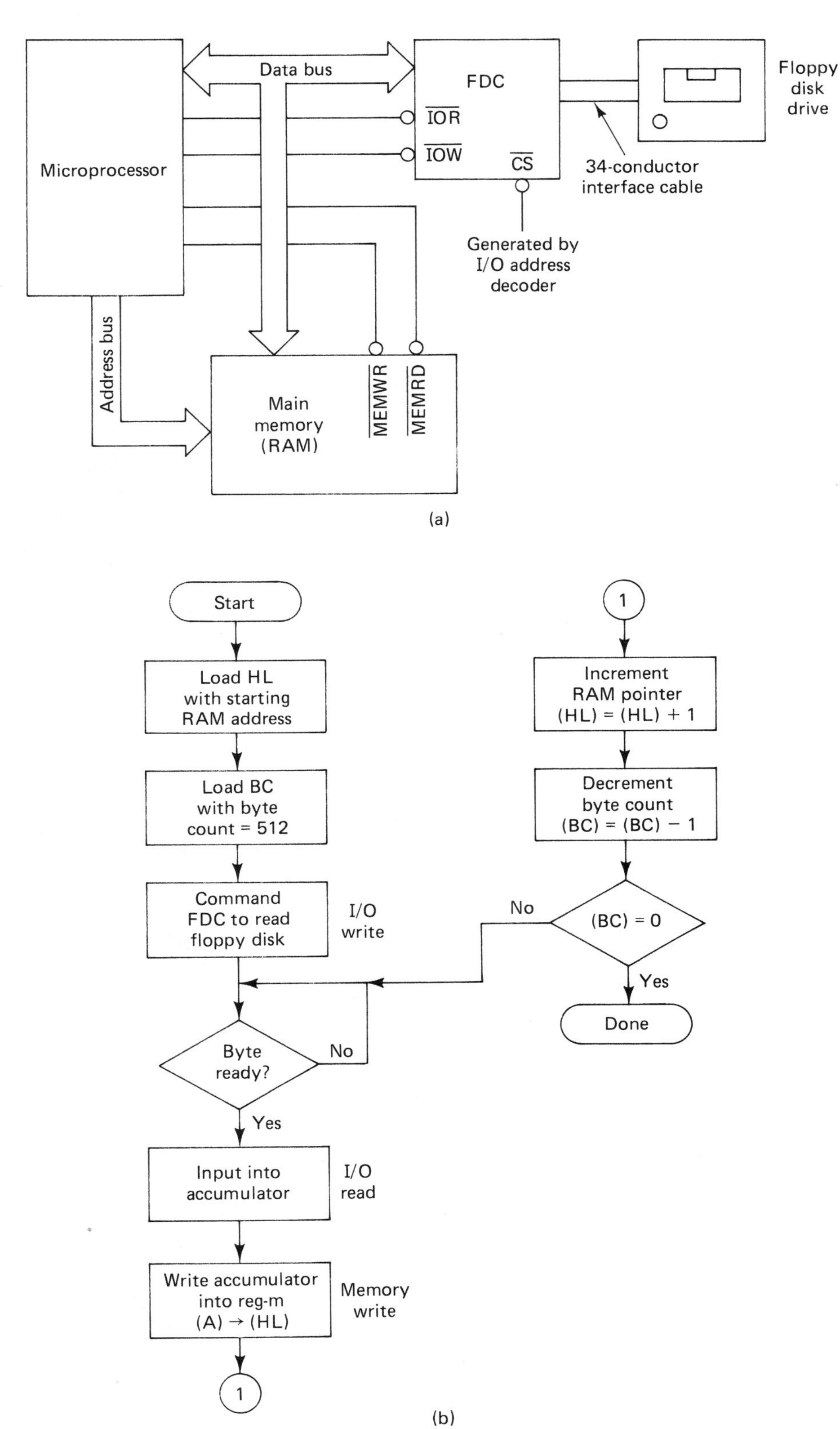

Figure 19.4 Programmed I/O (PIO): (a) system; (b) flow chart.

Block 8. The byte count must be decremented.

Block 9. The zero flag is checked to see if the complete sector has been transferred. If the BC is a nonzero value, the program will conditionally jump back to block 4. Here it will wait for another byte of data to become available.

Programmed I/O is a software-intensive operation. It requires that the microprocessor execute many instructions to transfer each byte of data. Notice that the FDC-to-RAM transfer is an indirect operation. The data byte is temporarily saved in the accumulator after the FDC data read instruction and then it must be written into memory. Although PIO is inexpensive because it relies exclusively on software, that software dependence makes PIO unbearably slow. The biggest bottleneck in microcomputer systems is disk I/O. When a large program or block of data must be written onto or read from disk storage, this process is extemely time consuming compared to the normal program execution.

An alternative to the software-intensive operation of PIO is DMA. DMA operations require a DMAC (direct memory access controller). The DMAC is a sophisticated, programmable LSI device that is capable of taking control of the system's data, address, and control buses. It is specifically designed to transfer large blocks of data between disk storage and system RAM.

Unlike a microprocessor, which relies on instructions that reside in ROM or RAM, the DMAC does not need to fetch external instructions. It accomplishes in hardware and internal microprograms what the microprocessor accomplishes in PIO through standard stored program execution. In addition to not fetching instructions from external memory, the DMAC speeds up disk-to-memory transfers by directly transferring a byte of data from the FDC to RAM (or vice versa). No temporary storage is needed, as occurred in programmed I/O.

The DMAC is another example of a specialized high-performance LSI device that is used to relive a difficult and time-consuming task from the microprocessor, with the effect of greatly increasing disk I/O throughput.

The DMAC employed in the PC is the 8237. The 8237 is a complex and powerful IC. It contains four independent DMA channels and 344 bits of RAM organized as internal 8- and 16-bit registers. The initial programming of the 8237 is mildly complex, but once programmed the 8237 operates independently of any further processor intervention. Figure 19.5 illustrates the functional pinout of the 8237. Because the DMAC can become the master of the system buses, it will have many similarities to a microprocessor: a bidirectional data bus, a 16-bit address output, and memory and I/O control signals.

Figure 19.5 indicates that A0 through A3 are bidirectional. As inputs, they are used to select one of the 16 registers in the

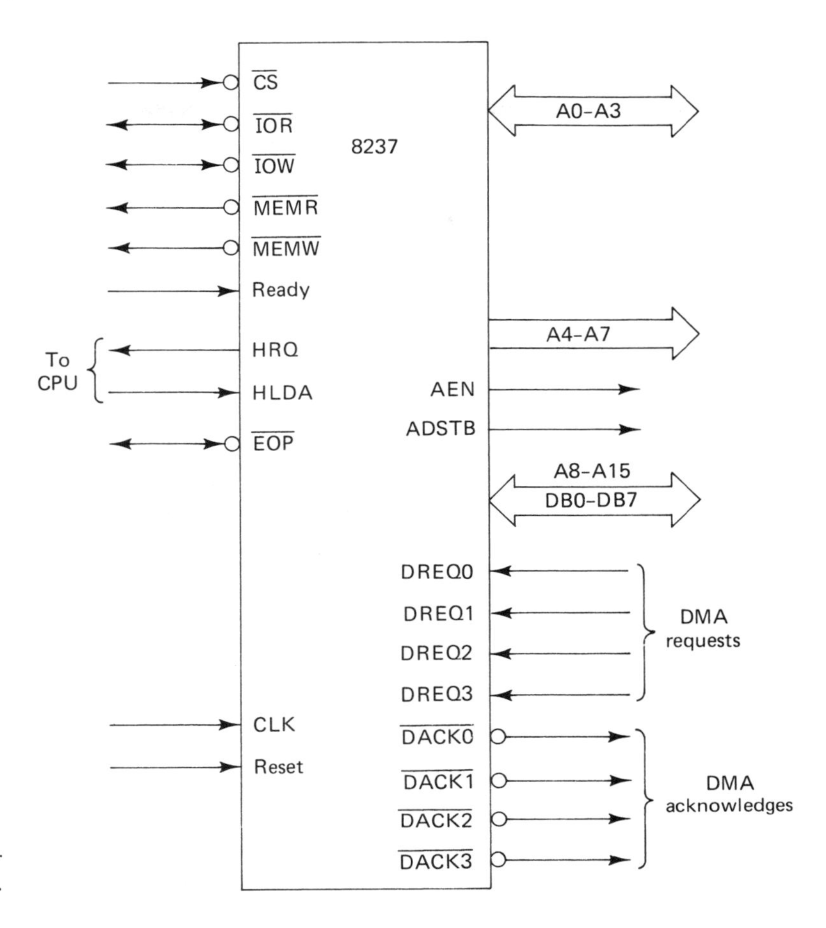

Figure 19.5 8237 direct-memory-access controller.

8237. As outputs, they provide the low nibble of a 16-bit memory address. Address bits A4 through A7 provide the high nibble of the low address byte. Unlike the 8085 and 8088 that multiplex the data bus with the low address byte, the 8237 multiplexes the data bus with the high byte of the 16-bit address.

A multiplexed address/data bus requires an address latch enable signal. ADSTB (address strobe) provides a demultiplexing signal for the 8237; ADSTB works in a similar fashion to ALE in the 8085 and 8088. The output AEN (address enable) is used to enable the DMAC's bus drivers and disable the microprocessor's bus drivers during the DMA cycle, as illustrated in Figure 19.1.

Now that we have established that the 8237 has a bidirectional data bus and can output a 16-bit address, we must turn our attention to the control bus. The $\overline{\text{MEMR}}$ and $\overline{\text{MEMW}}$ lines are outputs that function exactly as their counterparts in a microprocessor; the 8237 asserts $\overline{\text{MEMR}}$ during a memory-read operation and $\overline{\text{MEMW}}$ during a memory-write operation.

You may have noticed that the $\overline{\text{IOR}}$ (I/O read) and $\overline{\text{IOW}}$ (I/O write) pins are bidirectional. The 8237 DMAC must be initially programmed by the system microprocessor. This is accomplished by selecting the appropriate 8237 register with A0 through A3, and pulling $\overline{\text{CS}}$ and $\overline{\text{IOW}}$ active-low. The 8237 also has status registers that can be read by the system microprocessor by selecting the register with A0 through A3 and pulling $\overline{\text{CS}}$ and $\overline{\text{IOR}}$ active-low. In the write command and read status operations, the $\overline{\text{IOR}}$ and $\overline{\text{IOW}}$ pins are used as inputs. When the DMAC transfers bytes of data between memory and disk, the $\overline{\text{IOR}}$ and $\overline{\text{IOW}}$ lines become outputs.

A device that accesses memory and I/O must have a means of accommodating slow devices. Like every microprocessor we have seen, the 8237 has a ready input. If the ready line is taken low during a memory access, the 8237 will generate standard wait states.

The $\overline{\text{EOP}}$ (end of process) pin is a bidirectional line that is used to indicate when a DMA block transfer is complete. An external device can drive $\overline{\text{EOP}}$ to terminate a DMA transfer, or the 8237 will pull it active low after its word count is decremented to 0. CLK and Reset serve the same functions on a DMAC as they do on every other LSI device that we have studied.

Now we can consider how the 8237 handles DMA requests. The 8237 is capable of managing four independent DMA channels. There are four DMA request inputs (DREQ) and four DMA acknowledge outputs ($\overline{\text{DACK}}$). To add flexibility to the 8237, the active level of the DREQ and DACK lines are programmable as active-high or active-low signals. The HRQ (hold request) and HLDA (hold acknowledge) lines are used to interface the DMAC with the system processor's DMA request and DMA acknowledge lines.

DMACs are complex devices. The only way to truly understand how they function is to examine a typical DMA transfer. Consider the case when a block of data residing in RAM must be transferred onto a floppy disk. Assume that the 8237 DMAC has already been initialized by the microprocessor.

The floppy disk controller chip will request a DMA operation via one of the four DREQ inputs. The 8237 will respond by asserting its HRQ output. HDQ usually drives the DMA request input of the system microprocessor. After the microprocessor has finished its current instruction, it will bring its address, data, and control buses to high-Z and issue a DMA acknowledge that usually drives the HLDA input of the 8237. When HLDA goes active high, the 8237 is informed that it has control of the system's buses. The acknowledge is passed onto the FDC via the proper $\overline{\text{DACK}}$ output, and the DMA transfer will begin.

The 8237 can operate in many different modes. Assume that it is programmed to transfer a block of data starting at a given RAM location onto a particular track, sector, and side of the floppy disk.

The 8237 has an internal 16-bit address register and a 16-bit word count register for each channel. After each transfer the address register is automatically incremented and the byte count register is automatically decremented. The 8237 places the 16-bit address onto the address bus and simultaneously asserts $\overline{\text{MEMR}}$ and $\overline{\text{IOW}}$. The data byte read from the specified RAM location is written directly into the data register of the FDC. This results in an extremely fast transfer rate. The memory-read/IO-write process will continue until the word count register is decremented to 0. The $\overline{\text{EOP}}$ line will go low to signal the completion of the floppy disk store operation.

There is one major conflict involving the use of the 8237 in the 8088-based PC system. The 8088 has a 20-bit address bus, and the 8237 is capable of supplying only a 16-bit address. The solution to the problem is to have another device supply the most significant 4 bits of address. This is accomplished by a 74LS670 4 × 4 register file. You should think of the 74LS670 as a high-speed 4 × 4 static RAM. Each 4-bit "word" is used to provide the extra address bits for the four DMA channels of the 8237. The 8088 must write the high-order 4 bits of address before the DMA transfer begins. If the 8237 exceeds its 64K addressing space, the 8088 must interrupt the DMA process and increment the 4-bit extension word in the 74LS670.

DMA channels 1, 2, and 3 are run to the expansion slots. The PC's floppy disk control board uses DMA channel 2 for floppy disk I/O, while channel 3 is usually employed by hard disk systems. Channel 1 is available for other uses, such as networking PCs.

Channel 0 of the 8237 has a most important function in the PC system. It is used in conjunction with channel 1 of 8253 PIT to provide DRAM refresh.

Before examining the PC refresh scheme, let's take a moment to review the basic concepts of DRAM refresh. The address of a particular location in a DRAM is composed of an equal number of row and column address bits. A normal read or write operation latches the row address with the $\overline{\text{RAS}}$ signal, and then latches the column address with the $\overline{\text{CAS}}$ signal. The 64K × 1 DRAMs used in the PC are constructed from 256 rows and 256 columns of memory cells. Each row or column requires an 8-bit address. A row of memory cells in a DRAM is refreshed by reading any cell in the row. A refresh-only cycle consists of placing a row address on the address inputs of the DRAM and strobing $\overline{\text{RAS}}$. This refreshes every memory cell in the specified row.

If one could be sure that at least one memory cell in each row of every DRAM was accessed at least once every 4 ms, an external refresh operation would not be required. That concept is possible for a high-speed refresh memory in a video display board, but main memory in a microcomputer system is not accessed in such a uniform and predictable manner.

The PC refresh scheme is simple and effective. For 256 rows of DRAM to be refreshed a minimum of once every 4 ms, the 8088 programs channel 1 of the 8253 PIT to output a negative pulse every 15.2 μs. The rising edge of this pulse sets the Q output of a flip-flop. The output of the flip-flop is connected to DREQ0 of the 8237 DMAC. The ready input to the 8284 is taken to an inactive level, forcing the 8088 to idle by generating wait states, and the memory/IO command outputs of the 8288 and the address and data bus buffers are taken to high-Z. The system's address, data, and control buses can now be manipulated by the 8237.

The 8237 outputs a row address and strobes $\overline{RAS}$. Through a special refresh gate, the $\overline{RAS}$ lines on all the banks of DRAM are strobed and the selected row in each DRAM is refreshed. The 8237 increments the refresh counter register and returns control of the buses to the 8088. Notice that the 8237 is operating in a single byte mode during refresh, not a block mode as we have seen in disk transfers.

If a row of DRAM is refreshed every 15.2 μs, all 256 rows will be refreshed within 4 ms. This guarantees that main memory will not lose any data due to a missed refresh operation. Channel 0 of the 8237 was selected because it is the highest-priority DMA request. That means that the refresh operation will take priority over any floppy or hard-disk transfer operation that may be in progress via DREQ2 or DREQ3.

19.2 BLOCK DIAGRAM OF THE PC SYSTEM BOARD

Now that we have examined each major building block on the PC system board, it is time to see how these devices are connected to produce a working system. Figure 19.6 illustrates a block diagram of the PC system board. This figure does not illustrate the control bus, however, as the control bus would only add an extra layer of confusion to an already complex system. Be assured that many gates and decoders are used to generate the memory control signals, I/O chip selects, and other timing and qualification signals.

The CPU group includes the 8088, optional 8087, 8284, 8288, 8259 PIC, and a wait-state generator. The wait-state generator is a group of gates and flip-flops that are used to generate the ready input to the 8284 clock generator, various subsystem clocks, and DMA handshake signals. Notice that the 8259 PIC is connected to the local (unbuffered) address/data bus.

Three 74LS373 octal three-state transparent latches are used to buffer, latch, and demultiplex the address/data bus. A single 74LS245 octal three-state bus transceiver is used to buffer the 8-bit data bus. The buffered data and address buses are connected directly to the pins on the five I/O expansion connectors. All expansion boards will buffer the data and address buses. This practice of isolating buses ensures that the address and data bus buffers on the system board will not be excessively loaded when the I/O slots are filled with expansion boards.

The next group of devices is the system ROM (which is called ROS—read-only storage—by IBM), and the 8255 PPI and 8253 PIT. The address and data buses that are local to the ROM,

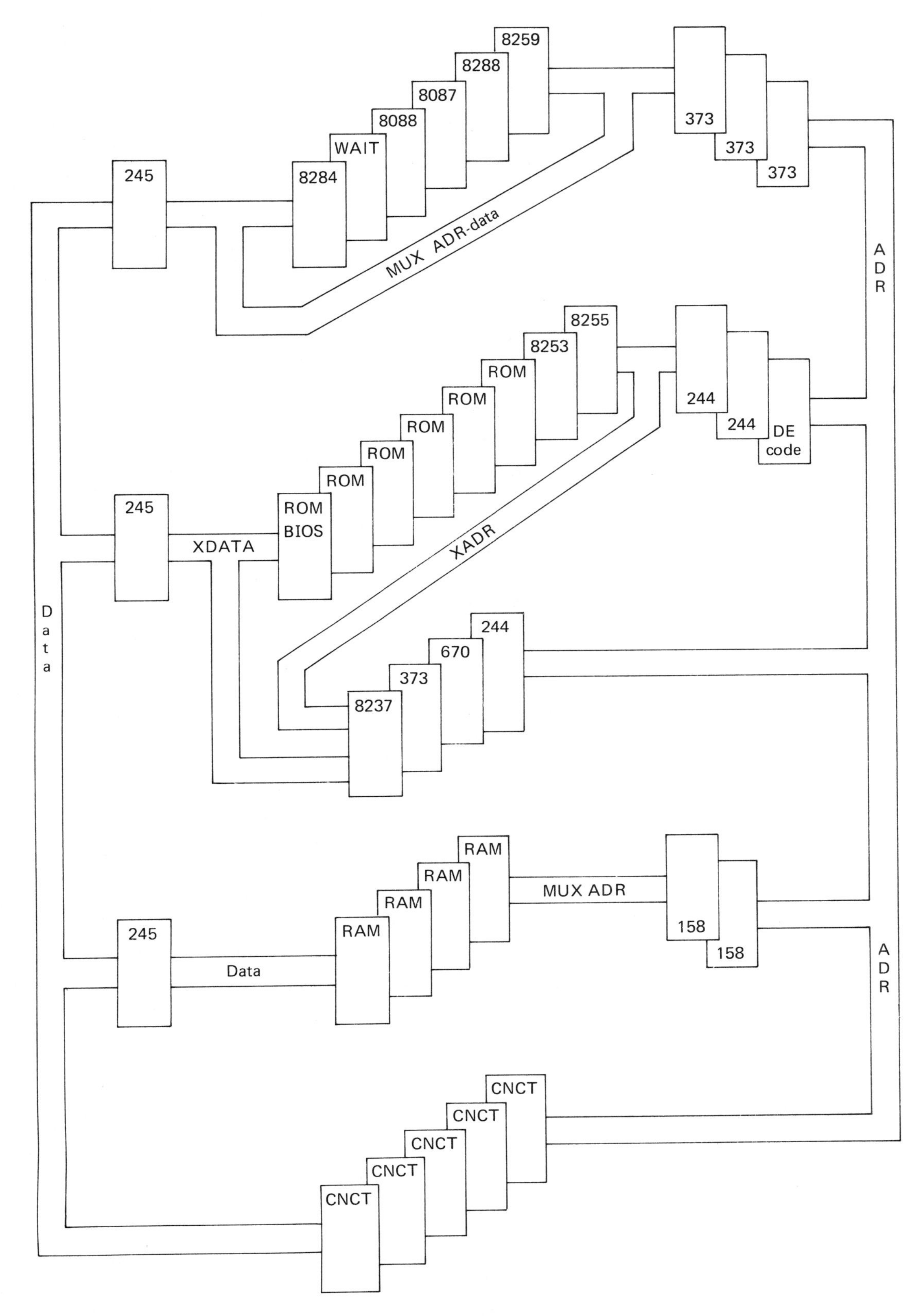

Figure 19.6 IBM-PC buses. (Courtesy of Dair Computers.)

PPI, and PIT subsystem are called external data and external address buses. The external address and data buses are isolated from the buffered buses that are run to the I/O connectors.

The DMA group provides some interesting circuitry. Remember that the 8237 DMAC can output only 16 bits of address, and that the upper byte of the 8237 is multiplexed with the data bus. When the 8237 has control of the system's buses during a DMA operation, the low 16 bits of memory address is provided by the 8237. The 74LS244 buffers the low address byte, and the 74LS373 octal latch is used to demultiplex the high address byte from the data bus. The most significant 4 bits of address are provided by the latched Q outputs of the 74LS670 4 × 4 register file. This 20-bit DMA address is output onto the buffered address bus.

The final group to examine in Figure 19.6 is the DRAM subsystem. It consists of four banks of nine 64K DRAMS. This provides the system board with a total of 256K bytes of parity-protected RAM. The 74LS158s multiplex the row and column addresses, and SSI/MSI devices are used to generate the $\overline{\text{RAS0}}$ through $\overline{\text{RAS3}}$ and $\overline{\text{CAS0}}$ through $\overline{\text{CAS3}}$ DRAM control signals.

The four banks of DRAM on the system board must be fully stuffed before memory expansion boards can be added to the PC system. Because more than one memory expansion board may be added to the system (up to a total of 640K), a means of identifying the range of addresses in which each board is active must be provided. This is accomplished by setting DIP switches on each memory expansion board. These DIP switches indicate the base address at which the expansion RAM will become active and the total number of bytes of RAM on the expansion board. The memory expansion board contains the actual DRAMs, row and column address multiplexers, buffers, and $\overline{\text{RAS}}/\overline{\text{CAS}}$ signal generation circuitry.

When memory expansion boards are added to the PC system, switches on the system board must be set to indicate the total system RAM. On newer versions of the PC, these switches are no longer used, and the total system RAM is calculated by a program during the power-up sequence.

19.3 TROUBLESHOOTING MICROCOMPUTERS

How does a technician troubleshoot a microcomputer system? The lowest level of troubleshooting is finding the bad circuit board or subassembly and swapping it with a new one. The highest level of troubleshooting is to repair a complete computer system to the component level.

The level of troubleshooting that a technician accomplishes is governed by many factors. Troubleshooting only to the bad board or subassembly level is a fairly easy and quick task. Thus the initial cost of technician's labor is low, and the computer is back on the job in a minimal length of time. The drawbacks of this shallow depth of troubleshooting are obvious:

a large and expensive inventory of computer boards, power suppliers, video monitors, keyboards, and disk drives must be kept. The bad subassembly must be sent to a repair location, which may take from two weeks to two months to be repaired and returned. The added costs of an extensive subassembly inventory, the services of the repair depot, and the freight charges may neutralize the advantages of troubleshooting only to the bad board level.

On the other hand, troubleshooting to the component level is a complex, time-consuming task. It requires an inventory of TTL SSI and MSI devices, standard LSI devices, custom ROMs, and other expensive proprietary parts. To efficiently troubleshoot microcomputers to the component level, expensive test equipment or extensive diagnostic programs must be utilized. The technician who troubleshoots to the component level must be highly skilled in the use of diagnostic programs and complex test equipment. This technician should command a much greater wage than the "board-swap tech."

The major advantages of this form of troubleshooting are that the control of the bad unit is not passed onto another repair depot or company. The cost of sending computer boards to and from a repair depot is an added expense. Furthermore, computer boards are often delayed, lost, or damaged in transit. Technicians also prefer a high-growth environment where they can be challenged by component-level repairs.

Most companies compromise between the two extreme levels of repair. They learn to fix relatively common component-level faults and pass the more difficult problems onto repair depots or to manufacturers.

Let's consider the typical PC repair environment. Two publications by IBM are required to perform any serious level of PC repair. The *Technical Reference Manual* describes the PC hardware in depth and supplies schematics of the system board and common IBM expansion boards. The *Hardware Maintenance Manual* provides tedious, detailed step-by-step procedures to locate bad subassemblies, important mechanical illustrations, parts inventory lists, and a floppy disk containing programs to test and exercise all the major systems and subsystems in the PC.

19.3.1 The POST: Power-On Self-Test

Almost every microprocessor based system contains a ROM-based self-test. These tests are automatically invoked each time the instrument is powered on. Typical parameters that self-tests examine are the ability of the RAM subsystem to store data, ROM check sums, and the read/write ability of registers in LSI support devices.

Self-tests are an important way to ensure the integrity of a system each time it is powered-on. If the self-test fails, the user is aware that the device requires maintenance and must not be used at the risk of malfunction and perhaps loss of important data. Various systems have different ways of indicating the results of a self-test. Some merely beep once for pass and twice for failure. Others display extensive fault messages on the system display.

A microprocessor-based instrument that does not have mass storage may also have extensive ROM-based diagnostic programs that are designed to exercise circuitry and isolate faults. Microcomputers provide complex, comprehensive diagnostics that are loaded into RAM from a floppy disk. These diagnostic programs are divided into tests that exercise a specific portion of the microcomputer's circuitry.

The important point here is: If a microcomputer system has a major circuit fault, such as a shorted data bus, it will not be able to run either the power-on self tests or the floppy-based diagnostics. These programs depend on the processor's ability to execute the self-test program instruction from ROM. Furthermore, if the self-test uses subroutines, we know that the return address must be pushed onto the stack. That implies that the RAM subsystem must also be functional or the self-test will pop an incorrect return address from the stack, and vector off to a random memory location to fetch the next instruction, crashing the system.

We can now derive the order in which a microcomputer should be tested. If a microcomputer has an internal cooling fan, it should begin to rotate after the ac power switch is turned on, which indicates that the machine is plugged into a functional power socket.

The most important (and most commonly forgotten) check is the power supply. Verify, preferably with an oscilloscope, that the dc voltages are within specified tolerances and have no appreciable switching noise. The switching noise is the product of a standard switching power supply. The more expensive and less efficient older-technology full-wave/filtered power supplies are usually employed only to power analog circuitry that requires much cleaner dc voltages than a switching power supply can produce. Remember to check dc voltages a the pins of the ICs, not on the output of the power supply. Bad connectors and intermittent wire/lug crimps are a common failure. Checking voltages at the ICs ensures the integrity of the power connectors and bus traces.

The next step is to check for valid system clocks. Microcomputer systems have many different clocks: processor clocks, peripheral clocks, and baud-rate clocks. All activity in the microprocessor system is synchronized to the clocks.

After the power and clocks have been verified, the address, data, and control buses must be checked for shorts or opens. The address and data bus traces on a PCB run in close proximity. It is easy to have two address bits or data bits shorted together.

Remember that a microprocessor system has many address and data bus buffers, so the integrity of all the local buses must also be ensured.

Assuming that no critical inputs, like a DMA request, wait input, or interrupt input are shorted to an active level, the processor should be up and running. If the correct chip select for the self-test ROM is generated, and the memory control lines are functional, the microprocessor should be able to perform the power-on self test.

Figure 19.7 describes the 29 power-on self tests contained in the PC's BIOS ROM. Under the second column is a brief description of the test. The third column lists the PC's indication that the POST is running correctly. The last column is of special interest. It indicates the PC's response to a test failure. The only external indication the PC will provide is beeping and displaying an error code.

Each subsystem in the PC has a unique error prefix. A DRAM error is indicated by the code "201." If the DRAM fault is a single 64K device, not a buffer or decode problem, the location of the bad chip can be easily derived from the "201" address error. This is one fix that is performed by technicians at all levels of repair.

A keyboard failure is indicated by the code "301." A known-good keyboard can easily be swapped into the system to verify the failure. The keyboard contains little repairable circuitry. The 8048 microcomputer chip in the keyboard can be swapped, but the most common fault is a bad connection at either the keyboard or PC end of the 6-ft coiled keyboard cable. The connections should be thoroughly checked for continuity with an ohmmeter.

To check the integrity of RAM the progam writes patterns into the RAM and reads them back. These patterns are extremely complex. They are designed to check for shorted bits, stuck bits, page boundary cross-checks, and other common errors. Because a ROM is read-only, how is it verified? When a ROM is constructed the contents of each memory location are added together to create a unique 16-bit quantity called a *ROM check sum.* To verify system ROM, the self-test reads all the bytes from the ROM and calculates the check sum. It then compares the calculated check sum with the check sum that is contained in the last two bytes of the ROM under test. If the two check sums do not match, the ROM is reported as bad.

Test 1AH tests the floppy disk drive interface and returns the error code "601" if a malfunction occurs. DRAM parity errors are reported to the 8088 via the NMI. Test 1CH enables external NMI enable circuitry. If a system board DRAM parity check error occurred, the PC will display "Parity Check 1." "Parity Check 2" indicates a parity error on a memory expansion board.

If the PC has passed the entire self-test, it will beep once

Test	Description	Working Indication	Failure Indication
1	8088 Internals		Halts at FE0AD
2	IO Initialization	Speaker clicks	
3	Verify ROM BIOS checksum		Halts at FE0AD
4	Verify 8253 Timer 1		Halts at FE0EE or FE103
5	Verify 8237 DMA		Halts at FE123
6	Initialize DMA		
7	Verify first 16K of RAM		Halts at FE164
8	Zero all RAM		
9	Initialize 8259 Interrupt Controller		
A	Read manufacturing test from keyboard		
B	Set temporary interrupt vectors		
C	Verify 8259 Interrupt Controller		Beeps long-short, halts at FE23E
D	Verify 8253 Timer		Beeps long-short, halts at FE23E
E	Set BIOS interrupt vectors		
F	Blink LED if manufacturing mode		
10	Initialize Video RAM		
11	Verify Video RAM		Beeps long-short-short
12	Verify horizontal synch timing		Beeps long-short-short
13	Scan Video ROM C0000-C7FFF	Screen clears	Beeps long-short-short
14	Verify expansion chassis		Displays "1801"
15	Verify RAM past first 16K		Displays "AA WW!!RR 201" where AA000=address, WW=byte written !!=exclusive OR, RR=byte read
16	Keyboard test		Displays "SS 301" where SS=scan code
17	Initialize Interrupt vectors		
18	Test cassette interface		Displays "131"
19	Verify and call ROMs C8000-F5FFF and BASIC at F6000-FDFFF		Displays "AAAA ROM" where AAAA0=address
1A	Test diskette interface	Motor starts	Displays "601"
1B	Init. parallel, serial, game interfaces		
1C	Enable NMI (non-maskable interrupt)		Displays "PARITY CHECK 1" (system board) or "PARITY CHECK 2" (plugin card) and/or halts at FE2DE
1D	Boot from diskette	Short beep, DOS loads	Short beep, displays "CAN'T LOAD"

The sequence of operations performed in the BIOS prior to loading DOS from disc is listed. The column "Working Indication" shows the observable result if that test passes. The "Failure Indication" shows the result of a detected error. If the BIOS halts, the halt address for ROM 1501476 is shown. Other releases perform much the same tests in much the same sequence but the halt addresses will differ from release to release. Consult the Technical Reference manual.

Figure 19.7 IBM-PC power-on self-test (POST).

and load DOS (the Disk Operating System) from the floppy in drive A. If the floppy does not contain DOS, it will notify the user to insert the correct diskette.

The PC's POST is an impressive group of checks, but to the computer service technician it can be extremely frustrating. If a fault occurs on the POST, the PC will give the indicated response and then "die." This indicated response is sufficient to troubleshoot simple DRAM and ROM errors and isolate video board and floppy disk controller board errors, but it does not enhance the technician's ability to troubleshoot system board faults. Notice that tests 1 through 12H simply halt or have an uninformative fault indication such as a long–short or long–short–short beep.

19.3.2 Microprocessor Emulators

The traditional test instrument that is used to troubleshoot microprocessor systems is the microprocessor *emulator*. The emulator is connected to the microcomputer via a flat ribbon cable and a DIP header. The microprocessor is removed from its socket and the DIP header is plugged in its place. The emulator can be used to dynamically exercise the microprocessor's bus systems. Emulator tests include routines for finding opens and shorts on the address and data buses. If an open or short is discovered, the emulator can be instructed to continually exercise the bad node, providing signals that enable the technician to troubleshoot the circuit with an oscilloscope.

Other emulator tests include calculating ROM check sums, testing RAM with standard patterns, and exercising serial and parallel I/O ports. The emulator can also be instructed to loop on any RAM failure. The address, data, and control signals generated by the emulator can be used to check the operation of buffers and decoders.

Sophisticated microprocessor emulators contain their own high-level programming language. A library of programs can be developed that are similar to the self-test. Unlike a normal self-test that halts the processor on a failure, the emulator tests can be used to generate helpful signals that will guide the technician in the troubleshooting process. The major drawback of emulators is their high price—typically $5000 to $10,000.

The Fluke 9010A illustrated in Figure 19.8 is an extremely sophisticated and flexible high-end microprocessor emulator. It contains many built-in tests, such as exercising the address, data, and control buses, RAM and I/O read/write tests, and ROM checksum tests. The 9010 also has its own high-level test development language. Comprehensive tests can be written to verify and troubleshoot complex microprocessor-based systems. The tests are loaded into the system via a microcassette mass-storage device.

A unique alternative approach to troubleshooting the IBM-

Figure 19.8 Fluke 9010A.

PC was recently developed by Dair Computer Systems of Palo Alto, California. The AID/88, illustrated in Figure 19.9, contains 8085 and 8088 microprocessors, a small keypad, a group of seven-segment displays, and associated support circuitry. We have seen that the major problem with the PC POST is that it halts after sensing an error, giving the technician little useful troubleshooting information. Instead of rewriting the POST in its own ROM, the Aid/88 employs and enhances the PC's POST.

A key labeled "BUS" is depressed to initiate a test that checks for functional address and data buses. If the bus test passes, the POST in the PC ROM can be accessed and executed under control of the Aid/88. If the bus test fails, the Aid/88 gives helpful information that is used to isolate the shorted or open address/data bits.

Another key on the 8088 is used to find the halted address of the PC. As we know from Figure 19.7, the halted address will indicate the type of fault that occurred. Other keys on the Aid/88 allow the technician to read data from memory, set hardware break points, and run and single step the system board through the POST.

At a price under $2600, the Aid/88 is a cost-effective alternative to the microprocessor emulator for troubleshooting the

Figure 19.9 AID/88—Dair Computer Systems.

IBM-PC. With the repair of just a few system boards, the Aid/88 can virtually pay for itself. Innovative and inexpensive test instruments like the Aid/88 can put the power of component-level troubleshooting into the hands of small microcomputer maintenance companies.

IBM provides a floppy disk with advanced diagnostic programs. These programs can be run on a PC that has a functioning system board, floppy disk controller board and floppy disk drive, and a video adapter board. The programs contained on the advanced diagnostics disk test the system board, DRAM, serial ports, parallel ports, and video adapter board.

19.4 PREVENTIVE MAINTENANCE ON MICROCOMPUTERS

Preventive maintenance is performed to ensure that the microcomputer is operating correctly and to prevent future malfunctions. The weak link in any computer system is not the electronics but the mechanical devices. A typical microcomputer system has only four mechanical devices: a cooling fan, floppy disk drives, a printer, and a keyboard.

A failure of a cooling fan can cause a microcomputer system to function intermittently or, in extreme cases, cause a massive failure of ICs. The quiet and smooth operation of fans should be checked on a periodic basis.

Printers used in microcomputer systems require periodic cleaning and testing. Typical printers have self-tests that print the entire character set. The self-test will indicate any local print failures. The new generation of dot-matrix printers contain a local processor that can be instructed to print various fonts and attributes. Although the vast majority of printer failures are mechanical in nature, the interface circuitry or internal processor circuitry should also be checked thoroughly during any printer malfunction.

19.4.1 Floppy Disk Drive Evaluation and Alignment

Floppy disk drives are the only part of a microcomputer system that require critical inspection and evaluation. One of the most important concepts of floppy disks is *media interchangeability.* Consider a school that has a classroom with 12 IBM-PCs. Over the course of a semester, a student may use a particular floppy disk on all 12 machines. This implies that a floppy may be formatted on one drive, have data written on it by another drive, and yet it must be readable by any floppy disk drive on any PC.

Because disk drives are mechanical devices, over time and use they come out of alignment. When they leave the factory, floppy disk drives must meet precise alignment specifications. As they age and come out of alignment, the data recorded on one floppy disk drive may not be readable on another floppy disk drive. To ensure that people can share data and be independent of a particular machine, it is vital that floppy disk drives be checked for alignment, and if needed, serviced every few months.

Many companies manufacture floppy disk drive evaluation packages. One such package is the Interrogator by Dysan of Santa Clara, California. The Interrogator contains an instruction manual and two floppy disks. The first floppy disk contains the application program used to exercise and evaluate the disk drives. The second floppy is the Dysan DDD (Digital Diagnostic Diskette). The DDD contains special test tracks of data that are miswritten in precise patterns. The application program derives the condition of the drives by the manner in which they read these divergent patterns.

Typical tests that are performed on floppy disk drives are spindle speed, centering, radial alignment, azimuth alignment, and hysteresis. Double-density $5\frac{1}{4}$-in. floppy disk drives rotate at 300 rpm. The spindle speed test checks for compliance with the 300-rpm specification. Floppy disks are clamped in the drive mechanism by a cone assembly. The centering test ensures that the cone assembly is clamping the floppy disk in the proper manner.

Figure 19.10 illustrates the meaning of the radial and azimuth alignment tests. The word "radial" is a reference to the circular shape of the floppy disk. In Figure 19.10 we can think of radial as meaning the up/down offset of the read/write head. "Azimuth" refers to the right/left offset of the read/write head. The head in Figure 19.10a is perfectly aligned on track 16.

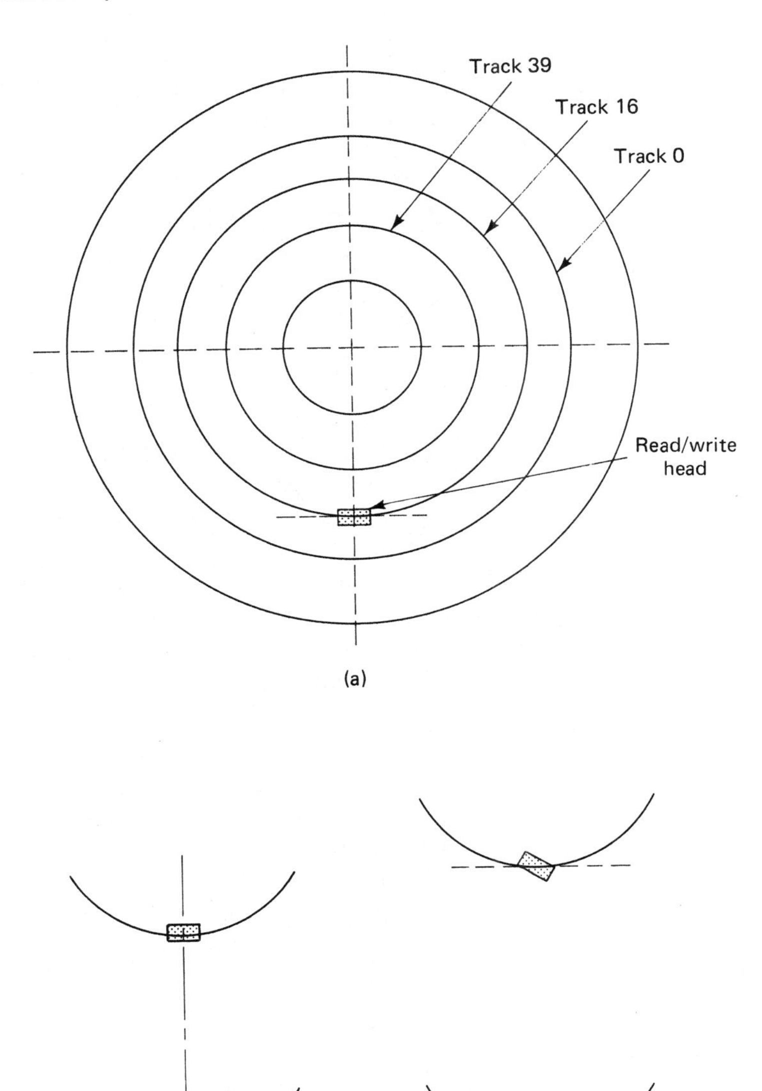

Figure 19.10 Radial and azimuth alignment on floppy disks.

Figure 19.10b illustrates a head with a radial misalignment. The top picture in Figure 19.10b illustrates a head that is too close to the center of the floppy. The bottom picture in Figure 19.10b illustrates a head that is too close to the outside edge of the floppy. If the two pictures in Figure 19.10b represent two different drives, both drives would probably appear to function in a normal manner. But a floppy disk recorded on one drive may not be readable by the second drive.

Figure 19.10c illustrates two drives with azimuth misalignment. Notice that the heads are aligned correctly in the up/down plane (radial), but they are tilted to the right or the left. The same media exchange problems would occur with the two misaligned drives in Figure 19.10c as occured in Figure 19.10b.

The hysteresis specification indicates the floppy disk drive's ability to repeatedly find the same location on a track as it is approached in different directions. Instruments called floppy disk alignment testers and exercisers are used with an analog alignment disk to realign floppy disk drives. The exerciser has circuitry that is used to position the read/write head on a selected track, and the analog alignment disk has analog patterns written on specified tracks.

The Lynx 470 Disk Drive Alignment Tester and Exerciser, illustrated in Figure 19.11, represents a new generation of disk drive service instruments. The older technology testers required the use of an oscilloscope to monitor a differential "cat's-eye" pattern for radial alignment. The Lynx 470- is a "scopeless" tester that is designed to be used with high-precision analog alignment disks. LEDs on the control panel are used to monitor all the major performance parameters of floppy disk drives. The 470 is both a floppy disk alignment tool and a powerful exerciser used to troubleshoot malfunctioning drives. The use of this type of tester/exerciser enables the technician to efficiently handle all the aspects of floppy disk drive maintenance.

As part of the preventive maintenance procedure, the read/write heads on a floppy disk drive can be cleaned by a special abrasive disk saturated with a cleaning chemical. This process is hotly debated throughout the industry. Many people feel that abrasive cleaning disks damage or shorten the life of the read/write heads, and that formatting a new, high-quality floppy disk actually cleans the heads without the abrasive side effects. Other groups feel that cleaning floppy disk drive read/write

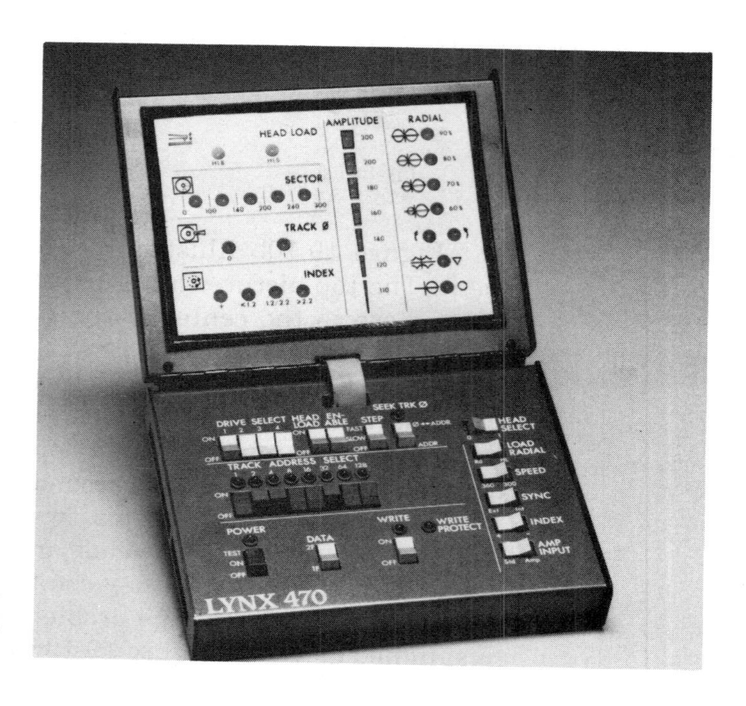

Figure 19.11 Lynx 470—alignment tester and exerciser.

heads with cleaning disks is an important aspect of microcomputer preventive maintenance.

19.5 THE LOGIC ANALYZER

Let's consider the limitations of the conventional oscilloscope. Most oscilloscopes can monitor a maximum of only two channels. This is perfectly acceptable for analog circuits, but digital circuits have many inputs and outputs that must be simultaneously monitored and checked for timing relationships.

Each sweep of the oscilloscope is initiated by a trigger signal. The trigger is a single event that can be provided by channel 1 or channel 2, an external trigger, or internally triggered on line frequency (60 Hz). Uniformly repetitive analog signals provide stable trigger sources. Stable triggers are almost impossible to derive from digital signals that have varying duty cycles and periods. This problem results in the oscilloscope display in Figure 19.12, which illustrates a bit on a microprocessor's data bus.

Remember that the horizontal axis of an oscilloscope represents a period of time equal to the setting of the time per division switch times the number of horizontal divisions. A microprocessor is continuously reading and writing data. Because this data vary with each operation and the oscilloscope is constantly triggering on new data, the display that results will be a mistriggered blur of overlapping signals.

The display in Figure 19.12 does provide some important information. First, and most important, it shows that there is activity on the data bus, and the signals appear to be at acceptable logic 1 and logic 0 levels. One extremely important question is: Why are there indeterminate voltage levels displayed in Figure 19.12? In Chapter 14 we examined the timing diagrams of the 8085 read and write operations. There are many points when the 8085 is not reading or writing data and its data bus is at high-Z. Even on data buses that are pulled up to +5 V with standard 2K to 10K pull-up resistors, the process of bringing a floating output to +5 V does take time, so the indeterminate level is visible on an oscilloscope display.

We have established that the oscilloscope suffers from two major faults that limits it effectiveness in the digital world: it can not display enough channels of information and it simply cannot trigger on aperiodic signals.

Figure 19.13 illustrates a digital signal that is sampled at an extremely high rate. Each sample, a logic 0 or logic 1 level, can be easily stored in a video refresh memory, where it will be displayed and examined at leisure. Although this process has

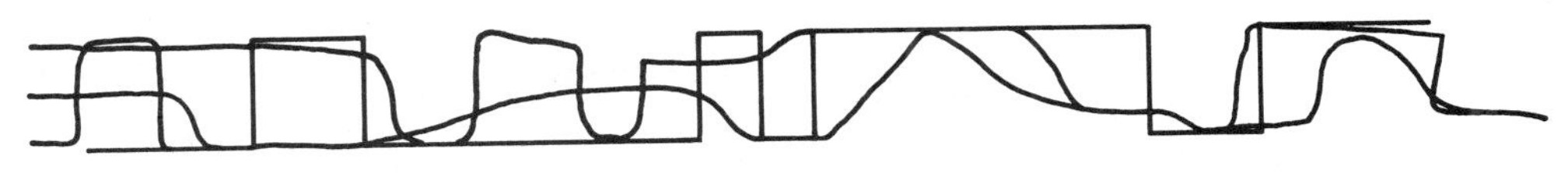

Figure 19.12 Data bus monitored with scope.

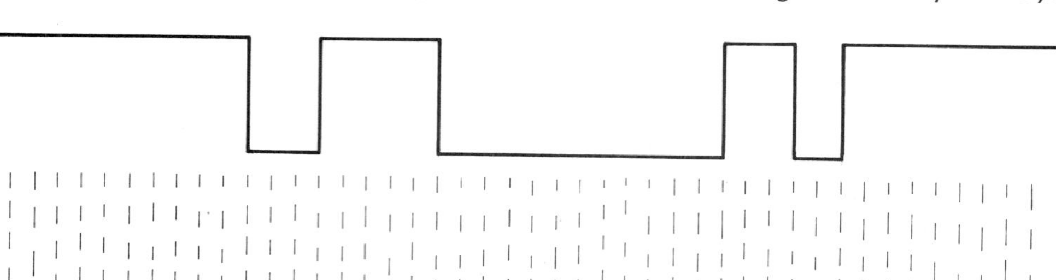

Figure 19.13 Sampling a digital signal.

much in common with an anlog-to-digital conversion, remember that the signal is already digital; it is merely being sampled and broken into discrete bits that can be stored in memory.

That concept is the basic idea behind a test instrument called a *logic analyzer.* The logic analyzer is the most powerful device available for analyzing the actions of digital circuits. The logic analyzer is often used in the R & D environment, but it is also an invaluable troubleshooting tool. Figure 19.14 represents the basic block diagram of a logic analyzer.

Typical analyzers have 8, 16, or 32 input channels. Each channel can monitor one digital signal. The system contoller block contains a microprocessor and associated circuitry that supervises and coordinates all system activity.

Display memory contains the semiconductor memory and support circuitry used to store the signal samples. The memory is constantly being updated at the rate defined by the sample clock. Typical sample clocks run between 20 MHz and 50 MHz.

Think of each channel in the display memory as a SIPO shift register. As each new sample is taken, the previous sample is shifted right by one bit. As the memory fills up, old samples are lost. The capacity of the display and the sampling rate of the system dictate the length of the digital signal that can be stored.

Triggering a logic analyzer is a widely misunderstood concept. The trigger block in Figure 19.14 can be thought of as consisting of an *n-bit comparator,* where *n* is equal to the number of channels of the analyzer. The operator sets a trigger word by toggling the appropriate switches. A channel can be set to trigger high, low, or a don't care. When the word in the switches matches the word on the channel inputs (ignoring any trigger bits set to don't care), the analyzer is "triggered" and information will be sent to the video display.

Figure 19.15 illustrates a typical address decoder. Assume that the decoder is in question and its operation must be verified. Fourteen channels of the logic analyzer can be used to monitor the enable inputs, select inputs, and decoded outputs. We will want to trigger the analyzer whenever the decoder is enabled: G1 is high, and G2a and G2b are low. The other trigger inputs will be set to don't care.

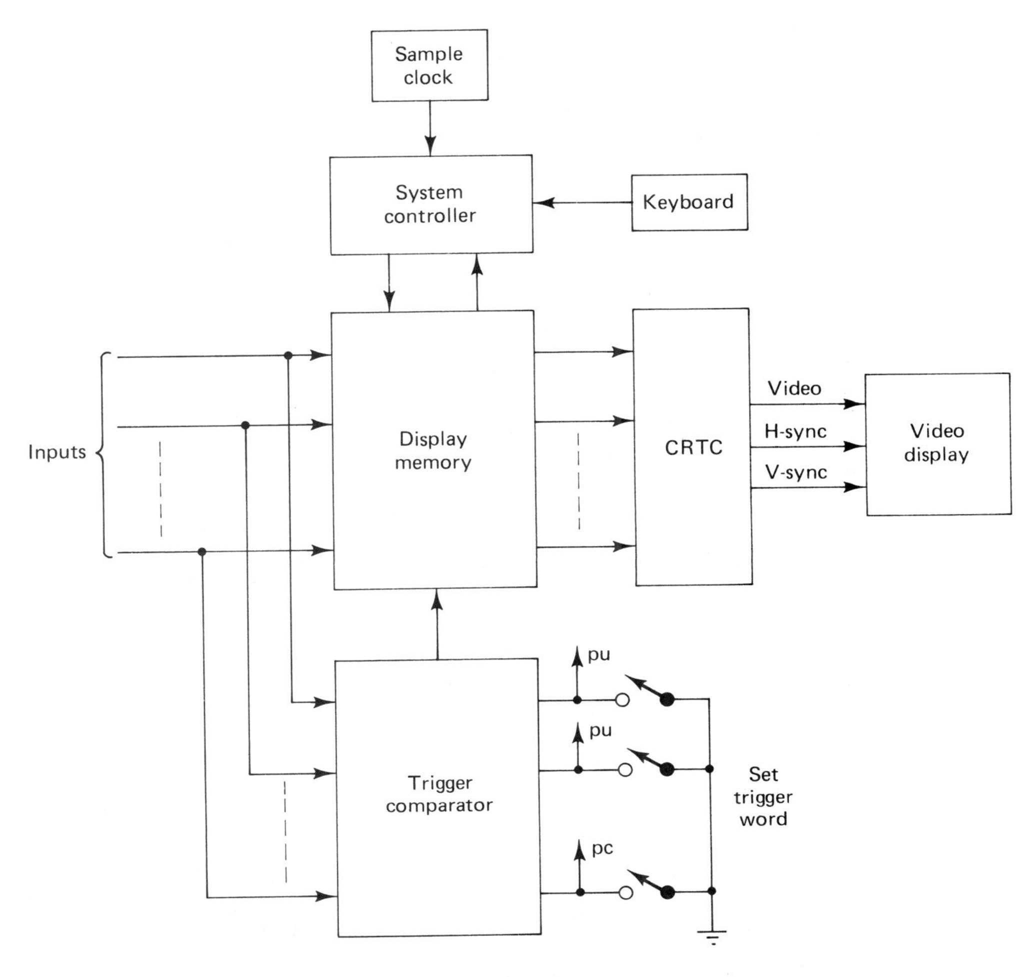

Figure 19.14 Logic analyzer block diagram.

Logic analyzers can be set to display the information that appeared before the trigger word, the information that occurs after a trigger word, or the analyzer will center the display around the trigger word. Logic analyzers are used to verify the operation of buffers, latches, decoders, and other multi-input sequential devices that reside in bus systems.

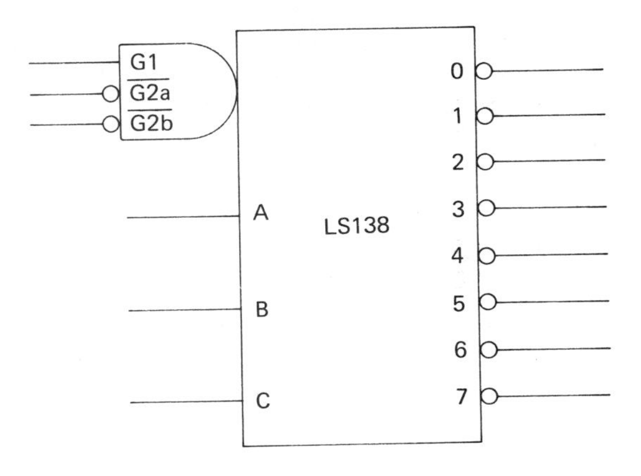

Figure 19.15 3-line-to-8-line decoder.

The greatest limitation of logic analyzers is their long setup time. It takes quite a while to attach the probes to the correct ICs and set the operational parameters of the analyzer. But many tough circuit faults cannot be isolated by any other means.

What does the output of a logic analyzer look like? If the device is functioning correctly, it looks like the timing diagrams that we have extensively examined throughout this text. That is why we have spent so much time understanding how circuits should work. The logic analyzer enables the technician to take a "snapshot" of the circuit's signals and analyze them for the correct levels and transitions.

Consider the possibility of connecting a "smart" logic analyzer to the address, data, and control buses of a microprocessor. If all these signals can be monitored, the logic analyzer should be able to display the actual assembly language instructions that the microprocessor is currently executing. This function is called *disassembly*. Advanced logic analyzers have a program for each popular microprocessor that is used to disassemble the

Figure 19.16 Omni-4 logic analyzer.

signals into instructions. This enables the technician to monitor the program as it is executed by the microprocessor and the memory and I/O locations accessed and the data values that are read and written.

A logic analyzer with the ability to disassemble 8088 instructions can be used to monitor the PC's POST as it is executed on power-up. The analyzer may be instructed to trigger whenever a halt instruction is executed and display the instructions that preceded the halt. These instructions will detail the location and cause of the POST failure.

Figure 19.16 illustrates the Omni-4 logic analyzer by OmniLogic, Inc., of Renton, Washington. The Omni-4 is an excellent topic in which to end this book. It embodies all the concepts that we have discussed involving digital and microprocessor systems. The Omni-4 is a high-performance logic analyzer that is built around a general-purpose, Z80-based microcomputer that runs the popular CP/M operating system. With the addition of a high-speed front-end logic board and extensive software controls, the microcomputer is transformed into a multifaceted logic analyzer.

It can monitor 16 channels (32 with the expansion option). With a maximum sampling speed of 20 MHz, the Omni-4 displays 1000 samples of 16 channels or 500 samples of 32 channels. The display can be standard digital waveforms or hexadecimal data. It can also be used to disassemble the instructions of the most popular 8-bit microprocessors and to analyze software performance. Yet it can still be used as a microcomputer running all the standard CP/M application programs.

The Omni-4 is a perfect example of the flexibility of software-intensive microprocessor-based systems. Because it depends on software rather then hardware, the Omni-4 is a low cost alternative to dedicated hardware-based logic analyzers.

QUESTIONS AND PROBLEMS

19.1. Describe the difference between a 8088 minimum and maximum mode.

19.2. How is the data/address bus demultiplexed?

19.3. Explain how the direction of the flow is controlled in data bus buffers.

19.4. How does the 8288 derive the memory and I/O control signals?

19.5. Why doesn't the PC have a standard *RC* time-delay reset circuit?

19.6. What signals are used to drive the buffered address and data bus to high-Z?

19.7. How are the 8088 maximum-mode system and the old 8080 system similar?

19.8. What are the major functions of the 8255 PPI in the PC system?

19.9. How are the system interrupts utilized?

19.10. Give a brief description of the 8253 PIT and how it is used in the PC.

19.11. What are the advantages of programmed I/O? DMA?

19.12. How are the four channels of DMA used in the PC system?

19.13. How is the DMAC similar to a microprocessor?

19.14. The 8088 has a 20-bit address bus and the 8237 DMAC has only a 16-bit address bus. How is this problem corrected in the PC?

19.15. Why are the address and data buses connected to the ROM and DMA subsystem called external buses?

19.16. What test facilities are usually built into microprocessor-based systems?

19.17. What is the greatest limitation of the PC's POST?

19.18. Why are single DRAM failures on the PC easy to repair?

19.19. Why is microprocessor emulation an effective method of troubleshooting "dead" microprocessor systems?

19.20. What is the only real preventive maintenance required on PC systems?

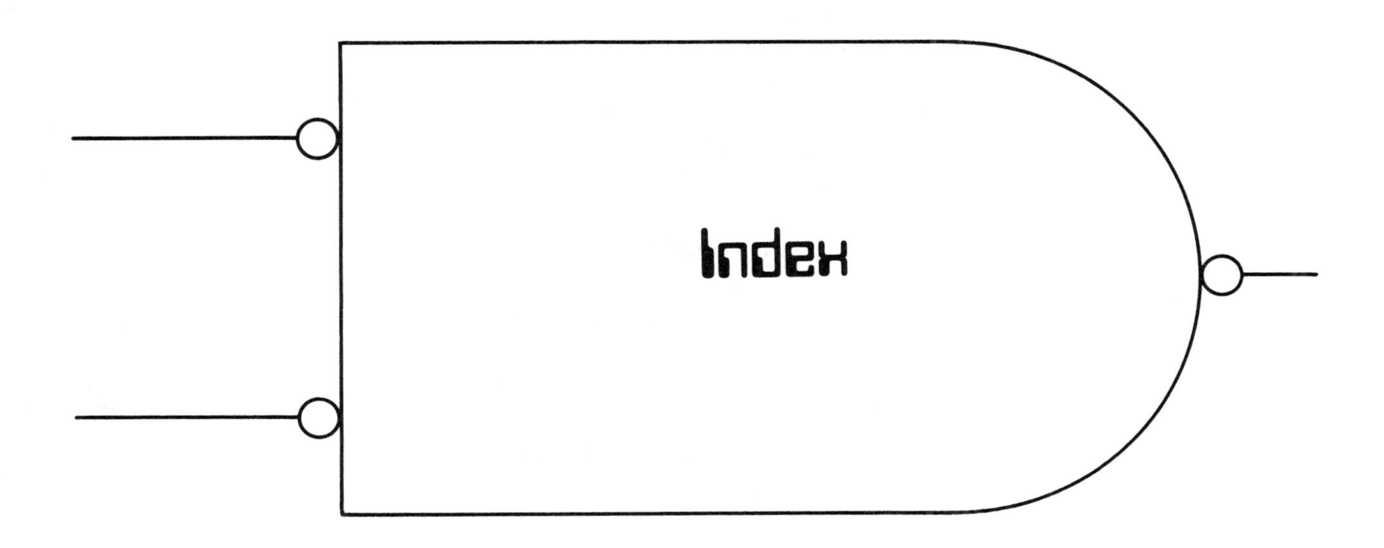
Index